Steel Science and Technology

Steel Science and Technology

Edited by
Felix Sampson

Willford Press
www.willfordpress.com

Published by Willford Press,
118-35 Queens Blvd., Suite 400,
Forest Hills, NY 11375, USA

ISBN: 978-1-68285-782-3

Cataloging-in-Publication Data

Steel science and technology / edited by Felix Sampson.
p. cm.
Includes bibliographical references and index.
ISBN 978-1-68285-782-3
1. Steel. 2. Steel--Metallurgy. 3. Materials science.
4. Steel industry and trade--Technological innovations. I. Sampson, Felix.
TN730 .S84 2020
669.142--dc23

For information on all Willford Press publications
visit our website at www.willfordpress.com

Willford Press

Printed in China.

Contents

Preface

Steel is an alloy that mostly consists of iron and carbon. It is the most important component used in buildings, infrastructure, ships, tools, automobiles and weapons. The major reason for its constant usage is its tensile strength and low cost. The base material of steel is iron which can take two forms- body-centered cubic and face-centered cubic. The body-centered cubic contains one atom in the center and 8 atoms at the vertices of each cubic unit cell. While, in the face-centered cubic, there is one atom at the center of each of the 6 faces of the cubic unit cell and 8 atoms at its vertices. Carbon may contribute up to 2.14% of the weight of a steel alloy. The varying amount of carbon and other elements affects the chemical and physical build-up of the final steel. It also affects few characteristics of steel such as its hardness, tempering behavior, quenching behavior and tensile strength. This book contains some path-breaking studies in the field of steel science and technology. As this field is emerging at a rapid pace, the contents of this book will help the readers understand the modern concepts and applications of the subject.

Significant researches are present in this book. Intensive efforts have been employed by authors to make this book an outstanding discourse. This book contains the enlightening chapters which have been written on the basis of significant researches done by the experts.

Finally, I would also like to thank all the members involved in this book for being a team and meeting all the deadlines for the submission of their respective works. I would also like to thank my friends and family for being supportive in my efforts.

Editor

1

High-Power Laser Cutting of Steel Plates: Heat Affected Zone Analysis

Imed Miraoui,[1,2] Mohamed Boujelbene,[3] and Mouna Zaied[2]

[1]*Department of Mechanical Engineering, College of Engineering, Aljouf University, Sakaka 2014, Saudi Arabia*
[2]*College of Sciences, Research Unit MEER, University of Gafsa, 2100 Gafsa, Tunisia*
[3]*Department of Industrial Engineering, College of Engineering, Hail University, Hail, Saudi Arabia*

Correspondence should be addressed to Imed Miraoui; aimed_mir@yahoo.fr

Academic Editor: Anna Richelli

The thermal effect of CO_2 high-power laser cutting on cut surface of steel plates is investigated. The effect of the input laser cutting parameters on the melted zone depth (MZ), the heat affected zone depth (HAZ), and the microhardness beneath the cut surface is analyzed. A mathematical model is developed to relate the output process parameters to the input laser cutting parameters. Three input process parameters such as laser beam diameter, cutting speed, and laser power are investigated. Mathematical models for the melted zone and the heat affected zone depth are developed by using design of experiment approach (DOE). The results indicate that the input laser cutting parameters have major effect on melted zone, heat affected zone, and microhardness beneath cut surface. The MZ depth, the HAZ depth, and the microhardness beneath cut surface increase as laser power increases, but they decrease with increasing cutting speed. Laser beam diameter has a negligible effect on HAZ depth but it has a remarkable effect on MZ depth and HAZ microhardness. The melted zone depth and the heat affected zone depth can be reduced by increasing laser cutting speed and decreasing laser power and laser beam diameter.

1. Introduction

Laser cutting process is used extensively in industries to achieve complex shapes with close tolerances. Compared to conventional cutting, laser cutting has more advantages such as small heat affected zone, accurate cutting, noncontact processing, and short time of processing [1]. The manufacturers are interested in using the laser cutting to optimize both the quality and the productivity of the items made by this process. The productivity and the quality are essentially related to the selection of the suitable parameters of laser cutting such as laser power, laser beam diameter, and cutting speed [2–4]. During laser cutting process, we have a thermal interaction and a mechanical interaction between matter and the laser cutting parameters. The heat affected zone (HAZ) is the zone of the base material which has not melted but whose microstructure and mechanical properties were affected by the heat generated during laser cutting. The heat generated by the laser cutting process and after cooling causes this change in the region near the cut surface. The heat affected zone may lead to undesirable effects such as fatigue resistance, surface cracking, and distortion. So, it is interesting to select the appropriate laser cutting parameters in order to minimize the heat affected zone.

Several studies were performed to examine the effect of the input laser cutting parameters on the quality of the cut surface and the heat affected zone.

Eltawahni et al. [5] have investigated the effect of laser cutting parameters on the kerf width and the cut surface roughness of MDF composite material. They have shown that the upper and lower kerf widths decrease as cutting speed increases and they increase as laser power increases. An experimental investigation of laser cutting of polymeric materials was done by Choudhury and Shirley [6]. They have concluded that the heat affected zone depth is proportional to laser power and inversely proportional to compressed air pressure and cutting speed. Radovanovic and Dasic [7] have examined the influence of laser cutting parameters on the roughness of cut surface for mild steel. They have concluded that the cut surface roughness increases as the sheet thickness increases and decreases as the laser power increases. Miraoui et al. [8, 9] have analyzed the surface roughness, microhardness, and microstructure of stainless steel sheet made by high-power laser cutting. They have indicated that the cutting of stainless steel sheets made by high-power laser had major effects on the microstructure and the roughness of the cut surface. They have showed that the laser cutting leads to the formation of cracks and periodic striations at the cut surface. Rajaram et al. [10] have investigated the CO_2 laser cut quality of 4130 steel and they have shown that the kerf and HAZ widths were influenced significantly by laser power, while cutting speed has played a minor role. Kumar Dubey and Yadava [11] have optimized the laser beam cutting process using a hybrid Taguchi method and response surface method with multiperformance characteristics for thin sheet of high silicon-alloy steel. An experimental investigation of the laser cutting parameters for ceramic composite was done by Quintero et al. [12]. They have indicated the influence of cutting speed, frequency, and assist gas pressure on heat affected zone. Prasad et al. [13] have discussed the laser cutting of metallic coated sheet steels of 1 mm as thickness. Miraoui et al. [14] have analyzed the effect of laser beam diameter on the melted zone and microhardness beneath the cut surface. They have concluded that the width of the melted zone depends on the laser beam diameter. Li et al. [15] have used the experimental method of Taguchi to obtain the optimal laser cutting parameters. They have concluded that the quality of laser cutting of QFN packages depends mainly on laser frequency, cutting speed, and laser driving current. Huehnlein et al. [16] have used the design of experiment (DOE) approach to optimize the laser cutting of thin ceramic layers. Miraoui et al. [17] have studied the effect of laser beam diameter and laser power on cut surface roughness of steel plates. They have indicated that the influence of laser beam diameter on cut surface roughness is negligible and the increasing of laser power improves the surface roughness. Grum and Zuljan [18] have analyzed the heat for laser cutting of steel. They have shown that the cutting quality was influenced by the high temperature zone developed in the cutting front.

In this study, the heat affected zones of low carbon steel plates obtained by CO_2 laser cutting were investigated. Laser power, cutting speed, and laser beam diameter were varied in order to analyze the effect of these input laser cutting parameters on heat affected zone depth and on microhardness beneath the cut surface. Design of experiment approach has been used in order to predict the relationship between the process parameters and heat affected zone depth.

2. Experimental Methods

The cutting operations were performed on a CO_2 laser machine type AMADA AS 4000 E. The dimensional capacity of this machine is 3000 × 1500 mm with an accuracy of cutting of ±0.1 mm. For experimentation, low carbon steel sheet with thickness of 8 mm was used. The dimensions of the workpiece were 15 mm in width and 30 mm in length (see Figure 1). Chemical composition of this steel structure is given in Table 1, in which the Max% levels of elements are indicated. The mechanical properties of this steel (S235) are presented in Table 2.

Figure 1: Low carbon steel plate carved by CO_2 laser.

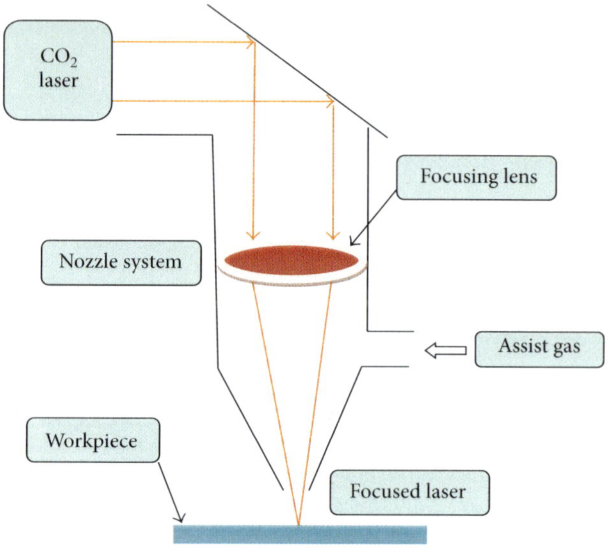

Figure 2: Experimental setup for laser cutting process.

Table 1: Chemical composition of S235 steel (wt%).

C	P	S	N	Mn	Si
0.18%	0.05%	0.05%	0.009%	1.40%	0.05%

Cutting specimens were investigated experimentally using nominal output laser power (P) ranged from 3 to 5 kW, laser beam diameter (D) ranged from 1 to 2 mm, and cutting speed (V) ranged 600–2200 mm/min. Figure 2 shows the experimental arrangement used in this study.

Vickers microhardness tester under a load of 50 g, during 10 s, was used to measure the microhardness beneath the cut surface of the specimens. The heat affected zone depths were measured by using optical microscopy. In order to examine the microstructure changes resulting from the laser cutting, we have used scanning electron microscopy, SEM.

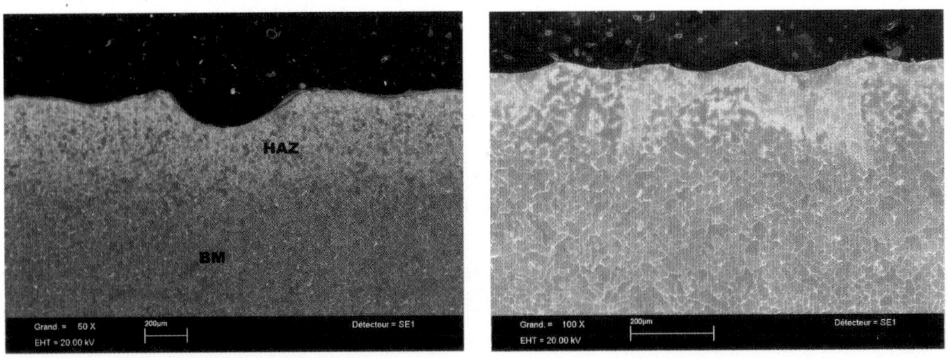

Figure 3: Microstructure of HAZ and base material (BM).

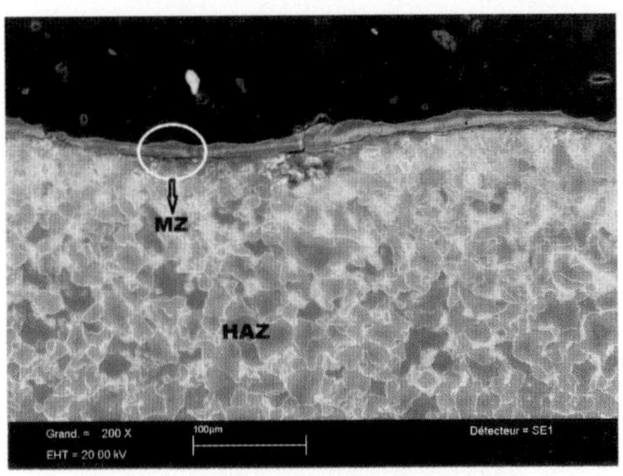

Figure 4: Microstructure of HAZ.

Table 2: Mechanical properties of S235 steel.

Elasticity modulus	Poisson ratio	Density	Tensile elastic limit
205000 MPa	0.3	7800 kg/m^3	235 MPa

3. Results and Discussion

3.1. Microstructure Analysis. The heat generated by the laser cutting process and after cooling has affected the microstructure of the region near the cut surface (HAZ). In the base material zone (BM), the material has not undergone any structural transformation (Figure 3). The result of microscopic analysis revealed the presence of a layered structure on the workpiece structure. Figure 4 shows the presence of a white layer called melted zone (MZ): zone of interaction beam-metal.

Full view and magnified image of cut surface microstructure is shown in Figure 5. It can be seen that laser cutting leads to the formation of cracks at the cut surface. Laser cutting had major effect on cut surface quality.

3.2. HAZ Depth Analysis

3.2.1. Experimental Study. The HAZ depth (zone beneath the cut surface) is one of the main performances of laser machining. The dimension of the heat affected zone (HAZ) was analyzed using optical microscope. HAZ is the area of

Table 3: Input laser cutting parameters and their levels.

Symbol	Input parameters	Unit	Level 1	Level 2
P	Laser power	W	3000	5000
V	Cutting speed	mm/min	600	2200
D	Laser beam diameter	mm	1	2

the work material which was not melted during cutting but its microstructure was affected by the intense heat during laser cutting. While in the base material zone, the material has not undergone any structural transformation (see Figures 3 and 6). The interface between HAZ and base material (BM) is the limit between the affected and unaffected microstructure. To study the influence of the input laser cutting parameters (laser power, laser beam diameter, and cutting speed) on the HAZ depth, seven HAZ values of each specimen were measured using optic microscope and an average was calculated for each specimen (see Figure 6).

Figure 7 presents the effect of laser power and cutting speed on HAZ depth. These results indicate that laser power and cutting speed had major effect on heat affected zone. HAZ depth increases as laser power increases and it decreases as cutting speed increases. The variations of HAZ depth depending on laser beam diameter and laser power are presented in the curves of Figure 8. It was shown that laser beam diameter has a negligible effect on HAZ depth.

3.2.2. Modelling. The statistical processing of the experimental results enabled development of model equations to determine the HAZ depth according to the cutting parameters (laser power, laser beam diameter, and cutting speed). The DOE approach (Taguchi method) has been used to develop the model equation. The input process parameters and their levels used in the designed experiment are given in Table 3. The measured values of HAZ corresponding to each experimental condition are listed in Table 4.

Analysis of variance (ANOVA) has been carried out to determine the significant model terms. This method has a gap that gives estimation value with an error 5%. Analysis of variance indicates that, on a significance level of 5%, laser

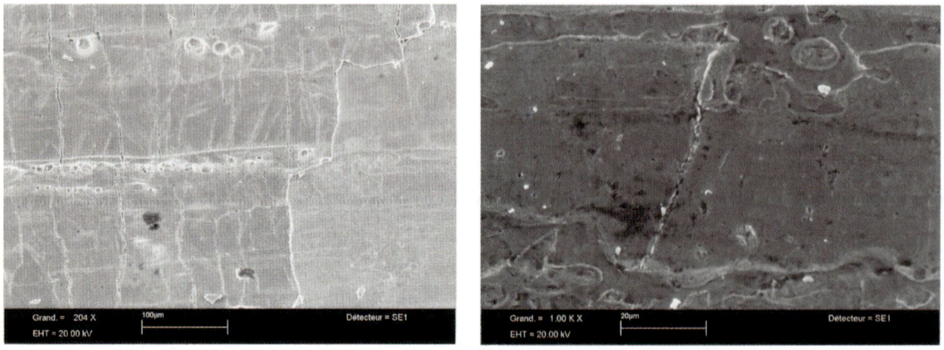

FIGURE 5: Full view and magnified images of cut surface microstructure.

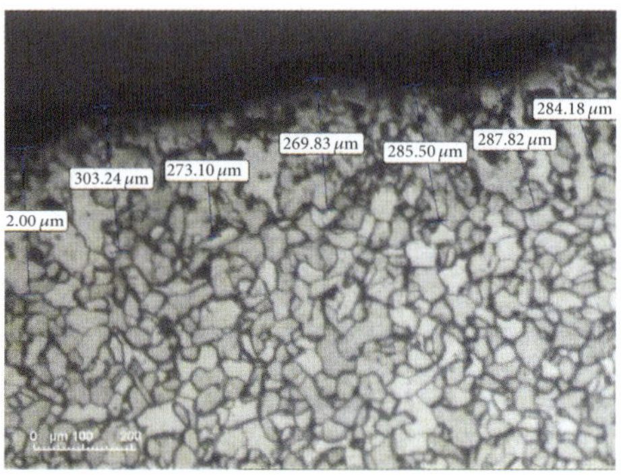

FIGURE 6: Measuring of HAZ depth.

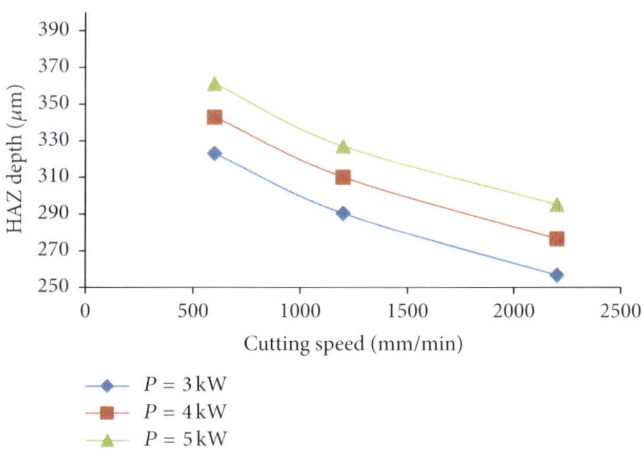

FIGURE 7: Effect of laser power and cutting speed on HAZ depth ($D = 1$ mm).

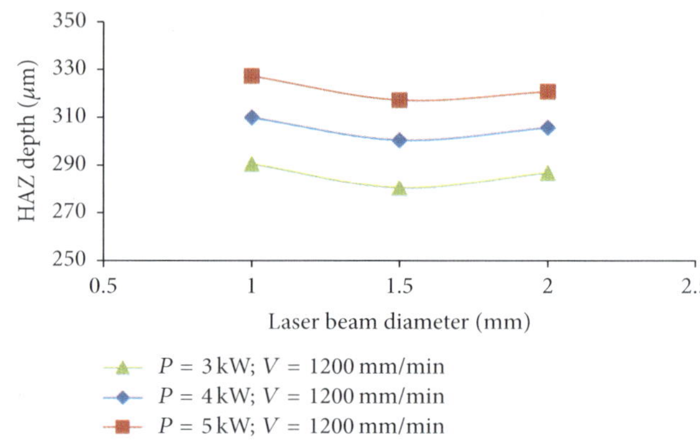

FIGURE 8: Effect of laser beam diameter on HAZ depth.

TABLE 4: Design matrix and experimentally measured response (HAZ).

Exp. number	P (W)	V (mm/min)	D (mm)	HAZ (μm)
1	3000	600	1	323
2	5000	600	1	360
3	3000	2200	1	257
4	5000	2200	1	295
5	3000	600	2	325
6	5000	600	2	361
7	3000	2200	2	266
8	5000	2200	2	317

power, cutting speed, and laser beam diameter are the most significant model terms, but their interaction does not have a significant effect on HAZ. It was observed that the laser power and the cutting speed are the most significant and the laser beam diameter is the least significant cutting parameters affecting the HAZ.

From the experimental data of HAZ depth, we have developed the following model equation:

$$\mathrm{HAZ} = 100.48 P^{0.259} V^{-0.146} D^{0.04}. \quad (1)$$

This equation shows that HAZ depth decreases with cutting speed and increases with laser power. The effect of laser power is more important than the effect of cutting speed on HAZ depth. It is clear that laser beam diameter has a negligible effect on HAZ depth. Laser power is the most important parameter affecting the HAZ. The increase in laser power generates the increase of the cut surface temperature which causes the increase of the heat affected zone. During the cooling period, it will have a change of phase in the solid state which is reflected in the increase of the heat affected zone depth. When the cutting speed is rapid, the propagation of heat in the sample is done in a shorter period of time which allows minimizing the cut surface damage and therefore

TABLE 5: Confirmation experiments—HAZ.

P (W)	V (mm/min)	D (mm)	Experimental HAZ	Modelled HAZ	Error%
4000	600	1	342.73	338.37	1.27
3000	1200	2	286.76	291.83	−1.76
5000	1200	1.5	316.98	329.30	−3.88

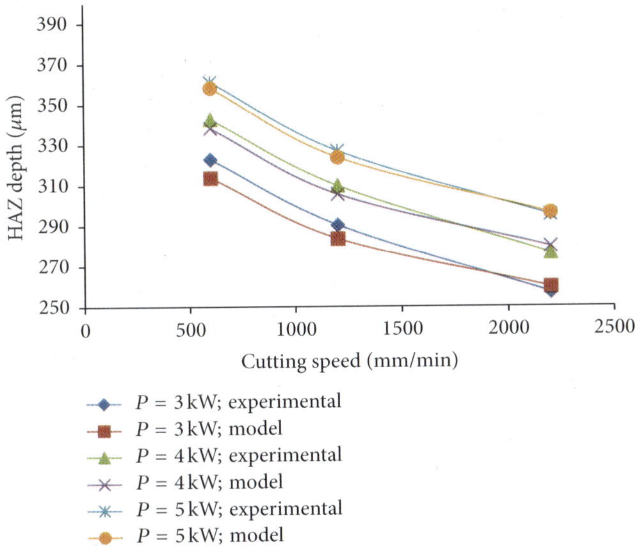

FIGURE 9: Comparison between experimental and modelled HAZ depth (D = 1 mm).

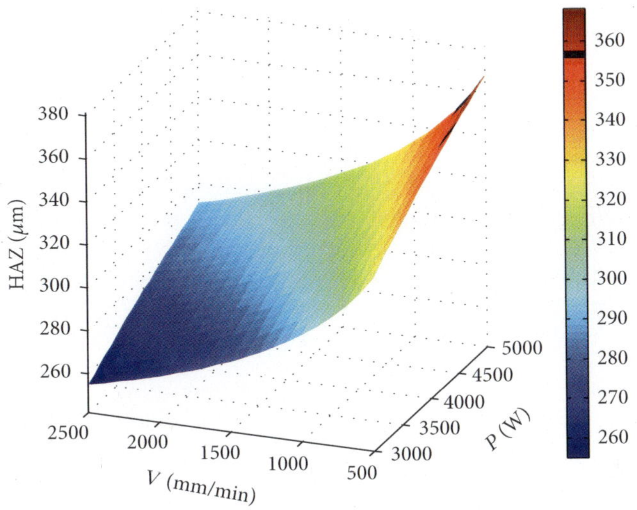

FIGURE 10: Response surface plot of HAZ depth with laser cutting speed and laser power (D = 1 mm).

has a thermally finer affected zone. All these results are in good agreement with the experimental study, as shown in Figure 9. The combined effects of laser cutting speed and laser power on HAZ depth are shown in Figure 10, where laser beam diameter (D) is kept at 1 mm. Figure 11 presents the effects of laser power and laser beam diameter on HAZ

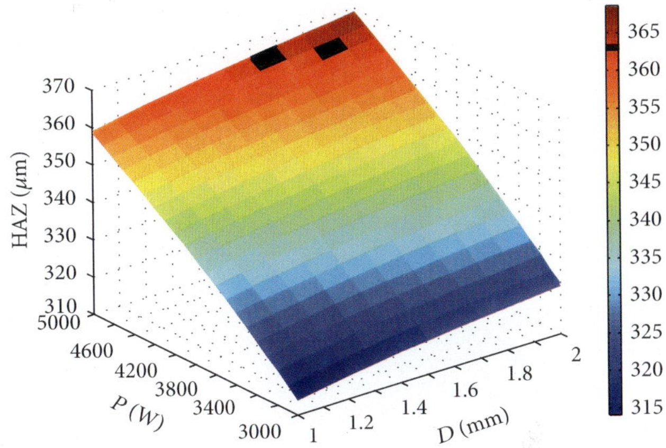

FIGURE 11: Response surface plot of HAZ depth with laser power and laser beam diameter (V = 600 mm/min).

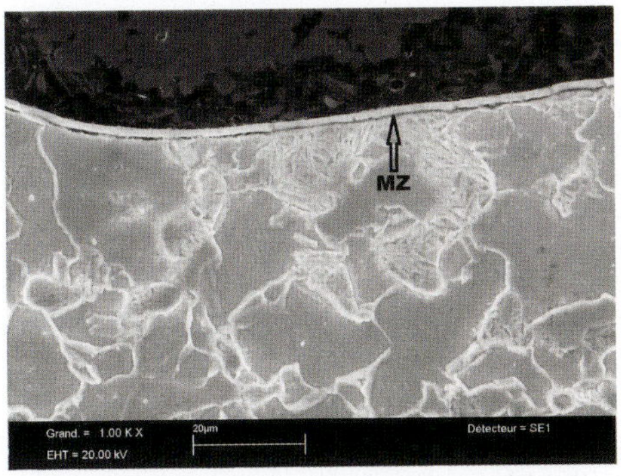

FIGURE 12: Measuring of MZ depth.

depth by keeping the laser cutting speed (V) as constant at 600 mm/min.

To confirm the adequacy of the developed models, three confirmation experiments were performed using new test conditions. The values of HAZ of the confirmation experiments which were determined using the developed model show good agreement with the experimental values (see Table 5).

3.3. Analysis of MZ Depth. To study the influence of the input laser cutting parameters (laser power, laser beam diameter, and cutting speed) on the MZ depth (the zone of interaction beam-metal), we have measured the average MZ depth from several measurements performed at the surface perpendicular to the cut surface (see Figure 12). A mathematical model for the melted zone depth is developed by using design of experiment approach (DOE). Taguchi

Table 6: Design matrix and experimentally measured response (MZ).

Exp. number	P (W)	V (mm/min)	D (mm)	MZ (μm)
1	3000	600	1	1.2
2	5000	600	1	2
3	3000	2200	1	1
4	5000	2200	1	1.75
5	3000	600	2	2
6	5000	600	2	3.20
7	3000	2200	2	1.50
8	5000	2200	2	2.50

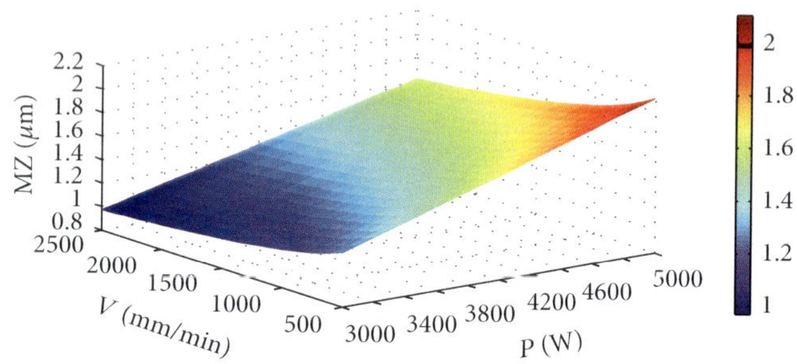

Figure 13: Response surface plot of MZ depth with laser cutting speed and laser power (D = 1 mm).

method was used to determine the mathematical relationship between the response (MZ depth) and the input laser cutting machining (laser power, laser beam diameter, and cutting speed). The input process parameters and their levels used in the designed experiment are given in Table 3. The measured values of MZ corresponding to each experimental condition are listed in Table 6. Analysis of variance (ANOVA) indicates that, on a significance level of 5%, laser power, cutting speed, and laser beam diameter are the most significant model terms, but their interaction does not have a significant effect on MZ. It was observed that laser power and laser beam diameter are the most significant and the cutting speed is the least significant cutting parameters affecting the MZ.

From the experimental data of MZ depth, we have developed the following model equation:

$$MZ = 0.00113P^{1.0039}V^{-0.164}D^{0.6281}. \quad (2)$$

It is clear that MZ depth decreases with cutting speed and increases with laser power. It can be seen that MZ depth increases with laser beam diameter, contrary to HAZ depth in which the diameter has a negligible effect. Laser power is the most important parameter affecting the MZ. The increase in laser power generates the increase of the cut surface temperature which causes the increase of the melted zone. If the cutting speed is rapid, the propagation of heat in the sample is done in a shorter period of time which allows minimizing the cut surface damage and therefore has a thermally finer MZ. The combined effects of laser cutting speed and laser power on MZ depth are shown in Figure 13, where laser beam diameter (D) is kept at 1 mm. Figure 14

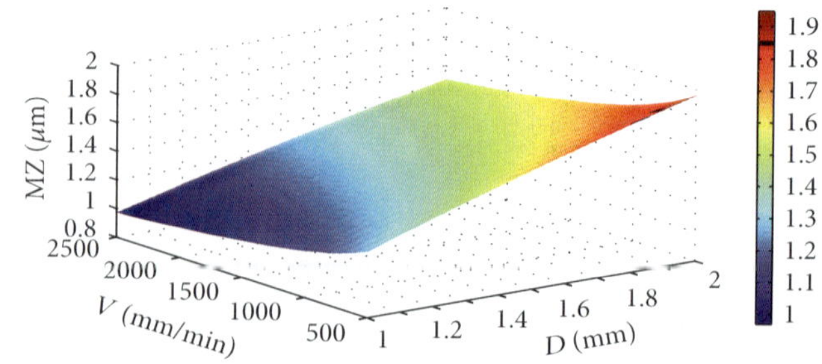

Figure 14: Response surface plot of MZ depth with laser cutting speed and laser beam diameter (P = 3000 W).

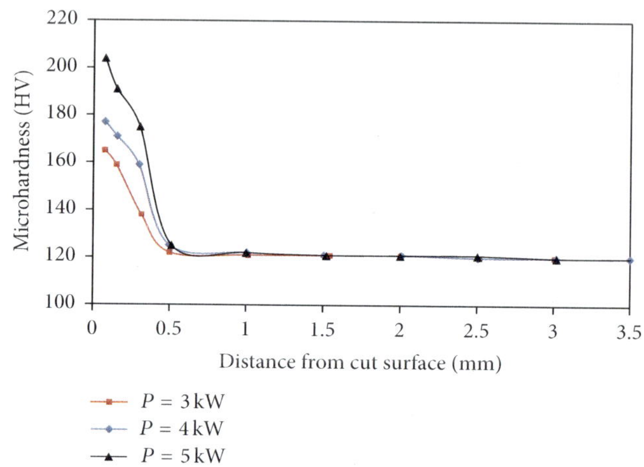

Figure 15: Effect of laser power on HAZ microhardness.

presents the effects of laser cutting speed and laser beam diameter on MZ depth by keeping the laser power (P) as constant at 3000 W.

To confirm the adequacy of the developed models, three confirmation experiments were carried out using new test conditions. The input laser cutting parameters, the experimental values, the modelled values, and the percentages of error are given in Table 7. These results indicate that all the values of the percentage of the error are acceptable.

3.4. Analysis of HAZ Microhardness. This section discusses the effects of the input laser cutting parameters (laser power, laser beam diameter, and cutting speed) on the HAZ microhardness. The microhardness profile across machined surface depths produced at different laser power (cutting speed and laser beam diameter are constant) is shown in Figure 13. It was shown that the maximum microhardness is observed near the cut surface and gradually decreases until it reaches the microhardness of the base material (120 HV: the ordinary microhardness of low carbon steel S235 used in this work). Laser cutting has a thermal effect on the original structure due to the high temperature reached at the cut edge followed by a high rapid cooling. This results in a hardening zone at the cut edge (HAZ). As shown in Figure 15, the microhardness in the HAZ increases as laser power increases.

Table 7: Confirmation experiments—MZ.

P (W)	V (mm/min)	D (mm)	Experimental MZ	Modelled MZ	Error%
5000	600	1.5	2.75	2.76	−0.36
4000	2200	1	1.40	1.36	2.9
3000	2200	2	1.50	1.57	−4.67

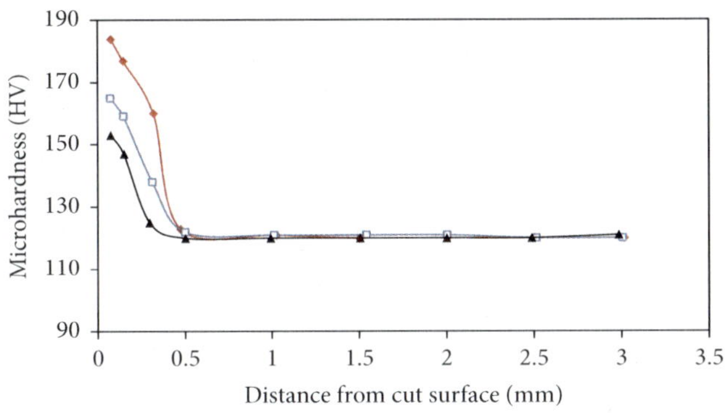

Figure 16: Effect of cutting speed on HAZ microhardness.

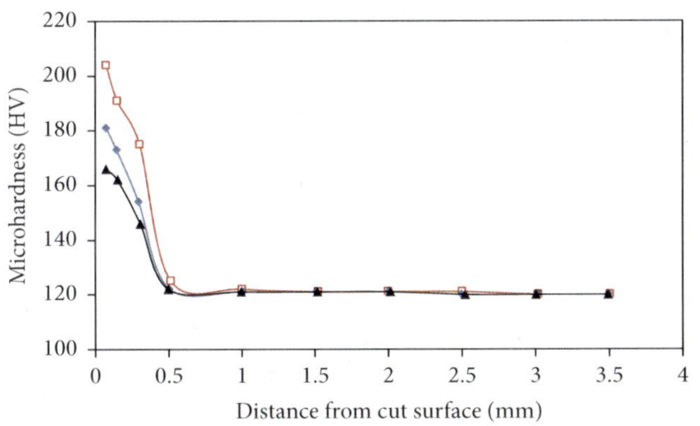

Figure 17: Effect of laser beam diameter on HAZ microhardness.

Figure 16 presents the effect of cutting speed on HAZ microhardness. It is clear that the microhardness in the HAZ decreases as cutting speed increases. The effect of laser beam diameter on HAZ microhardness is shown in Figure 17. These results indicate that laser beam diameter has an effect on HAZ microhardness. It decreases as laser beam diameter increases.

4. Conclusions

High-power CO_2 laser cutting of low carbon steel plates is examined and the effects of the input process parameters such as laser power, laser beam diameter, and cutting speed on heat affected zone and melted zone have been investigated. The scanning electron microscopy and the optical microscope are used to analyze the microstructure and the HAZ depth. The microhardness beneath the cut surface was measured by using Vickers microhardness. Design of experiment approach has been used to predict the relationship between the input process parameters and the response (heat affected zone and melted zone). From the present work, the following conclusions can be drawn. The depth of MZ, the depth of HAZ, and the microhardness of HAZ depend on the input process parameters. The laser power and the cutting speed are the most significant and the laser beam diameter is the least significant cutting parameters affecting the HAZ. However, the laser power and the laser beam diameter are the most significant and the cutting speed is the least significant cutting parameters affecting the MZ. The dimension of HAZ increases as laser power increases, and it decreases as cutting speed increases. The MZ depth decreases with cutting speed and increases with laser power. Results show that MZ depth increases with laser beam diameter, contrary to HAZ depth in which the diameter has a negligible effect. Laser power is the most important parameter affecting the MZ. The microhardness in the HAZ increases as laser power increases, and it decreases as cutting speed and laser beam diameter increase. The undesirable thermal effect of laser cutting of low carbon steel plates can be reduced by increasing laser cutting speed and decreasing laser power and laser beam diameter.

Conflict of Interests

The authors declare that there is no conflict of interests regarding the publication of this paper.

Acknowledgment

The authors acknowledge the financial support from Aljouf University, KSA, under Grant no. 35/294.

References

[1] B. S. Yilbas, S. Khan, K. Raza et al., "Laser cutting of 7050 Al alloy reinforced with Al_2O_3 and B_4C composites," *The International Journal of Advanced Manufacturing Technology*, vol. 50, no. 1–4, pp. 185–193, 2010.

[2] N. Jebbari, M. M. Jebari, F. Saadallah, A. Tarrats, R. Bennaceur, and J. P. Longuemard, "Themal affected zone obtained in machining steel XC42 by high-power CO_2 laser," *Optics and Laser Technology*, vol. 40, no. 6, pp. 864–873, 2008.

[3] R. Ragendram and M. B. Pate, "The effect of laser beam velocity on cut quality and surface temperature," *American Society of Mechanical Engineers, Heat Transfer Division*, vol. 104, pp. 121–127, 1998.

[4] I. Decker, J. Ruge, and U. Atzert, "Physical models and technological aspects of laser gas cutting," in *Proceedings of the Industrial Applications of High Power Lasers*, vol. 4 of *Proceeding of SPIE*, pp. 81–88, Linz, Austria, March 1984.

[5] H. A. Eltawahni, A. G. Olabi, and K. Y. Benyounis, "Investigating the CO_2 laser cutting parameters of MDF wood composite material," *Optics and Laser Technology*, vol. 43, no. 3, pp. 648–659, 2011.

[6] I. A. Choudhury and S. Shirley, "Laser cutting of polymeric materials: an experimental investigation," *Optics and Laser Technology*, vol. 42, no. 3, pp. 503–508, 2010.

[7] M. Radovanovic and P. Dasic, "Research on surface roughness by laser cut," *The Annals of University Dunarea de Jos of Galati Fasicule VIII, Tribology*, vol. 12, pp. 84–88, 2006.

[8] I. Miraoui, E. Elimi, M. Boujelbene, and E. Bayraktar, "Analysis of roughness and microstructure for high-power laser cutting of stainless steel," *Advanced Science Letters*, vol. 19, no. 2, pp. 483–486, 2013.

[9] I. Miraoui, E. Bayraktar, and E. Bayraktar, "Effects of laser cutting main parameters on microhardness and microstructure changes of stainless steel," *Advanced Materials Research*, vol. 664, pp. 811–816, 2013.

[10] N. Rajaram, J. Sheikh-Ahmad, and S. H. Cheraghi, "CO_2 laser cut quality of 4130 steel," *International Journal of Machine Tools and Manufacture*, vol. 43, no. 4, pp. 351–358, 2003.

[11] A. Kumar Dubey and V. Yadava, "Multi-objective optimisation of laser beam cutting process," *Optics & Laser Technology*, vol. 40, no. 3, pp. 562–570, 2008.

[12] F. Quintero, J. Pou, F. Lusquiños, M. Boutinguiza, R. Soto, and M. Pérez-Amor, "Quantitative evaluation of the quality of the cuts performed on mullite-alumina by Nd:YAG laser," *Optics and Lasers in Engineering*, vol. 42, no. 3, pp. 327–340, 2004.

[13] G. V. S. Prasad, E. Siores, and W. C. K. Wong, "Laser cutting of metallic coated sheet steels," *Journal of Materials Processing Technology*, vol. 74, no. 1–3, pp. 234–242, 1998.

[14] I. Miraoui, M. Zaied, and M. Boujelbene, "Effect of laser beam diameter on cut edge of steel plates obtained by laser machining," *Applied Mechanics and Materials*, vol. 467, pp. 227–232, 2014.

[15] C.-H. Li, M.-J. Tsai, and C.-D. Yang, "Study of optimal laser parameters for cutting QFN packages by Taguchi's matrix method," *Optics and Laser Technology*, vol. 39, no. 4, pp. 786–795, 2007.

[16] K. Huehnlein, K. Tschirpke, and R. Hellmann, "Optimization of laser cutting processes using design of experiments," *Physics Procedia*, vol. 5, pp. 243–252, 2010.

[17] I. Miraoui, M. Boujelbene, and E. Bayraktar, "Analysis of roughness and heat affected zone of steel plates obtained by laser cutting," *Advanced Materials Research*, vol. 974, pp. 169–173, 2014.

[18] J. Grum and D. Zuljan, "Analysis of heat effects in laser cutting of steels," *Journal of Materials Engineering and Performance*, vol. 5, no. 4, pp. 526–537, 1996.

2

Metallurgical and Stress State Factors Which Affect the Creep and Fracture Behavior of 9% Cr Steels

Jonathan Parker and John Siefert

Electric Power Research Institute, Charlotte, NC, USA

Correspondence should be addressed to Jonathan Parker; jparker@epri.com

Academic Editor: Paolo Ferro

EPRI-supported research has identified critical material information regarding the factors affecting the performance of creep strength-enhanced ferritic steels, in general, and Grade 91 steel, in particular. EPRI recommendations emphasize that a five-point, integrated strategy should be used for the effective life management of components fabricated from tempered martensitic steels. This integration promotes a balanced use of resources which, when properly focused, reduces uncertainty regarding creep and fracture behavior. Tighter control of processes from steel making, steel processing, and heat treatment ensures that alloys with deficient properties never enter service. One cornerstone of this proactive approach is the definition of 'Metallurgical Risk' which links the presence of inclusions and trace elements to the susceptibility for creep damage. The improved confidence in the high temperature performance of CSEF steel components promotes reliability, increases efficiency, and minimizes the risk of component fracture.

1. Background and Introduction

EPRI-coordinated, industry-sponsored research projects in creep strength-enhanced ferritic (CSEF) steels began in 2007 with a major effort to improve the life management of Grade 91 steel components. Since this initial project included more than 40 project participants and over $4 million in industry funding, several key follow-on projects have been completed or continue today. This series of related projects, illustrated in the schematic diagram shown in Figure 1, provides a critical base of information linking steel making and fabrication to microstructure and performance under simple and complex loading conditions.

These projects exemplify the EPRI mission statement "to provide thought leadership, industry expertise, and collaborative value to help the electricity sector . . . and addressed through effective research and development programs for the benefit of society."

It is important to emphasize that the learnings and findings were realized with direct input and perspective from stakeholders representing the entire electricity supply chain. A general outcome from this research is that for metallurgically complex steels, testing programmes should establish high temperature performance on steel sections which are carefully chosen and well characterized, and the full pedigree of the steel cast selected is known. Full details of the research performed and the reports, position papers, and documents are available from the EPRI website. Selected achievements directly linked to materials factors are summarized below.

(1) Detailed research examining the microstructure of tempered martensitic ferritic steels has provided key information concerning the rate controlling damage mechanisms for different conditions. Degradation during creep has been linked to many different factors; these include the formation of new phases, coarsening/dissolution of carbides, recovery of the dislocation substructure, and the nucleation, growth, and linking of microvoids [1–4]. Much of the prior research has been focused on the creep strength. However, from a standpoint of component performance assessment, it is important to evaluate both *when* a material fails (i.e., for laboratory testing, "time to rupture"), and also *how* it fails (i.e., for laboratory

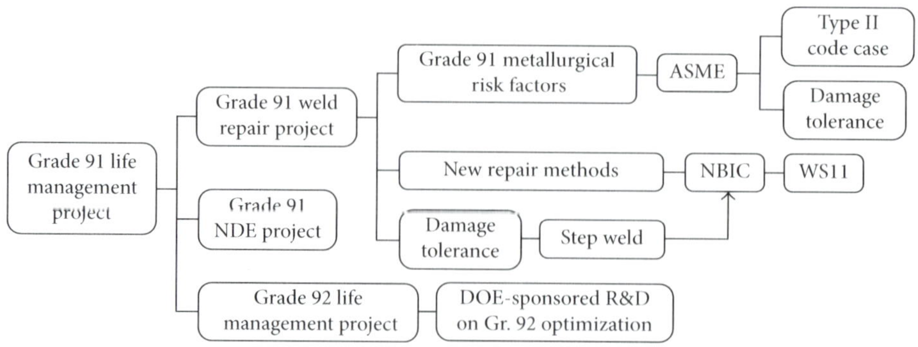

FIGURE 1: The evolution of major research and development projects specific to 9% Cr creep strength-enhanced ferritic (CSEF) steels within the Electric Power Research Institute since 2007.

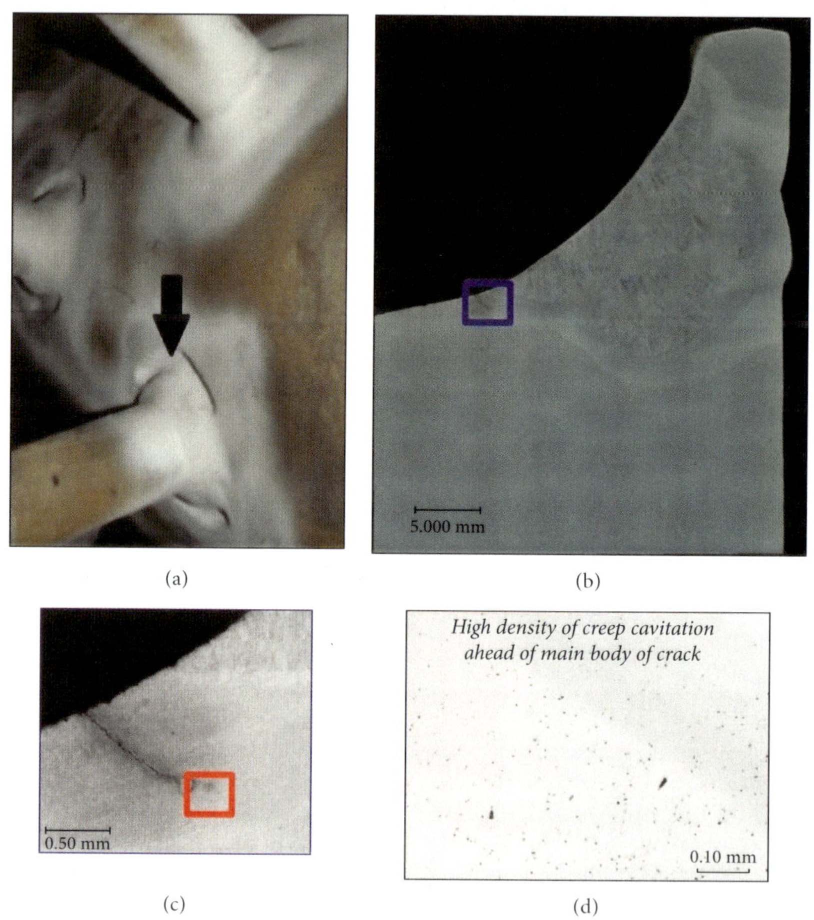

FIGURE 2: Example of in-service damage from Grade 91 heat "B2" [5]. The density of cavitation ahead of the crack varied from 700 to 1,000 voids/mm^2 and closely matches known maximum densities responsible for macrocracking in Grade 91 heat-affected zones (~1,000 voids/mm^2).

testing, "reduction of area (ROA)"). From assessment of in-service cracking, it is apparent that in the majority of cases, damage and failure are linked to the creep cavitation behavior. For example [5], the micrographs in Figure 2 show the very high number density of creep voids developed in the HAZ of a stub to header weld.

Root cause analysis established that the inherent cavity susceptibility of the Grade 91 base steel was a key factor linked directly to damage development in weld HAZs. Further details regarding the influence of specific trace elements on the nucleation and growth of cavities are presented later in this paper.

(2) The microstructure in the heat-affected zone (HAZ) of welds made from the 9% chromium martensitic steel is complex. Systematic microstructural investigations have defined the different regions of the

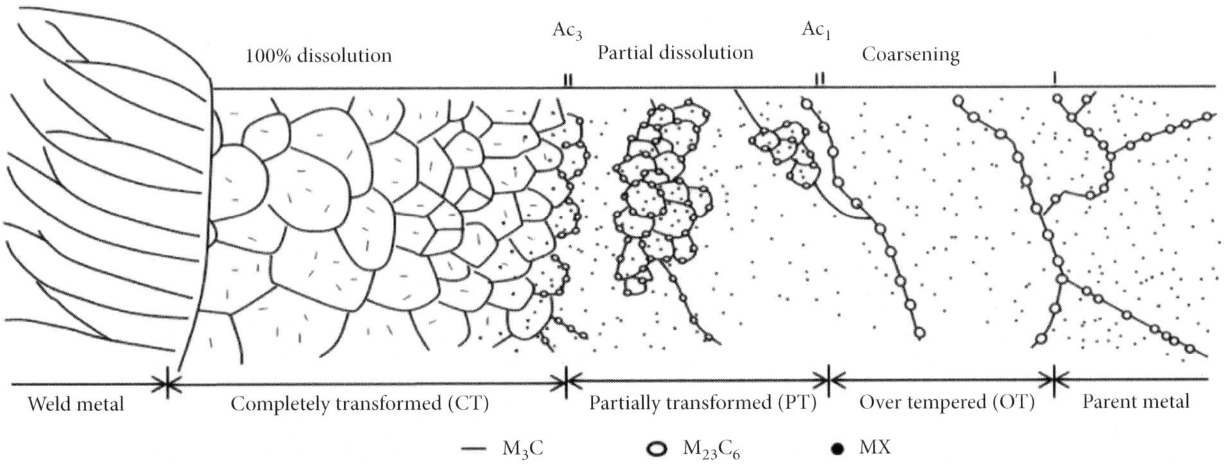

Figure 3: Schematic diagram illustrating the microstructural regions in the heat-affected zone of a fusion weld in a tempered martensitic steel [6, 7]. Meaningful characterization of microstructure in 9% Cr CSEF steels should take into account the phase dissolution and transformation behavior rather than simply describing the regions in the HAZ based on the size of the prior austenite grains.

microstructure across the HAZ as a function of the welding process [6, 7]. The microstructure in the HAZ of a single-pass, bead-on-plate weld on a Grade 92 parent was systematically investigated by using an extensive range of advanced electron and ion-microscopy-based techniques [6]. In addition, controlled thermal cycles were applied to simulate the microstructures in the different regions of the HAZ. It was found that the microstructure in the HAZ should be categorized in terms of the dissolution and transformation behavior as a function of a decreasing peak temperature during welding. The three primary microstructural regions, shown schematically in Figure 3, were defined as follows [6, 7]:

(a) The completely transformed (CT) region, in which the original matrix is completely reaustenitized with *complete* dissolution of the pre-existing secondary precipitate particles. The time at peak temperature is typically *sufficient* to dissolve inclusions present in the base steel.
(b) The partially transformed (PT) region, where the original matrix is partially reaustenitized along with a *partial* dissolution of the secondary precipitate particles from the original matrix. The time at peak temperature is typically *not sufficient* to dissolve inclusions present in the base steel.
(c) The over tempered (OT) region, where the pre-existing precipitate particles coarsen. No significant changes have been recorded to the size, number, and distribution of the inclusions present.

The PT region is the most susceptible area for creep damage base. This is consistent with evidence from the commonly reported HAZ failures in weldments constructed from these types of steels. This damage susceptibility arises because this location will always experience thermal cycles which degrade the local creep strength. However, the tendency for creep voids to form is directly linked to inclusions and trace elements. This may be present in following steel making and processing in steel where the level of control is below the Guidelines from EPRI. The peak temperatures and hold times achieved in the PT region are not sufficient to dissolve the inclusions which may be present in the base steel. Thus, damage susceptible HAZs will have low creep strength (from the weld thermal cycle) and poor resistance to creep void nucleation when large number densities of inclusions are present.

The link between creep ductility and each of these areas is critical in establishing the metallurgical risk factor(s) which control the susceptibility for low ductility failure in Grade 91 steels. It is clear that further evaluation of specific test results is of value to review potential trends in behavior. It is apparent that under conditions of low ROA, low ductility creep fracture is promoted as a result of the nucleation, growth, and link-up of creep voids.

The concept that creep cracking of in-service components is typically linked to the nucleation and growth of microvoids or creep "cavitation" is not new. This form of damage is well documented following root cause analysis of creep cracking during service in a large number of Grade 91 steel components. In addition, low creep ductility has been linked to cracking susceptibility of other established engineering steel systems. For example, in bainitic CrMoV steels [8–11], martensitic Grade 122 steel [12–14], and 316H stainless steel [15, 16], a link between low creep ductility and in-service cracking has been established. However, the specific materials factors which enhance cavitation vary and the key knowledge necessary to justify improved control of manufacturing process for CSEF steels has only recently been established.

Designers, manufacturers and operators of high energy equipment seek to use the most cost effective alloys which provide the required level of performance. Increasingly, this performance should be achieved without the use of extreme conservatism. Thus, it is important that the preferred alloys do not exhibit excessively large variations in properties. The present paper summarizes key information from recent research which has shown that for metallurgically complex alloys such as Grade 91 and Grade 92 steel, changes in composition and/or heat treatment have a significant influence on high temperature behavior. Key materials evidence for improved control of steel making and processing to minimize the risk of components entering service with deficient properties is reviewed and summarized.

2. Effect of Chemical Composition on Microstructure and Properties

Historically, the American Society of Mechanical Engineers Boiler and Pressure Vessel Code (ASME B&PV) for Grade 91 type steel specifies control of 14 elements. A specific example for SA-335 P91 is given in Table 1 [17].

In 2011, the Electric Power Research Institute published a "Best Practices" guideline for the specification and procurement of Grade 91 components [18], this was updated in 2015 [20]. In addition to the stated ASME requirements, this document recommended the control of four additional elements (namely As, Cu, Sb, and Sn). It is important to mention that one of the original ORNL reports on Grade 91 steel which laid the foundation for the acceptance of Grade 91 steel, recommended the control of three other elements (O, W, and B) [19]. It is therefore important to note that the traditional specification of the chemical composition mandated by ASME B&PV Section II and other codes (such as EN or ASTM) lack sufficient control of elements known to influence the properties in CrMo steels [9]. Direct evidence regarding the need for improved control of Grade 91 composition [21–26] suggests that steel makers around the world are monitoring a far wider range of elements than that is mandated by minimum requirements in code; see the example for SA 335 P91 in Table 1.

The need for control of additional elements in Grade 91 steel is directly linked to the susceptibility for nucleation of creep voids and the associated reduction in the creep ductility [26]. The problems associated with trace elements and creep brittle behavior are of increasing concern because of the growing desire for recycling of steel, that is, the use of scrap steel in the melting process. Poor control and inadequate sorting of scrap can introduce elements such as tin and copper into the steel. Once present these elements are very hard or even impossibly to remove. Although steel scrap is typically classified into various grades and steel makers may monitor scrap to obtain the required quality, it is clear that the influence of even trace levels of undesirable elements can have a detrimental effect either on strength or ductility or both. Furthermore, the nucleation of creep cavities can be linked to the local microstructure and composition. Thus, even relatively small (as measured in weight percent) additions of sulfur (S), copper (Cu), tin (Sn), antimony (Sb), lead (Pb), and arsenic (As) have been shown to be particularly damaging under creep conditions. There is circumstantial evidence from other martensitic CSEF steels to suggest that the explicit control of boron (B) is of benefit to mitigate the risk of formation of boron-nitride (BN).

In addition to the effects of chemical composition on creep ductility, there are known effects of specific elements in reducing the creep strength namely nickel (Ni), the nitrogen to aluminum (N : Al) ratio, and excessive additions of boron (B). The influence of these elements on performance has been reviewed and is summarized in the following sections.

2.1. Influence of Tin (Sn). A series of systematic studies evaluating the influence of Sn on the creep strength and fracture behavior has been published [27–29]. The steel compositions used by Masuyama et al. [27] are detailed in Table 2. Results showed that Sn appeared to be particularly damaging as it can have a detrimental effect on strength and creep ductility (Figures 4 and 5). As shown in Figure 4, the observed creep strength is clearly reduced from mean Grade 91 behavior (typical results are shown using a Larson Miller parameter comparison ($C = 31$)). The lives of the Sn-doped heat shown in Figure 4 have been used to make specific performance estimates for selected temperatures (Table 3).

As noted in Table 3, a reduction in creep life was not calculated until the anticipated creep temperature was above 550°C (1022°F). In Figure 5, Sn-doped Grade 91 steel exhibited a pronounced drop in the measured reduction in area in creep tests performed at 650°C (1202°F). This trend occurred even for creep test lives of less than 1000 hours [27].

The research outcomes published by Masuyama et al. [28] are consistent with the results of other research [30]. The recent work has established detailed information on the metallurgical behavior of Sn using field emission gun and scanning transmission electron microscopy (FEG-STEM). Song et al. [30] reported concentrations of Sn at subgrain boundaries that were over 10 times the values measured in the matrix. Based on these observations, it was suggested that the effect of Sn would be to reduce the boundary cohesion (as in creep) creating preferential sites for the nucleation of creep cavities. These effects would nucleate a higher number density of cavities leading to reduction in creep ductility in Grade 91 material where Sn was present. As reported in previous work [27–29], the authors also found a marked reduction in creep strength [30].

These effects were summarized by consideration of the results of creep tests at 600°C (1112°F) and 100 MPa (14.5 ksi) [30]:

(1) controlled Grade 91 composition life = 100,641 hours

(2) Sn-doped Grade 91 composition = 35,290 hours

It has been suggested (Table 1) that the maximum permissible value of Sn in Grade 91 steel should not exceed 0.010 weight percent [19]. In the early 1990s, a major supplier of Grade 91 material suggested a maximum value of 0.020 weight percent [21, 22]. For manufacture of thick-section Grade 22 CrMo steel, a value of 0.010 weight percent maximum has been recommended. In practice, where the Grade 91 composition has been controlled, Sn values are routinely <0.005 weight percent [19, 22, 26, 31].

Thus, EPRI recommend that the maximum limit for *Sn in Grade 91 steel should be ≤ 0.010 weight percent.*

Table 1: Comparison of recommended chemical composition ranges for Grade 91 steel products.

Element	SA-335 P91 [17]	ORNL 6303 spec. [18]	ORNL 6303 target [18]	1023199 [19]	P91 book [20]	V&M [21]	Arcelor Gr. 91 plate [22]	Arcelor Gr. 22 plate [23]	EPRI proposal for Grade 91 Cl. 2
C	0.08–0.12	0.08–0.12	0.1	0.08–0.12	0.08–0.12	0.08–0.12	—	—	0.08–0.12
Mn	0.30–0.60	0.30–0.50	0.4	0.30–0.60	0.30–0.50	0.30–0.50	—	—	0.30–0.50
P	0.020 max	0.020 max	<0.010	0.020 max	0.020 max	0.020 max	0.010 max	0.008	0.020 max
S	0.010 max	0.010 max	<0.010	0.010 max	0.005 max	0.005 max	0.005 max	0.005	0.005 max
Si	0.20–0.50	0.20–0.50	0.2	0.20–0.50	0.20–0.40	0.20–0.40	—	—	0.20–0.40
Cr	8.00–9.50	8.0–9.0	8.5	8.00–9.50	8.00–9.50	8.00–9.50	—	—	8.00–9.50
Mo	0.85–1.05	0.85–1.05	0.95	0.85–1.05	0.85–1.05	0.85–1.05	—	—	0.85–1.05
V	0.18–0.25	0.18–0.25	0.21	0.18–0.25	0.18–0.25	0.18–0.25	—	—	0.18–0.25
Nb	0.06–0.10	0.06–0.10	0.08	0.06–0.10	0.06–0.10	0.06–0.10	—	—	0.06–0.10
N	0.030–0.070	0.030–0.070	0.05	0.035–0.070	0.040–0.070	0.030–0.070	—	—	0.035–0.070
Ni	0.40 max	0.20 max	<0.10	0.20 max	0.40 max	0.20 max	—	—	0.20 max
Al	0.02 max	0.04 max	<0.02	0.020 max	0.025 max	0.025 max	—	—	0.015 max
Ti	0.01 max	0.01 max	<0.01	0.01 max	0.010 max	0.010 max	—	—	0.01 max
Zr	0.01 max	<0.01	<0.01	0.01 max	—	—	—	—	0.01 max
Cu	—	0.20 max	<0.10	0.25 max	0.10 max	0.10 max	—	—	0.10 max
As	—	—	—	0.012 max[1]			—	0.01	0.010 max[1]
Sn	—	—	—	0.010 max[1]	0.020 max	0.020 max	—	0.01	0.010 max[1]
Sb	—	<0.001	<0.001	0.003 max[1]	0.010 max	0.010 max	—	0.0025	0.003 max[1]
B	—	<0.001	<0.001	—	—	—	—	—	<0.001
W	—	<0.010	<0.010	—	—	—	—	—	<0.010
Pb	—	—	—	—	—	—	—	—	<0.001 max[1]
N : Al ratio	—	—	—	4.0 min	—	—	—	—	4.0 min

[1]The combined value of Sn + As + Sb + Pb < 0.010 weight percent.

Table 2: Chemical composition of examined Grade 91 heats with controlled additions of impurity/tramp elements P, S, Cu, Sn, and Sb [27].

	C	Si	Mn	P	S	Cu	Ni	Cr	Mo	V	Nb	Sn	Sb	N
A	0.10	0.42	0.40	0.012	0.001	—	0.06	8.53	1.00	0.21	0.080	—	—	0.054
P1	0.09	0.30	0.38	0.006	0.007	0.01	0.09	8.63	0.94	0.24	0.060	0.003	0.004	0.031
P2	0.10	0.30	0.38	0.021	0.007	0.01	0.08	8.66	0.94	0.23	0.061	0.003	0.004	0.032
P3	0.09	0.31	0.38	0.030	0.008	0.01	0.09	8.64	0.95	0.24	0.061	0.003	0.004	0.033
S1	0.10	0.31	0.38	0.010	0.002	0.01	0.09	8.61	0.95	0.24	0.058	0.003	0.004	0.032
S2	0.10	0.30	0.37	0.010	0.014	0.01	0.09	8.64	0.95	0.24	0.061	0.003	0.004	0.033
S3	0.10	0.30	0.38	0.009	0.024	0.01	0.09	8.62	0.95	0.24	0.059	0.003	0.004	0.034
Cu1	0.10	0.34	0.39	0.008	0.008	0.05	0.09	8.62	0.89	0.24	0.062	0.004	0.005	0.031
Cu2	0.10	0.34	0.39	0.009	0.008	0.15	0.09	8.58	0.90	0.24	0.064	0.005	0.005	0.031
Cu3	0.10	0.34	0.39	0.008	0.008	0.27	0.09	8.51	0.90	0.24	0.063	0.004	0.005	0.034
Sn1	0.10	0.33	0.39	0.008	0.008	0.01	0.09	8.63	0.93	0.24	0.063	0.011	0.005	0.033
Sn2	0.10	0.33	0.39	0.009	0.008	0.01	0.09	8.63	0.93	0.24	0.063	0.056	0.005	0.033
Sn3	0.10	0.33	0.39	0.008	0.008	0.01	0.10	8.64	0.94	0.24	0.063	0.110	0.005	0.037
Sb1	0.10	0.34	0.38	0.009	0.007	0.01	0.09	8.65	0.92	0.24	0.056	0.004	0.011	0.032
Sb2	0.10	0.34	0.38	0.009	0.007	0.01	0.09	8.73	0.92	0.24	0.057	0.004	0.053	0.032
Sb3	0.10	0.34	0.38	0.009	0.008	0.01	0.09	8.66	0.91	0.23	0.056	0.004	0.110	0.034

Note: the "A" heat is the control heat and may be identified in associated figures as "P0" or "Cu0", and so on.

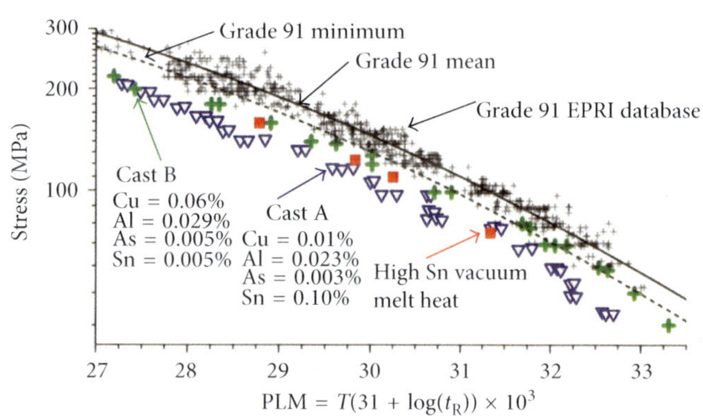

Figure 4: Analysis regarding the behavior of Tin (Sn), as plotted against the EPRI database of Grade 91 base metal using a Larson Miller parameter ($C = 31$), adapted from [27]. Note the reduction in strength for all casts with Sn. Even at the lowest value of 0.005 weight percent, the behavior is at the minimum (mean-20%) line. This reduction in strength may also be partially attributed to the presence of relatively high levels of Al in all examined heats.

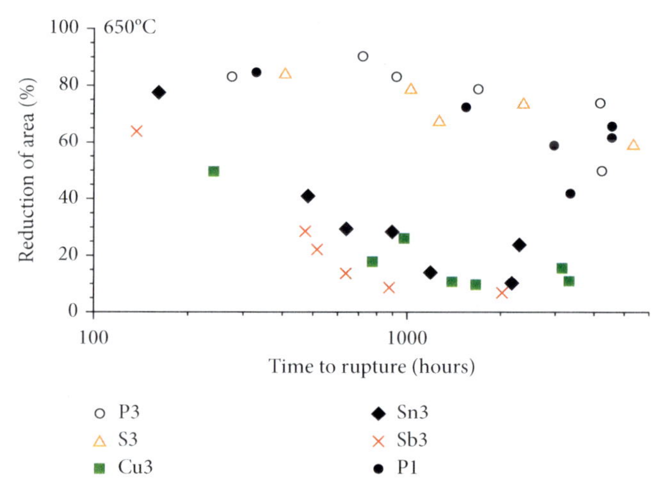

Figure 5: Analysis of Grade 91 behavior for experimental heats doped with P, S, Cu, Sn, and Sb and as compared to a control heat (P1) with low levels of tramp elements. Adapted from [27]. Note the marked decrease in reduction of area for heats doped in Cu, Sn, and Sb and the more gradual drop in behavior with time for heats doped with S and P.

2.2. Influence of Antimony (Sb). The effect of Sb, as analyzed by Masuyama et al. [27], has also been shown to influence the creep ductility. The steel compositions used in this research study [27] are detailed in Table 2. The ductility reduction was reported to initiate even at relatively low levels (Sb1 = 0.011 weight percent). The trend to lower ductility was accelerated by higher temperature and with increasing level of Sb (Figure 6).

As with the Sn-doped specimens, the reduction in the rupture life resulting from the higher amount of tramp elements at 650°C (1202°F) does not appear to be caused by the increase in the creep rate but rather by the decline in creep ductility. This is consistent with both the Sb- and Sn-doped samples showing enhanced creep microvoid nucleation leading to minimal reduction of area. As a consequence of the enhanced damage susceptibility, the rupture ductility reported with the Sb- and Sn-doped samples was lower than that of the P- and S-doped samples (Figure 5).

In the summary documentation prepared to support recommendations for Grade 91 steel, ORNL proposed limiting Sb to be less than 0.001 weight percent [19]. The more potent effect of Sb (e.g., as compared to Sn) is reflected in other guidance [25] where the Sb is recommended to be <0.010 weight percent. Furthermore, for Grade 22 CrMo steel, Sb should be <0.0025 weight percent. More recently, EPRI recommended controlling Sb to <0.003 weight percent in CSEF steels. In practice, the Sb level in commercial heats [22, 26] of Grade 91 steel has been highly controlled to values ≤ 0.0014 weight percent. This clearly reflects the more

TABLE 3: Analysis of the effect of Sn on performance as a function of service temperature for a given stress of 80 MPa (11.6 ksi).

Grade 91 material	Expected life (hours) at 80 MPa (11.6 ksi) and indicated temperature			
	550°C 1022°F	575°C 1067°F	600°C 1112°F	625°C 1157°F
Mean Grade 91	>1,000,000	>1,000,000	588,518	55,732
Min. Grade 91	>1,000,000	>1,000,000	204,913	19,983
Cast A (high Sn)	>1,000,000	360,250	32,340	3,320
Cast B (low Sn)	>1,000,000	>1,000,000	157,406	15,464

Data were taken from Figure 4 [27]. Note the dramatic drop off in anticipated life >550°C (1022°F).

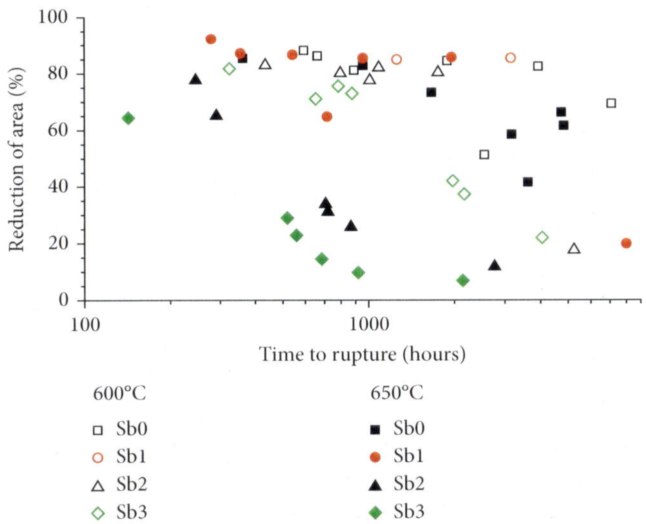

FIGURE 6: Analysis of Grade 91 behavior for experimental heats doped with Sb and tested at 600°C (1112°F) and 650°C (1202°F). Adapted from [27]. For even the heats with the low levels of Sb (Sb1 = 0.011 weight percent), there was a marked drop in the measured creep ductility, this drop was accelerated with increasing levels of Sb.

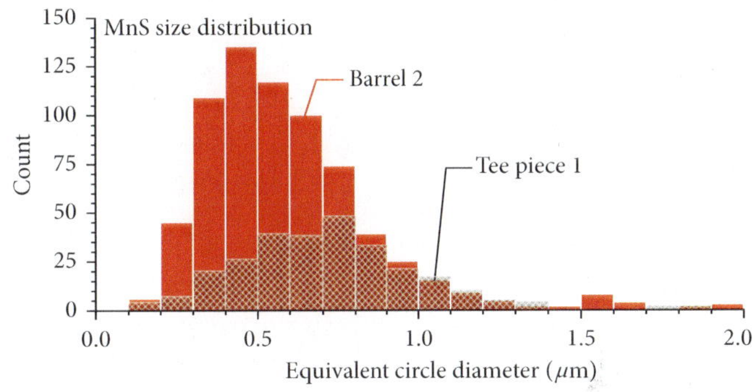

FIGURE 7: Number density of inclusions for two different but code acceptable Grade 91 steels. A higher density of MnS was measured in the creep damage susceptible steel which contained 0.009 wt.% S (labelled barrel 2) as compared to the cavity resistant steel which contained 0.002 wt.% S (labelled tee piece 1).

stringent pr+ocedures used by steel makers regarding the control of Sb. *Limiting Sb to < 0.003 weight percent* in Grade 91 steel, as originally suggested by ORNL [19], has been reinforced by the analysis presented here.

2.3. Influence of Copper (Cu). The influence of Cu on creep behavior has been examined [27]. It was reported that the influence of Cu on creep behavior was similar to that of Sn- and Sb-doped samples (Figure 5). Thus, Cu reduced the observed reduction of area even in plain bar uniaxial creep; the compositions used by Masuyama et al. [27] are detailed in Table 2. Interestingly, in a companion paper [29], the authors stated that, "the rupture life associated with the Cu-doped sample was seen to be independent of the content of Cu. This is because the Cu-doped samples exhibited poor ductility irrespective of Cu content."

With respect to the recommended composition limits for Cu in Grade 91 steel listed in Table 1, there were concerns reported by researchers at ORNL [19] regarding the effect of Cu on stabilizing the austenite phase at high temperature. Therefore, even at alloy introduction, it was recommended to keep the target chemistry at Cu < 0.10 weight percent [19]. This is the same value recommended in other early publications [21, 22]. Controlled composition Grade 91 type steels invariably contained less than 0.10 weight percent Cu [19, 23, 26, 31].

Thus, it is recommended that the *copper content of Grade 91 steel should be < 0.10 weight percent.*

2.4. Influence of Sulfur (S) and Phosphorous (P). In common with many engineering steels, these elements have been controlled from the first introduction of Grade 91 steel, [19]. The initial maximum levels of P and S appear to have been based on experience rather than detailed research study. The influence of S and P on the creep performance of Grade 91 steel has been examined in several subsequent research studies. It is noteworthy that in the work by Masuyama et al. [27], the levels of P and S used were below the specified maximum from typical design codes (Tables 1 and 3). A general trend of lower reduction in area was noted with increasing S for increasing time to rupture for creep tests conducted at 650°C (1202°F). Furthermore, and in the analysis of Grade 91 steel casts which exhibit excellent creep ductility and the stated maximum values quoted by some material suppliers, the values for S are specifically limited to values <0.005 weight percent.

EPRI-recommended methods for a component life management approaches identify that wherever possible root cause investigations should be performed to understand the factors influencing instances of service cracking. Indeed, several papers and reports document findings from investigations into cracking found during service in Grade 91 steel components. In particular, detailed laboratory

Table 4: Chemical composition (given in weight percent) for two exservice steels compared to the requirements in a typical standard specification [23].

Material	C	Cr	Mn	Mo	Ni	Si	Al	P	S	N	Nb	V	Ti	Zr	Cu	Pb	As	Sn	Sb	B	O
SA-387 P91	0.08 0.12	8.0 9.5	0.30 0.60	0.85 1.05	0.40 max	0.20 0.50	0.02 max	0.020 max	0.010 max	0.030 0.070	0.06 0.10	0.18 0.25	0.01 max	0.01 max	N/A	N/A	N/A	N/A	N/A	N/A	N/A
Damage susceptible heat	0.10	8.77	0.41	0.94	0.12	0.41	0.040	0.009	0.010	0.0454	0.071	0.21	<0.002	<0.002	0.19	0.00075	0.0128	0.008	0.0023	<0.0003	0.003
Damage resistant heat	0.10	8.30	0.40	0.94	0.19	0.33	0.020	0.012	0.002	0.0424	0.070	0.21	<0.002	<0.002	0.05	0.00003	0.0042	0.003	0.00063	<0.0003	0.0024

characterization and metallurgical evaluation have been carried out on sections from a secondary superheater header that was removed from a coal fired power plant after about 79,000 hours operation [5]. This replacement was required due to large numbers of cracks in which formed in the heat-affected zones of stub tube to header welds. The damage formed a pattern with different header sections showing different susceptibilities to creep crack formation. Study of the exservice steel thus permitted evidence to be obtained regarding the specific metallurgy of steel heats which had developed cracks (identified as damage susceptible) and those which were uncracked (identified as damage resistant). Comprehensive compositional analysis of these heats showed that the damage susceptible heat (identified as Barrel 2 in Figure 7) contained S at the top of the then code allowed specification (Table 4).

Further study was performed to understand details of how the sulfur level in the steel could influence susceptibility to formation of creep cavities. It was found that there was a consistent trend to increased number density of inclusions for Grade 91 steels with increasing S content. This increase was particularly notable for S values ≥0.003 wt.%. Additional analysis, undertaken for two heats of material using a refined and higher resolution approach developed by EPRI, confirms the presence of a higher density of MnS in creep damage susceptible steel (0.009 wt.% S) as compared to cavity resistant steel (0.002 wt.% S) (Figure 7). This figure shows the distribution of the size of MnS particles, represented as an equivalent circular diameter, as being centered at ~500 to 800 nm. For conventional assessment methods using basic SEM-EDS approaches, the detection of very small MnS inclusions would be unreliable. Thus, it is apparent that the minimum particle diameter identified reliably using conventional SEM-EDS is typically ~1,000 nm (1 μm). Only when a specialist procedure, optimized for detection of smaller particles, is used can accurate number densities below this size be recorded.

Review of the compositions in Grade 91 steel: it appears that the *S content should be limited to ≤ 0.005 weight percent*, and if possible it would appear to be beneficial to restrict S *to ≤ 0.003 weight percent.*

2.5. Influence of Boron (B). Boron has been used as an alloying addition in a number of CSEF steels, including Grades 92, 122, and SAVE 12 AD [32–36]. In controlled compositions, and with the correct heat treatment, boron can be beneficial to the creep behavior of tempered martensitic CSEF steels. Evidence suggests that one benefit of adding B is a reduced coarsening rate of $M_{23}C_6$ carbides. However, when nitrogen is present in sufficient quantities, excessive levels of B introduce a risk of forming BN [32]. The formation of BN can be problematic for two reasons. First, it reduces the amount of available nitrogen in the steel matrix and so can limit the formation of precipitation strengthening MX carbides. Secondly when of sufficient size, the BN inclusions can act as initiation sites for creep cavities, lowering the creep life and reducing the creep ductility (Figure 8).

Detailed metallographic characterization has documented the development of BN in Grade 92 and Grade 122 steels. Although BN has not been reported in Grade 91 steel, its formation could take place because B is not specifically controlled in most standards. Abe [33] has researched the formation of BN in Grade 92 steel in detail. The diagram shown in Figure 9 has been developed based on this previous work [33]. For the range of nitrogen in the Grade 91 specification, controlling the amount of $B \leq 0.001$ weight percent helps to prevent the formation of BN.

In practice, it appears that most suppliers of Grade 91 steel have controlled the levels of *B* to ≤0.001 *weight percent maximum*; this would be a sensible upper limit on the B content in alloy specifications for this steel.

2.6. Influence of Nickel (Ni). The original ORNL reports detailing the development for Grade 91 steel [24] recognized the risk that excessive Ni levels in Grade 91 steel could reduce the long-term creep strength. More recently, the results of long-term creep tests have confirmed that Ni level within the currently allowable range in alloy specifications can reduce the long-term creep strength [37]. Based on these results, estimated long-term creep lives for Ni at the maximum allowed (0.40 weight percent) would be below the minimum expected by the ASME code. The long-term detrimental effect of Ni is attributed to an increase in the rate of recovery of the microstructure and an increased tendency to form Z-phase. It has been reported that one of the main motivations behind having a specified limit for Ni of 0.40 weight percent maximum was the perceived benefit of being able to melt a wide range of steel scrap. Despite the recommendation by ORNL, when Grade 91 type steel was accepted by ASME [19], the nickel level was set at 0.40 weight percent maximum [17] as *it was believed to be a harmless element.*

The observations reported by NIMS [37] have recently been confirmed in work reported at the European Creep

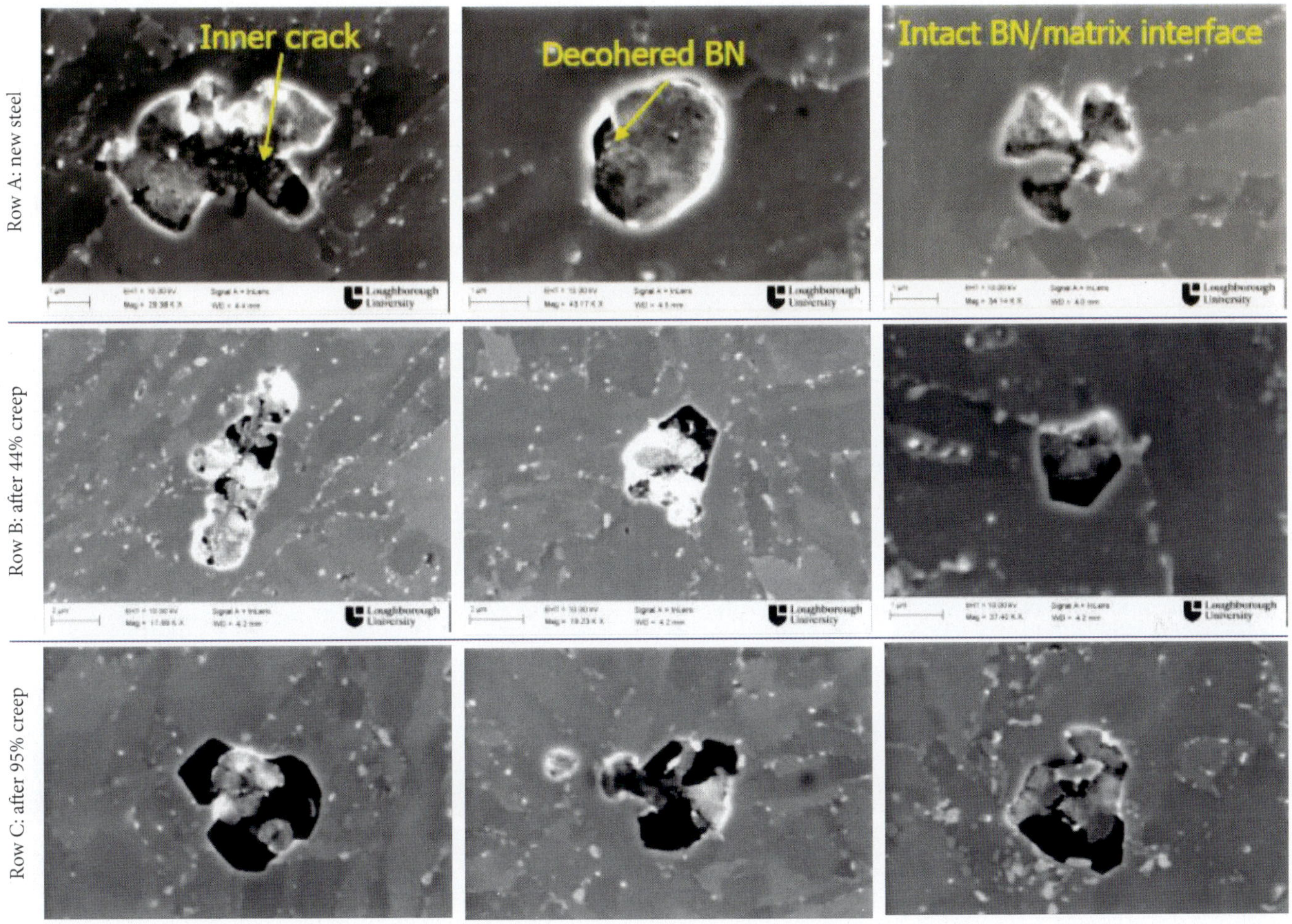

Figure 8: Micrographs from Grade 92 CSEF steel showing the development of creep cavities associated with BN particles, row A BN particles in new steel, row B after exposure to 44% creep life, and row C after exposure of 95% creep life.

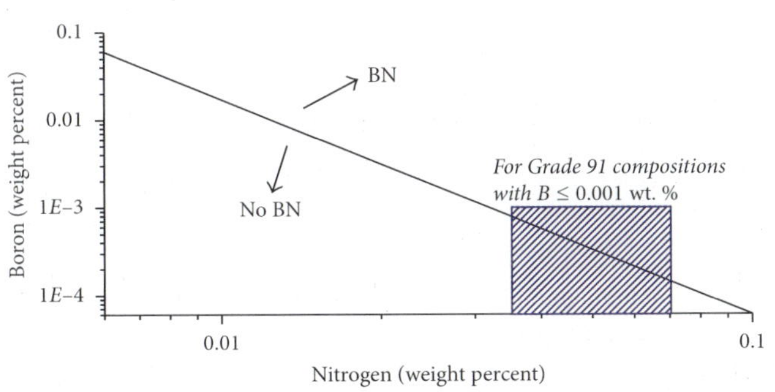

Figure 9: Relationship developed for the formation of BN in Grade 92 steel with different levels of B and N based on previous research [33]. The blue hashed box shows the range of N typically specified in Grade 91 steel for a B level of ≤0.001 weight percent.

Collaborative Committee (ECCC) [38]. It is noteworthy in that an early analysis of the ECCC database indicated no effect of Ni on performance of Grade 91 steel. More recently, however, the ECCC database was reevaluated using only long-term data for rupture lives >10,000 hours. Upon this reassessment of the ECCC database, the author remarked [38], "The long-term harmful influence of nickel on Grade 91 creep strength, proposed by Japanese researchers, is a real effect. . .. A restriction to 0.20 weight percent nickel maximum, as is now being introduced into many codes and standards is therefore to be strongly encouraged. . . Nevertheless, its effect [ref. nickel content] is comparable with that of aluminum, and should certainly be taken into account in component life management." It is thus concluded that excessive Ni can be a significant influence in long-term behavior with regard to strength.

A value of 0.20 weight percent maximum was first proposed in the ORNL summary report to ASME and ASTM in the early 1980s [19]. This recommendation was reinforced by the published maximum value in [21] and has more recently been recommended in [20] (Table 1). In practice, consideration of the maximum nickel value reported for components made to controlled compositions shows a value consistently below 0.20 weight percent [22–24, 26, 32].

There has never been any evidence reported to support the position that levels of Ni above 0.20 weight percent offer any benefit to the performance of Grade 91 steel components. It is thus concluded that this value should be regarded as an absolute maximum moving forward. Thus, in Grade 91

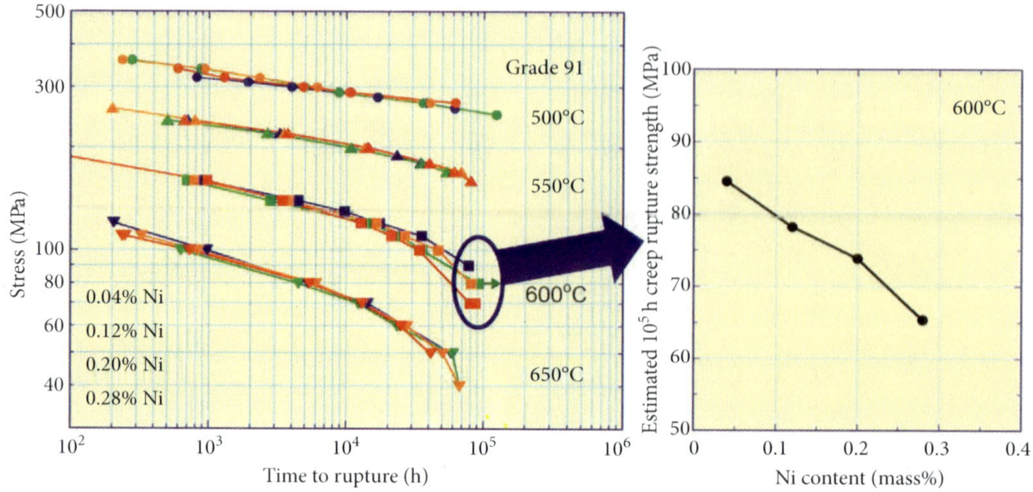

FIGURE 10: The influence of Ni on the creep strength in Grade 91 steel as reported by [37]. The detail at right shows that the 100,000 creep rupture strength decreases with increasing Ni content.

steel, a *maximum Ni content of 0.20 weight percent is recommended.*

2.7. Influence of the Nitrogen to Aluminum (N : Al) Ratio. As a result of postservice examination of Grade 91 steel components which cracked relatively early in life, a broad review of the composition of different Grade 91 components in UK power stations was undertaken. Because high levels of Al were commonly found in cracked components, this review also considered the challenges associated with accurate measurement of Al [39]. The results of this review supported the concept of a link between the ratio of N : Al present in the steel and the susceptibility for developing creep damage. This effect was explained on the basis that the formation of AlN was such that there was then insufficient nitrogen available to for the MX carbonitrides needed to ensure high temperature strength [39].

As a consequence of creep damage, a secondary superheater header was replaced after 79,000 hours. Background regarding the in-service experience with this header has been published previously [5]. Detailed root cause analysis was performed on sections of header which had developed hundreds of weld HAZ cracks in service. In summary, the HAZ cracking was found to be overwhelmingly associated with the three cylindrical sections (described as "barrels"). The two component sections with the highest percentage of cracks were identified as Barrel 2, >70% of stub welds cracked (N : Al = 1.42) and Barrel 5, >60% of stub welds cracked (N : Al = 1.57) (Figure 10). These two sections were reported to have been fabricated from the same ingot material. It is interesting to note that the third heavily damaged barrel, almost 50% cracking in the stub welds, Barrel 4 had an N : Al ratio of 3.67. An N : Al ratio of 3.67 is significantly higher than reported in earlier problem casts [8, 39]. Indeed, this ratio would be considered acceptable by today's ASME code requirements (Figure 11).

The reasons why an excessive N : Al ratio can be deleterious to performance of Grade 91 steel have been investigated [20]. One key finding was that when significant levels of Al are present, the AlN forms preferentially and so the amount of available N to form MX carbonitrides is reduced. In extreme cases, the number density of MX is not sufficient to stabilize the substructure.

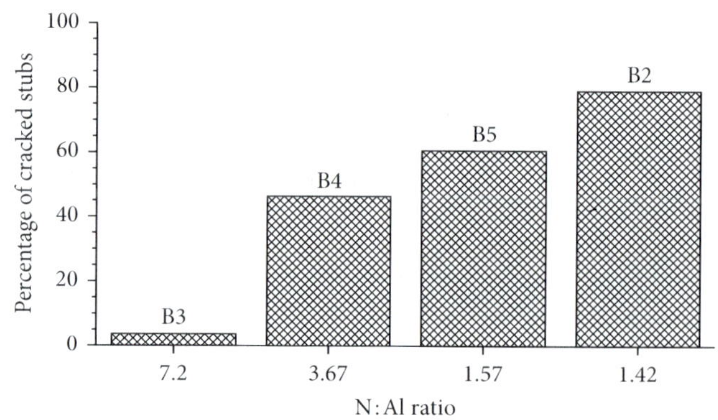

FIGURE 11: Analysis of the N : Al ratio for specific casts of Grade 91 steel as a function of the percentage of stubs cracked in a give barrel section. Adapted from [17].

The poorly stabilized substructure means that the steel has reduced resistance to tempering and shows a rapid degradation under creep conditions. The lower resistance to tempering has been shown in experiments tracking changes in hardness with increasing thermal exposure. These results which were obtained for Grade 91 steel samples with N : Al ratios of 1.3 G11, ~5 G8, and 13.5 J4 are summarized in Figure 12. For Grade 91 steel with very high N : Al ratios (J4), the resistance to tempering during exposure to elevated temperatures is remarkable. In this example, the examined material with a very high N : Al ratio (sample J4) exhibited a lower initial hardness value. However, for the most extreme condition, namely, tempering at 790°C (1450°F), that is, just below the A_1 temperature, for 50 hours, it can be seen that the J4 material has exceptional resistance to softening. Thus, even after a high degree of

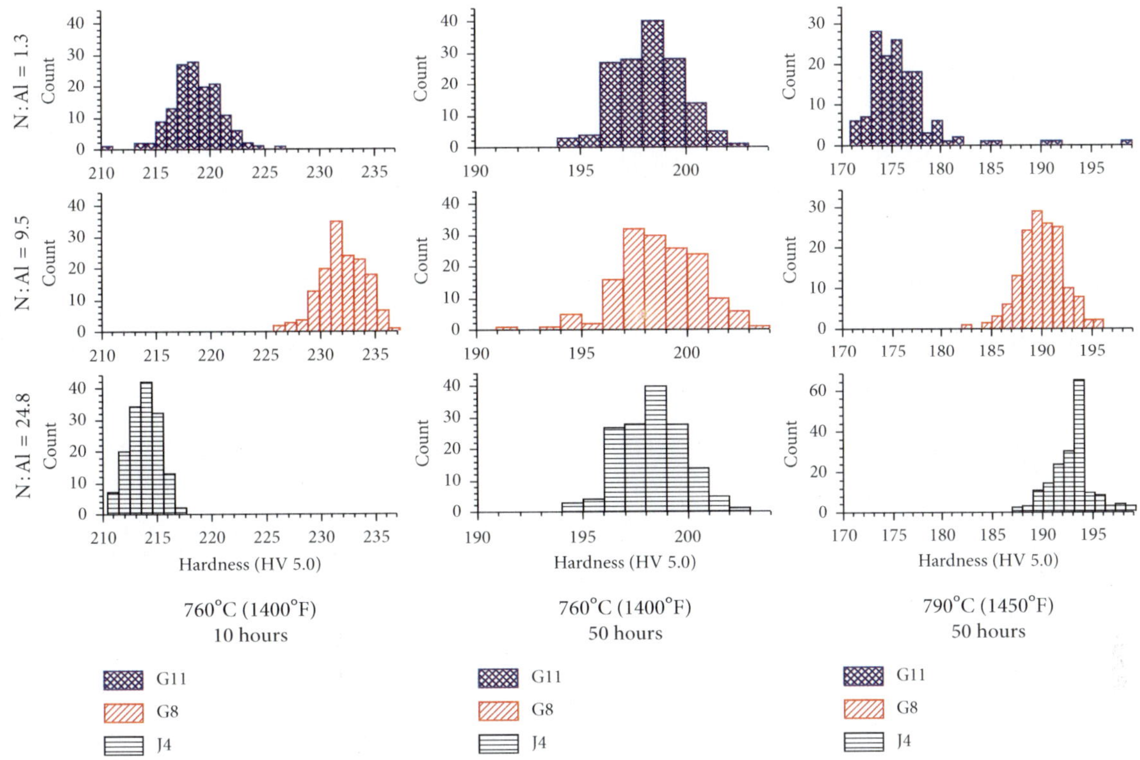

Figure 12: Analysis of the N : Al ratio on tempering behavior for specific casts of Grade 91 steel. Note that although steel J4 starts with the lowest initial hardness value, as demonstrated in the 760°C (1400°F) for 10 hours experiments, it exhibits the highest resistance to tempering and a more consistent (i.e., lower standard deviation) hardness for each examined tempering treatment.

tempering the measured hardness was above 190 Hv (Figure 12). In contrast, the steel with the lowest N : Al ratio, sample G11, has the least resistance to tempering. For this heat, tempering at 790°C (1450°F) for 50 hours reduced the vast majority of measured hardness values to less than 180 Hv (Figure 12).

No additional recommendations are suggested for control of the Al content in Grade 91 steels. Thus, to *reduce the risk of formation of AlN, it is recommended by EPRI that a minimum N : Al ratio of 4 should be adopted.*

3. Implications of Controlled Composition on Performance

The high temperature performance of CSEF steels is dependent on steel making, steel processing, composition, and heat treatment. It has been very well established that poor control of heat treatment leads to incorrect microstructures and poor high temperature performance. Thus, when irregularities of heat treatment result in a ferrite microstructure, the creep strength is significantly reduced compared to the creep strength of tempered martensite. In contrast, variability in the measured reduction in area during creep has been reported for steels exhibiting a tempered martensitic microstructure. This variability is such that at similar creep lifetimes, failure has been reported with either high ductility or low ductility (e.g., see Figure 13).

Posttest characterization of selected laboratory samples has shown that the low ductility failures were predominantly

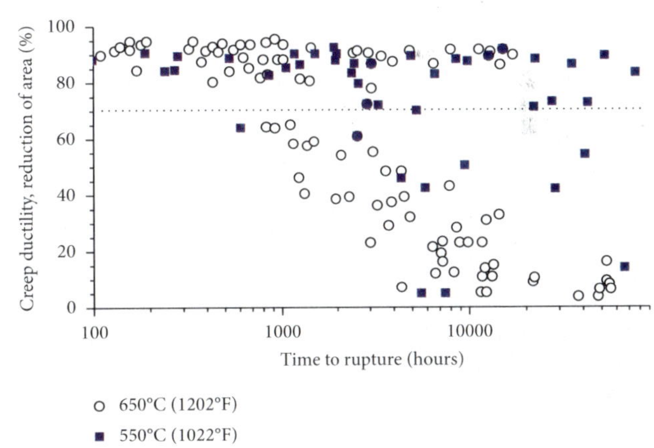

Figure 13: Creep ductility (reduction of area) for Grade 91 steel supplied by one manufacturer [40]. A dashed line is shown at 70% for the reduction of area since this was an acceptance criterion proposed by ORNL in the development of Grade 91 steel [20].

a consequence of the nucleation, growth, and link-up of cavities. The tendency for creep brittle behavior in CSEF steels is thus due to the formation of microvoids on prior austenite grain boundaries and at other microstructural features such as lath boundaries. The range of different sites for void nucleation means that in contrast to low allow steels, there are high number densities of creep voids present prior to microcrack formation. The diversity of void nucleation sites is also a challenge to tracking in-service component

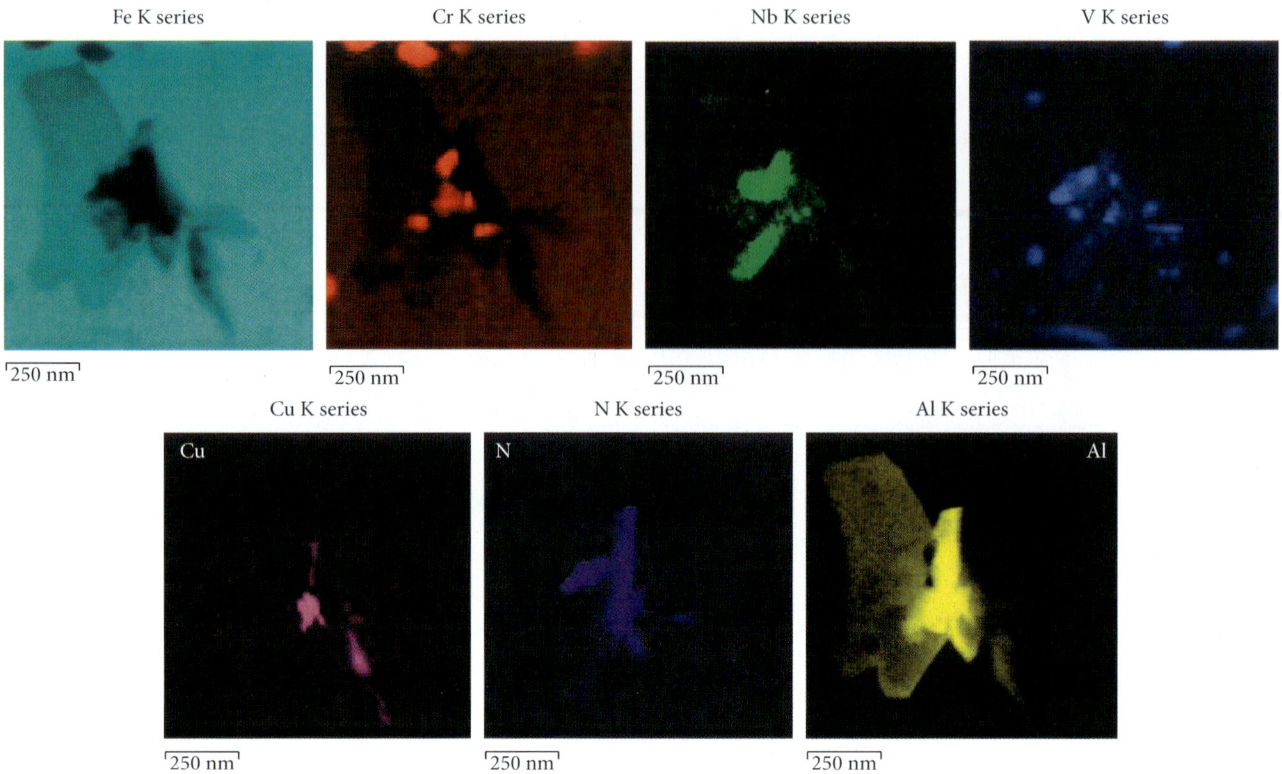

FIGURE 14: Detailed micrograph showing inclusions associated with an individual creep cavity as well as compositional maps showing the distribution of the elements Cr, Nb, V, Cu, N, and Al associated with this location.

damage using traditional inspection methods. While the details of the number of voids formed, and the tendency for reductions in strain to fracture, is different for the different CSEF steels, research to date shows that void nucleation is related to the presence of trace elements and hard non-metallic inclusions. An example of the link between the nucleation of a creep void, the presence of inclusions and segregation of copper is illustrated by the analysis results shown in Figure 14. A key factor in determining whether the inclusions nucleate voids is the particle size. Thus, only inclusions of a sufficient size (the critical inclusion size is directly linked to the creep stress) will be thermodynamically stable and thus able to act directly as nucleation sites.

It is also clear from the recent root cause analyses of Grade 91 steel components [5, 40] that steel composition and processing variables are linked to low creep ductility in base metal and weld HAZs. The observed in-service component cracking cannot be simply explained as a 'one off' anomaly. Examination of failures in several Grade 91 welded components [40] indicates a trend where an increasing number of failures of Grade 91 welds are occurring in times below that expected based on design analysis. The trend in the reduced creep ductility in Grade 91 steel and the link to very low HAZ creep life may be a characteristic of a significant number of components which entered service with a high susceptibility for cavity formation. Poor performance of weld HAZs is a key reason for the introduction of weld-efficiency factors. Further, reductions in weld creep performance could lead to increases in recommended weld strength reduction factors (WSRF) or weld-efficiency factors. In contrast, using Grade 91 steel with low densities of inclusions and controlled levels of deleterious trace elements should significantly reduce the risk of creep cracking associated with weldments.

It is thus clear from detailed examination that the creep ROA is vastly improved when the levels of inclusions and impurities in Grade 91 steels are low. Metallographic characterization using advanced techniques has established detailed trends in behavior. To consider a broader assessment of *Metallurgical Risk*, data were collated from the results of specific test programs. Test information was obtained from global sources including from EPRI projects, ORNL, Japan, and Europe. Although efforts were made to consider a broad range of research results, accurate and comprehensive compositions linked to long-term creep data were only available for a smaller number of heats than was expected. Thus, rather than being able to make a quantitative analysis of recorded values of creep ROA, the creep performance factor was simply organized as

(i) high susceptibility to damage (poor creep ductility),

(ii) low susceptibility to damage (high creep ductility).

A total of 36 detailed data sets were available for analysis, and the results are presented in Figure 15. As shown, this assessment covered the following elements–Cr, Mo, W, V, Nb, Ti, C, Mn, Ni, Si, Al, N, Cu, S, P, As, Sn, Sb, and Pb. In Figure 15, the heats which exhibited low susceptibility to creep damage are shown as blue box with heats which

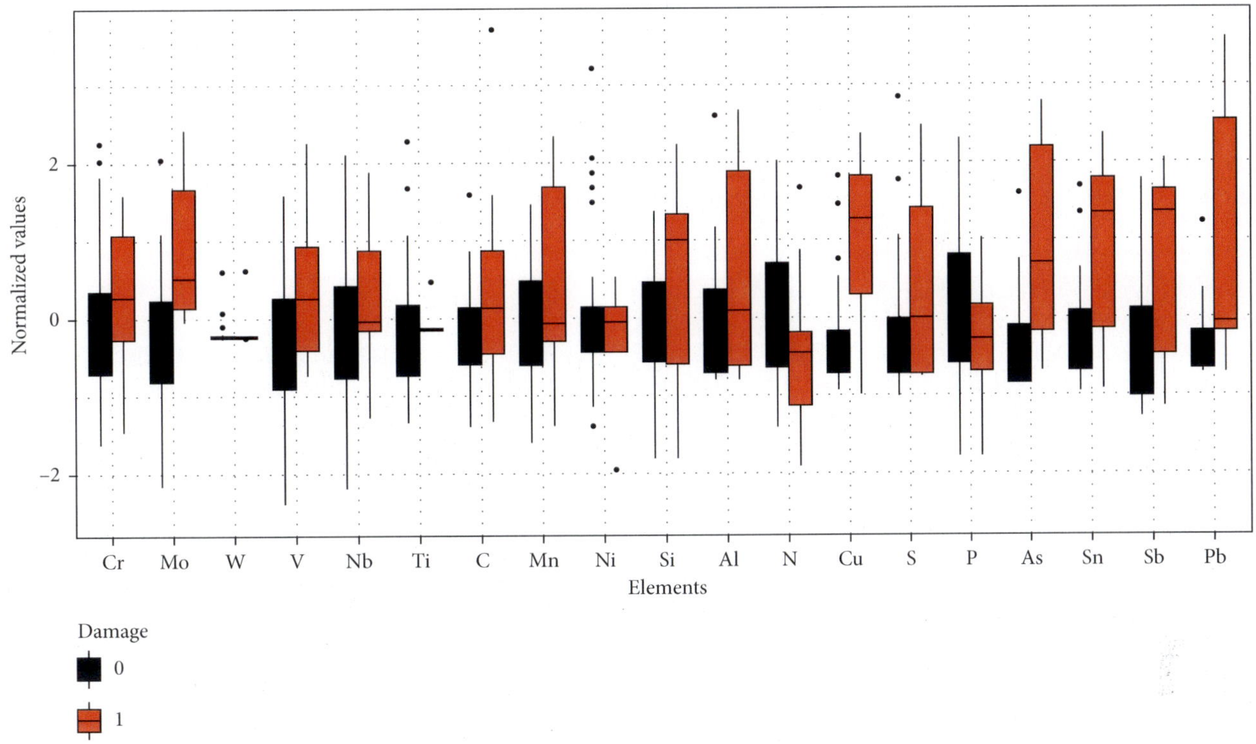

FIGURE 15: Trend showing how the level of the elements Mo, V, Nb, C, Mn, Si, Al, Cu, S, As, Sn, Sb, and Pb in Grade 91 steels varies for damage susceptible steels (red boxes) and damage resistant steels (blue boxes).

exhibited high susceptibility to creep damage shown as red boxes. The size of each of the boxes designates the range of each element in the data set with the horizontal line indicating the average value. It is apparent that the statistical trend shows that elements Mo, V, Nb, C, Mn, Si, Al, Cu, S, As, Sn, Sb, and Pb show a significantly higher level in the damage susceptible heats. An interpretation of these observations is that Mo, V, Nb, C, and Mn will be linked to creep strengthening and as such should be expected to enhance the deformation resistance of the steel. The elements Si, Al, Cu, S, As, Sn, Sb, and Pb are typically associated with the formation of inclusions and exhibit a tendency to promote the nucleation and growth of cavities. The combination of these trends is consistent with the previous metallographic examinations linked to observed decreases in creep ductility.

It is apparent that the results from the different assessment approaches show consistent outcomes; for example, for Grade 91 tempered martensitic steels, a controlled composition with minimal amounts of impurities and tramp elements can have a profound effect on the creep ductility. A controlled Grade 91 composition, specified as code case 2864 "Class 2" material, has recently been accepted by ASME. This Class 2 steel has a composition in line with the information presented in Table 1. It is believed these recommendations will have a dramatic effect on improving creep ductility and therefore provide significantly better in-service performance. These benefits will be particularly important for components operating under creep condition, that is, above about 550°C.

The trend showing that controlled composition and fabrication leads to improved behavior has been identified in other tempered martensitic steels. A comprehensive research program studying Grade 92 steel, from three distinct sources made in separate parts of the world, exhibited a range of behavior with regard to creep ductility. Clearly, the performance of "BM C" as opposed to "BM A" and "BM B" increased with respect to both creep strength and ductility. This improved creep performance was directly related to the controlled composition and variations in the size and distribution of inclusions in each of the steel heats.

For materials which may exhibit high levels of tramp elements, it would be advantageous to define a metallurgical risk factor. However, it must be emphasized that changes in ductility should be established for appropriate stress states, component geometry, and operational conditions. It is apparent that the component performance factors with respect to strength and ductility are different for changing operating regimes. For example, in the mid-2000s, the service experience with Grade 91 in the USA was particularly positive. It appears that this was mostly because the operating conditions were around 540°C (1004°F) and 240 bar (3,480 psi). Similarly, satisfactory performance has been reported from Europe. Clearly then the steam conditions of the power plant with respect to temperature and pressure are important in being able to anticipate future issues with performance of the Grade 91 material.

It is important to note that the chemical composition is only part of the overall proposal for more highly controlled Grade 91 steel. Research linked to understanding and modelling of the factors affecting creep behaviour should consider careful control and selection of the following metallurgical factors—steel composition, steel making, hot working method conditions and degree of hot reduction, normalizing temperature and time, cooling rate from normalizing, and tempering temperature, time, and controls. The influence of inclusion content and impurity elements on performance is recognized from the present results as being synergistic, that is, when both high levels of trace elements and inclusions are present, the creep performance is very poor. The evidence from the present work is supported by data from publications and the results of root cause analysis of cracking in-service. It is also important that research programs consider the role of stress state, particularly as constraint effects have an influence of creep damage development.

4. Conclusions

Design codes for components operating at high temperatures have traditionally assessed expected performance based on an evaluation of creep strength, underpinned by the expectation that alloys would be creep ductile. Thus, for CrMo type low alloy steels, relationships which combine defined trends in tensile and creep strength with Factors of Safety have typically ensured that the agreed allowable stress values provide the basis for achieving at least minimum performance. The usefulness of the traditional approaches is severely challenged when assessing the behavior of metallurgically complex steels. In particular, problems are encountered when steels exhibit poor creep ductility. In circumstances of low creep ductility, local stress relaxation can promote creep damage. It is now well established that the creep properties of CSEF steels are critically dependent on factors which include control of composition, steel making, processing, and all heat treatments. Lack of correct control of these factors can severely compromise strength, ductility, or both. Indeed, in-service failures of Grade 91 steel components have occurred relatively early in life. The very large range of creep behavior observed in Grade 91 steel components introduces many risks for life management and assessment. The greatest concern is that, as opposed to the expected leak-before-break failure, a crack in low ductility material will propagate rapidly and cause catastrophic failure. Moreover, the existence of significant densities of voids prior to crack initiation complicates in-service assessment of condition and weld repair of these steels.

Changes in composition and heat treatment clearly have a marked effect on creep strength. Tempered martensitic steels such as X20, Grades 91, 92, E911, and 122 all exhibit a trend to relatively low ductility as creep rupture times increase. The tendency for brittle behavior is, in all cases, due to the formation of creep voids on prior austenite grain boundaries and at other microstructural features such as lath boundaries. The detail of the number of voids formed and the tendency for reductions in strain to fracture is different for each of the steel grades. However, it appears that void nucleation is related to the presence of trace elements and other metallurgical risk factors such as nonmetallic inclusions or evolution of Z-phase. It is clear that proper control of the composition in Grade 91 steel can result in an increase in the performance of the material with respect to creep ductility.

The link between creep ductility and in-service damage has been shown for a wide range of materials. To ensure that more consistent Grade 91 steel can be procured through the relevant suppliers, it is suggested that the composition be further controlled by ASME and other Design codes.

5. Declaration

The Electric Power Research Institute (EPRI) conducts research, development, and demonstration projects for the benefit of the public in the United States and internationally. As an independent, nonprofit organization for public interest energy and environmental research, we focus on electricity generation, delivery, and use in collaboration with the electricity sector, its stakeholders, and others to enhance the quality of life by making electric power safe, reliable, affordable, and environmentally responsible.

Conflicts of Interest

The authors declare that they have no conflicts of interest.

References

[1] J. D. Parker, "Creep cavitation in CSEF steels," in *Proceedings of the 7th International Conference on Advances in Materials Technology for Fossil Power Plants*, Waikoloa, HI, USA, October 2013.

[2] L. Cipolla, A. Di Gianfrancesco, D. Venditti, G. Cumino, and S. Caminada, "Microstructural evolution during long term creep tests of 9%Cr steel grades," in *Proceedings of the 8th International Conference on Creep and Fatigue at Elevated Temperatures*, pp. 445–459, San Antonio, TX, USA, 2007.

[3] K. Kimura, "Creep rupture life prediction of creep resistant steels," *Journal of the Japan Institute of Metals*, vol. 73, no. 5, pp. 323–333, 2009.

[4] K. Sawada, H. Kushima, M. Tabuchi, and K. Kimura, "Microstructural degradation of Gr.91 steel during creep under low stress," *Materials Science and Engineering A*, vol. 528, no. 16-17, pp. 5511–5518, 2011.

[5] J. D. Parker and S. Brett, "Creep performance of a grade 91 header," *International Journal of Pressure Vessels and Piping*, vol. 111-112, pp. 82–88, 2013.

[6] X. Xu, G. D. West, J. A. Siefert, and R. C. Thomson, "The influence of thermal cycles on the microstructure of grade 92 steel," *Metallurgical Transactions A*, vol. 48, no. 11, pp. 5396–5414, 2017.

[7] X. Xu, G. D. West, J. A. Siefert, and R. C. Thomson, "Microstructural characterization of the heat affected zones in grade 92 steel welds: double-pass and multi-pass Welds," *Metallurgical and Materials Transactions A*, 2018.

[8] J. D. Parker and A. W. J. Parsons, "High temperature deformation and fracture processes in 214Cr-1Mo and 12Cr12Mo14V Weldments," *International Journal of Pressure Vessels and Piping*, vol. 63, no. 1, pp. 45–54, 1995.

[9] EPRI, *Embrittlement of Power Plant Steels*, EPRI, Palo Alto, CA, USA, 2013.

[10] B. L. King, "Intergranular embrittlement in CrMoV steels: an assessment of the effects of residual impurity elements on high

temperature ductility and crack growth," *Philosophical Transactions of the Royal Society A*, vol. 295, no. 1413, pp. 235–251, 1980.
[11] R. McPherson, "Compositional effects on reheat cracking of low-alloy ferritic steels," *Metals Forum*, vol. 3, no. 3, pp. 175–186, 1980.
[12] M. Yoshizawa, M. Igarashi, K. Moriguchi, A. Iseda, H. G. Armaki, and K. Maruyama, "Effect of precipitates on long-term creep deformation properties of P92 and P122 type advanced ferritic steels for USC power plants," *Materials Science and Engineering A*, vol. 510-511, pp. 162–168, 2009.
[13] H. K. Danielsen, P. E. DiNunzio, and J. Hald, "Kinetics of Z-phase precipitation in 9 to 12 pct Cr steels," *Metallurgical and Materials Transactions A–Physical Metallurgy and Materials Science*, vol. 44A, no. 5, pp. 2445–2452, 2013.
[14] K. Sawada, H. Kushima, M. Tabuchi, and K. Kimura, "Effect of creep deformation on Z phase formation in grade 91 steel," *Materials Science and Technology*, vol. 30, no. 1, pp. 12–16, 2014.
[15] EPRI, *Sigma Phase in 316 Stainless Piping, Main-steam Piping Failure: Eddystone-1 Plant*, EPRI, Palo Alto, CA, USA, 1989.
[16] J. Henry, T. Ward, and G. Zhou, "Lessons from the past: materials-related issues in an ultra-supercritical boiler at Eddystone plant," *Materials at High Temperatures*, vol. 24, no. 4, pp. 249–258, 2007.
[17] *SA-335/SA-335M Specification for Seamless Ferritic AlloySteel Pipe for High-Temperature Service, ASME Boiler and Pressure Vessel Code, Section II, Part A*, p. 567, 2013.
[18] J. D. Parker and K. Coleman, "EPRI Guidelines for Fabrication of Components Manufactured From Grade 91 Steel," in *Proceedings of the ASME 2012 Pressure Vessels and Piping Conference*, Toronto, ON, Canada, July 2012.
[19] J. R. DiStefano and V. K. Sikka, *Summary of Modified 9Cr-1Mo Steel Development Program, 1975-1986*, ORNL, Oak Ridge, TN, USA, 1986.
[20] EPRI, *Guidelines and Specifications for High-Reliability Fossil Power Plants: Best Practice Guideline for Manufacturing and Construction of Grade 91 Steel Components*, EPRI, Palo Alto, CA, USA, 2015.
[21] K. Haarmann, J. C. Vaillant, B. Vandenberghe, W. Bendick, and A. Arbab, *The P91 Book: Ferritic Pipe for High Temperature Use in Boilers and Petrochemical Applications*, Vallourec Industries, Aulnoye-Aymeries, France, 1992.
[22] P. A. Pellicani, J. C. Vaillant, J. Gabrel, and A. Vanderschaeghe, "Modified 9Cr (T91/P91) steel tubes and pipe: shop production and fabrication datas," in *Proceedings of the EPRI Third International Conference on Improved Coal-Fired Power Plants*, San Francisco, CA, USA, April 1991.
[23] S. Pillot, Z. Zhao, S. Corre, C. Chauvy, L. Coudreuse, and P. Toussaint, "Thick plates in grade 91 for fourth generation nuclear reactors," in *Proceedings of the ASME 2010 Pressure Vessels and Piping Division/K-PVP Conference*, Bellevue, WA, USA, July 2010.
[24] ArcelorMittal USA, *Plate A516 and A387 Pressure Vessel Steels: A Technical Overview*, Arcelor Mittal USA Publication, Chicago, IL, USA, 2013.
[25] Y. Tsuchida, K. Tokuno, and K. Hashimoto, *Development and BOF Manufacture of Modified 9Cr-1Mo Steel Plates with Excellent Strength and Toughness*, pp. 27–35, 1993, Nippon Steel Technical Report No. 58.
[26] J.D Parker and J.A. Siefert, "The effect of metallurgical factors and stress state on the performance of high energy components manufactured from creep strength enhanced steels," in *Proceedings of the ECCC Creep and Fracture 2017 Creep and Fracture in High Temperature Components Design and Life Assessment*, Düsseldorf, Germany, 2017.
[27] F. Masuyama, N. Nishimura, and M. Ozaki, "Creep properties of grade 91 castings and forgings with high residual elements," in *Proceedings of the Experience with Creep-Strength Enhanced Ferritic Steels*, San Diego, CA, USA, July 2004.
[28] F. Masuyama and T. Toyoda, "Creep strength and mechanical properties of air-melt mod. 9Cr-1Mo cast steel," in *Proceedings of the International Symposium on Improved Technology for Fossil Power Plants New and Retrofit Applications*, Washington, DC, USA, March 1993.
[29] N. Nishimura, M. Ozaki, and F. Masuyama, "The effect of the tramp elements on the creep property of the modified 9Cr1Mo steel," in *Proceedings of the 5th International Conference on Clean Steel*, vol. 2, pp. 117–125, Vienna, Austria, July 1997.
[30] S. H. Song, Y. W. Xu, and H. F. Yang, "Effect of impurity tin on the creep properties of a p91 heat-resistant steel," *Metallurgical and Materials Transactions A*, vol. 45, no. 10, pp. 4361–4370, 2014.
[31] R. F. Hanus, K. H. Schonfeld, and H. Wagner, "Transformation of knowledge and technology from research and development to the commercial production of heavy steel castings and forgings for power engineering, made of advanced creep resistant steels," in *Proceedings of the 5th International Charles Parsons Turbine Conference*, Churchill College, Cambridge, UK, July 2000.
[32] M. Rauch, K. Maile, P. Seliger, and A. Reuter, *Creep Damage Development in Martensitic 9Cr Steels*, VGB Power Tech, Essen, Germany, 2004.
[33] F. Abe, "Creep deformation behavior and alloy design philosophy of creep-resistant tempered martensitic 9Cr steel," in *Proceedings of the Sixth International Conference on Advances in Materials Technology*, Santa Fe, NM, USA, September 2011.
[34] U. A. Sachadel, P. F. Morris, and P. D. Clarke, "Design of 10% Cr martensitic steels for improved creep resistance in power plant applications," *Materials Science and Technology*, vol. 29, no. 7, pp. 767–774, 2013.
[35] F. Abe, H. Semba, M. Igarashi, M. Yoshizawa, N. Komai, and A. Fujita, "Feasibility of MARBN steel for application to thick section boiler components in USC power plant at 650°C," in *Proceedings of the Fifth International Conference on Advances in Materials Technology*, pp. 92–106, Marco Island, FL, USA, October 2008.
[36] EPRI, *State of Knowledge: Grades 92 and 122 Steel for Fossil Power Plants*, EPRI, Palo Alto, CA, USA, 2008.
[37] K. Kimura, H. Kushima, and K. Sawada, "Long-term creep strength property of advanced ferritic creep resistant steels," in *Proceedings of the Sixth International Conference on Advances in Materials Technology*, pp. 732–751, Santa Fe, NM, USA, September 2011.
[38] D. J. Allen, "Analysis of ECCC creep rupture data on grade 91 steel–the influence of nickel," in *Proceedings of the 3rd International Conference on Creep and Fracture*, Rome, Italy, May 2014.
[39] S. J. Brett, J. S. Bates, and R. C. Thomson, "Aluminium nitride precipitation in low strength grade 91 power plant steels," in *Proceedings of the Fourth International Conference on Advances in Materials Technology for Fossil Power Plants*, pp. 1183–1197, Hilton Head Island, SC, USA, October 2005.
[40] J. D Parker and J. A. Siefert, "The creep and fracture behaviour of tempered martensitic steels," *Materials at High Temperatures*, pp. 1–13, 2017.

3

Experimental Study on Autogenous and Drying Shrinkage of Steel Fiber Reinforced Lightweight-Aggregate Concrete

Shunbo Zhao, Changyong Li, Mingshuang Zhao, and Xiaoyan Zhang

School of Civil Engineering and Communication, North China University of Water Resources and Electric Power, No. 36 Beihuan Road, Zhengzhou 450045, China

Correspondence should be addressed to Shunbo Zhao; sbzhao@ncwu.edu.cn

Academic Editor: Osman Gencel

Steel fiber reinforced lightweight-aggregate concrete (SFRLAC) has many advantages applied in structural engineering. In this paper, the autogenous shrinkage and drying shrinkage of SFRLAC for up to 270 days were measured, considering the effects of types of coarse and fine aggregates with the changes of water-to-binder ratio and volume fraction of steel fiber, respectively. The properties of mix workability, apparent density, and compressive strength of SFRLAC were also reported and discussed in relation to above factors. Test results show that the development of autogenous and drying shrinkage of SFRLAC was fast within 28 days and tended to be steady after 90 days. The development of autogenous shrinkage of SFRLAC reduced with the increasing water-to-binder ratio and by using the expanded shale with higher soundness and good water absorption, especially at early age within 28 days; the later drying shrinkage was reduced and the development of drying shrinkage was slowed down with the increasing volume fraction of steel fiber obviously; manufactured sand led to less autogenous shrinkage but greater drying shrinkage than fine lightweight aggregate in SFRLAC. The regularities of autogenous shrinkage and drying shrinkage of SFRLAC expressed as the series of hyperbola are analyzed.

1. Introduction

Shrinkage is a common phenomenon of concrete, which can induce tensile stress when the shrinkage is restrained, and may lead to cracking and finally reduce the service life and durability of concrete structures [1]. With the development of long span, large size, and high rise buildings, the design for considering the influence of concrete shrinkage becomes more and more important [2]. Based on the forming mechanisms, shrinkage can be divided as autogenous shrinkage and drying shrinkage. As the volume of cement hydration products is less than the sum of volumes of water and binders, pores and capillaries formed in the hardened cement paste. Autogenous shrinkage takes place because of the self-desiccation of pores and capillaries, resulting from the continuous hydration of binder materials after initial setting. Drying shrinkage takes place mainly due to the free water evaporated from the pores and capillaries near element surfaces to ambient air. The main factors influencing drying shrinkage are concrete strength, cement type, geometry, curing condition, and starting age [1–4] as well as some properties of concrete mix such as cement dosage, fine aggregate percentage, air content, and slump [1, 3, 4]. The above properties of concrete mix are also the primary factors influencing the autogenous shrinkage of concrete, which determines the composite microstructure of cement paste in concrete. Meanwhile, the coarse aggregate is the bone of concrete, which constrains the shrinkage of concrete in different degrees depending on the type, grading, maximum size, fine aggregate percentage, and volume ratio of aggregate in concrete [5].

In view of the marked shrinkage of normal weight concrete (NWC), especially of high-strength concrete, the internal curing provided by a partial replacement of normal weight aggregate (NWA) by lightweight aggregate (LWA) has been developed as an effective technique for shrinkage mitigation in recent years [6–15]. In this method, the use of prewetted LWAs as internal water sources of concrete is the key of study, as the unsaturated LWAs may lead to an opposite effect due to the absorption of water from matrix pores into aggregate pores [7–11]. The saturated LWAs can absorb significant amounts of water before mixing and release the water gradually during concrete hardening. The released water will contribute to restoring part of the water which has been lost through internal self-desiccation or external drying of concrete. However, the efficiency of internal curing by LWAs depends not only on their water absorption but also on their desorption characteristics, as not all kinds of LWA can release a sufficient amount of absorbed water during hardening of concrete, although most of them are able to absorb significant amounts of water because of their high porosity. Generally, the desorption characteristics depend on the microstructures of pores [13], the surface open porosity, and the particle size [14]; a coarse pore structure with a high proportion of well-interconnected pores leads to better desorption behavior for internal curing [15]. Meanwhile, the capability of internal curing of LWAs reduces with the decrease of water-to-binder ratio and the increase of concrete strength [16–18].

Theoretically, if the coarse NWA was fully replaced by the coarse LWA, NWC will become the lightweight aggregate concrete (LAC). As in most cases, structural LAC is made with a LWA as coarse aggregate and normal weight sand (natural river sand or manufactured sand) as the fine aggregate [19]. Therefore, the internal curing of LAC by saturated LWAs should be developed to the best; the difference of shrinkage of LAC to NWC should mainly depend on the difference of properties of coarse aggregates. As reported [20, 21] that with the same strength, LAC with higher moisture content LWAs has smaller shrinkage at early age but is finally larger than NWC; however the shrinkage of LAC with lower moisture content LWAs is always larger than that of NWC. When part of the lower water absorption LWAs was replaced by crushed limestone, the shrinkage of LAC could be reduced. This is because, for the same strength, LAC has larger volume of cement paste, and the modulus of elasticity of LWA is lower than that of NWA. However, with the different compositional parameters such as types of LWA (fly ash based artificial aggregate, sintering expanded clay, pumice aggregate, or oil palm shell), amount and type of binder, water-to-binder ratio, mineral admixture (silica fume, fly ash), and initial prewetting method for LWA, the shrinkage of LAC may be lower or greater than that of NWC [22–27]. Generally, the shrinkage of LACs is reduced by the rigid LWA with higher modulus of elasticity in water saturated condition prewetted and with the greater volume of LWA in the same volume of paste [24–26]. And most importantly, in current standard methods, the shrinkage of LAC is usually estimated from the expressions defined for NWC and always gets higher value than those measured in practice [24, 28]. For the proper prediction of shrinkage, the special characteristics of LAC should be considered.

With the application of fiber reinforced LAC in structure engineering, the effects of steel fibers or hybrid fibers on shrinkage of LAC were also studied. Among them, LACs were made by using cold-bonded fly ash aggregate [29], sintered fly ash aggregate [30], and expanded clay aggregate [31–33] as coarse aggregate and river sand as fine aggregate. Generally, the shrinkage of LACs did not change much at early age; however it reduced later to some extent. Fibers (carbon fiber, steel fiber, and polypropylene fiber) in LAC could not change the shrinkage behavior of base LAC affected by LWAs and water-to-binder ratio; their effect is mainly on the restraining of drying shrinkage. As the type and content of fibers are new parameters affecting the shrinkage of LAC, the research become complex and should be carried out comprehensively with other parameters changed.

In view of the wide application of sintering expanded shale as a LWA [20, 21] and the inexorable trend of natural sand replaced by manufactured sand [5, 34], this paper is focused on the autogenous and drying shrinkages of SFRLAC (steel fiber reinforced LAC) with the change of expanded shale and fine aggregate. The water-to-binder ratio and the volume fraction of steel fiber are taken into consideration to study their effects on autogenous shrinkage and drying shrinkage, respectively. Meanwhile, the mix workability, apparent density, and compressive strength are also given out. The aim of this study is to quantify the effects of aggregates, water-to-binder ratio, and volume fraction on the shrinkage of SFRLAC and finally gives a reference for the practical analysis of structural design.

2. Experiment

2.1. Raw Materials

Cement: grade P.O. 52.5 ordinary silicate cement produced in accordance with China Standards [35]; the physical and mechanical properties are listed in Table 1.

Admixture: class-II fly ash with physical properties listed in Table 2.

Coarse aggregate: two kinds of ceramsite of sintering expanded shale in continuous gradation with maximum size of 20 mm. They were sieved based on the maximum density principle. The physical and mechanical properties are listed in Table 3.

Fine aggregate: lightweight sand of fine sintering expanded shale and manufactured sand in continuous gradation with size of 1.6–5 mm. The physical properties are listed in Table 4.

Steel fiber: milling type, 36 mm long, 1.35 mm equivalent diameter, and 27.1 aspect ratio.

Water reducer: polycarboxylic acid superplasticizer with water-reducing rate of 19%.

Water: tap water.

The water absorption of lightweight aggregates was measured in accordance with China Standard GB/T 17431.2 [36]; the results are drawn in Figure 1.

2.2. Test Parameters and Mix Proportion of Concrete. Test parameters of this study were considered as volume fraction of steel fiber (ρ_f), water-to-binder ratio (W/B), two kinds of coarse lightweight aggregate, and fine lightweight aggregate and manufactured sand [37, 38]. Their combinations for

Table 1: Physical and mechanical properties of cement.

Strength grade	Water requirement of normal consistency (%)	Setting time (min)		Compressive strength (MPa)		Tensile strength (MPa)	
		Initial	Final	3 d	28 d	3 d	28 d
52.5	27.4	150	255	33.6	51.7	6.8	9.2

Table 2: Physical properties of fly ash.

Fineness modulus (pass 0.045 mm) (%)	Water demand ratio (%)	Apparent density (kg/m^3)	Water content (%)
26.4	104.2	2049	0.4

Table 3: Physical and mechanical properties of expanded shale.

Identifier	Apparent density (kg/m^3)	Bulk density (kg/m^3)	1 h water absorption (%)	Mud content (%)	Cylinder compressive strength (MPa)
H	1444	888	7.65	0.7	8.3
N	1307	816	8.50	0.9	7.3

Table 4: Physical properties of lightweight sand and manufactured sand.

Identifier	Grade (mm)	Fineness modulus	Apparent density (kg/m^3)	Bulk density (kg/m^3)	1 h water absorption (%)	Mud content (%)	Stone powder (%)
L	0.16~5	3.60	1350	850	9.02	0.11	—
M	0.16~5	2.82	2730	2320	0.70	2.45	5.0

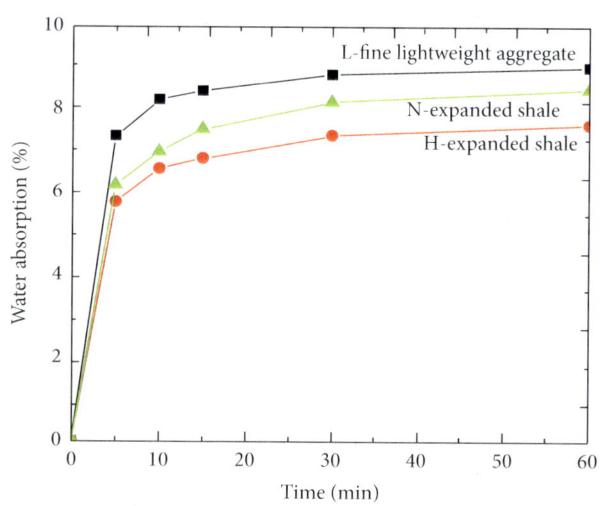

Figure 1: Water absorption of lightweight aggregates changing with time.

testing of autogenous shrinkage and drying shrinkage are listed in Table 5, where the double letters of mix number are the identifiers of expanded shale and sand; the following digits represent W/B and ρ_f.

The mix proportion of SFRLAC was designed in accordance with the specifications in China Standard JGJ 51 [19], where the absolute volume method was selected. For all mixes, the dosages of cement and fly ash were 440 kg/m^3 and 110 kg/m^3; the dosage of water reducer was 5.5% cement.

2.3. Test Methods. From Figure 1, we can know that the water absorption of lightweight sand at 5 min reached 81.7% at 1 hour; the water absorption of ordinary and high-strength expanded shales at 10 min reached 82.4% and 86.3% at 1 hour. Therefore, the lightweight sand was prewetted for 5 min; the expanded shales were prewetted for 10 min, before they were put into concrete mixer. The dosage of prewetted water was calculated by the water absorption at 1 hour.

Test for mix slumps was in accordance with the specification of China Standard GB/T50080 [39], which is identical to British Standard BS EN 12350-2-2009 [40].

The apparent density of SFRLAC was measured in accordance with the specification of China Standard JGJ51 [19]. Three cubes were used as a group for each SFRLAC. They were dried to be a steady weight in drying box at (105~110)°C, and then the apparent density was gotten from the weight divided by the volume of cube.

The cubic compressive strength at 28 d of each group was tested by using the cubes in dimension of 150 mm; each group had three cubes. The cubes were tested on YE-2000A hydraulic compression machine for compressive strength of SFRLAC in accordance with the specifications of China Standard GB/T50081 [41], which is identical to British Standard BS EN 12390-3-2009 [42].

The shrinkage of SFRLAC was measured in accordance with the specification of China Standards GB/T50082 [43], JG/T472 [44], and CECS 13:2009 [45]. As listed in Table 6, six groups of SFRLAC were used to study the autogenous shrinkage; each group had two specimens; eight groups of SFRLAC were used to study the drying shrinkage; each group had three specimens. The measuring device for autogenous

TABLE 5: Parameters of shrinkage tests and mix proportion of concrete.

Mix number	W/B	ρ_f (%)	Mix water (kg/m^3)	Expanded shale (kg/m^3)		Sand (kg/m^3)		Shrinkage test	
				N	H	L	M	Autogenous	Dry
NL0.30/0.8	0.30	0.8	134.8	482.5	0	368.7	0	√	√
NM0.30/0.8	0.30	0.8	134.8	482.5	0	0	745.6	√	√
HM0.30/0.8	0.30	0.8	134.8	0	533.1	0	745.6	√	√
HL0.30/0.4	0.30	0.4	134.8	0	542.2	370.9	0		√
HL0.30/0.8	0.30	0.8	134.8	0	533.0	368.7	0	√	√
HL0.30/1.2	0.30	1.2	134.8	0	523.9	366.4	0		√
HL0.30/1.6	0.30	1.6	134.8	0	514.8	364.1	0		√
HL0.30/2.0	0.30	2.0	134.8	0	505.6	361.9	0		√
HL0.35/0.8	0.35	0.8	157.2	0	514.2	355.9	0	√	
HL0.25/0.8	0.25	0.8	112.3	0	551.8	381.4	0	√	

TABLE 6: Test results of mix slump and apparent density of SFRLAC.

Mix number	Slump (mm)	Apparent density (kg/m^3)		Cubic compressive strength (MPa)
		Wet	Dry	
HL0.3/0.4	165	1754	1695	47.2
HL0.3/0.8	197	1809	1753	49.3
HL0.3/1.2	170	1842	1784	57.6
HL0.3/1.6	98	1843	1787	60.9
HL0.3/2.0	46	1919	1843	64.8
HL0.25/0.8	35	1820	1783	59.3
HL0.35/0.8	175	1796	1696	47.8
HM0.3/0.8	85	2069	1946	51.7
NL0.3/0.8	150	1682	1651	45.0
NM0.3/0.8	75	2000	1869	54.1

shrinkage is shown in Figure 2; the specimen is Ø150 mm × 450 mm cylinder. Two micrometer gauges were fixed symmetrically on the specimen accompanying the copper rods by using the embedded gaging heads with standard datum length of 250 mm. The specimens were placed in the standard curing room with (20 ± 2)°C temperature and (65 ± 5)% RH. The cast surface was covered by wet cloth. After cast for 12 hours, they were demolded and closely wrapped with polypropylene plastic film. The measuring started at 24 hours after specimen cast.

The prisms in dimension of 100 mm × 100 mm × 550 mm were used to measure the drying shrinkage. The copper ends were embedded at specimen ends along the longitudinal centroidal axis. The prisms were demolded after casting for 12 hours and placed into the standard curing box shown in Figure 3(a) with (20±2)°C temperature and (60±5)% RH. The length comparator shown in Figure 3(b) was used to measure the specimen length due to drying shrinkage. The measuring started at 24 hours after specimen cast.

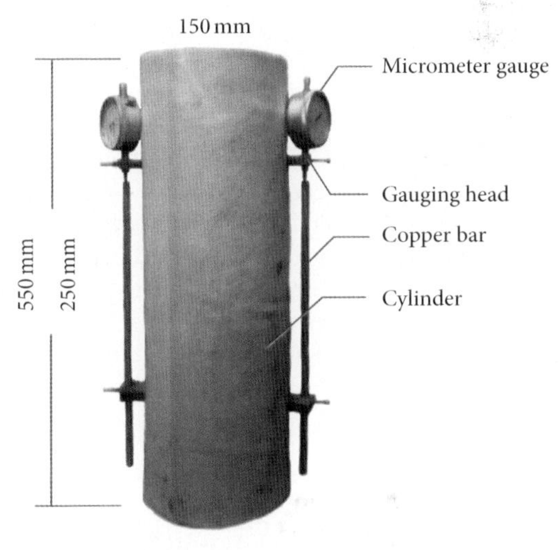

FIGURE 2: Measuring device for autogenous deformation.

3. Results and Analysis

3.1. Mix Workability. Test results of mix slump are listed in Table 6. As tests reported [37], the mix slump tended to reduce with the increase of volume fraction of steel fiber. This is mainly due to the bridging effect of steel fiber; the mix flowability increased normally when the mix was vibrated. The mix slump was affected by water-to-binder ratio obviously; the mix with $W/B = 0.25$ did not satisfy the requirement of flowing concrete. Meanwhile, the mixes with manufactured sand had lower slump; this is reasonable because of the peculiarity of manufactured sand with rough

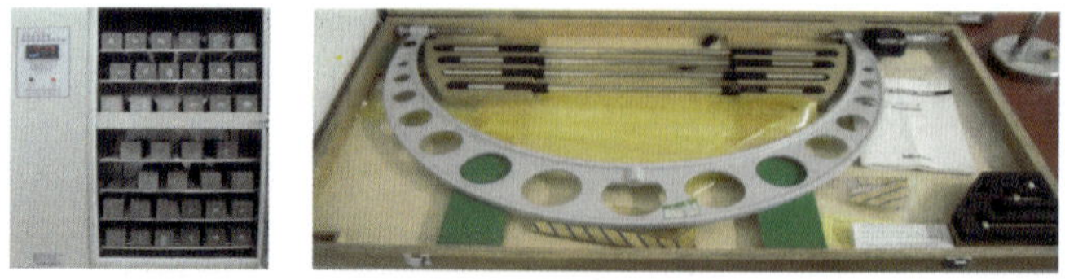

(a) Standard curing box (b) Length comparator

Figure 3: Devices for drying shrinkage of specimen.

surface, corner angles, and stone powder [34, 38]. Compared with the mix with H-expanded shale, the mix with N-expanded shale had lower slump. This is because of the larger specific surface area of N-expanded shale with lower density.

3.2. Apparent Density and Compressive Strength. As can be seen from Table 6, due to the high density of steel fiber and design by the absolute volume method, the apparent density of SFRLAC increased with the volume fraction of steel fiber. With the 0.4% increment of volume fraction of steel fiber, the apparent density increased by 1~3% with an average of 2.2%; however the dry apparent density of SFRLAC with lightweight sand was lower than 1850 kg/m^3. With the increase of W/B, the density of SFRLAC reduced somewhat. The density of SFRLAC with manufactured sand was heavy; fortunately it was also within the specified range of LAC less than 1950 kg/m^3 [19].

As tests reported [37], the compressive strength of SFRLAC increased with the volume fraction of steel fiber and decreased with the increase of W/B. SFRLAC with manufactured sand had higher compressive strength than that with lightweight sand.

3.3. Autogenous Shrinkage. Test results of autogenous shrinkage of SFRLAC for up to 270 days are listed in Table 7 and shown in Figure 4. As can be seen in this table and figure, the development of autogenous shrinkage of SFRLAC was fast within 28 days and tended to be steady after 90 days. For the SFRLAC with H-expanded shale and L-fine aggregate, the autogenous shrinkage grew slowly with the increasing W/B, especially at early age within 28 days; this is because of the cement hydration water supplied by sufficient free water in pores of SFRLAC with larger W/B, resulting in the moist pores and capillaries with less pore shrinkage stress. With the same W/B and L-fine lightweight aggregate, the autogenous shrinkage of SFRLAC with N-expanded shale developed fast within 28 days and was overall larger than that of SFRLAC with H-expanded shale; this is mainly due to the less compressive strength of N-expanded shale resulting from the weaker particle soundness, resulting in the bad resisting to the shrinkage deformation. Compared to SFRLAC mixed with L-fine lightweight aggregate, SFRLAC mixed with manufactured sand (M-sand) had smaller autogenous shrinkage; this may be due to the higher particle soundness of M-sand filed into the pores of expanded shales to constrain the autogenous deformation and the better interfacial bond property between cement paste and particles

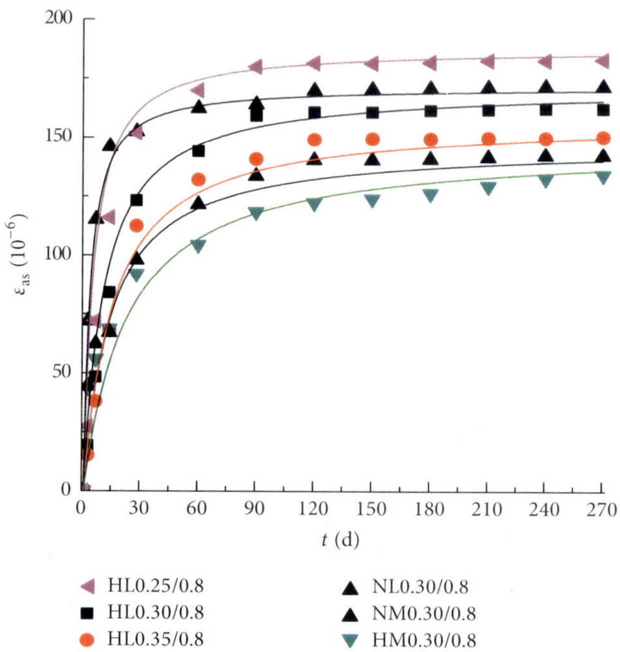

Figure 4: Autogenous shrinkage of SFRLAC.

of manufactured sand [5, 38]; all of these are beneficial to reduce the autogenous shrinkage of SFRLAC.

Based on the statistical analysis of test data, the regularity of autogenous shrinkage of SFRLAC can be expressed well as a series of hyperbola:

$$\varepsilon_{as}(t, t_s) = \frac{360(t - t_s) \times 10^{-6}}{52.5(\alpha_1\beta_1 \cdot 4.68(W/B) - 1) + (1 + \alpha_2\beta_2 \cdot 3.65(W/B))(t - t_s)}, \quad (1)$$

where $\varepsilon_{as}(t, t_s)$ is the autogenous shrinkage of SFRLAC at t age and t_s is the initial age for the testing of autogenous shrinkage, here taken as 1 day after molding. α_1 (α_2) and β_1 (β_2) are the shape factors related to coarse aggregate and fine aggregate, respectively; they are listed in Table 8.

The comparison of formula (1) with test results is also shown in Figure 4, which fitted better with correlation coefficient greater than 0.90.

From formula (1), it can be concluded that W/B is a main factor influencing not only the development rate at early age, but also the later developing state of autogenous shrinkage. As can be seen from Table 8, the effects of

TABLE 7: Test results of autogenous shrinkage of SFRLAC.

Mix number	Curing age (days)												
	1	3	7	14	28	60	90	120	150	180	210	240	270
HL0.25/0.8	0	27	72	116	152	170	180	182	182	182	183	183	183
HL0.30/0.8	0	19	48	84	123	144	159	161	161	162	162	162	162
HL0.35/0.8	0	15	38	68	112	132	141	149	150	150	150	150	150
NL0.30/0.8	0	72	115	146	152	162	164	170	170	171	171	172	172
HM0.30/0.8	0	44	56	69	92	104	118	122	124	126	130	133	134
NM0.30/0.8	0	44	62	67	98	121	134	140	140	141	142	142	142

TABLE 8: Values of influencing factors in formula (1).

Aggregate type	Coarse aggregate		Fine aggregate	
	H	N	L	M
α_1 (α_2)	1.0 (1.0)	0.8 (1.0)	—	—
β_1 (β_2)	—	—	1.0 (1.0)	1.45 (1.32)

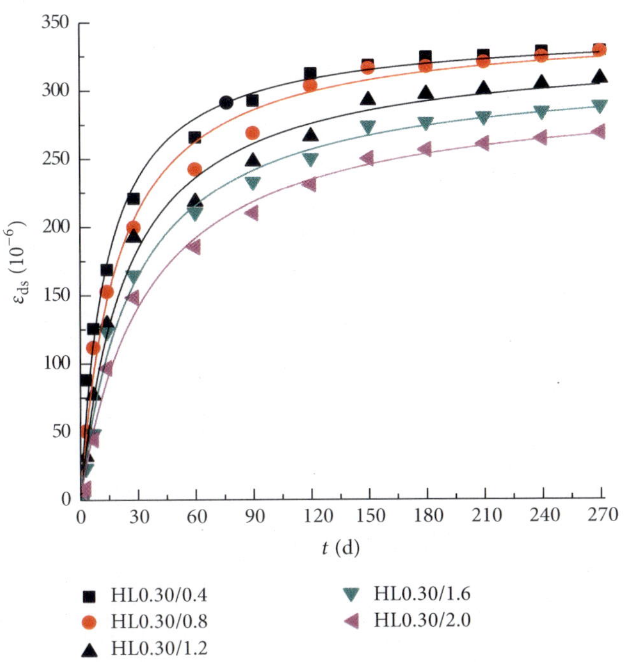

FIGURE 5: Drying shrinkage of SFRLAC with different volume fraction of steel fiber.

different type of aggregate on the autogenous shrinkage of SFRLAC are well reflected by the changes of influencing factors. Under the same condition of other parameters, and compared to the SFRLAC with H-expanded shale and L-fine lightweight aggregate, N-expanded shale increases the development of autogenous shrinkage at early age; M-sand decreases the development rate at early age but prolongs the later developing stage.

3.4. Drying Shrinkage. Test results of drying shrinkage of SFRLAC for up to 270 days are listed in Table 9. As can be seen in this table and Figure 5, the development of drying shrinkage of SFRLAC was slowed down with the increasing volume fraction of steel fiber. The three-dimensional distribution of steel fibers in concrete builds the frame constraining the shrinkage deformation of concrete; this effect will be stronger with the increasing volume fraction of steel fiber due to the better and better net structure bridged among steel fibers. As the large elastic modulus and tensile strength of steel fibers, the constraining effects of steel fiber on the drying shrinkage of SFRLAC were obvious.

Based on the statistical analysis of test data, the regularity of drying shrinkage of SFRLAC with different volume fraction of steel fiber can be expressed as a series of hyperbola:

$$\varepsilon_{ds}(t, t_s) = \frac{a(t - t_s)}{b + (t - t_s)} \times 10^{-6}, \quad (2)$$

where $\varepsilon_{ds}(t, t_s)$ is the drying shrinkage of SFRLAC at t age and t_s is the initial age for the testing of autogenous shrinkage, here taken as 1 day after molding. a and b are the shape factors related to the volume fraction of steel fiber.

From formula (2), the shape factors a and b can be obtained by the fitting analysis of test data shown in Figure 5. After that, the linear relation of them with volume fraction of steel fiber can be obtained as follows and is shown in Figure 6:

$$a = 5(73.4 - 5.7\rho_f), \quad (3)$$

$$b = 8.91 + 12.53\rho_f. \quad (4)$$

Combined with formula (2), factor a decreases and b increases with the increase of volume fraction of steel fiber. This means that the later drying shrinkage will be reduced and the development of drying shrinkage will be slowed down by the strong restraint effect of steel fibers, respectively. The comparison of formula (2) with test results of SFRLAC with different volume fraction of steel fiber is drawn in Figure 7, which fitted better with correlation coefficient greater than 0.90.

Similar to the development of autogenous shrinkage, with the same water-to-binder ratio and L-fine lightweight aggregate, the drying shrinkage of SFRLAC with N-expanded shale developed fast within 28 days and was overall greater than that of SFRLAC with H-expanded shale. This showed that the constraining effect of steel fibers on shrinkage deformation of SFRLAC could not change the inherent shrinkage of base LAC formed by cement paste and aggregate. Compared to SFRLAC mixed with L-fine lightweight aggregate, SFRLAC mixed with manufactured sand (M-sand) had greater drying

TABLE 9: Test results of drying shrinkage of SFRLAC.

Mix number	Curing age (days)												
	1	3	7	14	28	60	90	120	150	180	210	240	270
HL0.30/0.4	0	88	126	169	221	266	292	312	318	320	325	328	329
HL0.30/0.8	0	50	111	152	199	242	268	303	316	317	320	324	328
HL0.30/1.2	0	32	77	129	192	218	248	266	292	297	300	304	308
HL0.30/1.6	0	23	48	123	165	210	233	250	274	276	280	284	288
HL0.30/2.0	0	8	44	96	148	185	210	230	250	256	260	264	268
NL0.30/0.8	0	109	124	184	225	272	298	323	343	344	345	345	348
HM0.30/0.8	0	84	137	202	283	329	355	376	382	392	400	400	400
NM0.30/0.8	0	129	257	325	361	408	436	464	480	484	488	496	505

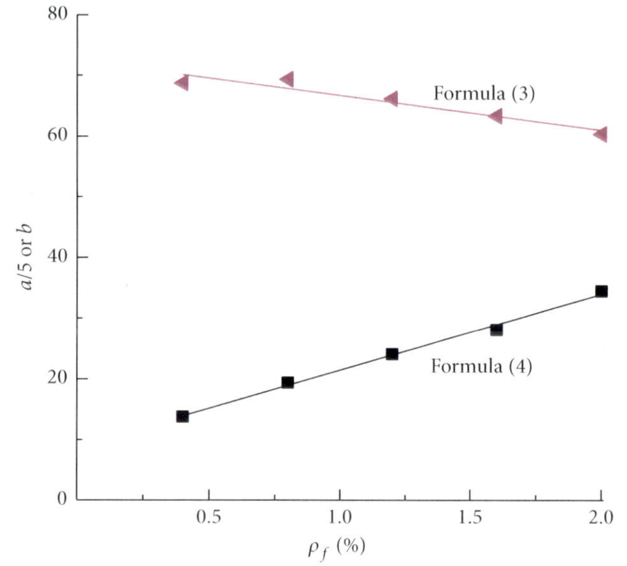

FIGURE 6: Relation of shape factors a and b with volume fraction of steel fiber.

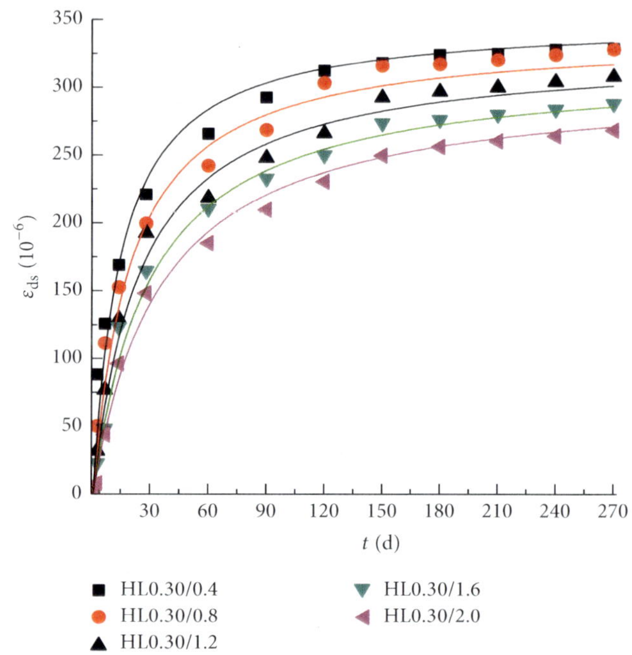

FIGURE 7: Comparison of test data with computation of formula (2).

shrinkage. This is because of the larger water absorption of L-fine lightweight aggregate leading to the beneficial internal curing for cement hydration. In this case, the beneficial effects such as the rigid particle of M-sand filling into the pores of expanded shales to constrain the autogenous deformation and the better interfacial bond property between cement paste and particles of M-sand become secondary.

Based on the statistical analysis of test data, the regularity of drying shrinkage of SFRLAC with different aggregates can also be expressed as a series of hyperbola. On the basis of formula (2), it is

$$\varepsilon_{\mathrm{ds}}(t, t_s) = \frac{\alpha a (t - t_s)}{\beta b + (t - t_s)} \times 10^{-6}, \qquad (5)$$

where α and β are the influencing factors related to coarse aggregate and fine aggregate.

Results of statistical analysis for values of influencing factors in formula (5) are listed in Table 10. The comparisons of test data with computation of formula (5) are shown in Figure 8. As can be seen from Table 10, the values of influencing factors α and β for SFRLAC with N-expanded shale/M-sand are not equal to the product of those for SFR-LAC, respectively, with N-expanded shale/L-fine lightweight aggregate and with H-expanded shale/M-sand. This means that the influences of expanded shale and fine aggregate (sand) on drying shrinkage of SFRLAC were correlated. In this study, the drying shrinkage of SFRLAC with N-expanded shale and M-sand was largest.

TABLE 10: Values of influencing factors in formula (5).

Aggregate type	Coarse aggregate ∗ fine aggregate			
	H ∗ L	N ∗ L	H ∗ M	N ∗ M
α	1.0	1.04	1.22	$1.45 > 1.04 \times 1.22 = 1.27$
β	1.0	0.71	0.77	$0.45 < 0.71 \times 0.77 = 0.55$

4. Conclusions

From this study, the main conclusions can be given as follows.

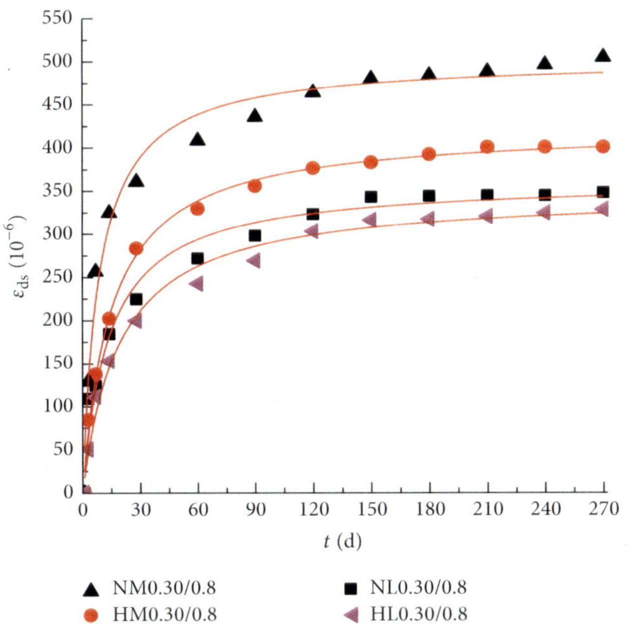

FIGURE 8: Comparison of test data with computation of formula (5).

(1) The mix slump was affected by water-to-binder ratio obviously and tended to reduce with the increase of volume fraction of steel fiber. Mixes with N-expanded shale or manufactured sand had lower slump.

(2) The apparent density of SFRLAC increased with the volume fraction of steel fiber and reduced somewhat with the increase of W/B; however it was less than 1850 kg/m^3 or 1950 kg/m^3 by using lightweight sand or manufactured sand, respectively. The compressive strength of SFRLAC increased with the volume fraction of steel fiber and decreased with the increasing W/B, while it was higher with manufactured sand than that with lightweight sand.

(3) Test results of autogenous shrinkage of SFRLAC for up to 270 days are given out. The development of autogenous shrinkage of SFRLAC was fast within 28 days and tended to be steady after 90 days. The autogenous shrinkage of SFRLAC with H-expanded shale and L-fine aggregate grew slowly with the increasing W/B, especially at early age within 28 days. With lower compressive strength than that of H-expanded shale, N-expanded shale resulted in the fast development within 28 days and overall greater autogenous shrinkage of SFRLAC. However, L-fine lightweight aggregate led to greater autogenous shrinkage than manufactured sand in SFRLAC.

Based on the test data, the regularity of autogenous shrinkage of SFRLAC can be expressed well as a series of hyperbola, in which the effects of W/B, coarse aggregate, and fine aggregate are independently qualified on the development rate at early age and the later developing state of autogenous shrinkage.

(4) Test results of drying shrinkage of SFRLAC for up to 270 days are presented. The later drying shrinkage was reduced and the development of drying shrinkage was slowed down with the increasing volume fraction of steel fiber obviously. Similar to the development of autogenous shrinkage, N-expanded shale leads to the fast development within 28 days and overall greater drying shrinkage than H-expanded shale in SFRLAC. But the internal curing of L-fine lightweight aggregate was more beneficial to reduce the drying shrinkage compared with manufactured sand in SFRLAC.

Based on the test data, the regularity of drying shrinkage of SFRLAC can be also expressed as a series of hyperbola in which the shape factors are linear with the volume fraction of steel fiber, and the influencing factors of coarse aggregate and fine aggregate are qualified. The values of influencing factors show that the effects of these two expanded shales, fine lightweight aggregate and manufactured sand, on drying shrinkage of SFRLAC were correlated. In this study, the drying shrinkage of SFRLAC with N-expanded shale and M-sand was largest.

Competing Interests

The authors declare that they have no competing interests.

Acknowledgments

This study was financially supported by the Science and Technology Innovation Team of Eco-Building Material and Structural Engineering in the University of Henan Province, China (13IRTSTHN002), and the Key Research Project of University in Henan Province, China (16A560024).

References

[1] ACI Committee 209, Prediction of creep, shrinkage and temperature effects in concrete structures, ACI, 209R: Manual of concrete practice, Part 1, 1992.

[2] GB 50010-2010, *Code for Design of Concrete Structures*, China Building Industry Press, Beijing, China, 2010.

[3] EN, "Eurocode 2: design of concrete structures—part 1.1: general rules and rules for buildings," EN 1992-1-1, CEN, Brussels, Belgium, 2004.

[4] CEB-FIP Model Code 2010-Fib Bulletin 55, First Complete Draft, Vol. 1, Fib-International Federation for Structural Concrete, Lausanne, Switzerland, 2010.

[5] F. L. Li, C. J. Liu, L. Y. Pan, and C. Y. Li, *Machine-Made Sand Concrete*, China Water Power Press, Beijing, China, 2013.

[6] S. Weber and H. W. Reinhardt, "New generation of high performance concrete: concrete with autogeneous curing," *Advanced Cement Based Materials*, vol. 6, no. 2, pp. 59–68, 1997.

[7] T. Merikallio, R. Mannonen, and V. Penttala, "Drying of lightweight concrete produced from crushed expanded clay aggregates," *Cement and Concrete Research*, vol. 26, no. 9, pp. 1423–1433, 1996.

[8] B. Akcay and M. A. Tasdemir, "Effects of distribution of lightweight aggregates on internal curing of concrete," *Cement and Concrete Composites*, vol. 32, no. 8, pp. 611–616, 2010.

[9] M. Golias, J. Castro, and J. Weiss, "The influence of the initial moisture content of lightweight aggregate on internal curing," *Construction and Building Materials*, vol. 35, pp. 52–62, 2012.

[10] A. Bentur, S.-I. Igarashi, and K. Kovler, "Prevention of autogenous shrinkage in high-strength concrete by internal curing using wet lightweight aggregates," *Cement and Concrete Research*, vol. 31, no. 11, pp. 1587–1591, 2001.

[11] T. Ji, D.-D. Zheng, X.-F. Chen, X.-J. Lin, and H.-C. Wu, "Effect of prewetting degree of ceramsite on the early-age autogenous shrinkage of lightweight aggregate concrete," *Construction and Building Materials*, vol. 98, pp. 102–111, 2015.

[12] D. Cusson and T. Hoogeveen, "Internal curing of high-performance concrete with pre-soaked fine lightweight aggregate for prevention of autogenous shrinkage cracking," *Cement and Concrete Research*, vol. 38, no. 6, pp. 757–765, 2008.

[13] S. Ghourchian, M. Wyrzykowski, P. Lura, M. Shekarchi, and B. Ahmadi, "An investigation on the use of zeolite aggregates for internal curing of concrete," *Construction and Building Materials*, vol. 40, pp. 135–144, 2013.

[14] Y. Wei, Y. Xiang, and Q. Zhang, "Internal curing efficiency of prewetted LWFAs on concrete humidity and autogenous shrinkage development," *Journal of Materials in Civil Engineering*, vol. 26, no. 5, pp. 947–954, 2014.

[15] S.-D. Hwang, K. H. Khayat, and D. Youssef, "Effect of moist curing and use of lightweight sand on characteristics of high-performance concrete," *Materials and Structures*, vol. 46, no. 1-2, pp. 35–46, 2013.

[16] J. Zhang, Y. D. Han, Y. Gao, and Y. Luosun, "Integrative study on the effect of internal curing on autogenous and drying shrinkage of high-strength concrete," *Drying Technology*, vol. 31, no. 5, pp. 565–575, 2013.

[17] J. Zhang, J. H. Wang, and Y. D. Han, "Simulation of moisture field of concrete with pre-soaked lightweight aggregate addition," *Construction and Building Materials*, vol. 96, pp. 599–614, 2015.

[18] I. Maruyama, M. Kanematsu, T. Noguchi, H. Iikura, A. Teramoto, and H. Hayano, "Evaluation of water transfer from saturated lightweight aggregate to cement paste matrix by neutron radiography," *Nuclear Instruments and Methods in Physics Research Section A: Accelerators, Spectrometers, Detectors and Associated Equipment*, vol. 605, no. 1-2, pp. 159–162, 2009.

[19] JGJ, "Technical specification for lightweight concrete," JGJ 51-2002, China Building Industry Press, Beijing, China, 2002.

[20] H. Sun, L. Ye, J. Ding, and Y. Guo, "Shrinkage and creep of high-strength lightweight aggregate concrete," *Journal of Tsinghua University*, vol. 47, no. 6, pp. 765–767, 2007.

[21] P.-J. Song, J.-T. Ding, and Y.-S. Guo, "Shrinkage of high strength lightweight aggregate concrete and factors influencing it," *Journal of Building Materials*, vol. 7, no. 2, pp. 138–144, 2004.

[22] A. Nilsen and P. C. Aitcin, "Properties of high-strength concrete containing light-, normal-, and heavyweight aggregate," *Cement, Concrete and Aggregates*, vol. 14, no. 1, pp. 8–12, 1992.

[23] A. Terzić, L. Pezo, V. Mitić, and Z. Radojević, "Artificial fly ash based aggregates properties influence on lightweight concrete performances," *Ceramics International*, vol. 41, no. 2, pp. 2714–2726, 2015.

[24] J. A. Bogas, R. Nogueira, and N. G. Almeida, "Influence of mineral additions and different compositional parameters on the shrinkage of structural expanded clay lightweight concrete," *Materials and Design*, vol. 56, pp. 1039–1048, 2014.

[25] P. Shafigh, H. Ghafari, H. B. Mahmud, and M. Z. Jumaat, "A comparison study of the mechanical properties and drying shrinkage of oil palm shell and expanded clay lightweight aggregate concretes," *Materials and Design*, vol. 60, pp. 320–327, 2014.

[26] N. Kabay and F. Aköz, "Effect of prewetting methods on some fresh and hardened properties of concrete with pumice aggregate," *Cement and Concrete Composites*, vol. 34, no. 4, pp. 503–507, 2012.

[27] A. I. M. Ismail, M. S. Elmaghraby, and H. S. Mekky, "Engineering properties, microstructure and strength development of lightweight concrete containing pumice aggregates," *Geotechnical and Geological Engineering*, vol. 31, no. 5, pp. 1465–1476, 2013.

[28] H. Costa, E. Júlio, and J. Loureno, "New approach for shrinkage prediction of high-strength lightweight aggregate concrete," *Construction and Building Materials*, vol. 35, pp. 84–91, 2012.

[29] E. Güneyisi, M. Gesoğlu, A. Mohamadameen, R. Alzeebaree, Z. Algın, and K. Mermerdaş, "Enhancement of shrinkage behavior of lightweight aggregate concretes by shrinkage reducing admixture and fiber reinforcement," *Construction and Building Materials*, vol. 54, pp. 91–98, 2014.

[30] L. Domagała, "Modification of properties of structural lightweight concrete with steel fibres," *Journal of Civil Engineering and Management*, vol. 17, no. 1, pp. 36–44, 2011.

[31] O. Kayali, M. N. Haque, and B. Zhu, "Drying shrinkage of fibre-reinforced lightweight aggregate concrete containing fly ash," *Cement and Concrete Research*, vol. 29, no. 11, pp. 1835–1840, 1999.

[32] B. Chen and J. Liu, "Contribution of hybrid fibers on the properties of the high-strength lightweight concrete having good workability," *Cement and Concrete Research*, vol. 35, no. 5, pp. 913–917, 2005.

[33] B. Chen and J. Liu, "Properties of lightweight expanded polystyrene concrete reinforced with steel fiber," *Cement and Concrete Research*, vol. 34, no. 7, pp. 1259–1263, 2004.

[34] X. X. Ding, C. Y. Li, Y. Y. Xu, F. L. Li, and S. B. Zhao, "Experimental study on long-term compressive strength of concrete with manufactured sand," *Construction and Building Materials*, no. 108, pp. 67–73, 2016.

[35] SAC, "Common Portland cement," GB 175-2007, Standardization Administration of China, Beijing, China, 2007.

[36] GB/T, "Test method for lightweight aggregate," GB/T 17431.2-2010, China Standard Press, Beijing, China, 2010.

[37] Z. Shen, M. H. Chen, M. S. Zhao, and X. K. Li. Experimental study, "Experimental study on tensile properties of steel fiber reinforced lightweight-aggregate fly-ash flowable concrete," in *Architectural Engineering and New Materials*, pp. 1–8, DEStech, 2015.

[38] S. Zhao, Y. Xu, and C. Li, "Study on workability of plastic fresh concrete with machine-made sand," in *Proceedings of the International Conference on Transportation, Mechanical, and Electrical Engineering (TMEE '11)*, pp. 2013–2016, Changchun, China, December 2011.

[39] GB/T, "Standard for test method of performance on ordinary fresh concrete," GB/T 50080-2002, China Building Industry Press, Beijing, China, 2002.

[40] BS EN, "Testing fresh concrete—slump-test," BS EN 12350-2-2009, 2009.

[41] CABP, "Standard for test method of mechanical properties on ordinary concrete," GB/T 50081-2002, China Architecture and Building Press, Beijing, China, 2002.

[42] "Testing hardened concrete—compressive strength of test specimens," BS EN 12390-3-2009, 2009.

[43] GB/T, "Standard for test method of long-term performance and durability of ordinary concrete," GB/T 50082-2009, China Building Industry Press, Beijing, China, 2010.

[44] "Steel fiber reinforced concrete," JG/T 472-2015, China Building Industry Press, Beijing, China, 2015.

[45] CECS, "Standard test methods for fiber reinforced concrete," CECS 13:2009, China Plan Press, Beijing, China, 2010.

4

Combating Wear of ASTM A36 Steel by Surface Modification using Thermally Sprayed Cermet Coatings

Vineet Shibe[1] and Vikas Chawla[2]

[1] *Research Scholar, IKG Punjab Technical University, Kapurthala, Punjab 144603, India*
[2] *Ferozepur College of Engineering and Technology, Ferozepur, Punjab 142052, India*

Correspondence should be addressed to Vineet Shibe; shibevineet@gmail.com

Academic Editor: Katsuyuki Kida

Thermal spray coatings can be applied economically on machine parts to enhance their requisite surface properties like wear, corrosion, erosion resistance, and so forth. Detonation gun (D-Gun) thermal spray coatings can be applied on the surface of carbon steels to improve their wear resistance. In the present study, alloy powder cermet coatings WC-12% Co and Cr_3C_2-25% NiCr have been deposited on ASTM A36 steel with D-Gun thermal spray technique. Sliding wear behavior of uncoated ASTM A36 steel and D-Gun sprayed WC-12% Co and Cr_3C_2-25% NiCr coatings on base material is observed on a Pin-On-Disc Wear Tester. Sliding wear performance of WC-12% Co coating is found to be better than the Cr_3C_2-25% NiCr coating. Wear performance of both these cermet coatings is found to be better than uncoated ASTM A36 steel. Thermally sprayed WC-12% Co and Cr_3C_2-25% NiCr cermet coatings using D-Gun thermal spray technique is found to be very useful in improving the sliding wear resistance of ASTM A36 steel.

1. Introduction

Wear controls the service life of machine components. Most of the components fail due to the combination of different types or modes of wear, such as corrosion, adhesion, impact, surface fatigue, abrasion, and erosion. Wear is the removal of material from the surfaces in contact when they slide or roll relative to each other [1]. It is a very slow process, but metal removal takes place steadily and continuously [2]. Wear can be minimized or controlled by using the costly, superior wear resistant materials or by the application of surface modification techniques on the low cost existing, inferior substrate materials [3]. The service life of the machine parts depends upon their surface properties and hence should be considered in design [4]. Surface engineering deals with different techniques that can be applied to achieve desired surface characteristics [5]. Surface engineering improves the performance and service life of machine parts. Its effectiveness on the performance and service life depends upon the material of the surface, alloy, type of service conditions, and the application process used [6].

Wear and friction are responsible for many problems and costs more in a modern civilization. Engineers and designers must take these factors into account while designing and constructing different equipment [7]. The economic losses due to friction and wear related issues are about 1 to 2.5% of the gross national product [8]. The estimated costs of losses due to abrasive wear range from 1 to 4% of the gross national product of an industrialized nation [9]. So, it is important to focus on improving wear resistance of machine parts, as it impacts the economy the most.

Although wear cannot be eliminated completely, it can be minimized to some extent by using different wear prevention methodologies. Some of the methods used for the prevention of wear are [10] selection of better material, lubrication, temperature control, alignment control, environmental control, contact pressure control, preventive maintenance, and use of surface coatings. Surface coating by thermal spraying is an optimum preventive method which is mostly used in the practical field of application.

The desired technical or decorative properties can be enhanced by the application of surface coatings. A suitable

Table 1: Nominal and actual percentage composition of ASTM A36 steel.

	Chemical composition (weight % age) of ASTM A36 steel substrate (IS 2062)														
	% C	% Si	% Mn	% P	% S	% Al	% Cu	% Cr	% Mo	% Ni	% Pb	% Ti	% V	% W	% Fe
Nominal	0.16	0.17	0.46	0.026	0.019	0.007	0.048	0.084	0.018	0.039	0.007	<0.001	0.003	<0.001	98.89
Actual	0.19	0.18	0.92	0.019	0.022	0.01	0.01	0.01	0.002	0.01	—	—	0.001	—	98.626

coating can be applied on the surface of existing low cost inferior substrate material in the form of a superior layer of material [11]. The purpose of applying coatings to the surface of the substrate or base material is to prevent its surface from different types of wear and to improve the appearance of its surface [12].

A number of thermal spray coating techniques are available such as D-Gun, plasma spray, and high velocity oxy-fuel (HVOF) [13]. It is a group of coating techniques in which a superior material to be coated is heated and accelerated towards the base metal to form a surface coating. A large number of materials can be deposited by thermal spraying to produce surface coatings.

D-Gun thermal spraying technique finds a variety of applications like in the aircraft industries of the United States and Japan [13]. A D-Gun consists of a long barrel, a powder feeder, and a chamber in which the combustion is initiated and controlled. Gas flows are regulated and controlled with the help of control panels provided in D-Gun [14]. It includes a manipulator to hold and control the movements of the work piece. A measured mixture of oxygen and acetylene gases is fed into the barrel along with the coating powder to be deposited. The gas mixture is ignited with a spark and the combustion produces detonation waves. The detonation waves thus produced heats and accelerated the coating powder to be deposited. Some advantages of D-Gun spray coating process is increase in wear resistance by several times, dense microstructure, lower porosity, smooth surface finish, higher microhardness, lower oxidation of in-flight particles, improved wear and erosion wear resistance, and so forth.

2. Materials and Methods

2.1. Selection of the Substrate Material. Selection of the base material is made after consultation with Pressure and Process Boilers, Saharanpur (India). Nominal and actual chemical composition (weight % age) of ASTM A36 steel substrate is given in Table 1.

2.2. Preparation of Test Specimens from ASTM A36 Steel Substrate Material. Small cylindrical pins of circular cross-section having a diameter of 8 mm and length 30 mm are prepared from ASTM A36 steel material. Manufacturing process such as machining or metal cutting/removal is used for manufacturing cylindrical pins. The pins are machined to required dimensions on lathe machine by using operations like simple turning, facing, and so forth. The end faces of the specimens are ground using emery papers and subsequently polished by polishing papers before the application of surface coating.

Table 2: Detonation spray parameters.

	WC-12% Co	Cr_3C_2-25% NiCr
Gases	Flow rate (std. liters per hour)	
Oxygen	2960	2720
Acetylene	2400	2320
Nitrogen	720	720

2.3. Thermal Spray Powders for Coatings. Two types of cermet coating powders, namely, WC-12% Co and Cr_3C_2-25% NiCr, are selected for detonation spray deposition on the ASTM A36 steel substrate specimens after the through literature review. Scanning electron microscope (SEM) images and energy-dispersive X-ray spectroscopy (EDS) patterns showing the morphologies, microstructure, and elemental composition of WC-12% Co and Cr_3C_2-25% NiCr feedstock powders have been shown in Figure 1. The results obtained from EDS confirm that WC-12% Co powder contains C, Co, and W elements and Cr_3C_2-25% NiCr powder contains C, Ni, and Cr elements.

2.4. Formulation of the Coating. WC-12% Co and Cr_3C_2-25% NiCr powders are effectively deposited on an ASTM A36 steel substrate by the D-Gun thermal spray coating technique. These coatings are deposited at SVX Powder M Surface Engineering Private Limited, Greater Noida, U.P. (India). Very hard, dense coatings with high bond strength ranging from 70 MPa to 75 MPa are produced by depositing WC-12% Co and Cr_3C_2-25% NiCr coating powders by D-Gun technique on an ASTM A36 steel substrate. The average bond strength of WC-12% Co and Cr_3C_2-25% NiCr coating on an ASTM A36 steel substrate is tested using the ASTM standard C633-01 at Metallizing Equipment Company Private Limited, Jodhpur, India. A total of three specimens are tested for WC-12% Co and Cr_3C_2-25% NiCr coating and the average value of the bond strength is found to be 74 MPa and 71 MPa, respectively. The WC-12% Co powder is supplied by H. C. Starck, Germany, under the product code Amperit 518.074. The powder is agglomerated and sintered having a particle size as 10–45 microns. Cr_3C_2-25% NiCr powder is supplied by H. C. Starck, Germany, under the product code Amperit 588.074. The powder is agglomerated and sintered having a particle size of 10–45 microns. The parameters used for D-Gun thermal spraying for the two coatings at a spraying distance of 165 mm are shown in Table 2.

The front view and top view (macrographs) of uncoated or bare test specimens, WC-12% Co coated pins and Cr_3C_2-25% NiCr coated pins, have been shown in Figures 2(a), 2(b), and 2(c) respectively.

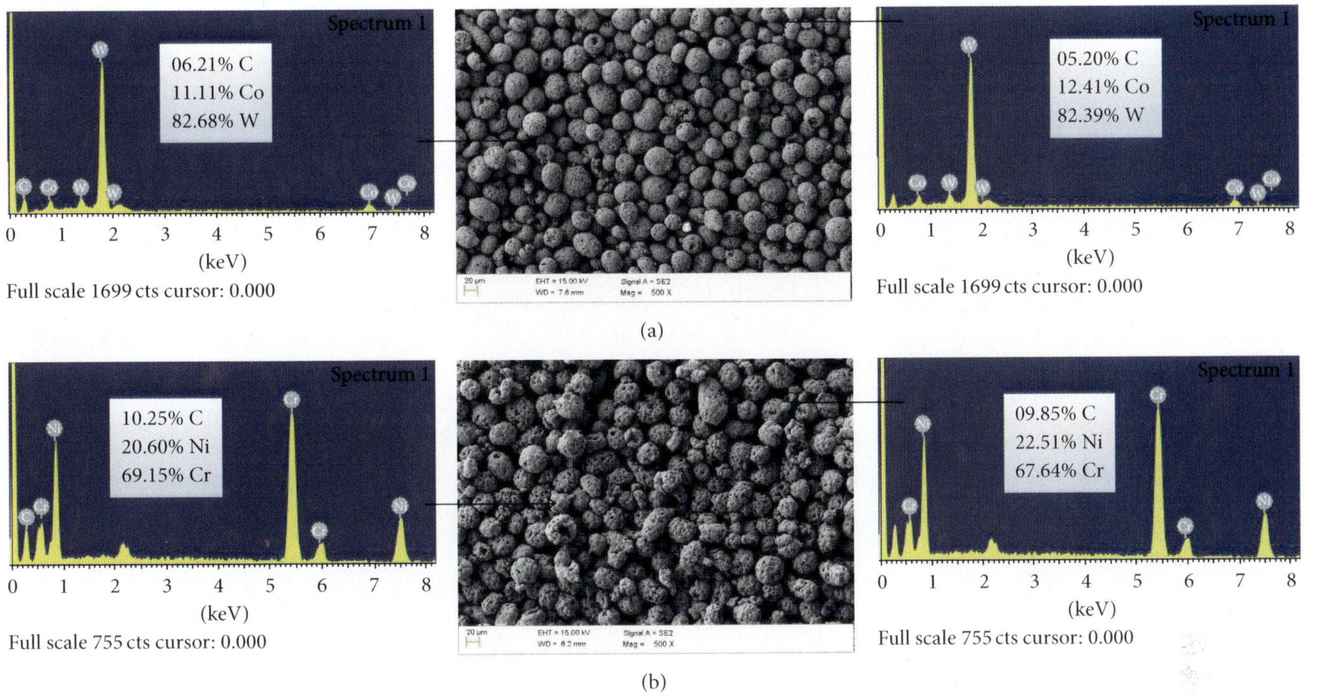

Figure 1: SEM images and EDS patterns showing the morphologies, microstructure, and elemental composition of the (a) WC-12% Co; (b) Cr_3C_2-25% NiCr feedstock powders.

Table 3: Chemical composition (weight % age) of the EN-31 steel disc.

Chemical composition (weight % age) of EN31 steel						
C	Si	Mn	Cr	Ni	S	P
0.90–1.20	0.10–0.35	0.30–0.75	1.00–1.60	1.46	0.04	0.04

2.5. Measurement of Coating Thickness. Average thickness of WC-12% Co and Cr_3C_2-25% NiCr coatings on ASTM A36 steel is measured with a thickness gauge Minitest-2000 during D-gun spraying process. Efforts are made to deposit coatings of uniform thickness of 250 ± 10 μm in all the cases of WC-12% Co and Cr_3C_2-25% NiCr powders for ASTM A36 steel substrate.

2.6. Experimental Procedures. Sliding wear tests are conducted on bare and cermet coated ASTM A36 steel samples using a pin-on-disc Tester. The model of the wear and friction monitor tester used is TR-20-PHM-CHM-600 of Ducom make (Ducom Instruments Pvt. Ltd., Bangalore, India) conforming to ASTM G 99 standard. Wear tests are performed on the flat end faces of the pins at room temperature in air having a relative humidity of 51.5%. During wear testing the specimen pin is held stationary against the counter face of a rotating disc at fixed track diameter. Rotating disc is made of an alloy steel (EN-31) which was case hardened (63 to 65 HRC). The composition of EN-31 steel material is given in Table 3.

2.7. Sliding Wear Studies. The specifications of the pin-on-disc tester used for wear testing include wear track diameter ranging from 10 to 100 mm, speed ranging from 200 to 2000 rpm, normal load ranging from 5 to 200 N, and sliding velocity ranging from 0.5 to 10 ms^{-1}. Both the disc and the pin are thoroughly cleaned and dried, before conducting the wear tests. The pin is loaded against the rotating disc through a dead weight loading system. The test parameters taken in the present investigation include wear track diameter (D) of 80 mm, speed of rotation of the disc (N) at 477 rpm, and three different normal applied loads of 40 N, 50 N and 60 N. The specimen pins are held at a fixed wear track diameter of 80 mm in order to have a constant sliding velocity (v) of 2 ms^{-1} throughout the wear testing, as $v = (\pi DN/60)\ ms^{-1}$. Thus, the wear tests are conducted for bare and coated specimens at a constant sliding velocity of 2 ms^{-1} at three different normal applied loads of 40 N, 50 N, and 60 N. A constant linear sliding velocity of 2 ms^{-1} is maintained by adjusting the speed of rotation of the disc. Wear tests are conducted for a total sliding distance of 10,800 m for each specimen. Weight losses of all the samples are measured after 5, 5, 10, 10, 20, and 40 minutes to find the wear loss. The weight is measured on a microbalance to an accuracy of 0.001 g. This weight loss data is used to determine the cumulative wear rate (C.W.R). Results of C.W.R. (in Bowden) with the sliding distance (m) are reported in the results and discussion.

3. Results

3.1. Microstructure of the Coatings. The Scanning electron microscope (SEM) micrographs for D-Gun sprayed WC-12% Co coating on ASTM A36 is shown in Figure 3(a) and for Cr_3C_2-25% NiCr coating on the same substrate is shown in Figure 3(b). Figure 3(a)(A) depicts the small size splats in

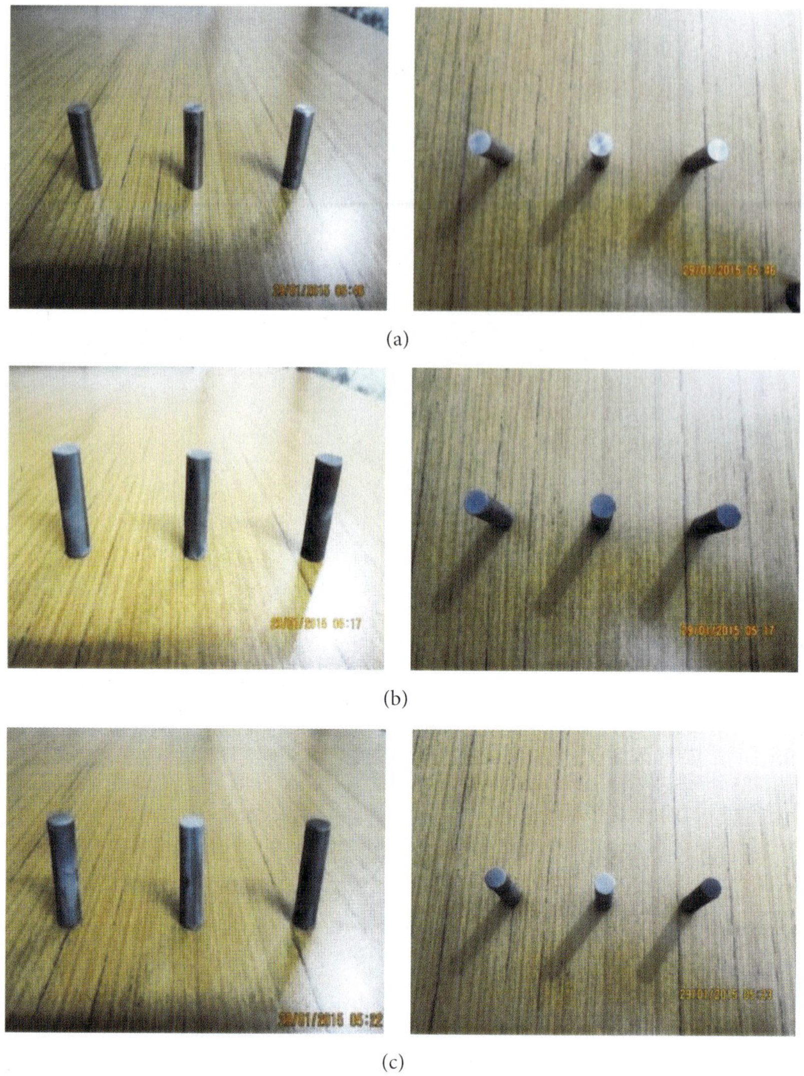

FIGURE 2: The figure shows the front view and top view (macrographs) of (a) uncoated or bare (ASTM A36) test specimens; (b) WC-12% Co coated pins; (c) Cr_3C_2-25% NiCr coated pins prepared for the sliding wear testing on pin-on-disc wear tester.

case of WC-12% Co coating whereas Figure 3(b)(A) shows the coarse splats in case of Cr_3C_2-25% NiCr coating. Figures 3(a)(B), 3(a)(C), 3(a)(D), 3(b)(B), 3(b)(C), and 3(b)(D) depict the SEM images of the worn samples of WC-12% Co and Cr_3C_2-25% NiCr coating after the conduct of wear tests at the normal applied loads of 40 N, 50 N, and 60 N, respectively. It is quite clear from these SEM images of worn samples that, at higher applied load, there is more extensive loss of material.

3.2. X-Ray Diffraction (XRD) Analysis. XRD patterns for D-Gun sprayed WC-12% Co on an ASTM A36 steel substrate is shown in Figure 4(a). XRD patterns for D-gun sprayed Cr_3C_2-25% NiCr coating on an ASTM A36 steel substrate are shown in Figure 4(b). From the XRD patterns for D-Gun sprayed WC-12% Co coating on ASTM A36 steel substrate, it is found that WC-12% Co coated specimen showed the major phases of WC and the minor phases of Co.

From the XRD patterns for D-gun sprayed Cr_3C_2-25% NiCr coating on ASTM A36 steel substrate, it is found that Cr_3C_2-25% NiCr coated specimen showed the major phases of Cr_3C_2, Cr_7C_3, and Ni which are desired phases of this coating.

3.3. Evaluation of Microhardness. The microhardness of the coatings is measured on the surface and along the cross-section. The cross-sectional analysis reading is taken with reference to the substrate-coating interface at intervals of 50-micron approximate distances. Miniload 2 Microhardness Tester (Leitz, Germany) fitted with a Vicker's pyramidal diamond indenter is used for measuring the microhardness of the coatings at 200 g load at Central Tool Room, Ludhiana, Govt. of India Society, Ministry of Micro, Small and Medium Enterprises. The critical microhardness values of the substrate are found to be in the range 211–220 $HV_{0.2}$. The WC-12%

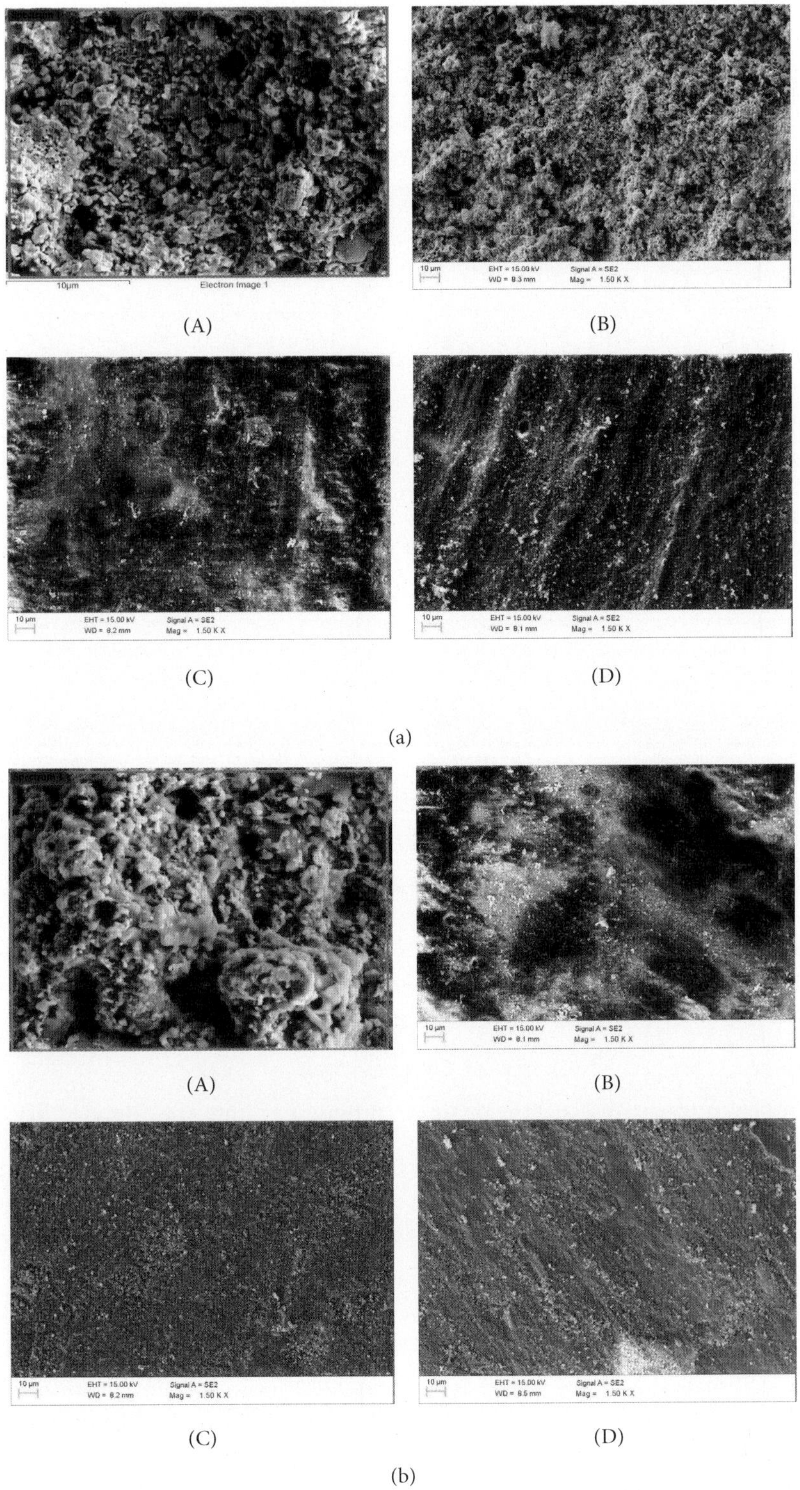

FIGURE 3: (a) SEM images of D-Gun sprayed WC-12% Co coating showing: (A) Unworn surface; (B) Surface worn under a load of 40 N; (C) Surface worn under a load of 50 N; (D) Surface worn under a load of 60 N. (b) SEM images of D-Gun sprayed Cr_3C_2-25% NiCr coating showing: (A) Unworn surface; (B) Surface worn under a load of 40 N; (C) Surface worn under a load of 50 N; (D) Surface worn under a load of 60 N.

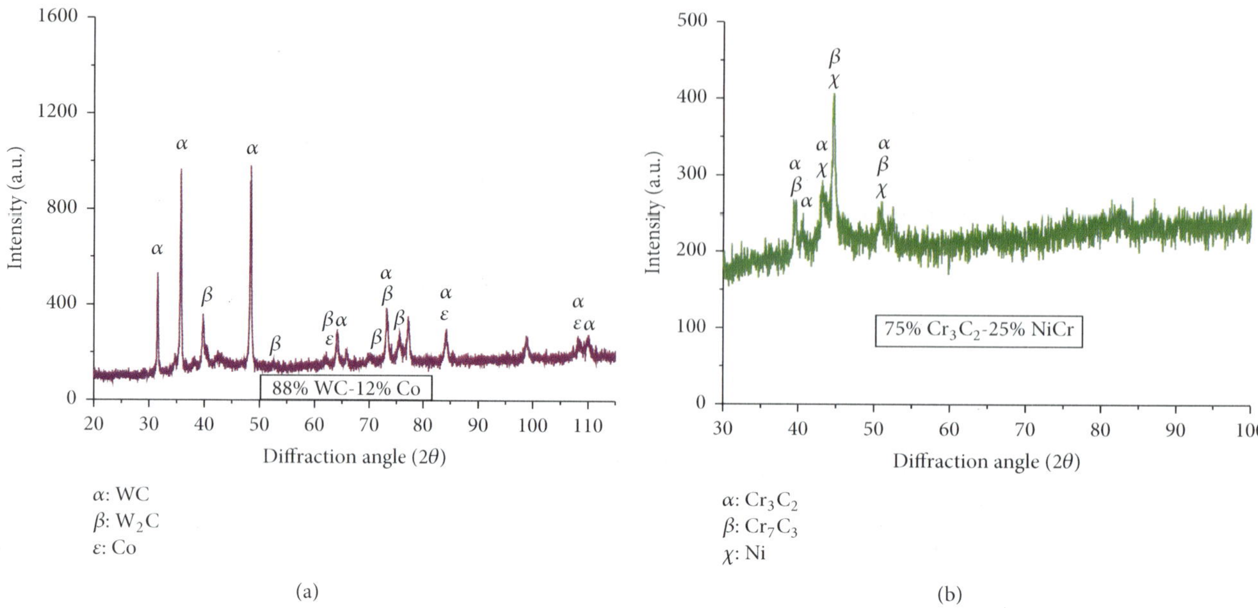

FIGURE 4: X-ray diffraction patterns of D-Gun sprayed (a) WC-12% Co; (b) Cr_3C_2-25% NiCr coating on an ASTM A36 steel substrate.

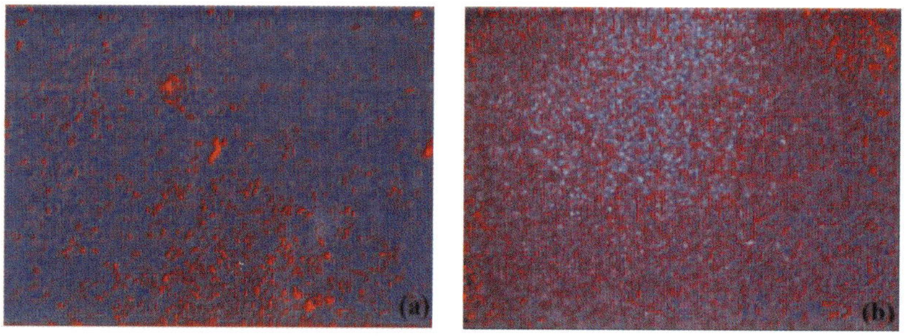

(a) (b)

FIGURE 5: Optical images showing the porosity in D-Gun sprayed (a) WC-12% Co; (b) Cr_3C_2-25% NiCr coating on an ASTM A36 steel substrate.

Co coating has shown maximum microhardness of the order of 1178–1192 $HV_{0.2}$. Cr_3C_2-25% NiCr coating has shown microhardness in the order of 716–726 $HV_{0.2}$.

3.4. Measurement of Porosity. Few pores and voids appear in the D-gun sprayed coatings, as it is an intermittent process during metal spraying. The porosity measurements are made by PMP3 inverted metallurgical microscope with stereographic imaging. The magnification is selected in such a way that the coating microstructure image covers the screen and permits the resolution of the voids that provides remarkably to the total porosity area percentage. The method of choosing the suitable range of light gray contrast spots is accomplished systematically by stereographic imaging to guarantee that only voids are chosen. The analysis using an image processing software decides the pore area size in the sight range by changing over the pore areas into a background color; for example, red despite whatever is left of the microstructure remains in its actual color. The area of one element is mathematically related to the total area of the image, as the program counts the number of one-color-type pixels, for example, red, and sets that as a proportion of the total number of pixels in the image, that is, the total area. Around twenty discrete locations are chosen to prevent the overlap between two locations and decide the area percent porosity [15]. The optical images depicting the porosity in D-Gun sprayed WC-12% Co and Cr_3C_2-25% NiCr coating on an ASTM A36 steel substrate are shown in Figures 5(a) and 5(b), respectively. The average porosity of WC-12% Co and Cr_3C_2-25% NiCr coating is found to be 0.28776% and 0.84625%, respectively, on an ASTM A36 steel substrate. Thus, the average porosity of all the as sprayed coatings is found to be less than 1.0%.

3.5. Measurement of Surface Roughness (R_a) Values. Veeco Optical Profilometer (NT 1100, USA make) is used to measure the surface roughness (R_a values) at IIT, Roorkee. It has the three-dimensional profiling capability with excellent precision and accuracy. The roughness is found to be in the range of 2.96–4.27 μm for WC-12% Co coatings and 3.96–4.64 μm for Cr_3C_2-25% NiCr, respectively.

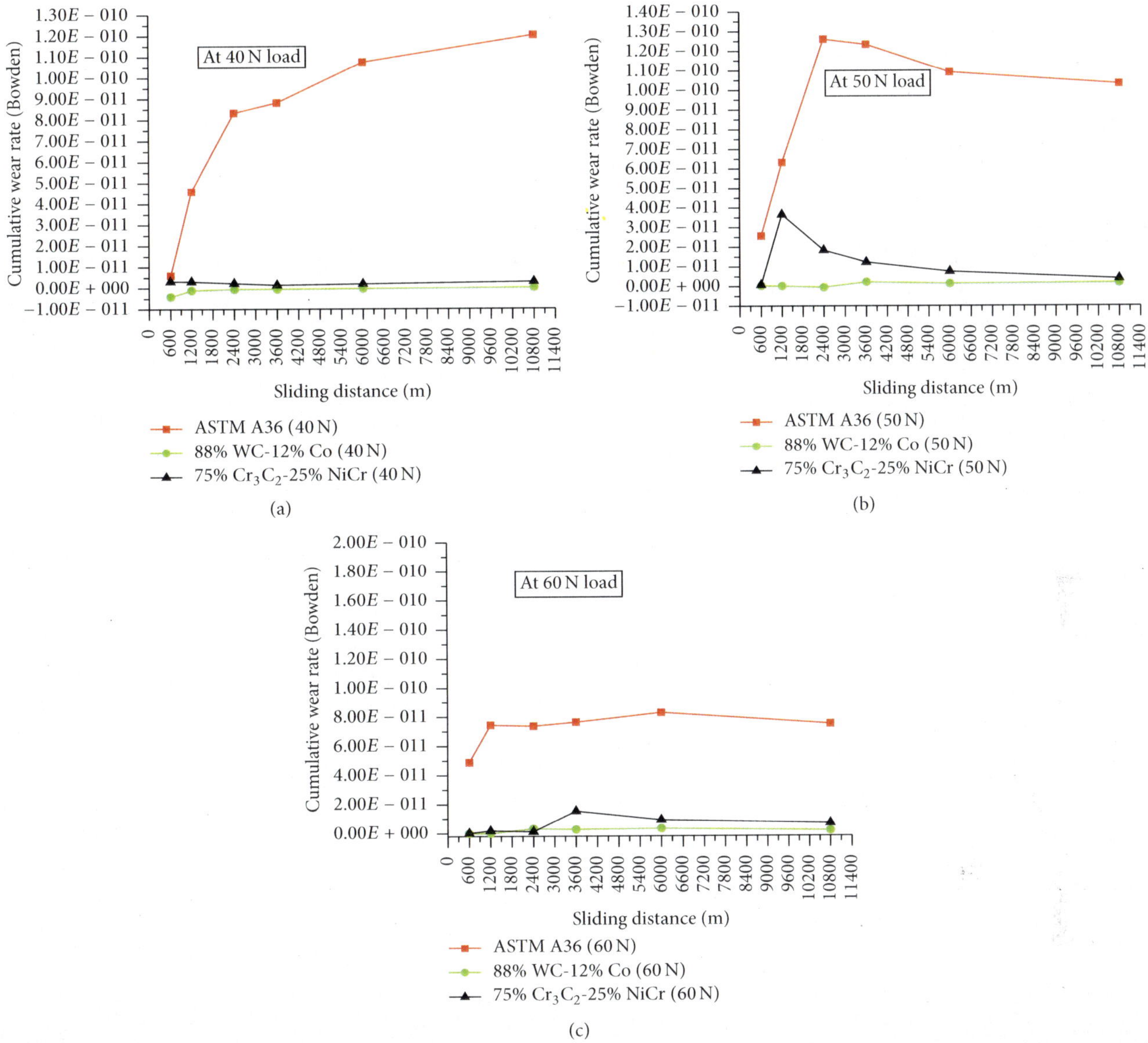

FIGURE 6: Variation of C.W.R. for uncoated and D-Gun spray coated ASTM A36 steel at normal applied load of (a) 40 N; (b) 50 N; (c) 60 N at sliding velocity of 2 ms^{-1}.

3.6. Wear Behavior. Wear testing of bare as well as cermet coated specimens is conducted on a Pin-On-Disc wear tester [16] as per the standard procedure conforming to ASTM G99 standard. Variation of C.W.R. (in Bowden) with the sliding distance has been discussed in this section for the different cases under investigation. Variation of the C.W.R. with the sliding distance for bare and WC-12% Co, Cr_3C_2-25% NiCr coated ASTM A36 steel at a normal applied load of 40 N, 50 N, and 60 N has been plotted in Figures 6(a), 6(b), and 6(c), respectively.

It is clear from Figure 6 that the WC-12% Co coated samples have shown considerable wear resistance at the three different applied loads of 40 N, 50 N, and 60 N in comparison with Cr_3C_2-25% NiCr coated and bare samples.

Bar charts as shown in Figures 7(a), 7(b), and 7(c) clearly depict the C.W.R. for bare and D-Gun spray coated ASTM A36 steel at normal applied load of 40 N, 50 N, and 60 N at a sliding velocity of 2 ms^{-1}, after a sliding distance of 10,800 m.

These plots and bar charts clearly show that the bare ASTM steel specimen has shown much higher C.W.R as compared to cermet coated specimens at all the different applied loads of 40 N, 50 N, and 60 N. Therefore, it can be concluded that at different loads of 40 N, 50 N, and 60 N the D-Gun sprayed WC-12% Co coating has shown considerable wear resistance as compared to the Cr_3C_2-25% NiCr coated specimens and bare ASTM A36 steel.

4. Discussion

4.1. Microstructure of the Coatings. The WC-12% Co coating has small size splats whereas the splats are coarse for Cr_3C_2-25% NiCr. From the SEM images as shown in Figures 3(a)(B),

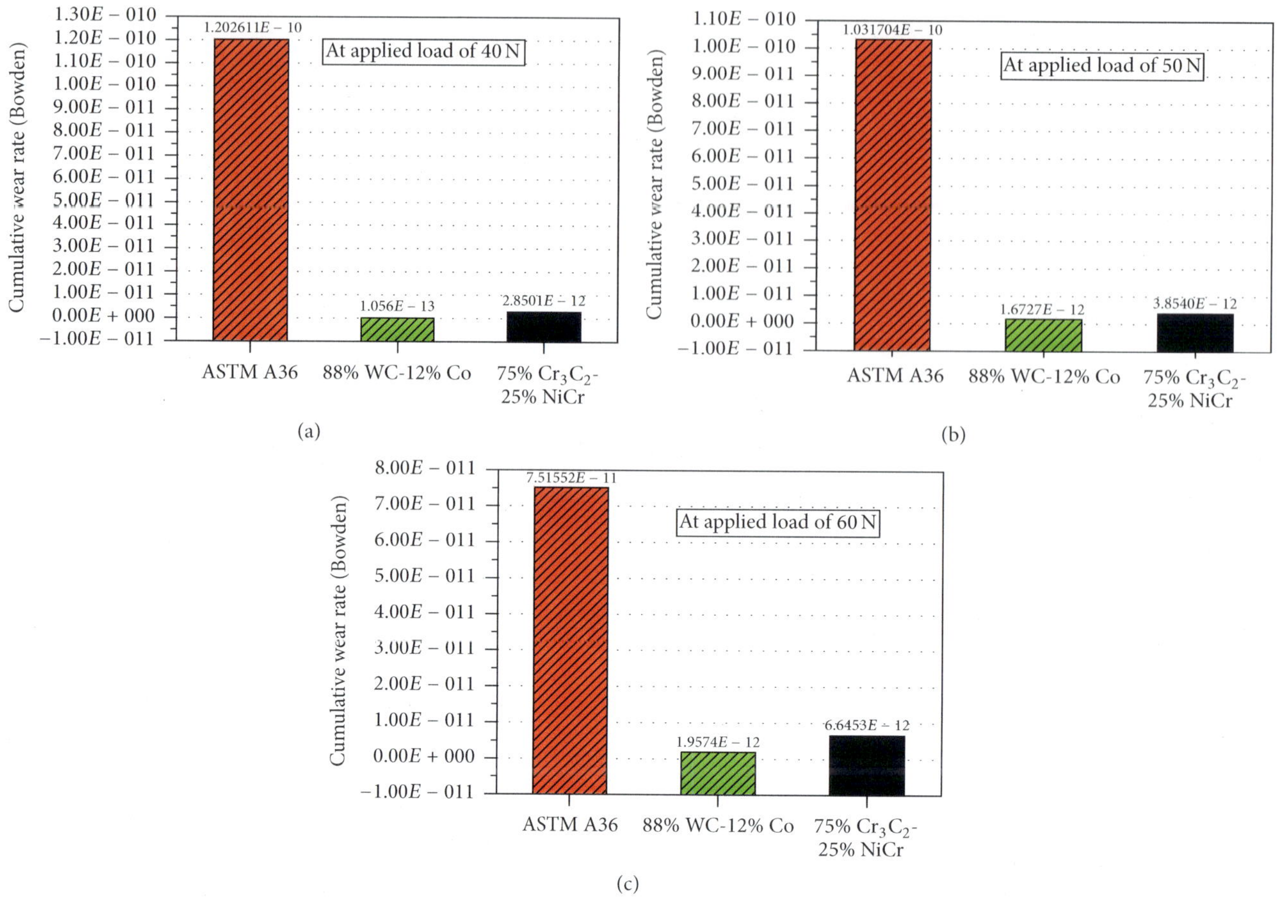

FIGURE 7: Bar chart showing C.W.R. for uncoated (bare) and D-Gun sprayed coated ASTM A36 steel at normal applied load of (a) 40 N; (b) 50 N; (c) 60 N at a sliding velocity of 2 ms^{-1}, after a sliding distance of 10,800 m.

3(a)(C), 3(a)(D), 3(b)(B), 3(b)(C), and 3(b)(D) it is clear that, at higher applied load, there is more extensive loss of material. There is no subsurface cracking in the case of the coated specimens. WC-12% Co coated samples were found to be more wear resistant at three different applied loads of 40 N, 50 N, and 60 N than Cr_3C_2-25% NiCr. In the present investigation, superior wear resistance of WC-12% Co coating is found which may be due to the presence of W on its surface. Similar results and microstructure are obtained by Chivavibul et al. [17] for WC-Co coatings. The microstructure obtained for Cr_3C_2-25% NiCr coatings are similar to the findings of Kamal et al. [18].

4.2. X-Ray Diffraction (XRD) Analysis. The number of peaks corresponding to different phases of WC-12% Co and Cr_3C_2-25% NiCr coatings on ASTM A36 material can be seen from diffraction patterns as shown in Figures 4(a) and 4(b), respectively. Theses identified phases for WC-Co coatings are in agreement with those observed by Suresh Babu et al. [19]. XRD phases identified for Cr_3C_2-25% NiCr coatings are in agreement with the findings of Mohanty et al. [20].

4.3. Evaluation of Microhardness. The higher microhardness values of WC-12% Co and Cr_3C_2-25% NiCr cermet coatings may be one of the reasons of high wear resistance of these coatings, which is in agreement with those observed by Sahraoui et al. [21].

4.4. Measurement of Porosity. The lower porosity of WC-12% Co and Cr_3C_2-25% NiCr cermet coatings may be one of the reasons of high wear resistance of these coatings, which is also supported by the findings of Sahraoui et al. [21].

4.5. Surface Roughness (R_a) Values. Coatings are differentiated on the basis of their properties like porosity, microhardness, microstructure, surface roughness and residual stresses. The surface of the coating is rough due to the due to the presence of fully melted and partially melted particles.

4.6. Wear Behavior. From the plots showing the variation of the C.W.R. with the sliding distance for bare and WC-12% Co, Cr_3C_2-25% NiCr coated ASTM A36 steel at a normal applied load of 40 N, 50 N, and 60 N, it is evident that the WC-12% Co and Cr_3C_2-25% NiCr cermet coated samples have shown lower C.W.R than uncoated ASTM A36 steel samples at all these three different applied loads.

From the bar charts showing C.W.R. for uncoated and D-Gun sprayed coated ASTM A36 steel at normal applied load of 40 N, 50 N, and 60 N at a constant sliding velocity of 2 ms^{-1}, after a sliding distance of 10,800 m, it is clear that the bare ASTM steel specimen has shown much higher C.W.R as compared to cermet coated specimens at all these three different applied loads. D-Gun sprayed WC-12% Co coating has shown considerable wear resistance as compared to the Cr_3C_2-25% NiCr coated specimens and bare ASTM A36 steel at all these three different applied loads. Also, it is found that the C.W.R for WC-12% Co and Cr_3C_2-25% NiCr cermet coated samples and bare ASTM A36 steel samples increases with an increase in the applied normal load.

D-Gun sprayed coatings are more wear resistant than bare or uncoated material which is in agreement with what was observed by Murthy et al. [22], Sundararajan et al. [23], and Wang et al. [24]. The wear rate is observed to increase with increase in applied load which is in agreement with what was observed by Qureshi and Sheikh [25]. Glaeser [26] opinioned that the wear rate is proportional to load and sliding distance during wear testing which fully supports the test results. Identical to the findings of the wear tests in the present study, enhancement of wear resistance of steel substrates is also observed after the deposition of coatings by Kulu and Phil [27]. The improvement of wear resistance after the deposition of coatings is also supported by the findings of Sahraoui et al. [21].

5. Conclusions

In the present investigation, the main conclusions on the basis of the results obtained are as follows:

(1) WC-12% Co and Cr_3C_2-25% NiCr cermet coatings have successfully been deposited on ASTM A36 steel substrate.

(2) WC-12% Co and Cr_3C_2-25% NiCr cermet coated samples have shown lower C.W.R than uncoated ASTM A36 steel samples.

(3) Enhancement in sliding wear resistance of ASTM A36 steel is observed after the deposition of WC-12% Co and Cr_3C_2-25% NiCr cermet coatings.

(4) C.W.R for cermet coated specimens and uncoated specimens increases with an increase in the applied normal load.

(5) Minimum C.W.R is observed for WC-12% Co coated specimens. WC-12% Co is found to be the best coating to be deposited on an ASTM A36 steel substrate in the present study.

(6) Sliding wear resistance of WC-12% Co coating is found to be better than Cr_3C_2-25% NiCr coating.

(7) Sliding wear resistance for all the investigated cases followed the trend as given below:

the trend is WC-12% Co coated ASTM A36 steel > Cr_3C_2-25% NiCr coated ASTM A36 steel > Bare ASTM A36 steel,

therefore, out of these combinations WC-12% Co–ASTM A36 coating-substrate combination is found to be the best combination.

Competing Interests

The authors declare that there is no conflict of interests regarding the publication of this paper.

Acknowledgments

The authors wish to thank the Pressure and Process Boilers, Saharanpur, India, for providing the substrate material; SVX Powder M Surface Engineering Private Limited, Greater Noida, U.P., India, for providing the facility of the detonation gun (D-Gun) thermal spray coating; Ducom Instruments Pvt. Ltd., Bangalore, India, for providing the wear and friction monitor tester (Model: TR-20-PHM-CHM-600) for the conduct of wear testing; IIT, Roorkee, India, for extending the facility of SEM/EDS and XRD analysis during this work. The authors are highly grateful to the officials, staff, and supervisor appointed by the IKG Punjab Technical University, Kapurthala, for their guidance and support.

References

[1] B. Bhushan and B. K. Gupta, *Handbook of Tribology: Material Coating and Surface Treatments*, McGraw-Hill, New York, NY, USA, 1991.

[2] J. F. Archard, *Wear Control Handbook: Wear Theory and Mechanisms*, Edited by M. B. Peterson and W. O. Winer, ASME, New York, NY, USA, 1980.

[3] H. Singh, M. S. Grewal, H. S. Sekhon, and R. G. Rao, "Sliding wear performance of high-velocity-oxy-fuel spray Al_2O_3/TiO_2 and Cr_2O_3 coatings," *Proceedings of the Institution of Mechanical Engineers, Part J: Journal of Engineering Tribology*, vol. 222, no. 4, pp. 601–610, 2008.

[4] J. Stokes, *Thermal Spray Process: Theory and Application of the High-Velocity-Oxy-Fuel (HVOF)*, Dublin City University, Dublin, Ireland, 2008.

[5] J. Halling, *Introduction: Recent Development in Surface Coating and Modification Processes*, Mechanical Engineering Publications Limited, London, UK, 1985.

[6] R. Arulmani and S. Pandey, "Surfacing applications: a review," in *Proceedings of the National Workshop on Welding Technology*, Department of Mechanical Engineering, S.L.I.E.T., Longowal, India, 2003.

[7] H. P. Jost, "Tribology—origin and future," *Wear*, vol. 136, no. 1, pp. 1–17, 1990.

[8] Z. G. Karl-Heinz, *Microstructure and Wear of Materials*, Tribology Series 10, Elsevier, Tokyo, Japan, 1987.

[9] A. Vencl, "Abrasive wear resistance of the iron and WC-based hardfaced coatings evaluated with scratch test method," in *Proceedings of the 13th International Conference on Tribology (SerbiaTrib '13)*, Kragujevac, Serbia, 2013.

[10] S. K. Srivastava, *Tribology in Industries*, S. Chand and Company Limited, New Delhi, India, 2001.

[11] B. Q. Wang and S. W. Lee, "Elevated temperature erosion of several thermal-sprayed coatings under the simulated erosion conditions of in-bed tubes in a fluidized bed combustor," *Wear*, vol. 203-204, pp. 580–587, 1997.

[12] J. Stokes, *Theory and Application of HVOF Thermal Spray Process*, Dublin City University, Dublin, Ireland, 2003.

[13] R. C. Tucker Jr., *Handbook of Deposition Technologies for Film and Coatings: Advanced Thermal Spray Deposition Technologies*, Edited by R. F. Bunshah, Noyes, Park Ridge, Ill, USA, 2nd edition, 1994.

[14] K. Rao, D. A. Somerville, and D. A. Lee, "Properties and characterization of coating made using jet kote thermal spray technique," in *Proceedings of the 11th International Conference on Thermal Spray (Thermal Spray '86)*, Welding Institute of Canada, Montreal, Canada, September 1986.

[15] N. Kumar, G. Prashar, and R. K. Dhawan, "To check the feasibility of Cr_2O_3 coating on boiler steel tubes simulated coal fired boiler conditions to prevent the erosion," *International Journal on Emerging Technologies*, vol. 3, no. 1, pp. 126–129, 2012.

[16] Test Method for Wear Testing with a Pin-on-Disk Apparatus, "American Society for Testing and Materials International Standards," ASTM Standard G99, ASTM, 2003.

[17] P. Chivavibul, M. Watanabe, S. Kuroda, and K. Shinoda, "Effects of carbide size and Co content on the microstructure and mechanical properties of HVOF-sprayed WC-Co coatings," *Surface and Coatings Technology*, vol. 202, no. 3, pp. 509–521, 2007.

[18] S. Kamal, R. Jayaganthan, S. Prakash, and S. Kumar, "Hot corrosion behavior of detonation gun sprayed Cr_3C_2-NiCr coatings on Ni and Fe-based superalloys in Na_2SO_4-60% V_2O_5 environment at 900○C," *Journal of Alloys and Compounds*, vol. 463, no. 1-2, pp. 358–372, 2008.

[19] P. Suresh Babu, B. Basu, and G. Sundararajan, "Processing–structure–property correlation and decarburization phenomenon in detonation sprayed WC–12Co coatings," *Acta Materialia*, vol. 56, no. 18, pp. 5012–5026, 2008.

[20] M. Mohanty, R. W. Smith, M. De Bonte, J. P. Celis, and E. Lugscheider, "Sliding wear behavior of thermally sprayed 75/25 Cr_3C_2/NiCr wear resistant coatings," *Wear*, vol. 198, no. 1-2, pp. 251–266, 1996.

[21] T. Sahraoui, N.-E. Fenineche, G. Montavon, and C. Coddet, "Structure and wear behaviour of HVOF sprayed Cr_3C_2–NiCr and WC–Co coatings," *Materials and Design*, vol. 24, no. 5, pp. 309–313, 2003.

[22] J. K. N. Murthy, D. S. Rao, and B. Venkataraman, "Effect of grinding on the erosion behaviour of a WC–Co–Cr coating deposited by HVOF and detonation gun spray processes," *Wear*, vol. 249, no. 7, pp. 592–600, 2001.

[23] G. Sundararajan, D. Sen, and G. Sivakumar, "The tribological behaviour of detonation sprayed coatings: the importance of coating process parameters," *Wear*, vol. 258, no. 1–4, pp. 377–391, 2005.

[24] J. Wang, B. Sun, Q. Guo, M. Nishio, and H. Ogawa, "Wear resistance of a Cr_3C_2-NiCr detonation spray coating," *Journal of Thermal Spray Technology*, vol. 11, no. 2, pp. 261–265, 2002.

[25] F. S. Qureshi and A. K. Sheikh, "A probabilistic characterization of adhesive wear in metals," *IEEE Transactions on Reliability*, vol. 46, no. 1, pp. 38–44, 1997.

[26] W. A. Glaeser, "Friction and wear," *IEEE Transactions on Parts, Hybrids, and Packaging*, vol. 7, no. 2, pp. 99–105, 1971.

[27] P. Kulu and T. Phil, "Selection criteria for wear resistant powder coatings under extreme erosive wear conditions," *Journal of Thermal Spray Technology*, vol. 11, no. 4, pp. 517–522, 2002.

5

Microstructure of Al-Si Slurry Coatings on Austenitic High-Temperature Creep Resisting Cast Steel

Agnieszka E. Kochmańska

West Pomeranian University of Technology in Szczecin, Piastow 17, 70-310 Szczecin, Poland

Correspondence should be addressed to Agnieszka E. Kochmańska; agnieszka.kochmanska@zut.edu.pl

Academic Editor: Yuanshi Li

This paper presents the results of microstructural examinations on slurry aluminide coatings using scanning electron microscopy, X-ray microanalysis, and X-ray diffraction. Aluminide coatings were produced in air atmosphere on austenitic high-temperature creep resisting cast steel. The function of aluminide coatings is the protection of the equipment components against the high-temperature corrosion in a carburising atmosphere under thermal shock conditions. The obtained coatings had a multilayered structure composed of intermetallic compounds. The composition of newly developed slurry was powders of aluminium and silicon; NaCl, KCl, and NaF halide salts; and a water solution of a soluble glass as an inorganic binder. The application of the inorganic binder in the slurry allowed to produce the coatings in one single step without additional annealing at an intermediate temperature as it is when applied organic binder. The coatings were formed on both: the ground surface and on the raw cast surface. The main technological parameters were temperature (732–1068°C) and time of annealing (3.3–11.7 h) and the Al/Si ratio (4–14) in the slurry. The rotatable design was used to evaluate the effect of the production parameters on the coatings thickness. The correlation between the technological parameters and the coating structure was determined.

1. Introduction

High-temperature corrosion is a common problem in various industrial processes, such as heat treatment, chemical processing, refining, etc. in, for example, petrochemical, aerospace, and gas turbine industries. Furnaces for thermochemical treatment, for example, carburising or nitriding furnaces, are often made of high-temperature creep resisting cast steels, especially Fe-Cr-Ni based alloys. A high mechanical strength at elevated temperatures and a resistance to thermal shocks are required from these materials. It is difficult to keep a satisfactory compromise between the mechanical strength and the surface degradation resistance in such severe working conditions. On the surface of Cr-rich materials, a protective chromium oxide layer (Cr_2O_3) is formed and provides the carburisation resistance at temperatures below 1050°C [1]. The chromia is also liable to spalling and cracking during creep and thermal shocks. Therefore, the other protective layers such SiO_2 and Al_2O_3 are considered. For oxidation resistance above 1200°C, alloys that rely on the protective Al_2O_3 scale formation are preferred over those forming chromia. Unfortunately, the addition of aluminium and silicon to the high-temperature creep resisting alloys to develop full protection involves metallurgical compromise in strength and ductility. Therefore, silicon is usually limited to 2% and aluminium to less than 4% in alloys [2].

One of the most frequent solutions to introduce aluminium and silicon is an application of appropriate coatings. Inward diffusion of aluminium and outward diffusion of iron, nickel, and chromium results in the formation of intermetallic phases with very interesting properties. They are widely used for the improvement of the corrosion resistance of iron- and nickel-based alloys. Among these intermetallics, Fe and Ni aluminides improve the resistance to many aggressive environments including oxidising [3, 4], sulphidising [5], and carburising [6, 7] agents. These coatings extend the lifetime of the equipment as a result of the carburisation reduction of the surface layer. Coatings, that protect the base material in industrial processes, need not only to inhibit the diffusion of elements, but also must offer resistance to thermal shocks that occur during quenching following the carburising process [8]. These shocks cause accelerated destruction of the protective coatings. In addition to aluminium in the coating, very important is the presence of silicon, which decreases the carbon solubility in austenite and is one of the main elements improving the resistance to carbon absorption at high temperatures. Moreover, it has been observed that SiO_2 is present as a subscale beneath chromium oxide [1].

The manufacturing method of the coating should be easy to apply on complex-shaped parts, and the cost should be as low as possible. One of these methods is the slurry cementation. Slurry aluminisation can provide advantages in the process, that is, simplicity and economical efficiency. Typical slurries manufactured on the base of aluminium powder (as an active component) and of the solution of an organic binder are widely used [9–12]. When the polyvinyl alcohol is used as the organic binder in the slurry, slurry-coated specimens have to be held at a temperature of 300°C for 4 h [12] or 400°C for 3 h [10] to expel the binder and only after that heat treated. The use of the other type of binder instead of the organic one could allow to produce the coatings in an one single step without additional annealing at an intermediate temperature.

In this work, a single-step slurry cementation was applied to create the protective Al-Si coatings. The method used by the author is a modification of the conventional one. In this study, the inorganic binder, that is, a water solution of sodium silicate, was applied instead of the organic one. The simultaneous use of the sodium silicate and the silicon powder in the slurry causes dehydration and self-hardening of the slurry. This means that silicon can be beneficial in the slurry both from a technological point of view and later in the diffusion-forming process. The innovative composition of the slurry has forced the determination of optimum production parameters of the protective coatings on austenitic high-temperature creep resisting cast steel.

The slurry usually comprises metal powders with a grain size below 40 μm [13], but very often the grain size is in the order of a few micrometers [14, 15]. In this work, powders of larger granularity were used. This has reduced the cost of the slurry. In addition, a larger grain of aluminium powder reduces the risk of an explosion during production.

Additionally, in this work, a mixture of three selected salts was used. These salts applied in the slurry are a flux and have good properties from an ecological and economic point of view. The existence of oxides on both aluminium particles in the slurry and the surface of samples is a barrier for diffusion processes. The oxide shell is dissolved by this flux during annealing. The coating is formed in a liquid state (the temperature of annealing is above the temperature of aluminium melting). The liquid state guarantees high rates of the diffusion reaction, and the interruption of the oxide shell accelerates processes in the liquid state.

2. Materials and Methods

2.1. Samples and Slurry Composition. The substrate used for the investigation was austenitic high-temperature creep resistant cast steel (GX20NiCrSi30-18-1) containing approximately 0.2 wt.% C, 18% Cr, 30% Ni, 0.7% Mn, 1.5% Si, 0.03% P, 0.03% S, and Fe balance. Dimensions of the samples were $10 \times 20 \times 6$ mm. The surface of samples used in the experiment was finished by grinding. However, several samples of the raw cast surface were subjected to produce the coatings as well (to verify the possibility of the coating formation). The wall thickness of the casting was about 9 mm; therefore, the dimensions of this sample were $10 \times 20 \times 9$ mm. The active slurry was prepared from silicon and aluminium powders. The grain size was approximately 100 μm for the aluminium powder and within the range 200–250 μm for the silicon powder. The binder used in the slurry consisted of an aqueous solution of sodium silicate (soluble glass). The flux was used of composition of sodium and potassium salts (NaF, NaCl, and KCl). The ratio of the sodium silicate to (Al + Si) powders was 15/100, and the ratio of the salts to (Al + Si) powders was 9.5/100. The quantitative composition of the slurry was determined experimentally and patented [16].

The samples after degreasing with acetone were immersed several times in the water slurry and dried until the value of the covering mass amounted to 0.3 $g{\cdot}cm^{-2}$ and then annealed in air atmosphere (without any protective atmosphere). The slurry residues were removed after annealing, and the samples were washed in an ultrasonic bath in acetone.

2.2. Design of Experiment. The purpose of the application of the plan is the determination of the optimum parameters of coatings producing. The annealing temperature, time, and the ratio Al/Si in the slurry were adopted as three independent variables in the design of the experiment. The determination of the equations describing the thickness as a function of these three parameters was the objective in the chosen experiment design. A rotatable central composite design was chosen. Thus, rotatable designs do not favour one direction over another during exploring. This is reasonable when the knowledge about the response surface before experimentation is insufficient. This design requires twenty interactions (total points in the design) if there are three variable parameters [17]. The various combinations of levels of these interactions lie on the edge (factorial) points ($\pm 1, \pm 1, \pm 1$) and the centre point (0, 0, 0) of a cubic experimental region, as illustrated schematically in Figure 1.

The plan assumes the existence of additional (axial) points ($\pm\alpha$, 0, 0), (0, $\pm\alpha$, 0), and (0, 0, $\pm\alpha$). In the rotatable design, the variance of the predicted response is the same at all points that have the same distance from the centre of the design. The plan of the rotatable design for the coded variables (x_i) together with all the real values (X_i) of the interaction between the three variables is shown in Table 1. It was assumed that the unknown characteristics can be approximated by the second-order model which describes

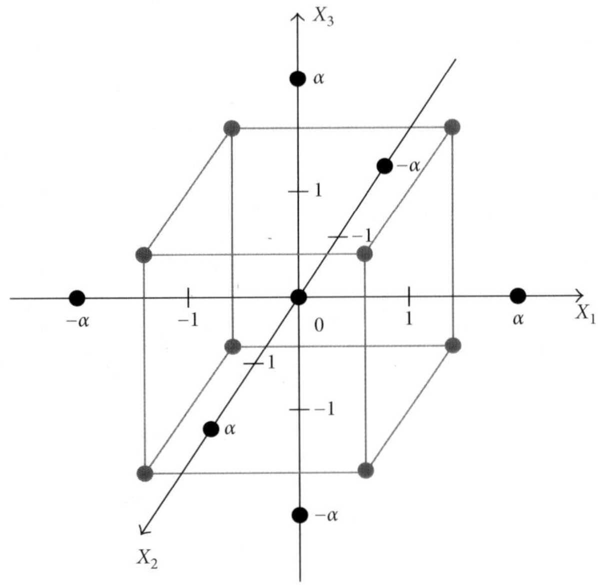

FIGURE 1: Graphical representation of the rotatable design for the three variables.

TABLE 1: The rotatable design for three-factor interactions with coded and real variables.

Run number/number of coating	Temperature		Time		Ratio Al/Si	
	x_1	X_1/°C	x_2	X_2/h	x_3	X_3
1	−1	800	−1	5	−1	6
2	+1	1000	−1	5	−1	6
3	−1	800	+1	10	−1	6
4	+1	1000	+1	10	−1	6
5	−1	800	−1	5	+1	12
6	+1	1000	−1	5	+1	12
7	−1	800	+1	10	+1	12
8	+1	1000	+1	10	+1	12
9	−1.682	732	0	7.5	0	9
10	+1.682	1068	0	7.5	0	9
11	0	900	−1.682	3.3	0	9
12	0	900	+1.682	11.7	0	9
13	0	900	0	7.5	−1.682	4
14	0	900	0	7.5	+1.682	14
15	0	900	0	7.5	0	9
16	0	900	0	7.5	0	9
17	0	900	0	7.5	0	9
18	0	900	0	7.5	0	9
19	0	900	0	7.5	0	9
20	0	900	0	7.5	0	9

quadratic surfaces, and quadratic surfaces can take several shapes [17]. Second-order model for three independent variables takes the form

$$Y = \beta_0 + \beta_1 X_1 + \beta_2 X_2 + \beta_3 X_3 + \beta_4 X_1^2 + \beta_5 X_2^2 + \beta_6 X_3^2 + \beta_7 X_1 X_2 + \beta_8 X_1 X_3 + \beta_9 X_2 X_3. \quad (1)$$

A second-order design has a uniform precision which is a reasonable criterion, because we want to know precision how close to the origin a maximum or other surface feature may be [17].

The central temperature, the central annealing time, and the central ratio Al/Si were correspondingly 900°C, 7.5 hours and 9 : 1, respectively. The remaining points (± 1, $\pm\alpha$) were

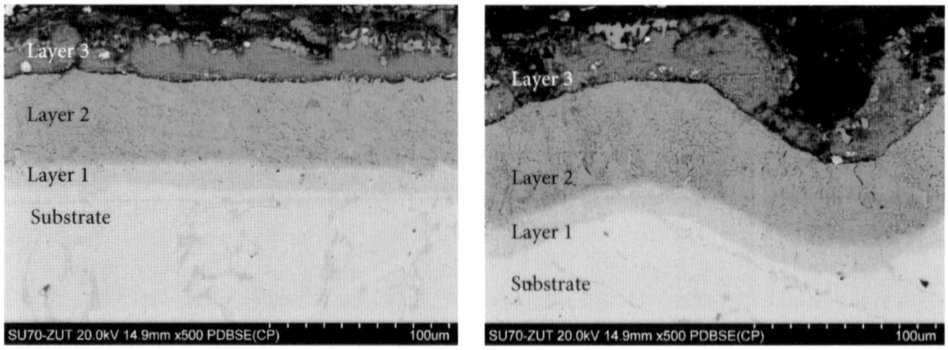

(a) (b)

Figure 2: Microstructure (BEI) of Al-Si coating on sample (a) number 16 (surface after grinding), (b) with a rough cast surface (technological parameters in both cases: 900°C, 7.5 h, Al/Si = 9; central point of the experiment).

calculated assuming an appropriate step. The steps (levels coded as ±1) for the temperature, time, and ratio Al/Si were correspondingly ±100, ±2.5, and ±3, respectively. The ratio Al/Si in the slurry was changed from 4 : 1 to 14 : 1. The samples covered with the slurry were annealed at corresponding temperatures between 732°C and 1068°C for corresponding times between 3.3 and 11.7 hours, respectively, in the air atmosphere. The design predicted the coating properties throughout the area, even at parameter combinations actually not working. The properties of coatings are predicted for the whole range of the three parameters using only the parameter combinations considered.

2.3. Microstructural Characterization. Metallographic preparations of the cross sections of all coatings were conducted according to the standard methods. All samples have been examined using the field emission scanning electron microscopy (FESEM) (Hitachi SU-70) equipped with the energy dispersive X-ray microanalysis (EDS; Thermo Scientific NORAN System 7). Accelerating voltage of 20 kV was applied for SEM and EDS analyses. The COMPASS software was used by collecting a complete, deadtime-corrected spectrum at every point of the sample. COMPASS is a statistical-based software applied in the Thermo NORAN 7 software which enables quantitative analysis of each elemental phase [18]. The COMPASS software finds image locations that have the same spectral "fingerprint," with no assumptions about the chemical composition of the sample. Such created elemental phase maps show the spatial distribution of regions that have statistically similar spectral features (peak locations and relative peak heights) and cumulative spectra of these regions.

The XRD analysis was conducted with Panalytical X'Pert-PRO X-ray diffractometer using Co-Kα radiation and the HighScore software to analyse the diffraction patterns. Consecutive layers of the coating material were removed by grinding. Such a protocol enabled to identify phase components present in particular layers.

The microhardness tests were performed using a Buehler hardness tester using a Knoop indenter at the load of 0,098 N during 15 s.

3. Results and Discussion

The coatings were successfully produced for each set of assumed technological parameters. The microscopic observations proved that the microstructures of the coatings depended on the production parameters. In general, the typical coating consisted of three layers: the internal layer (1), the middle layer (2), and the external layer (3) (Figure 2(a)). The total thickness of the resulting coatings for particular cases was rather uniform, and the standard deviations did not exceed 5%, except the coatings produced at a low temperature, especially in a long time, which were less uniform (the standard deviations did not exceed 10%).

The microstructure of the coating obtained with the central technological parameters (900°C, 7.5 h, 9 Al/Si) is shown in Figure 2. Figure 2(a) shows the Al-Si coating obtained on the surface after grinding and Figure 2(b) on the raw cast surface. No significant differences in microstructure between the coatings on the machined and the raw cast surface were observed. Thus, the coatings can be produced on the cast surface without further mechanical processing like grinding [19]. The possibility to obtain coatings on the surface of the raw cast is very important from the technological point of view. The presence of the flux in the slurry promotes the formation of the coating despite the existence of a thick oxide layer on the surface of the casting.

Different structures and properties of the coatings were obtained depending on the technological parameters in the established plan of the experiment. For example, the structures of the two coatings prepared with the parameters belonging to set numbers 7 and 2 are shown in Figures 3(a) and 3(b), respectively. The main differences in the structure of the coatings were the thickness and the chemical composition.

The relations (as regression models) between the thickness and various parameters were determined using Statistica 12 software, and the equations thus obtained are presented below. The following notation was used: Y_x, the thickness of the x sublayer (μm); Y_t, the total thickness (μm); X_1, the annealing temperature (°C); X_2, the annealing time (h); X_3, the Al/Si ratio; R, multiple correlation coefficient; R^2, unbiased estimator; F, value of Snedecor test.

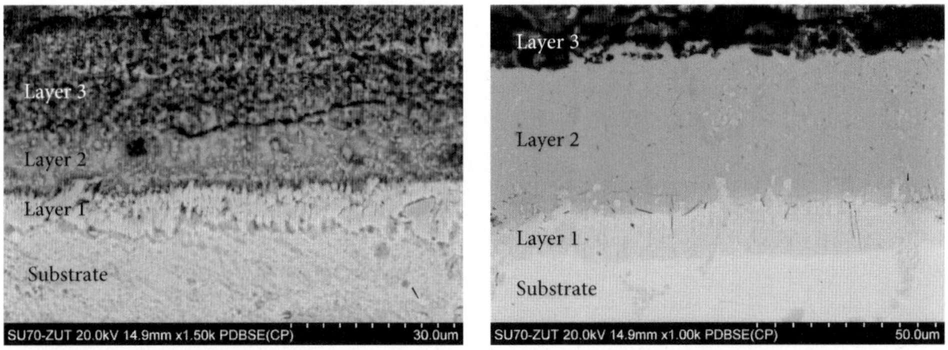

(a) (b)

FIGURE 3: Microstructure (BEI) of Al–Si coating on sample (a) number 7 (technological parameters: 800°C, 10 h, Al/Si = 12) and (b) number 2 (technological parameters: 1000°C, 5 h, Al/Si = 6).

$$Y_1 = 523 - 1.07X_1 - 30X_2 + 0.0006X_1^2 + 0.56X_2^2 + 0.026X_1X_2 \quad (2)$$
$$R = 0.96;\ R^2 = 0.93;\ F = 36.78$$

$$Y_2 = 43.3 - 3.89X_2X_3 + 0.005X_1X_2X_3 \quad (3)$$
$$R = 0.72;\ R^2 = 0.52;\ F = 9.23$$

$$Y_3 = -6.68 + 0.008X_1X_2 \quad (4)$$
$$R = 0.65;\ R^2 = 0.42;\ F = 13.42$$

$$Y_t = -206.94 + 0.00035X_1^2 + 0.966X_2^2 \quad (5)$$
$$R = 0.78;\ R^2 = 0.60;\ F = 12.80.$$

The determined regression models for the thickness were used to show the dependence on the annealing time and the temperature in the following graphs (Figure 4). The total thickness (5) of the coatings and the thickness of each layer depended mainly on the temperature and the time of the annealing. The ratio of aluminium to silicon in the slurry only affected the thickness of the second layer (3). Hence, the Al/Si ratio had no significant influence on the thickness of the first and third layers (in the applied range). The graph of the second layer thickness has been drawn for the Al/Si ratio $X_3 = 6$. It was found that the total thickness of the coating increased with increasing temperature and time of annealing (Figure 4(d)). The same holds for the thickness of the third layer (Figure 4(c)). The situation was different using the regression models representing the thickness of the first and the second layer. The increase in the thickness of these layers is apparent with increasing temperature and the time, but only at temperatures above 900°C. The thickness of the first and the second layer decreased significantly with the increasing times at temperatures below 800°C.

The Knoop hardness measurement results performed in layers 1 and 2 and in the base material were correspondingly 784 ± 137 HK0.01 in the first layer, 1251 ± 224 HK0.01 in the second layer, and 436 ± 68 HK0.01 in the base material. The results of the coating hardness measurements showed that the first layer has an intermediate hardness value in between those of the substrate and the second layer reducing the stress gradient between the substrate and the coating. The third layer was brittle at room temperature, and hardness tests have not been successfully performed. The spalling of the third layer occurring probably during metallographic preparation was observed, especially in coatings where the thickness of first and the second layer was small and third layer was big (Figure 4(c)—in a long time of annealing).

Figure 5 shows the elemental mapping of the coating obtained at central parameters of the experiment design. The results of a quantitative point analysis (EDS) for the first layer and the second layer of the coatings are presented in Figures 6(a) and 6(b) and Table 2.

Quasiphase maps by the COMPASS procedure showed the spatial distribution of regions that have statistically similar spectral features (Figure 7). The results of the chemical quantitative analysis of each phase suggested by the COMPASS software are given in Table 3.

XRD phase analysis was performed on sections parallel to the surface of the coating, and subsequent XRD patterns are shown in Figure 8.

AlN, Al_5FeNi, and β-NiAl phases were found near the surface of the coating formed at 1000°C. The β-NiAl phase was identified in the second layer of this coating, and the α-(Fe, Cr, Si) solution and $CrSi_2$ were identified in the first layer. Alloyed austenite was identified in the substrate. XRD studies were also carried out for the coatings formed at a temperature of 800°C, because the morphology of these coatings (formed at 800°C) was slightly different from those formed at 1000°C (Figures 3(a) and 3(b)). It was also observed that the kinetics of growth of the first and the second layer at a temperature of 800°C differs from the kinetics of growth of these layers annealed at higher temperature (Figures 4(a) and 4(b)). A decrease in thickness with time was observed at annealing temperature of 800°C. Figures 9 and 10 show XRD diffraction patterns of two coatings produced at 800°C for 10 hours but using different slurry compositions.

AlN phase was observed in the surface layer of the coating produced from the slurry with the ratio of Al/Si = 6 (the amount of the silicon in the slurry is higher). In the

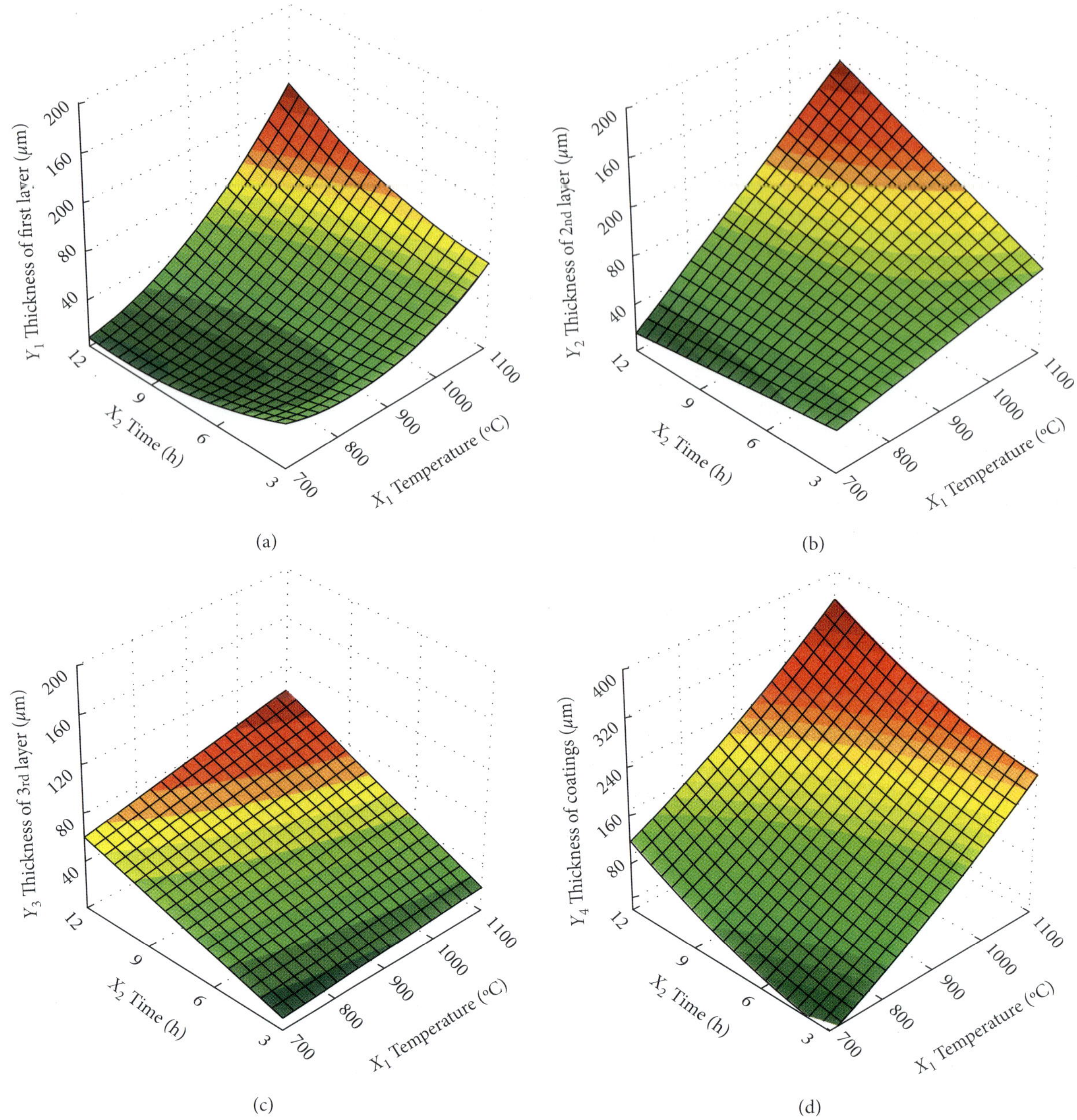

FIGURE 4: Graph of the correlation between the technological parameters and the thickness of (a) first layer, (b) second layer, (c) third layer, and (d) coatings.

second and the third layers, Al_5FeNi, Cr(Al, Si)$_2$, Al_3Ni_2, and β-AlNi phases were identified. The Al_5FeNi phase dominated in the third layer and the Al_3Ni_2 and β-AlNi phases in the second layer. The α-(Fe, Cr, Si) solution in the first layer was identified. Cr(Al, Si)$_2$, Al_8Cr_5, Al_3Ni_2, and β-AlNi phases were observed near the surface of the coating produced from the slurry with the ratio of Al/Si = 12. Cr(Al, Si)$_2$, Al_3Ni_2, and β-AlNi were in the third and second layers. The α-(Fe, Cr, Si) solution was observed in the first layer similar to the previously described cases. The presence of M_7C_3 carbides from the substrate was also visible in the coating.

The first layer was a diffusion layer formed by outward iron diffusion and inward diffusion of silicon. Nickel diffusion from this area to the second layer led to Ni content decrease in the first layer, consequently promoting a transition of f.c.c. to b.c.c. crystal lattice. The γ-(Fe, Ni, Cr) solution transformed into the α-(Fe, Cr, Si) solution [20], with a certain content of nickel and aluminium. The solubility of aluminium in γ-Fe is rather small contrary to that of α-(Fe, Cr) [21]. The aluminised coatings obtained at 1000°C in this work consisted of similar phase compositions to coatings with interfacial layers of the α-(Cr, Fe) and β-FeAl layers and the β-FeAl layer and a thick outer layer of δ-Fe_2Al_5 reported by Hu et al. [22]. They also observed [22] spherical β-FeAl precipitates present in the α-(Cr, Fe) grains. In the current work, a fine-grained intermetallic β-Al(Ni, Fe) phase was formed in the matrix of the first layer containing

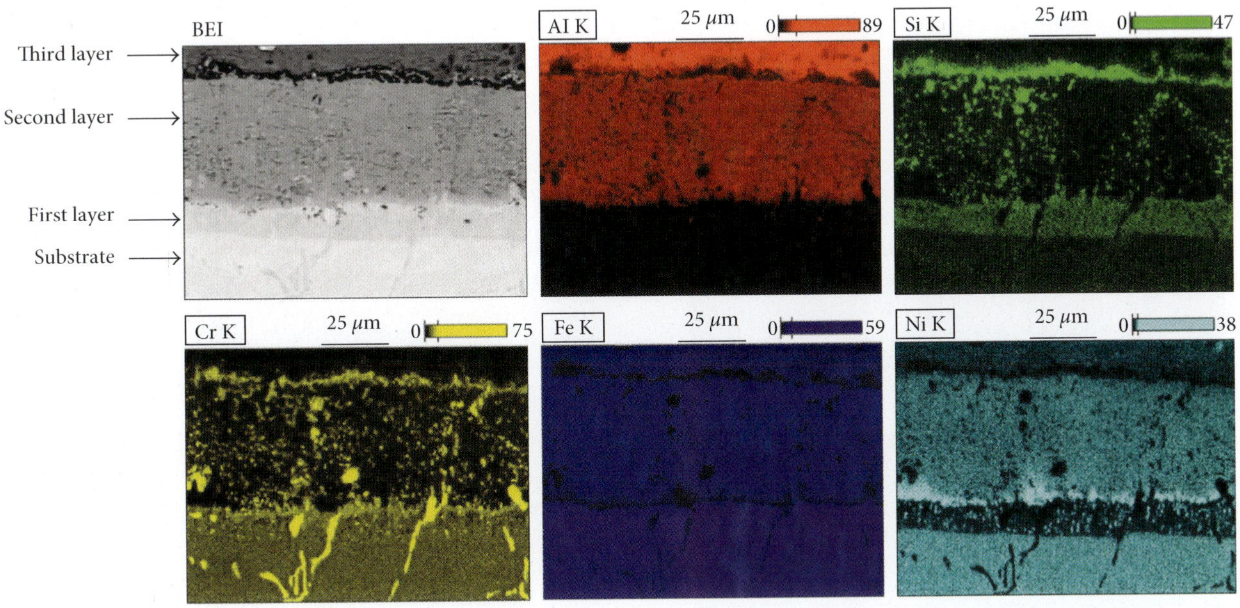

Figure 5: Elemental mapping in a cross section in coating number 17 (technological parameters 900°C, 7.5 h, Al/Si = 9; central point of the experiment).

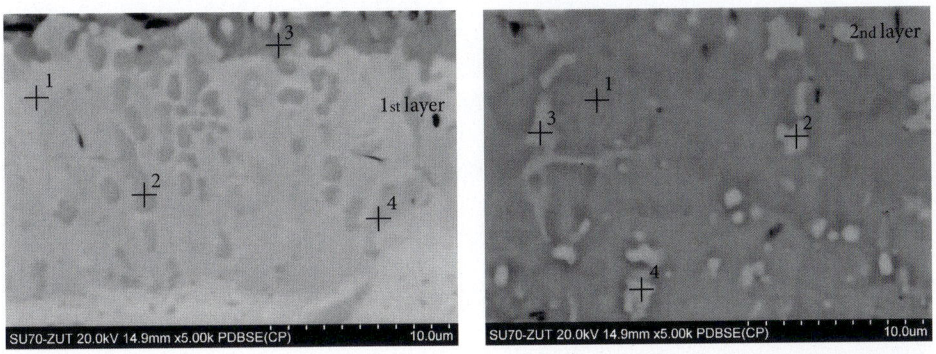

(a) (b)

Figure 6: Microstructure (BEI) of (a) the first layer and (b) the second layer of the coating no. 4 obtained at temperature 1000°C in 10 h (Al/Si = 6) with points of EDS X-ray microanalysis.

Table 2: Chemical composition according to Figures 6(a) and 6(b).

Element	Chemical composition (%at.), according to Figure 6(a)				Chemical composition (% at.), according to Figure 6(b)			
	Point 1	Point 2	Point 3	Point 4	Point 1	Point 2	Point 3	Point 4
Al	5.7	39.5	44.0	10.1	47.0	16.5	6.1	16.7
Si	10.1	2.7	0.2	1.4	5.2	28.1	33.0	19.1
Cr	29.9	6.3	8.0	48.8	4.0	32.4	48.6	42.7
Fe	44.9	17.3	12.8	21.6	24.4	15.5	10.4	13.8
Ni	9.5	34.2	35.0	18.2	19.4	7.6	1.9	7.5

the α-(Fe, Cr, Si) phase, as well. The β-Al(Ni, Fe) precipitates had spherical shapes (Figure 6(a), Table 2, point 2). The amount of nickel in the β-Al(Ni, Fe) precipitates from first layer was significantly higher than that in the β-Al(Fe, Ni) phase from the second layer (Table 3, phase 3 and phase 4).

It was found that, with increasing production temperature, the thickness of the first layer increased, and simultaneously, a thin sublayer rich in nickel was formed at the boundary with the second layer (Figure 7, Table 3, phase 4). This sublayer had nearly the same chemical and phase

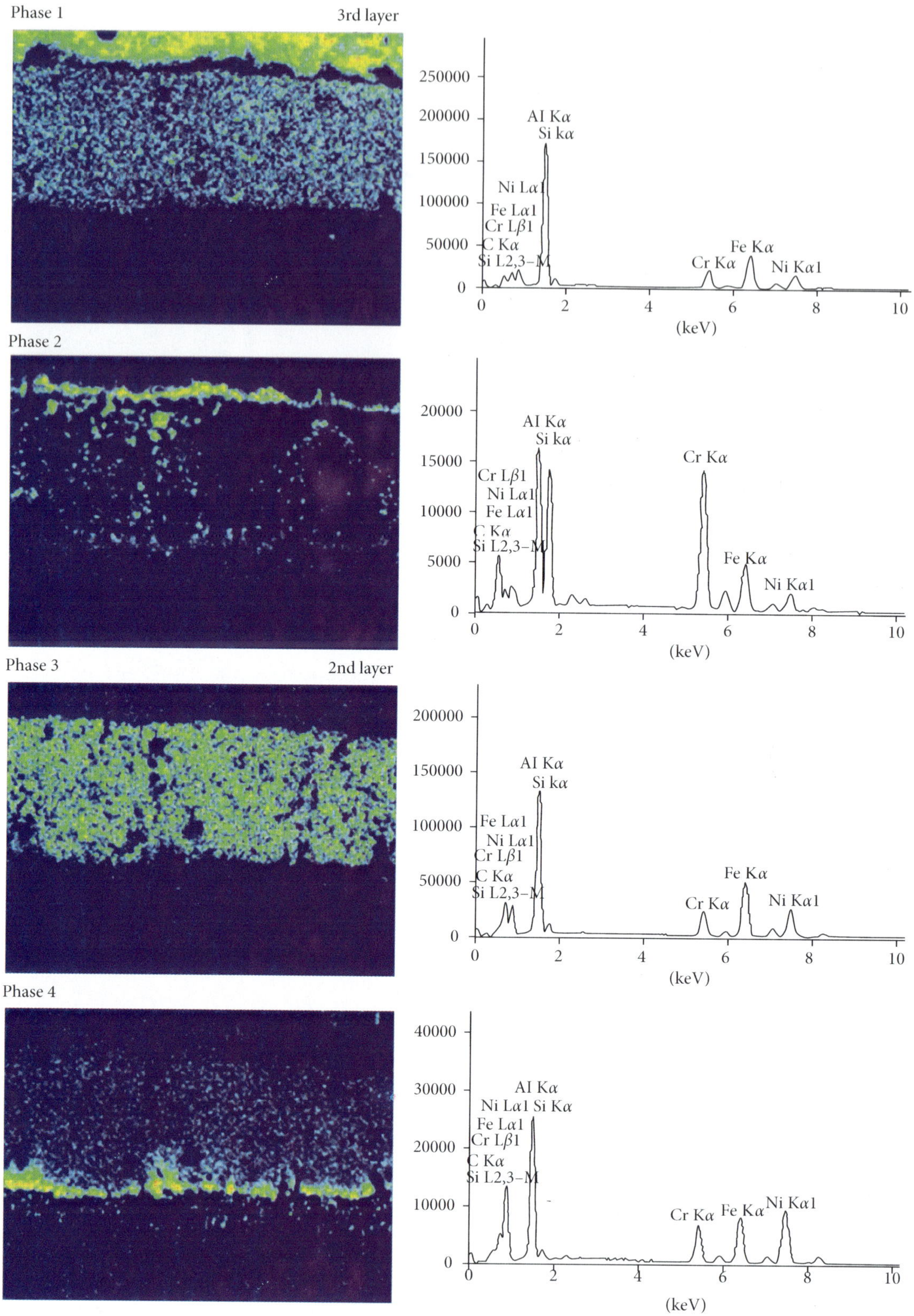

Figure 7: Continued.

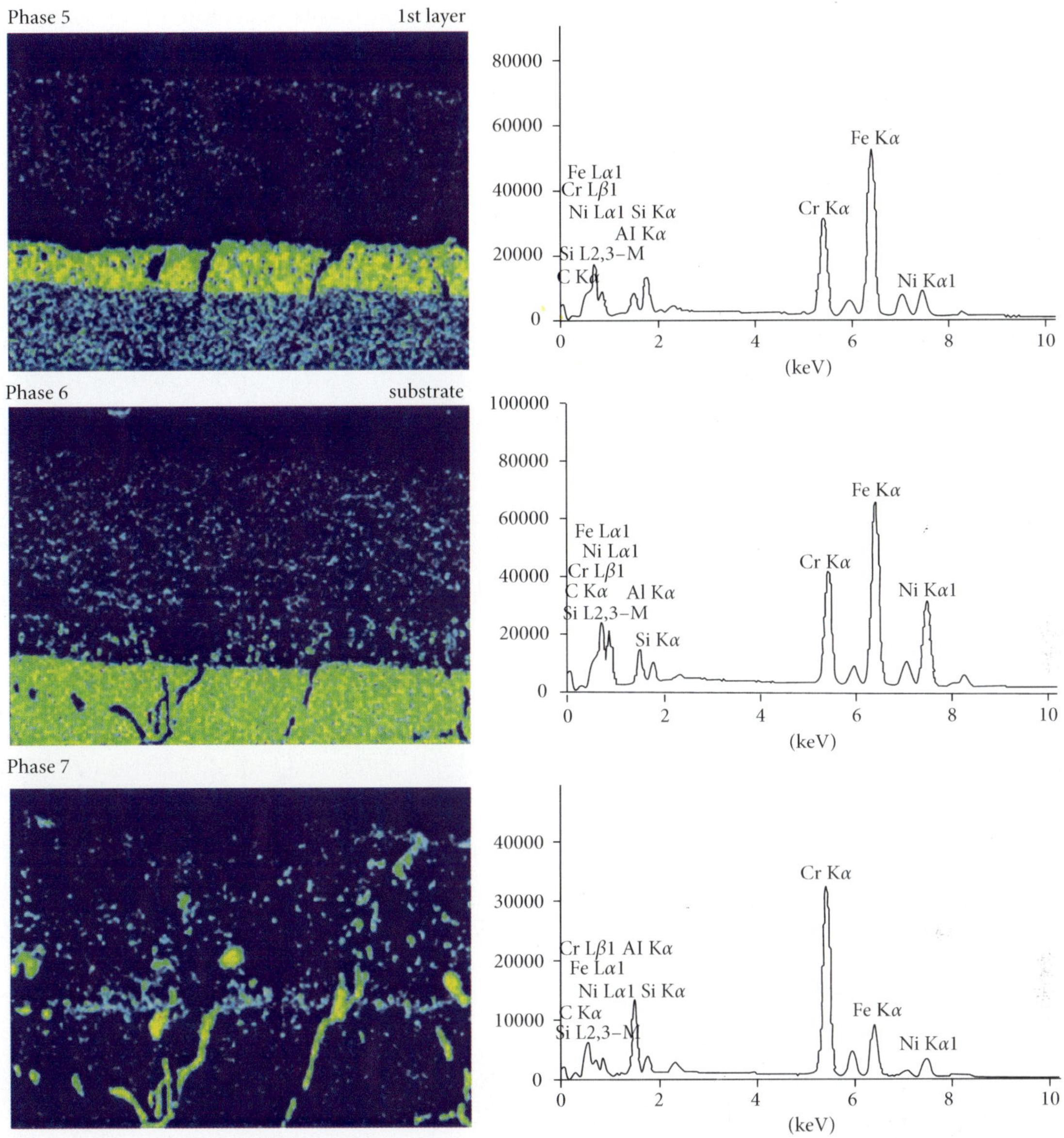

FIGURE 7: Mapping of phase components (according to Figure 5) with cumulative EDS spectrums.

compositions as those of β-Al(Ni, Fe) precipitates from the first layer (Figure 6(a), Table 2, point 2 and point 3). In the vicinity of this Ni-rich sublayer, a self-forming Cr-rich layer (Figure 6(a), Table 2, point 1) was observed. This sublayer can be used as a barrier blocking aluminium diffusion, because of a low solid solubility of aluminium in Cr-rich layer, as reported by Gao et al. [23].

Alumina (α-Al_2O_3) and AlN were also formed near the surface. Additionally, aluminide coatings have an ability to form a protective oxide Al_2O_3 scale at higher temperatures during oxidation [24]. Aluminium oxide has good protective properties at high temperature, which is advantageous. AlN phase is oxidized to α-Al_2O_3, but not directly. The transition oxides are transformed into amorphous Al_2O_3 gradually and finally to α-Al_2O_3 [25]. Therefore, the presence of the AlN phase on the surface could be also very advantageous.

4. Summary

The aluminide protective coatings by the slurry method were successfully formed on austenitic creep-resisting cast steel. The coatings consisted of three-layers having different thicknesses and properties. The thickness of the coatings from the high-temperature processes were very uniform, contrary to the coatings produced using the low temperature, especially in a long time, which were less uniform. The use of the long time of annealing caused growing of the third layer, which was very brittle. The structure and, as a consequence, the properties of the coatings depended on the production parameters. The thickness was the function of time and temperature of the annealing.

The growth of the second layer is mainly induced by the aluminium diffusion from an active mixture and nickel and

TABLE 3: Average content of chemical elements in phases according to Figure 7.

Element	Content (% at.)						
	Phase 1	Phase 2	Phase 3	Phase 4	Phase 5	Phase 6	Phase 7
Al	62.0	28.2	49.5	40.7	4.8	3.0	20.4
Si	3.1	14.0	1.6	2.3	8.4	3.5	2.6
Cr	6.6	31.1	8.4	12.5	22.2	19.5	53.5
Fe	17.8	15.6	22.2	16.2	54.7	46.3	16.5
Ni	10.5	11.2	18.3	28.3	10.0	22.7	7.1
Total							100.0
The main phase	Al_5FeNi	$CrSi_2$	β-Al (Fe, Ni)	β-Al (Ni, Fe)	A (Fe, Cr, Si)	Γ (Fe, Ni, Cr)	$(Cr, Fe)_7C_3$

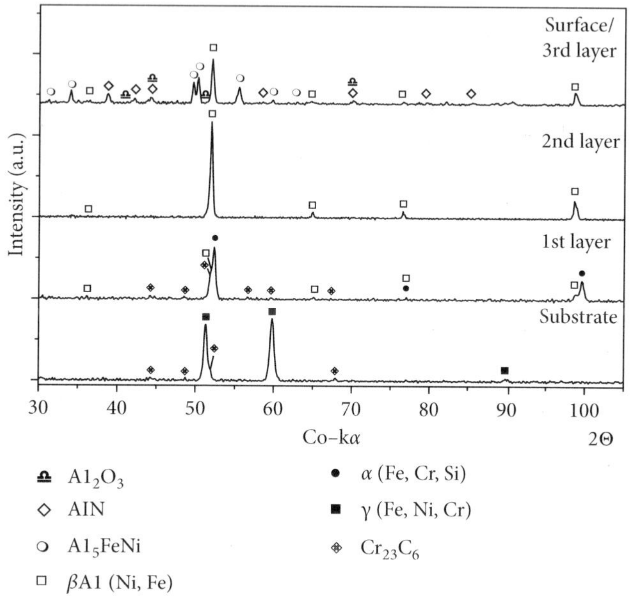

FIGURE 8: XRD patterns (the coating number 4 obtained at 1000°C, 10 hours, Al/Si = 6).

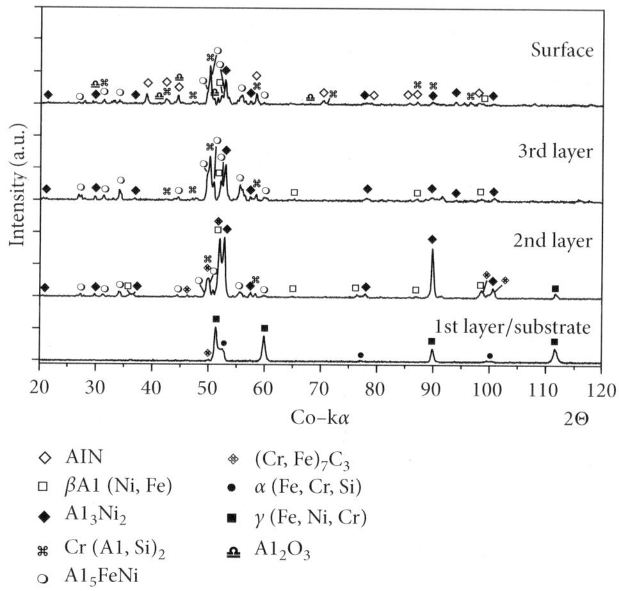

FIGURE 9: XRD patterns (the coating number 3 obtained at 800°C, 10 hours, Al/Si = 6).

iron diffusion from the substrate. The activity of the diffusing elements during the heat treatment is the parameter which determines the type of coating that is obtained. There are two variants: outwards and inwards diffusion coatings. The first one is produced in a low activity of aluminium and leads to the β-Al(Fe, Ni) phase forming. This case was observed in coatings produced at 900, 1000, and 1068°C (Figure 8). In the second layer also, the $Cr(Al, Si)_2$ phase formed in the matrix of the β-Al(Fe, Ni) phase was observed (Figures 3(b) and 5).

The inwards diffusion coating is produced in a high activity of aluminium and leads to the Ni_2Al_3 or Al_5FeNi phase creation. This case was observed in coatings produced at 732 and 800°C. Phases which contain a higher amount of aluminium were present, for example Al_5FeNi (sample no. 3, Figure 9) and Ni_3Al_2 (sample no. 3 and no. 7, Figure 10). The $Cr(Al, Si)_2$ phase was also observed in both the investigated cases.

For various Al to Si ratios in the slurry, differences in the structure of the coatings were significant in coatings manufactured at the lower temperature, that is, 800°C. The Al_8Cr_5 phase was created, when the amount of silicon was smaller in the slurry (Al/Si = 12), and Al_5FeNi phase was observed, when the amount of silicon in the slurry was higher (Al/Si = 6). Analysis of the results allowed to develop a simplified model of the coatings showing the influence of the silicon in the slurry (Figure 11).

The Al-Si coatings were designed to protect the high-temperature creep resisting cast steel against carburisation. Assessment of the coatings' behaviour in the carburising atmosphere under thermal shock conditions will be the final verification of their protective properties. The promising results of the preliminary investigations of carburisation and oxidation of coated cast steel have been carried out and presented in the article [26]. The selection of the production parameters leads to the formation of the coating having a suitable structure and properties. The phase composition of the coatings, which is Al/Si ratio dependent, can have a significant influence on the resistance against carburisation.

Coatings can be produced on the raw cast surface without any further surface finishing processes. This is a very

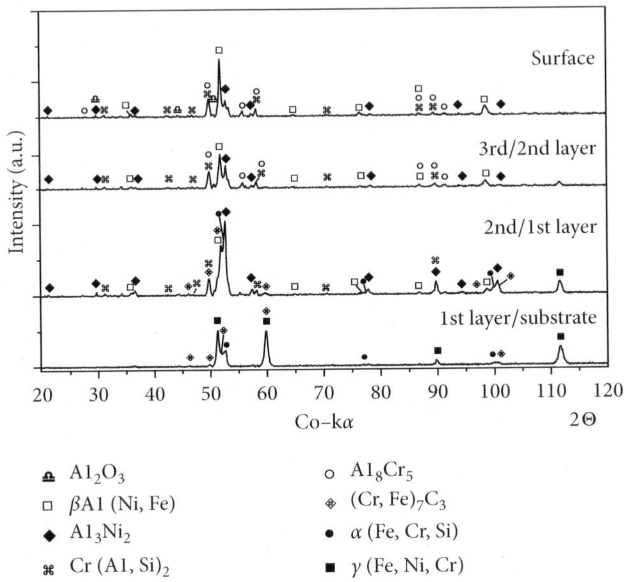

Figure 10: XRD patterns (the coating number 7 obtained at 800°C, 10 hours, Al/Si = 12).

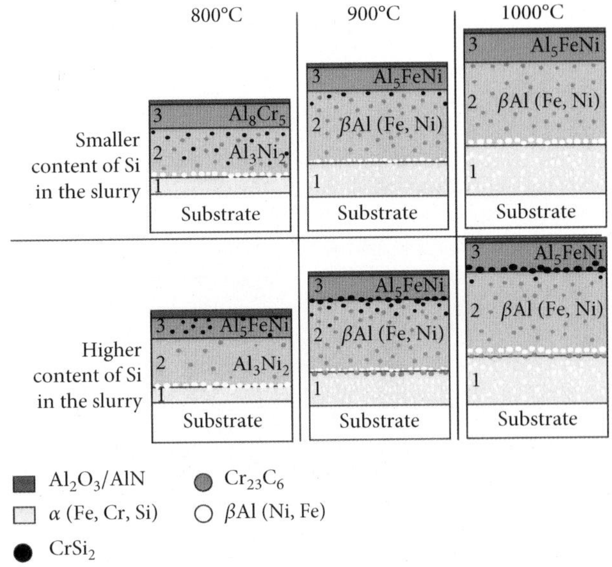

Figure 11: A simplified model of the coatings. Effect of the silicon content in the slurry on the phase composition of the coatings.

important advantage of the applied method, because it simplifies the process of equipment manufacturing and reduces costs.

The advantageous slurry method is more convenient and less harmful than the classic powder methods. This method is also easy to apply and could be used for a local application. Other advantages of this method are a low expenditure of the slurry materials and possibility of producing uniform and high-quality coatings on equipment with a large and complex shape. A very important aspect from an industrial conditions perspective is the increase of the mechanical strength combined with the decreased brittleness of the dried slurry (before annealing), caused by the soluble glass additive. Moreover, the use of the powder of coarser granularity reduces the risk of an explosion.

5. Conclusions

(1) Experimental rotatable designs are useful tools for identifying significant process parameters for the aluminide coatings by the slurry method.

(2) Three-layer Si-rich aluminide coatings were obtained on high-temperature creep resisting cast steel using a modified slurry method with soluble glass as the binder in the air atmosphere. Application of the aqueous solution of sodium silicate in the slurry allowed for annealing in an one single step.

(3) The outer layer of the coatings contained Al-rich phases. The middle layer contained intermetallic phases, mainly β-Al(Ni, Fe). The crystal lattice of the first layer was different from the substrate and from the other layers of the coating. Ferrite was formed from Ni-depleted austenite. This occurred because nickel diffuses from the substrate to the second layer, and Ni content was lower in the first layer than in the substrate.

(4) Thickness of the coatings depended on the production parameters (mainly on time and temperature). The ratio Al/Si in the slurry (in the applied range) did not have a significant impact on the thickness of the coating. The content of aluminium and silicon in the slurry affected the phase components formed in the coatings.

(5) Hardness value of internal layer was intermediate between the hardness of the middle layer and the base material, which is favourable from the thermal shock resistance viewpoint.

(6) The slurry method is also applicable for materials of a raw cast surface thanks to the presence of a flux in the slurry.

Conflicts of Interest

The author declares that there are no conflicts of interest regarding the publication of this paper.

References

[1] J. F. Norton and W. T. Bakker, "Materials for coal gasification plant," *Materials at High Temperature*, vol. 11, pp. 1–4, 1993.

[2] P. R. Roberge, *Handbook of Corrosion Engineering*, McGraw-Hill, New York, NY, USA, 2000.

[3] Z. Zhan, Y. He, L. Li, H. Liu, and Y. Dai, "Low-temperature formation and oxidation resistance of ultrafine aluminide coatings on Ni-base superalloy," *Surface and Coatings Technology*, vol. 203, no. 16, pp. 2337–2342, 2009.

[4] Z. Xu, J. Dai, J. Niu, L. He, R. Mu, and Z. Wang, "Isothermal oxidation and hot corrosion behaviors of diffusion aluminide coatings deposited by chemical vapor deposition," *Journal of Alloys and Compounds*, vol. 637, pp. 343–349, 2015.

[5] D. B. Lee, H. Mitsui, H. Habazaki, A. Kawashima, and K. Hashimoto, "The high temperature sulfidation behavior of Nb-Al-Si coatings sputter-deposited on a stainless steel," *Corrosion Science*, vol. 38, no. 11, pp. 2031–2042, 1996.
[6] K. L. Wang, F. S. Chen, and G. S. Leu, "The aluminizing and Al-Si codeposition on AISI HP alloy and the evaluation of their carburizing resistance," *Materials Science and Engineering: A*, vol. 357, no. 1-2, pp. 27–38, 2003.
[7] Y. Wang and W. Chen, "Microstructures, properties and high-temperature carburization resistances of HVOF thermal sprayed NiAl intermetallic-based alloy coatings," *Surface and Coatings Technology*, vol. 183, no. 1, pp. 18–28, 2004.
[8] A. Kochmańska and J. Kubicki, "Efficiency of protective coatings on high creep resistant cast steel," *Archives of Foundry Engineering*, vol. 9, no. 2, pp. 129–132, 2009.
[9] A. J. Rasmussen, A. Agüero, M. Gutierrez, and M. J. L. Østergård, "Microstructures of thin and thick slurry aluminide coatings on Inconel 690," *Surface and Coatings Technology*, vol. 202, no. 8, pp. 1479–1485, 2008.
[10] C. Boulesteix, B. Grégoire, and F. Pedraza, "Oxidation performance of repaired aluminide coatings on austenitic steel substrates," *Surface and Coatings Technology*, vol. 326, pp. 224–237, 2017.
[11] J. Bermejo Sanz, R. Roussel Garcia, V. Kolarik, and M. Juez Lorenzo, "Influence of the slurry thickness and heat treatment parameters on the formation of aluminium diffusion coating," *Oxidation of Metals*, vol. 88, pp. 179–190, 2017.
[12] X. Montero, M. C. Galetz, and M. Schutze, "A novel type of environmentally friendly slurry coatings," *JOM*, vol. 67, no. 1, pp. 77–86, 2015.
[13] Y. Tamarin, *Protective Coatings for Turbine Blades*, ASM International, Novelty, OH, USA, 2002.
[14] A. Agüero, M. Gutiérrez, R. Muelas, D. Plana, A. Román, and M. Hernández, "Laboratory corrosion testing of coatings and substrates simulating coal combustion under a low NO_x burner atmosphere," *Materials and Corrosion*, vol. 65, no. 2, pp. 149–160, 2014.
[15] A. Agüero, M. Gutiérrez, R. Muelas, and K. Spiradek-Hahn, "Overview of steam oxidation behaviour of Al protective oxide precursor coatings on P92," *Surface Engineering*, vol. 34, no. 1, pp. 30–39, 2016.
[16] West Pomeranian University of Technology in Szczecin, "Method of manufacturing coatings protecting metal parts exposed to carburization, oxidation and thermal shock," PL Patent 209678, 2011.
[17] G. W. Oehlert, *A First Course in Design and Analysis of Experiments*, University of Minnesota, Minneapolis, MN, USA, 2010.
[18] K. Thompson, *Unleashing the Power of COMPASS*, 2015, http://www.thermoscientific.com/content/dam/tfs/ATG/CAD/CAD%20Documents/Videos%20&%20Webinars/Microanalysis%20and%20Electron%20Microscopy/WS52733-Power-of-COMPASS.pdf.
[19] A. Kochmańska, J. Kubicki, and P. Kochmański, "Protective Al–Si coatings obtained on high–temperature creep resistance cast steel," *Materials Science*, vol. 12, no. 2, pp. 120–123, 2006.
[20] A. Kochmańska and M. Garbiak, "High-temperature diffusion barrier for Ni-Cr cast steel," *Defect and Diffusion Forum*, vol. 312–315, pp. 595–600, 2011.
[21] H. C. Akuezue and D. P. Whittle, "Interdiffusion in Fe-Al system: aluminizing," *Metal Science*, vol. 17, no. 1, pp. 27–33, 1983.
[22] T. L. Hu, H. L. Huang, D. Gan, and T. Y. Lee, "The microstructure of aluminized type 310 stainless steel," *Surface and Coatings Technology*, vol. 201, no. 6, pp. 3502–3509, 2006.
[23] F. Gao, Q. Yang, R. Liu, and X. Huang, "Oxidation resistance of novel multilayer coatings for gas turbine components," *Surface Engineering*, vol. 30, no. 9, pp. 624–635, 2014.
[24] H. Svensson, J. Angenete, and K. Stiller, "Microstructure of oxide scales on aluminide diffusion coatings after short time oxidation at 1050°C," *Surface and Coatings Technology*, vol. 177–178, pp. 152–157, 2004.
[25] R. Yue, Y. Wang, Y. Wang, and C. Chen, "SIMS study on the initial oxidation process of AlN ceramic substrate in the air," *Applied Surface Science*, vol. 148, no. 1-2, pp. 73–78, 1999.
[26] A. Kochmańska, "Hot corrosion resistance properties of Al–Si coatings obtained by slurry method," *Defect and Diffusion Forum*, vol. 326–328, pp. 273–278, 2012.

6

Organic Phosphorus Compounds as Inhibitors of Corrosion of Carbon Steel in Circulating Cooling Water: Weight Loss Method and Thermodynamic and Quantum Chemical Studies

Liang Liu, Ting-Ting Cao, Qi-Wei Zhang, and Chong-Wei Cui

Harbin Institute of Technology, School of Environment, Harbin 150090, China

Correspondence should be addressed to Chong-Wei Cui; cuichongwei1991@126.com

Academic Editor: Alfredo Juan

Circulating cooling water plays an important role in industrial water use. In this study, the corrosion inhibition effects of PBTCA, HEDP, and ATMP organic phosphorus inhibitors were investigated using the weight loss method by varying the dosage of inhibitors, ClO_2 concentrations, and pH values on carbon steel in recirculating cooling water with a low concentration of ClO_2 solution. The results showed that the three corrosion inhibitors had a satisfactory corrosion inhibition effect and that corrosion inhibition efficiency is positively correlated with the concentration of organic phosphorus inhibitors and pH. The average corrosion inhibition efficiency of the three inhibitors was about 80% at the concentration of inhibitors = 35 mg/L, pH = 9.0, and the concentration of ClO_2 = 7.0 mg/L, of which the single-phosphorus molecular corrosion inhibitor proved to be the best inhibitor. When the ClO_2 concentration was 7 mg/L, the corrosion inhibition efficiencies of the three corrosion inhibitors were relatively stable. Using the density functional theory (DFT) algorithm in the Gaussian 09 program, the optimization calculation was completed by the B3LYP/6-31G (d, p) method at the microlevel. The molecular structures of the three organic phosphorus inhibitors and the number of phosphorus-containing atoms were compared to the sustained-release properties. Organic phosphorus inhibitors, as an electronic buffer, not only provided electrons but also received electrons. They formed a complex with iron and zinc ions in water in order to attach to the surface of the carbon steel and to alleviate corrosion. In addition, the adsorption with a metal surface followed the Langmuir adsorption isotherm.

1. Introduction

Circulating cooling water accounts for 60–90% of total industrial water use [1–3]. Because of continuous recycling, the water quality of circulating cooling water becomes more concentrated leading to corrosion, fouling, and microbial breeding [4–6], which results in lower heat transfer efficiency of heat exchangers, corrosion and perforation of metal pipelines, and safety issues [7, 8].

Since the 1930s, the use of organic phosphorus inhibitors in circulating cooling water has alleviated the corrosion of carbon steel [9, 10]. To inhibit microbial breeding in circulating cooling water, oxidizing or nonoxidizing fungicides are added to the water [9, 11, 12]. The combined use of the corrosion inhibitor and the bactericide has relieved the impact of the challenges mentioned above and has promoted the development of the industrial economy. Organic phosphorus inhibitors, and also in combination with ClO_2, have good corrosion inhibition properties, but there are only a few articles about their corrosion inhibition mechanism with carbon steel [13, 14]. A large number of studies have shown that the efficiency of corrosion inhibitors is closely related to their structures [15–17]. The quantum chemical method can be used to calculate the molecular structure parameters and to explore the mechanism of corrosion inhibition at the microlevel [18, 19]. Kumar et al. [20] used the AM1 semiempirical quantum chemical method to study the corrosion inhibition mechanism of synthetic organic matter for N80 carbon steel and found that the adsorption of inhibitors followed the Langmuir adsorption isotherm. Gece and Bilgi [21] used the B3LYP/6-311 ++ G (d, p) method to study the mechanism of a nitrogen corrosion inhibitor in acidic media and concluded that the nitrogen atom belongs to the active centre and has good corrosion inhibition performance. Shi et al. [22] studied the adsorption behaviour of four pyrazine inhibitors on a copper surface using the density functional theory and the difference in charge density and distribution analysis. The vertical adsorption of the corrosion inhibitor molecules on the metal surface was by chemical adsorption where the coordination bond was formed. By analyzing the frontier orbitals, it was verified that the active site of the inhibitor was the N atom on the pyrazine ring.

Hitherto, good results have been obtained in the research of nitrogen corrosion inhibitors, but few reports have described the mechanism of action between phosphorus corrosion inhibitors and metal surfaces. In this paper, three kinds of organic phosphorus corrosion inhibitors: 2-phosphate group-1,2,4-tricarboxylate butane (PBTCA), hydroxyethylidene disphosphonic acid (HEDP), and aminotrimethylene phosphonic acid (ATMP), were studied. The corrosion inhibition performance of the organophosphorus corrosion inhibitors in combination with ClO_2 on the carbon steel was analyzed experimentally. The charge distribution, frontier orbit, and adsorption model of stable molecules were investigated using ChemBioOffice 2014, Gaussian 09, and Origin 2017 software to explore the corrosion inhibition mechanism of the organic phosphorus inhibitor on the metal surface and provide theoretical support for its application in circulating cooling water.

2. Experimental Materials and Research Methods

2.1. Experimental Material. The composition and content of the standard water used in the experiment are shown in Table 1.

2.2. Weight Loss Method. The effect of different inhibitor concentrations, ClO_2 concentrations, and pH values on the corrosion inhibition efficiency of the corrosion inhibitors on carbon steel in circulating cooling water with a low concentration of ClO_2 and zinc ion solution was studied by using the weight loss method. Determination methods and procedures were referenced to the regulation of corrosion resistance of water treatment agents (GB/t 18175-2014) [23, 24]. The corrosion rate V and corrosion inhibition efficiency E were determined using the following equations:

$$V = K\frac{W1 - W2}{F \times t \times \gamma},$$
$$E = \frac{C_0 - C_t}{C_0} \times 100, \quad (1)$$

where $W1$ and $W2$ are the weight of specimens before and after corrosion, respectively; F is the surface area of the specimen exposed to cooling water; K is a constant, whose value is 87.6; t is the specimen immersion corrosion time; γ is the metal density; and C_0 and C_t are corrosion rates in the absence and presence of the inhibitor, respectively.

Table 1: The composition and content of the standard water used in the experiment.

Ion	Ca^{2+}	Mg^{2+}	Na^{+}	Zn^{2+}	Cl^{-}	SO_4^{2-}	HCO_3^{-}
Concentration (mg/L)	200	48	305	5	426	192	122

2.3. Quantum Chemical Calculation. The molecular structures of organic phosphorus inhibitors PBTCA, HEDP, and ATMP were constructed using ChemBioDraw Ultra 14.0, and the RHF/6-31G method of GAMESS was used to preliminarily optimize the structure. Density functional theory (DFT) helps to study the electronic structure of a multi-electron system with high computational accuracy [25], so the DFT algorithm in the Gaussian 09 program was used, without any symmetry restriction being imposed, in the B3LYP/6-31G (d, p) method [26]. The quantum chemical parameters and related parameters of each molecule were obtained from the output results by optimizing the three inhibitors. The stable composition of the three corrosion inhibitors after optimization is shown in Figure 1. The quantum chemical parameters included the highest occupied molecular orbital energy (E_{HOMO}), the lowest unoccupied molecular orbital energy E_{LUMO}, the energy gap (ΔE), the distribution of the atomic Mulliken charge, and the molecular dipole moment (μ). Other relevant parameters included the ionization potential (I), electronegativity χ, electron affinity (A), hardness (η), and affinity index (ω). These parameters were calculated using the following equations [27]:

$$I = -E_{HOMO},$$
$$A = -E_{LUMO},$$
$$\chi = \frac{I + A}{2}, \quad (3)$$
$$\eta = \frac{I - A}{2},$$
$$\omega = \frac{\chi^2}{2\eta}.$$

3. Results and Discussion

3.1. Evaluation of Corrosion Inhibition Performance of Organophosphorus Agents. PBTCA, HEDP, and ATMP are commonly used water treatment agents, and they have excellent corrosion and scale inhibition and a moderate price and are suitable for large-scale use. In this section, we studied the influence of three factors: corrosion inhibitor concentration, ClO_2 concentration, and pH value, on the corrosion inhibition efficiency of corrosion of carbon steel.

3.1.1. Effect of Inhibitor Concentration on the Inhibition of Corrosion of Carbon Steel. The effect of the concentration of PBTCA, HEDP, and ATMP on the inhibition of corrosion of

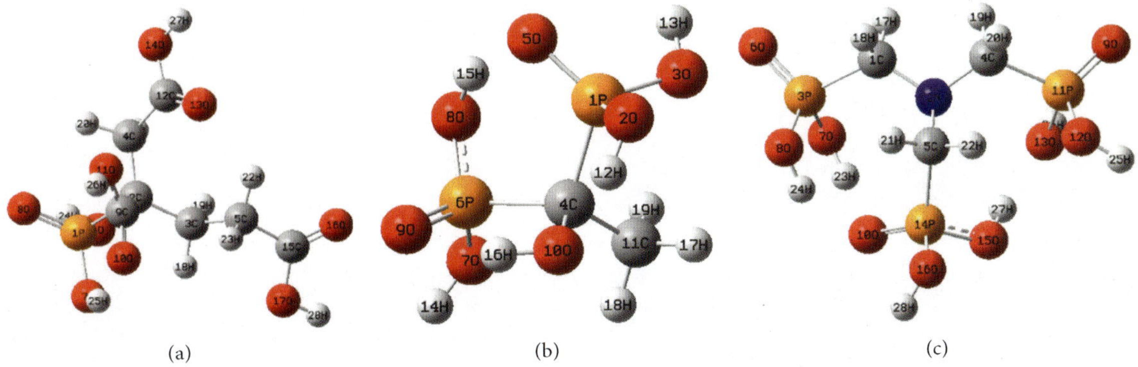

Figure 1: Molecular model diagram of three organophosphorus inhibitors. (a) PBTCA. (b) HEDP. (c) ATMP.

carbon steel was studied using the carbon steel test, and the test liquid was a mixture of standard water, ClO_2, and corrosion inhibitor. The concentration of ClO_2 was 7.0 mg/L, the temperature was (40.0 ± 1.0)°C, the pH was 9.0, and the rotation reaction time was 72 h. The experiment was repeated three times, and the average values are shown in Figure 2.

The above figure shows that, with the increasing organic phosphorus concentration, the inhibition rate increased gradually and the average inhibition rate of the three corrosion inhibitors reached above 80%. When the concentration was increased to 35.0 mg/L, the corrosion inhibition rate of ATMP was more than 70%. Figure 2 shows that the corrosion inhibition rate of three corrosion inhibitors is in the order of PBTCA > HEDP > ATMP based on the number of phosphorus atoms in the inhibitor. The results also show that the inhibition performance is related to the number of P atoms, and the single-phosphorus molecule inhibitor exhibits a preferable corrosion inhibition effect.

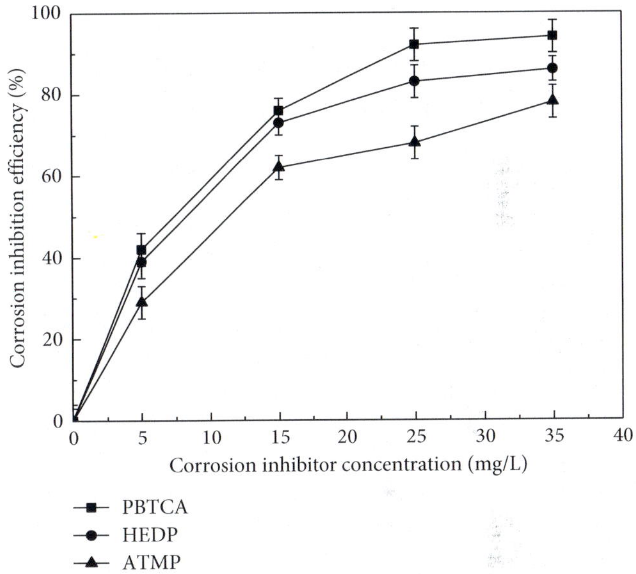

Figure 2: Effect of concentration of the three corrosion inhibitors on the inhibition of corrosion of carbon steel.

3.1.2. Effect of ClO_2 Concentration on Corrosion Inhibition of Carbon Steel. ClO_2 is a good fungicide and can be combined with a corrosion inhibitor by the compounding technology. The test solution was a mixture of standard water, ClO_2, and corrosion inhibitor. The effect of the ClO_2 concentration on the inhibition of corrosion of carbon steel was studied. The corrosion inhibitor concentration was 15.0 mg/L, the temperature was (40.0 ± 1.0)°C, the pH was 9.0, and the rotation reaction time was 72 h. The experiment was repeated three times, and the average results are shown in Figure 3.

From Figure 3(a), it can be seen that the addition of ClO_2 had a certain inhibitory effect on the slow-release performance of the three inhibitors, but the compound of an inhibitor and ClO_2 still had good corrosion inhibition effect. When the ClO_2 concentration was increased, the slow-release performance of HEDP steadily decreased. Generally, the average slow-release performance of HEDP and ATMP was lower than that of PBTCA. The slow-release performance of PBTCA showed fewer fluctuations, and its slow-release performance was more stable with ClO_2 concentration changes as compared with the inhibitors HEDP and ATMP (Figure 3(b)). However, with the increase in ClO_2 concentration, the inhibition rate of ATMP decreased, and the lowest inhibition rate recorded was only about 30%. Once the concentration of ClO_2 was 7 mg/l, the slow-release performance of PBTCA and HEDP was significantly higher than that of ATMP. Overall, the average corrosion inhibition rate of the three corrosion inhibitors in the presence of ClO_2 was PBTCA > HEDP > ATMP.

3.1.3. Effect of pH on Inhibition of Corrosion of Carbon Steel. The test solution in this test was a mixture of standard water, ClO_2, and corrosion inhibitor, and the effect of pH on corrosion inhibition was determined. The initial concentration of the corrosion inhibitor was 15.0 mg/L, the ClO_2 concentration was 7.0 mg/L, the temperature was (40.0 ± 1.0)°C, and the rotation reaction time was 72 h. The experiment was repeated three times, and the average results are shown in Figure 4.

The inhibition rate of the three organic phosphorus reagents increased with increasing pH, indicating that the protective film was more favourable to the reagent formation in the alkaline environment. This is likely because the concentration of hydrogen ions in water decreases with the

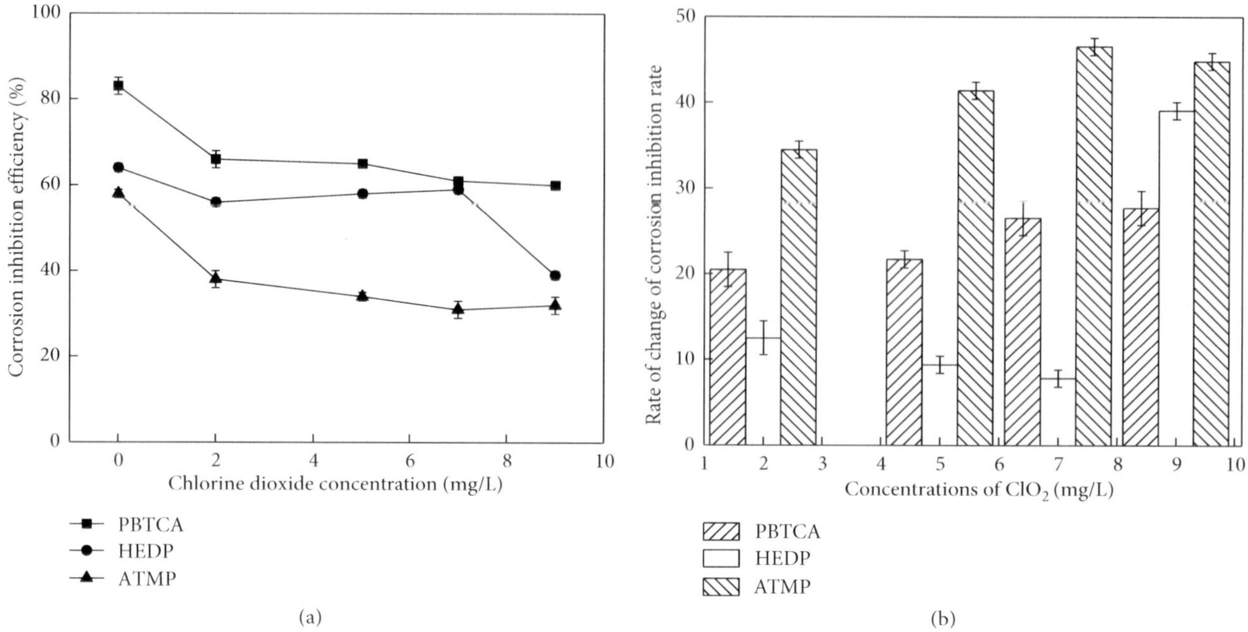

Figure 3: (a, b) Corrosion inhibition rate of carbon steels with different concentrations of ClO_2.

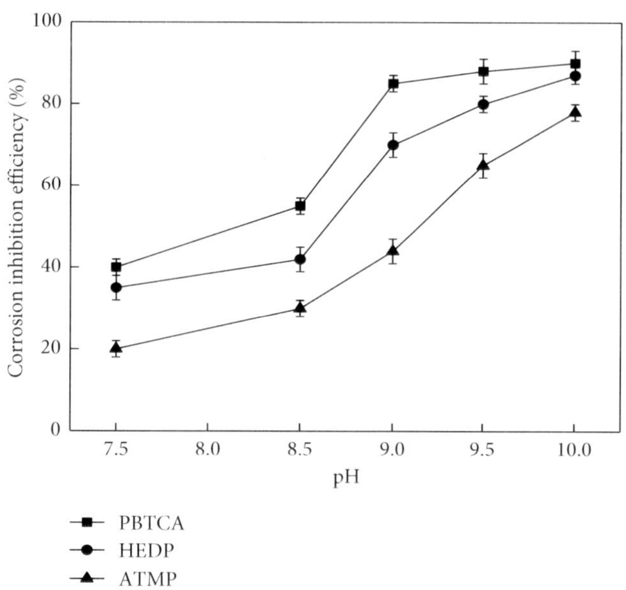

Figure 4: Corrosion inhibition rate of the corrosion inhibitor to carbon steel at different pH.

increasing pH, and thus, the depolarization of hydrogen ions in the process of metal corrosion is inhibited at high pH. Therefore, the tendency to form a protection film on the surface of carbon steel increases, as well as the corrosion inhibition rate. As shown in Figure 4, the corrosion inhibition rate of the three organic phosphorus reagents can be presented as PBTCA > HEDP > ATMP.

The influence of three factors on the inhibition of corrosion of carbon steel was studied by adding different organic phosphorus inhibitor concentrations, ClO_2 concentrations, and pH. The corrosion inhibition rate of the three organic phosphorus inhibitors was above 80%, and the average corrosion inhibition effect was satisfactory. The order of corrosion inhibition effect was PBTCA > HEDP > ATMP. In addition, with the increase in the number of P atoms, the inhibition effect of the corrosion inhibitor decreased; i.e., the corrosion inhibitor with a single phosphorus could exhibit a more effective corrosion inhibition effect.

3.2. Quantum Chemical Calculation. In order to investigate the corrosion inhibition mechanism of organic phosphorus corrosion inhibitors on carbon steel at the microscopic level, we used the density functional theory in quantum chemistry to quantify the molecular structure of organic phosphorus corrosion inhibitors, Fe, Fe^{2+}, and Zn^{2+}. Figure 1 and Table 2 show the geometric structure models and the optimized Mulliken atomic charge distributions of the three organic phosphorus inhibitors by the B3LYP/6-31G (d, p) method. As can be seen from Table 2, the P atoms were positively charged, indicating that they could accept electrons and form chemical bonds, while both N and O have a high negative charge, indicating that they could also be adsorbed onto the surface of the metal atom by the loss of electron interaction. In addition, all the C atoms in the PBTCA molecule that bonded to the C=O double bond were positively charged, while the other alkyl carbon atoms were all negatively charged; alkyl carbon and oxygen atoms in HEDP were also negatively charged; the alkyl carbon atoms in the ATMP were also negatively charged, and all the H atoms were positively charged. All these data are consistent with the results of the induced effect.

Figure 5 shows the frontier orbital diagram of the three organic phosphorus inhibitor molecules after optimization. The HOMO orbitals of the PBTCA were essentially distributed on alkyl chains and carboxyl groups linked to the

TABLE 2: The charge distribution of the Mulliken atom in the three corrosion inhibitors.

PBTCA		HEDP		ATMP	
Atom	Mulliken charge	Atom	Mulliken charge	Atom	Mulliken charge
P1	1.1658	P1	1.1134	C1	−0.2667
C2	−0.2905	O2	−0.5764	N2	−0.4657
C3	−0.1777	O3	−0.5953	P3	1.1046
C4	−0.2663	C4	−0.1195	C4	−0.2575
C5	−0.2588	O5	−0.6431	C5	−0.2861
O6	−0.5877	P6	1.1237	O6	−0.6589
O7	−0.5932	O7	−0.5694	O7	−0.6391
O8	−0.6405	O8	−0.6027	O8	−0.6129
C9	0.6524	O9	−0.6360	O9	−0.6436
O10	−0.5367	O10	−0.5683	O10	−0.6735
O11	−0.4980	C11	−0.3356	P11	1.1320
C12	0.6226	H12	0.3999	O12	−0.6159
O13	−0.5344	H13	0.3958	O13	−0.5825
O14	−0.5040	H14	0.4016	P14	1.1323
C15	0.5876	H15	0.4087	O15	−0.6002
O16	−0.5478	H16	0.3616	O16	−0.5800
O17	−0.4948	H17	0.1451	H17	0.1858
H18	0.1462	H18	0.1447	H18	0.1650
H19	0.1433	H19	0.1517	H19	0.1936
H20	0.1970			H20	0.1751
H21	0.1839			H21	0.1911
H22	0.1469			H22	0.1927
H23	0.1654			H23	0.4111
H24	0.3972			H24	0.3915
H25	0.4024			H25	0.4062
H26	0.3761			H26	0.3995
H27	0.3741			H27	0.4033
H28	0.3690			H28	0.3987

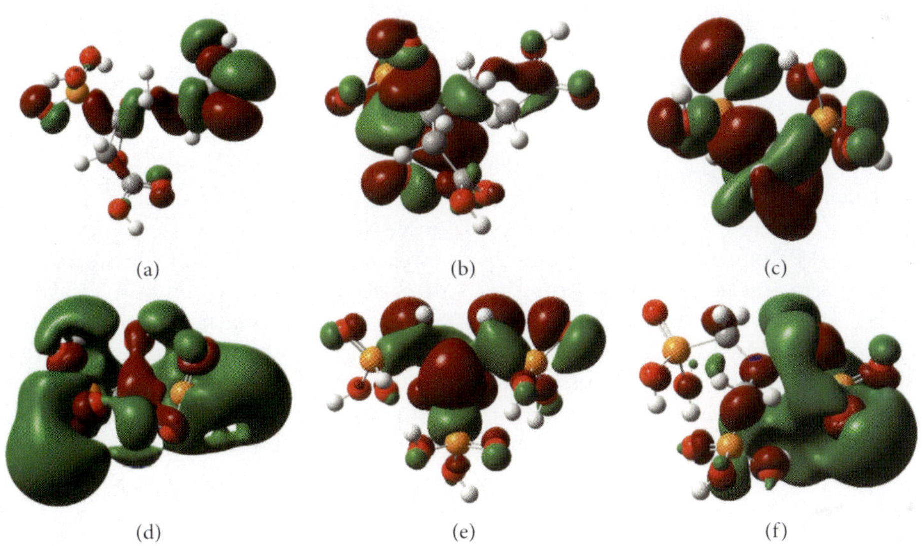

FIGURE 5: Optimized frontier orbitals of organic phosphorus inhibitors. (a) HOMO of PBTCA. (b) LUMO of PBTCA. (c) HOMO of HEDP. (d) LUMO of HEDP. (e) HOMO of ATMP. (f) LUMO of ATMP.

central carbon atom, while some of them were distributed near the phosphorus-oxygen double bonds belonging to the electron loss region. The LUMO orbitals were distributed near P atoms and positively charged carbon atoms which are an easy-to-accept electron region. The HOMO orbits of HEDP were located near oxygen and carbon atoms, while the LUMO orbits were located near P atoms. The HOMO orbitals of ATMP were mainly distributed near the central nitrogen and oxygen atoms, even though the LUMO orbitals were mainly distributed near the P11 atomic group.

Table 3 shows a list of quantum chemical parameters optimized for the three organic phosphorus inhibitors. The

Table 3: Quantum chemical parameters of the three organic phosphorus inhibitors at the B3LYP/6-31G (d, p) theoretical level.

Quantum chemical parameters	Corrosion inhibitor PBTCA	HEDP	ATMP
E_{HOMO} (eV)	−7.5811	−7.5321	−6.4845
E_{LUMO} (eV)	−0.3320	0.6857	0.9960
ΔE (eV)	−7.2491	−8.2178	7.4805
Dipole moment (debye)	3.7949	5.7964	12.0357
Electron affinity (A)	0.3320	−0.6857	−0.9960
Electronegativity (χ)	3.9566	3.4232	2.7443
Hardness (η)	3.6246	4.1089	3.7403
Affinity index (ω)	2.1595	1.4260	1.0068

size of E_{HOMO} represents the ability of molecules to provide electrons, while the size of E_{LUMO} represents the electron-accepting capability, and the energy gap, ΔE, represents the stability of the molecule. The smaller the value of ΔE, the more efficient the inhibitor molecule's corrosion inhibition performance. The electron-donating capacity order of the three organic phosphorus inhibitors was HEDP > PBTCA > ATMP, the electron-accepting capacity order was PBTCA > HEDP > ATMP, while the energy gap order was PBTCA < ATMP < HEDP. This reflects the high eletrophilicity of PBTCA and high nucleophilicity of ATMP. According to the three principles of orbital bonding (the orbital symmetric principle, the energy approximation principle, and the maximum overlap principle), the HOMO energies of all three corrosion inhibitors are lower than the minimum orbital energy of iron ions (0.6912 eV), which disagrees with the orbital symmetric principle. Therefore, the corrosion inhibitor molecules cannot provide electrons to empty orbits of iron atoms. The HOMO energies of all three corrosion inhibitor molecules were lower than the minimum empty orbital energy of Fe^{2+} (−1.9048 eV), which conforms to the orbital symmetric principle. Thus, HOMO energies can provide electrons to empty orbits of Fe^{2+}. The difference between HOMO orbital energy of HEDP and LUMO orbital energy of Fe^{2+} is the highest (6.1013 eV), followed by those of PBTCA (5.5092 eV) and ATMP (4.9903 eV). According to the principle of energy similarity, ATMP shows the strongest bonding capacity with Fe^{2+}, followed by PBTCA and HEDP. On the contrary, the LUMO energies of HEDP, PBTCA, and HEDP are all higher than the maximum orbital energy of iron ions (−3.0749 eV). According to the orbital symmetric principle, iron ions cannot provide electrons to HEDP and ATMP, but they can offer electrons to PBTCA. The difference between LUMO energy of PBTCA and HOMO energy of iron ions is 2.1005 eV. Although the LUMO orbital energy of the three corrosion inhibitors is higher than HOMO orbital energy of Fe^{2+}, electrons are provided to PBTCA and ATMP but not to HEDP. The differences between the LUMO orbital energies of PBTCA and ATMP and the HOMO orbital energy of Fe^{2+} are 9.1877 eV and 8.2258 eV, respectively. All three corrosion inhibitors can provide electrons to Fe^{2+} according to the comparison of the differences in HOMO orbital energy of inhibitors and LUMO orbital energy of Fe^{2+}. In conclusion, the PBTCA is superior to the other two inhibitors in terms of slow-release performance, which is similar to previous experimental results. Specifically, PBTCA can provide electrons to Fe^{2+} and receive electrons from iron atoms and can form chemical bonds with the corrosion medium in cyclic cooling water, thus avoiding corrosion.

Zn^{2+} has been proved to be an effective index of the corrosion effect in cyclic cooling water [28]. It is necessary to add corrosion inhibitors to water to inhibit corrosion upon the increase of Zn^{2+} concentration. Moreover, the composite corrosion inhibitor formed by PBTCA, HEDP, and ClO_2 presents good corrosion-relieving performance [15]. Table 3 shows that all three inhibitors can provide electrons to Zn^{2+} according to the orbital symmetric principle although their HOMO energies are lower than the LUMO energy of Zn^{2+}. The difference between the HOMO orbital energy of HEDP and the LUMO orbital energy of Zn^{2+} (−2.2177 eV) is 5.7884 eV, which is the highest relative to that of PBTCA (5.1963 eV) and that of ATMP (4.6774 eV). According to the energy similarity principle, the descending order of the three inhibitors with respect to the bonding capacity with Zn^{2+} is ATMP > PBTCA > HEDP. With abundant negative loads, oxygen atoms and nitrogen atoms in molecules of all three inhibitors easily contribute surplus electrons (Table 2). As a result, water can receive surplus electrons from the three inhibitors, which is attributed to the small content of Zn^{2+} in the water. Consequently, chemical bonds are formed and attached on the iron surface, thus further relieving the corrosion of iron. As electron buffers, all three organic phosphorus inhibitors receive electrons from iron atoms due to the positively charged phosphorus atoms. Since oxygen atoms and nitrogen atoms are rich in negative charges and provide electrons to Zn^{2+} in water, a two-layer complex is formed in the cyclic cooling water and attaches on the iron surface to protect carbon steel from corrosion. The flow graph of electrons is shown in Figure 6.

The dipole moment was caused by the uneven distribution of the atomic charges in the molecule and can, therefore, be used to measure the polarity of the inhibitor molecules. The greater the value of the dipole moment, the stronger the polarities of the molecules and the more likely the electrons are to be lost, clearly indicating the weak electron absorption between the corrosion inhibitor molecules and the metal surface. From Table 3, it can be concluded that the order of dipole moment of the three organic phosphorus molecules was ATMP > HEDP > PBTCA. The corrosion inhibitor ATMP had a better electron-donating capability. Hence, PBTCA had a better electron-receiving capability. Because of the high number of P atoms in ATMP, the steric effect was great, which was not conducive to the formation of the dense membrane. Therefore, PBTCA showed better corrosion inhibition performance.

According to Koopmans' theorem, quantum chemical parameters are related to the frontier orbital energy, such as ionization potential (I), electron affinity (A), electronegativity (χ), hardness (η), and affinity index (ω). The results are shown in Table 3.

The electron affinity potential indicates the electronic trend of the molecules, and the higher its value, the easier it is to obtain the electron. The order of electron affinity of the

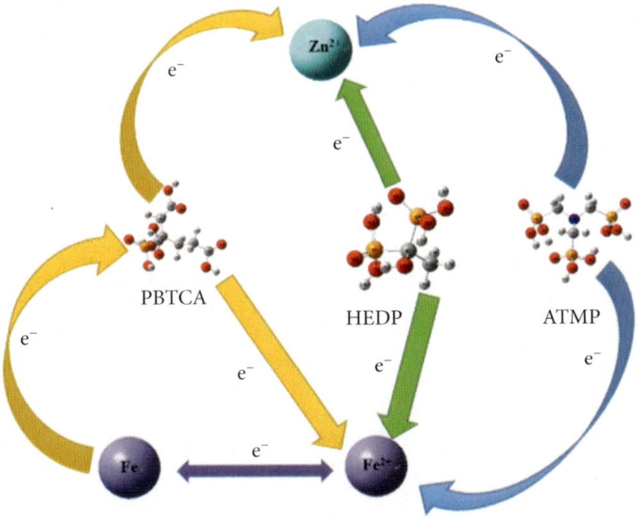

FIGURE 6: Electron-flow diagram of the corrosion inhibition mechanism between the inhibitor and carbon steel.

three organic phosphorus molecules was PBTCA > HEDP > ATMP. PBTCA molecules can easily obtain electrons. Electronegativity represents the ability of molecules to attract electrons. The electronegativity of the three organic phosphorus molecules was PBTCA > HEDP> ATMP, where PBTCA can easily accept electrons and react with metal atoms. Hardness can reflect stability and reactivity of molecules. The hardness of the three organic phosphorus molecules is in the order HEDP > ATMP > PBTCA. According to the principle of "hard close to hard and soft close to soft," HEDP can easily form stable products with ions. The affinity index also represents the affinity of the molecule, and the greater that value, the easier it is to attract the electrons. The three organic phosphorus corrosion inhibitors' affinity index was ranked as PBTCA > HEDP > ATMP. This clearly shows that PBTCA could accept the electrons easily, which was the main component of the three organic phosphorus corrosion inhibitors. The analysis of electron affinity, electronegativity, hardness, and electron-affinity index of the three organic phosphorus molecules also shows that PBTCA had a better corrosion inhibition performance and it was more likely to accept the free electrons outside the iron atoms and form coordination bonds. The smaller the number of P atoms contained in the molecule, the lesser its steric effect. As a result, PBTCA formed a dense protective film on the surface of the metal, thus hindering corrosion, which satisfies the requirements of prefilm in circulating cooling water.

3.3. Study on Corrosion Inhibitor Adsorption Behaviour. The adsorption isotherm of a corrosion inhibitor can describe the adsorption behaviour at a certain temperature. In order to study the adsorption models of three organic phosphorus inhibitors, we assumed that they were adsorbed on a single molecular layer on the surface of carbon steel, such that corrosion rate y was approximated by a metal surface coverage θ.

By the Langmuir adsorption isotherm, the equation is as follows:

$$\frac{\theta}{1-\theta} = KC_i. \quad (4)$$

The formula can be concluded as follows:

$$\frac{C}{\theta} = C + \frac{1}{K}, \quad (5)$$

where C is the concentration of the inhibitor (mg/L) and K is the adsorption equilibrium constant [29]. The result is shown in Figure 7.

It can be seen from Figure 7 that the regression coefficients of the organic phosphorus inhibitors were $R_{\mathrm{PBTCA}}{}^2 = 0.9931$, $R_{\mathrm{HEDP}}{}^2 = 0.9927$, and $R_{\mathrm{ATMP}}{}^2 = 0.9825$, respectively, indicating that the inhibitors were Langmuir adsorbed on the surface of carbon steel. The adsorption equilibrium constants were calculated using intercepts as $K_{\mathrm{PBTCA}} = 0.0732$ L/mg, $K_{\mathrm{HEDP}} = 0.0737$ L/mg, and $K_{\mathrm{ATMP}} = 0.1164$ L/mg. According to the adsorption theory, the relationship between the standard molar adsorption free energy and the adsorption equilibrium constant K is

$$K = \frac{1}{C_{\mathrm{solvent}}} \exp \frac{-\Delta G^{\theta}}{RT}, \quad (6)$$

where $C_{\mathrm{solvent}} = 55.5$, T stands for thermodynamic temperature (K), and R is the gas constant (8.314 J/mol·K). The higher the absolute value of the standard molar adsorption free energy, the higher the stability of the adsorption layer. ΔG^{θ} is negative throughout the calculation, which means that the adsorption was spontaneous. When the absolute value of ΔG^{θ} was below 20 kJ/mol, the adsorption of the inhibitor molecules on the metal surface was physical adsorption. At an absolute value of ΔG^{θ} greater than 40 kJ/mol, the corrosion inhibitor molecule formed a covalent bond with the metal surface, which shows the chemical adsorption [18]. The standard molar adsorption free energy of the three organic phosphorus molecules was obtained by introducing the adsorption equilibrium constant conversion unit into the formula, and the results were $\Delta G_{\mathrm{PBTCA}}{}^{\theta} = 43.414$ kJ/mol, $\Delta G_{\mathrm{HEDP}}{}^{\theta} = 42.743$ kJ/mol, and $\Delta G_{\mathrm{ATMP}}{}^{\theta} = 43.667$ kJ/mol, respectively. All values were greater than 40 kJ/mol, demonstrating that the three organic phosphorus corrosion inhibitors formed coordination bonds with the metal surface by chemical adsorption, which impedes metal corrosion.

4. Conclusion

In this paper, three organophosphorus inhibitors, PBTCA, HEDP, and ATMP, were used as the research objects, and the influence factors of corrosion inhibition efficiency and corrosion inhibition mechanism on carbon steel in circulating cooling water were discussed. The following conclusions can be drawn:

(1) The weight loss method proved that the average inhibition rate of the three organic phosphorus corrosion inhibitors was more than 80%, the inhibition rate increases with the increasing concentration of

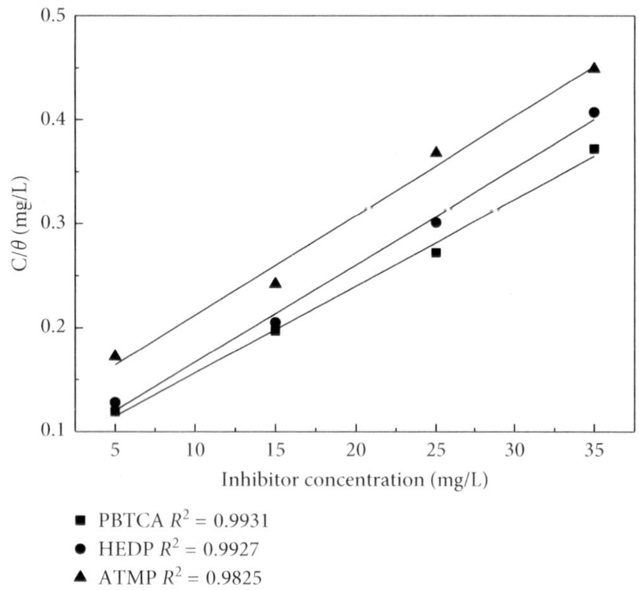

FIGURE 7: Isotherm fitting of three organic phosphorus inhibitors to Langmuir adsorption.

inhibitors and decreases with the increase in the number of P atoms, and the corrosion inhibition effect of single-phosphorus inhibitor PBTCA was better.

(2) The adsorption of the organophosphorus inhibitor onto the metal surface followed the Langmuir monolayer adsorption.

(3) The DFT algorithm in Gaussian 09 software was used to optimize the inhibitors at the theoretical level by the B3LYP/6-31G (d, p) method. According to molecular layout analysis and frontier orbital theory, the three organophosphorus corrosion inhibitors were used as electron buffers. By receiving the free electron of the outer orbit of the iron atom, the inhibitor and iron atom can form the stable co-ordination bond and form a protective film on the metal surface to obstruct the metal corrosion. At the same time, the inhibitor can provide electrons to zinc ions present in the water, forming complex compounds and meeting the requirements of the pre-membrane in circulating cooling water.

Conflicts of Interest

The authors declare that they have no conflicts of interest.

Acknowledgments

We would like to thank Harbin Gas Works for making available the corrosion agents for the study. Furthermore, we wish to thank the Harbin Institute of Technology. Liu Liang is grateful for a funded PhD stipend provided by the Nanqi Ren Studio. The research was supported by the Nanqi Ren Studio, Academy of Environment and Ecology, Harbin Institute of Technology (grant number HSCJ201704).

References

[1] T. F. Tang, "Optimization of compound scale corrosion inhibitor formula for circulating cooling water," *Environmental Science and Technology*, vol. 27, pp. 126–128, 2004.

[2] C. F. Wei, L. Hong, and B. F. Liang, "Status and development of corrosion and scaleinhibitors for circulating cooling water," *Science and Technology of Icic*, vol. 5, pp. 126–128, 2000.

[3] D. M. Wei, "Treatment of circulating cooling water in Refrigeration system," *Technology and Development of Chemical Industry*, vol. 1, pp. 42–57, 2011.

[4] R. Touir, N. Dkhireche, M. Ebn Touhami et al., "Study of phosphonate addition and hydrodynamic conditions on ordinary steel corrosion inhibition in simulated cooling water," *Materials Chemistry and Physics*, vol. 122, no. 1, pp. 1–9, 2010.

[5] Z. Quan, C. Wang, B. Li, Y. Chen, and C. Ma, "Experimental study of circulating cooling water scale inhibition by low voltage electrostatic," *Water Purification Technology*, vol. 6, pp. 1–9, 2007.

[6] F. Tang, A. P. Zheng, and Z. Z. Song, "Research progress and application of scale inhibitors for industrial circulating cooling water," *Science and Technology in Chemical Industry*, vol. 3, pp. 1–9, 2010.

[7] Y. Wang, "Mechanism and control countermeasures of scale and corrosion in circulating water," *Petrochemical Technology*, vol. 7, p. 43, 2017.

[8] E. Ilhan-Sungur and A. Otuk, "Microbial corrosion of galvanized steel in a simulated recirculating cooling tower system," *Corrosion Science*, vol. 52, no. 1, pp. 161–171, 2010.

[9] F. Liu, X. Lu, W. Yang, J. Lu, and H. Zhong, "Optimizations of inhibitors compounding and applied conditions in the simulated circulating cooling water system," *Desalination*, vol. 313, pp. 18–27, 2013.

[10] R. Liang, J. Li, M. Liu, and Z. Y. Huang, "Influence of inhibitors on the adhesion of SRB to the stainless steel in circulating cooling water," *Colloids and Surfaces B: Biointerfaces*.vol. 172, pp. 1–9, 2018.

[11] K. H. Rahmani, R. Jadidian, and S. Haghtalab, "Evaluation of inhibitors and biocides on the corrosion, scaling and biofouling control of carbon steel and copper-nickel alloys in a power plant cooling water system," *Desalination*, vol. 393, pp. 1174–1185, 2016.

[12] C. Hou, "Corrosion control, fouling deposit control and biofouling control in the seawater circulating cooling system," *Ocean Technology*, vol. 4, pp. 1174–1185, 2002.

[13] K. Rahmani, "Reducing water consumption by increasing the cycles of concentration and Considerations of corrosion and scaling in a cooling system," *Applied Thermal Engineering*, vol. 114, pp. 849–856, 2017.

[14] Y. J. Teng, S. H. Shao, and C. W. Cui, "Study on compatibility of chlorine dioxide with organic phosphine reagents," *Journal of Harbin Commercial University (Natural Science Edition)*, vol. 3, pp. 313–315, 2008.

[15] X. H. He, J. W. Mao, Q. Ma, and Y. M. Tang, "Corrosion inhibition of perimidine derivatives for mild steel in acidic media: electrochemical and computational studies," *Journal of Molecular Liquids*, vol. 268, pp. 260–268, 2018.

[16] E. S. H. E. Ashry and S. A. Senior, "QSAR of lauric hydrazide and its salts as corrosion inhibitors by using the quantum chemical and topological descriptors," *Corrosion Science*, vol. 53, pp. 1025–1034, 2011.
[17] G. Gece, "The use of quantum chemical methods in corrosion inhibitor studies," *Journal of Molecular Liquids*, vol. 50, no. 11, pp. 2981–2992, 2008.
[18] H. Hamani, T. Douadi, D. Daoud, M. Ai-Noaimi, and R. A. Rikkouh, "1-(4-Nitrophenyl-imino)-1-(phenylhydrazone)-propane-2-one as a corrosion inhibitor for mild steel in 1 M HCl solution: weight loss, electrochemical, thermodynamic and quantum chemical studies," *Journal of Electroanalytical Chemistry*, vol. 801, pp. 425–438, 2017.
[19] A. A. Olajire, "Corrosion inhibition of offshore oil and gas production facilities using organic compound inhibitors - a review," *Journal of Molecular Liquids*, vol. 248, pp. 775–808, 2017.
[20] S. Kumar, D. Sharma, and P. Yadav, "Experimental and quantum chemical studies on corrosion inhibition effect of synthesized organic compounds on N80 steel in hydrochloric acid," *Industrial and Engineering Chemistry Research*, vol. 52, no. 39, pp. 14019–14029, 2013.
[21] G. Gece and S. Bilgi, "Molecular-level understanding of the inhibition efficiency of some inhibitors of zinc corrosion by quantum chemical approach," *Industria and Engineering Chemistry Research*, vol. 51, no. 43, pp. 14115–14120, 2012.
[22] X. Shi, Y. Jiang, H. Wang, Y. Hao, and S. Chen, "Density functional theory study on adsorption behavior of four pyrazine inhibitors on Cu (111) surface," *Journal of Chemical Engineering*, vol. 68, pp. 3211–3217, 2017.
[23] C. W. Cui, S. F. Li, H. Yang, W. T. Feng, and Y. Liu, "Theoretical study of corrosion inhibition of PBTCA, HEDP, ATMP," *Material Science and Technology*, vol. 14, pp. 608–611, 2006.
[24] Y. M. Chen, C. X. Sun, Y. S. Cheng, M. Zhang, and M. Chen, "Study on corrosion inhibition of carbon steel, copper and stainless steel by the polyaspartic acid compound," *Application of Chemical Industry*, vol. 43, pp. 647–650, 2014.
[25] I. B. Obot, D. D. Macdonald, and Z. M. Gasem, "Density functional theory (DFT) as a powerful tool for designing new organic corrosion inhibitors. Part 1: an overview," *Corrosion Science*, vol. 99, pp. 1–30, 2015.
[26] P. Geerlings, F. De Proft, and W. Langenaeker, "Conceptual density functional theory," *Chemical Reviews*, vol. 103, no. 5, pp. 1793–1873, 2003.
[27] D. B. Hmamou, R. Salghi, A. Zarrouk, M. R. Aouad, and O. Benali, "Weight loss, electrochemical, quantum chemical calculation, and molecular dynamics simulation studies on 2-(Benzylthio)-1,4,5-triphenyl-1H-imidazole as an inhibitor for carbon steel corrosion in hydrochloric acid," *Industrial and Engineering Chemistry Research*, vol. 52, no. 40, pp. 14315–14327, 2013.
[28] L. Liu, T. T. Cao, Q. W. Zhang, and C. W. Cui, "Optimization of associative model between chlorine dioxide and organo-phosphorus corrosion inhibitor by RSM-BBD," *Industrial Water Treatment*, vol. 12, 2018.
[29] H. Gerengi, M. Mielniczek, G. Gece, and M. M. Solomon, "Experimental and quantum chemical evaluation of 8-hydroxyquinoline as a corrosion inhibitor for copper in 0.1 M HCl," *Industrial and Engineering Chemistry Research*, vol. 55, no. 36, pp. 9614–9624, 2016.

7

Ageing Effects on Microstructure, Mechanical Properties, and Fracture Behaviour of 9Cr-1.5Mo-1Co-VNbBN Martensitic Steel Welded Joint for High Temperature Application

Ladislav Falat,[1] Viera Homolová,[1] Lucia Čiripová,[1] Peter Ševc,[1] and Milan Svoboda[2]

[1]*Institute of Materials Research, Slovak Academy of Sciences, Watsonova 47, 04001 Košice, Slovakia*
[2]*Institute of Physics of Materials, Academy of Sciences of the Czech Republic, Žižkova 22, 61662 Brno, Czech Republic*

Correspondence should be addressed to Ladislav Falat; lfalat@imr.saske.sk

Academic Editor: Akihiko Kimura

This paper deals with long-term ageing effects of 9Cr-1.5Mo-1Co-VNbBN (CB2) steel weldment on its impact toughness, creep rupture behaviour, and hardness in relation to microstructure and fracture characteristics. The weldment was studied in PWHT state and after isothermal expositions at 625°C for 10000 and 30000 hours. Microstructure evolution was studied using analytical scanning and transmission electron microscopy. Charpy V-notch impact toughness tests were performed for all heat-treated states with a notch location in distinct weld regions such as weld metal (WM), heat-affected zone (HAZ), and base material (BM). The overheated HAZ region exhibited the lowest impact toughness as a result of severe welding induced microstructure degradation. Creep tests were performed at 625°C in the stress range between 80 and 120 MPa. At the highest applied stress, creep fracture occurred in WM, whereas at lower stresses the failure position shifted towards fusion zone at WM/HAZ interface. The hardness profiles experienced significant scattering due to weld microstructural heterogeneity. The major fracture mechanisms involved transgranular quasi cleavage and intergranular creep cracking in impact and creep loading conditions, respectively.

1. Introduction

The welded joints of high chromium tempered martensitic creep-resistant steel are frequently employed in construction of modern fossil-fired thermal power plants operating at supercritical and ultrasupercritical conditions [1]. Their main application area covers structural parts of thick section boiler components including high temperature header section and main steam pipelines [2–4]. It generally holds that local microstructure formation of individual weld regions, namely, the weld metal (WM), fusion zone (FZ), and heat-affected zone (HAZ) depends on their location within the weldment and reached peak temperature during the weld preparation. Due to the presence of microstructural gradient and welding-related residual stresses in welded joints of martensitic/ferritic power plant steel, it is necessary to perform their postweld heat treatment (PWHT). Beside thermal stabilization of initial weld microstructure and residual stress relief, the application of PWHT decreases unallowably high hardness in FZ and increases the weldments toughness [2, 5, 6]. Worldwide constant efforts are still being put into developments aiming for creep strength improvement of these materials. Apart from classical strengthening mechanisms, such as solid solution, precipitation, and subboundary strengthening, Abe et al. [7] revealed specific boron strengthening mechanism in weldments of Japanese boron alloyed 9Cr-3W-3Co-VNbBN (MARBN) martensitic steel based on the HAZ microstructure modification via boron related effects on the retardation of secondary precipitates coarsening and suppression of diffusive ferrite-to-austenite phase transformation during welding. The European research initiatives in the frame of international collaboration within COST Action 536 resulted in the development of a series 9Cr-Mo-Co-B-VNbBN model alloys for production of thick castings and forgings intended for the use at supercritical steam conditions [8]. One of them is represented by 9Cr-1.5Mo-1Co-VNbBN (CB2) steel with controlled boron and nitrogen content.

Table 1: Chemical composition in wt.% of CB2 steel BM and FM (Thermanit MTS 5 Co 1).

	C	Si	Mn	Cr	Ni	Mo	V	Ti	Al	Co	Nb	B	N	Fe
BM	0.12	0.29	0.86	9.14	0.22	1.5	0.19	0.002	0.006	0.95	0.06	0.012	0.020	Balance
FM	0.11	0.45	0.65	9.10	0.7	1.3	0.24	0.01	0.01	1.09	0.06	0.003	0.028	Balance

Our previous works [9, 10] were focused on preliminary studies of ageing effects of thick-walled CB2 weldments with respect to thermal stability of their microstructures and local mechanical properties in different loading conditions. The present study represents an extended and continuing research work aiming for deeper investigation and interpretation of the effects of long-term ageing expositions on mechanical and creep properties variations of the same CB2 steel weldment in relation to its detailed microstructural and fracture characteristics.

2. Experimental Material and Procedure

The welded joint of cast 9Cr-1.5Mo-1Co-VNbBN (CB2) steel was produced by tungsten inert gas (TIG) welding using electrode weaving technique and modified filler metal Thermanit MTS 5 Co1 at PHB Stahlguss International. The thickness of welded plates was around 98 mm. Both the base material (BM) and filler metal (FM) chemical composition is given in Table 1. After the weld preparation, subsequent postweld heat treatment (PWHT) was carried out industrially at 730°C for 24 h by the weld producer. Cross-weld segments obtained within COST Action 536 collaboration were afterwards long-term isothermally aged at 625°C for 10 and 30 thousand hours in our laboratory conditions. Charpy V-notch (CVN) samples with dimensions of 10 × 10 × 55 mm and the notch individually located in the BM, WM, and overheated HAZ next to FZ were used to test impact toughness after long-term ageing expositions and in the PWHT state, as well. The tests were performed at room temperature (RT) for all the material states. After the preageing at 625°C for 30 kh, the samples were impact tested also at 100°C. Creep rupture tests were carried out at 625°C with applied stresses of 80, 100, and 120 MPa using round M8 tensile creep specimens with gauge length of 40 mm and body diameter of 5.5 mm. Cross-weld creep tests were performed for the weldments in initial PWHT state and separately also for the welds preaged at 625°C for 30 kh. The BM preaged in the same way was also creep tested for reference. Hardness HV10 profiles were measured for as-crept material states across the welded joint samples. The microstructure of individual experimental states was observed and analyzed by means of light microscopy (LM) and scanning and transmission electron microscopy (SEM, TEM) equipped with energy dispersive X-ray (EDX) analyzer. Carbon extraction replicas were used for semiquantitative chemical microanalyses of metallic elements in extracted secondary phase precipitates. Experimental results of the phase analyses have been compared with those obtained by thermodynamic calculations using Thermo-Calc Classic software [11] and noncommercial database Steel 16 formulated by Kroupa et al. [12]. Fractographic studies included fracture path and fracture surface observations using LM and SEM.

3. Results and Discussion

3.1. Microstructure and Phase Analysis. Overall microstructural features of individual weld regions of studied CB2 weldment in the initial PWHT state are shown on light-optical micrographs in Figure 1. Generally, all the microstructures consist of tempered martensite with precipitated particles of secondary phases occurring within matrix as well as on all types of inner microstructural interfaces such as the primary austenitic grain boundaries, block, packet, and lath subboundaries. Apart from these common features, the microstructures of individual weld regions possess their own specific microstructural characteristics. The original cast, normalized and tempered martensitic microstructure of BM shows noticeably coarse grains and laths (Figure 1(a)), whereas the remelted microstructure of WM (Figure 1(b)) is significantly refined due to its fast cooling during the welding. Transition region between the BM and WM is represented by notably heterogeneous HAZ microstructure following the locality of weld FZ (Figure 1(c)). Beside its heterogeneous grain size, the transition zone exhibits some minor occurrence of nonequilibrium δ-ferrite grains appearing like small precipitate-free areas (see Figure 1(c)). More detailed microstructures of BM and WM in PWHT state and also in long-term aged states at 625°C for 10 and 30 thousand hours are visualized on SEM-micrographs in Figures 2–4. In comparison to the microstructures in PWHT state (Figure 2), the microstructures after long-term ageing expositions (Figures 3 and 4) can be characterized by additional precipitation and coarsening of secondary phase particles accompanied by proceeding recovery processes of highly tempered martensitic matrix. The differences in size, morphology, and distribution of secondary precipitates occurring in PWHT state and after the 625°C ageing for 30 kh are shown on TEM carbon extraction replicas for BM (Figure 5) and WM (Figure 6).

Phase analyses of individual precipitates on the TEM replicas were performed by their semiquantitative chemical microanalyses via local EDX measurements. The results are given in Table 2 in the form of mean chemical composition (metallic part) in at.% of individual precipitated phases. In addition, thermodynamic calculations of equilibrium phase compositions were calculated to support the experimental findings. According to the results presented in Table 2, only two types of the precipitates were revealed in PWHT state (Figure 2), namely, fine intragranular MX carbonitrides (M = V, Nb, Cr; X = C, N) at tempered martensitic lath boundaries and large intergranular Cr-rich $M_{23}C_6$ carbides at grain, block, and packet boundaries. Both of these precipitate types exhibit similar greyish appearance by their SEM visualization using back-scattered electrons contrast (see Figures 2(b) and 2(d)). On the other hand, during subsequent long-term

FIGURE 1: Light-optical microstructures of individual regions of studied CB2 weldment in PWHT state: BM (a), WM (b), and HAZ (c).

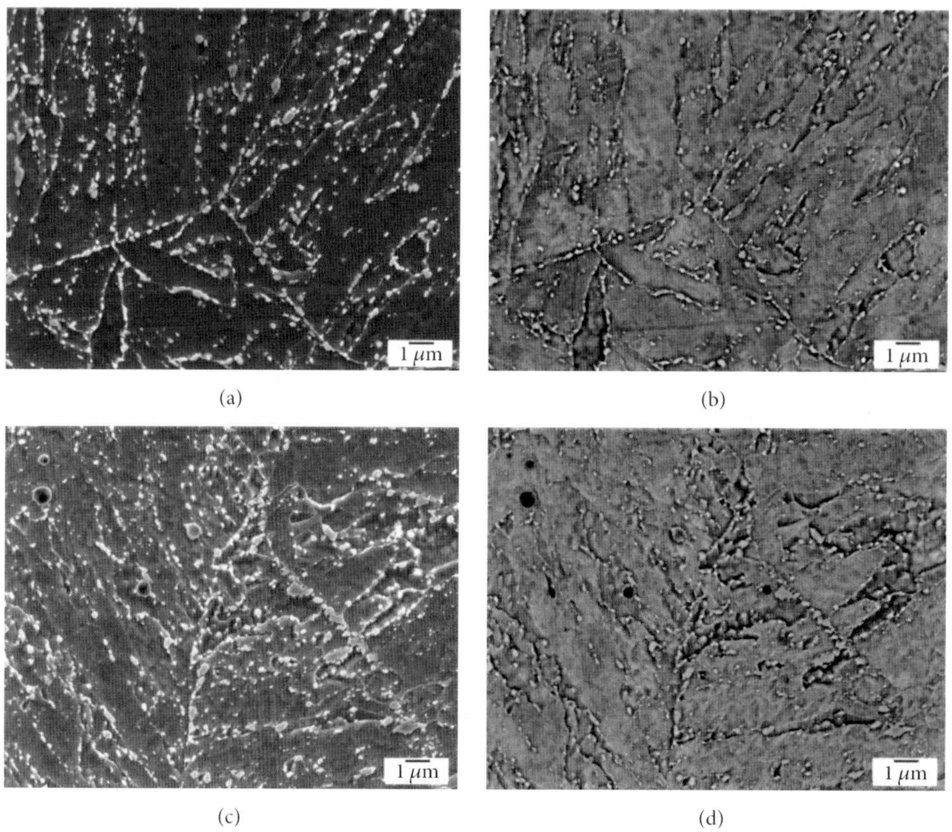

FIGURE 2: SEM-micrographs of initial PWHT microstructures of BM (a, b) and WM (c, d) of the CB2 weldment visualized using secondary electrons (a, c) and back-scattered electrons (b, d).

ageing at 625°C for both time durations, the particles of $(Fe,Cr)_2Mo$ based Laves phase have been revealed, often in the vicinity of Cr-rich $M_{23}C_6$ carbides (see Figures 3 and 4 and Table 2). These intermetallic precipitates show clear bright appearance in the SEM visualization using back-scattered electrons contrast (Figures 3(b), 3(d), 4(b), and 4(d)) due to their notable molybdenum content and thus much higher average atomic number compared to other precipitated phases in the microstructures. The precipitation of Laves phase belongs to the most significant microstructural changes in 9Cr creep-resistant steel during its long-term thermal or creep exposure [13–15]. Furthermore, as indicated in present study by both the experimental measurements and thermodynamic calculations (Table 2), beside the Laves phase also some other newly precipitated phases, namely, the Cr(V,Nb)N type Z-phase and Nb-rich MC carbides, have been revealed in thermally exposed state at 625°C for 30 kh hours. As visible in Table 2, several thermally induced changes in chemical composition of individual phases took place during the ageing: the MX precipitates represented by (V,Nb,Cr) rich carbonitrides show certain small enrichment in vanadium and iron at the expense of

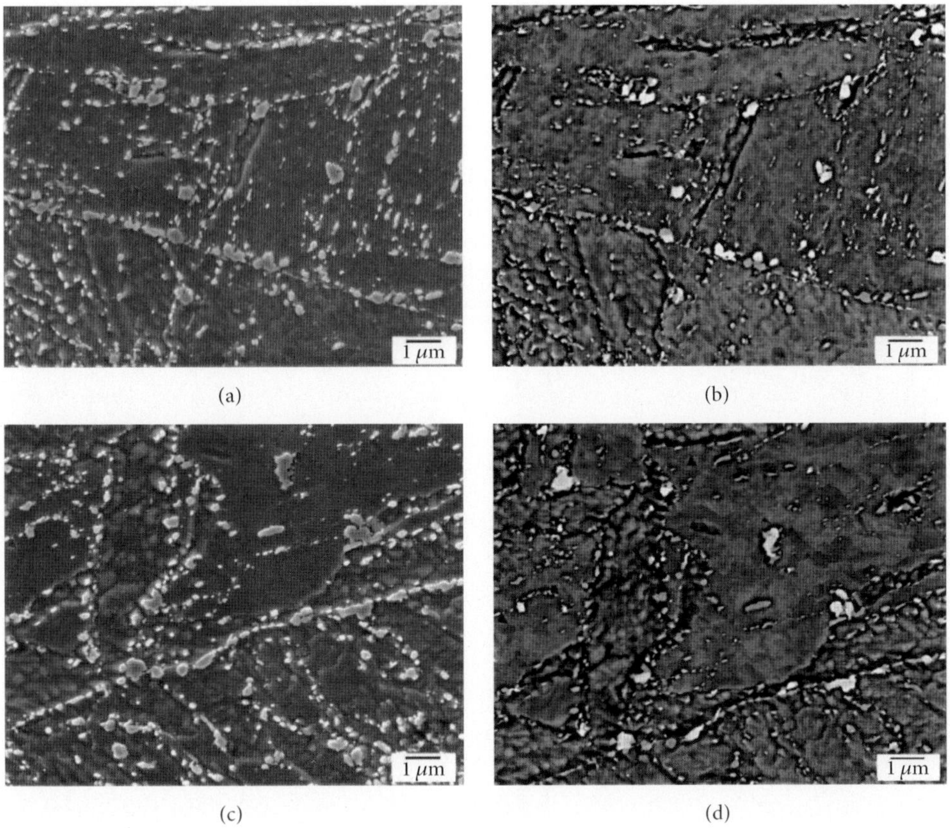

FIGURE 3: SEM-micrographs of thermally exposed microstructures at 625°C for 10 kh of BM (a, b) and WM (c, d) of the CB2 weldment visualized using secondary electrons (a, c) and back-scattered electrons (b, d).

TABLE 2: Mean chemical composition (metallic part) in at.% of precipitate phases in CB2 weld regions and material states and calculated equilibrium phase composition at 625°C.

Material state	Phase	BM					WM				
		V	Cr	Fe	Nb	Mo	V	Cr	Fe	Nb	Mo
PWHT (730°C/24 h)	MX	62	17	5	14	2	60	15	1	22	2
	$M_{23}C_6$	3	65	23	1	8	2	63	27	1	7
Isothermally aged (625°C/30 kh)	MX	64	14	8	12	2	65	15	3	15	2
	$M_{23}C_6$	2	67	23	1	7	2	69	21	1	7
	Laves phase	1	14	50	2	33	1	18	46	2	33
	Z-phase	39	33	13	11	4	39	40	10	6	5
	MC	22	18	3	53	4	22	17	2	54	5
Equilibrium calculatedat 625°C	$M_{23}C_6$	3	74	17	0	6	3.3	73	17.7	0	6
	Laves phase*	0	16	46	0	33.3	0	15.3	46	0	33.5
	Z-phase	35	43.5	6.5	15	0	38	43.7	6	12.3	0
	MC	10.1	0.2	0	89.2	0.5	12.8	0.2	0	86.5	0.5

*Containing also 4.7 at.% Si in BM and 5.2 at.% Si in WM.

niobium and chromium. Starting from the initial PWHT state followed by long-term ageing up to achievement of equilibrium condition, the $M_{23}C_6$ precipitates exhibit gradual increasing in their chromium content at the expense of other alloying elements. Experimentally determined Laves phase composition is almost identical with the calculated one (Table 2). As already shown before (Figures 3–6), the long-term ageing expositions of CB2 weldment at 625°C resulted in significant occurrence of coarse Laves phase precipitates. In contrast, only a few Z-phase particles were revealed in studied weldment after the ageing at 625°C for 30 kh. The observation of Z-phase was also reported by Jandová et al. [16] for the CB2 steel material after its long-term creep exposure at 650°C. Depending on the occurrence in either BM or WM, the Z-phase precipitates exhibited some small variations in their chemical composition (Table 2). The calculated Z-phase

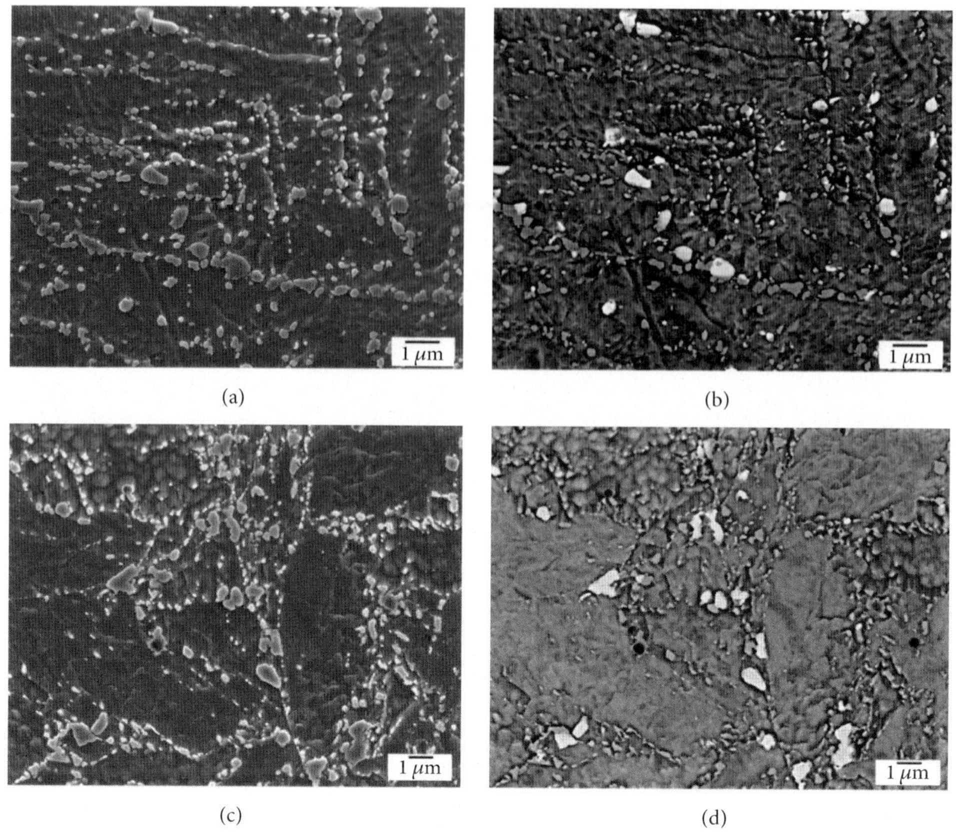

FIGURE 4: SEM-micrographs of thermally exposed microstructures at 625°C for 30 kh of BM (a, b) and WM (c, d) of the CB2 weldment visualized using secondary electrons (a, c) and back-scattered electrons (b, d).

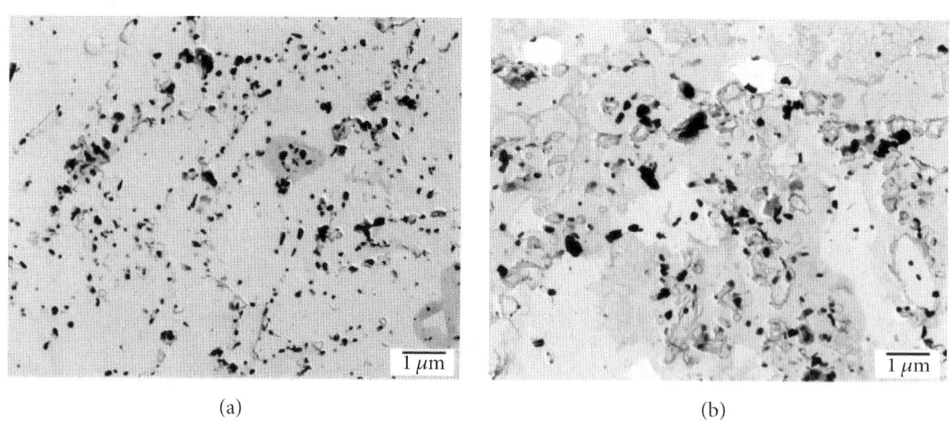

FIGURE 5: TEM carbon extraction replica visualization of secondary phase precipitates in BM of studied CB2 weldment in PWHT state (a) and after 30 kh ageing at 625°C (b).

equilibrium compositions expressed by the metallic elements ratio to be 43.5Cr-35V-15Nb-6.5Fe for CB2 BM and 43.7Cr-38V-12.3Nb-6Fe for CB2 WM (Table 2) correspond well with the findings by Panait et al. [17] for P91 steel base material. In present study, the experimentally determined Z-phase metallic elements ratios were 33Cr-39V-11Nb-13Fe-4Mo and 40Cr-39V-6Nb-10Fe-5Mo for the long-term aged BM and WM, respectively. Such closely similar vanadium content of experimentally analyzed Z-phase precipitates like that of the calculated ones may indicate their early stage of precipitation at the V-rich MX precipitates which also agrees with other observations performed on thermally exposed 9–12Cr steel [18, 19]. Figure 7 shows temperature dependencies of equilibrium mole fractions of individual phases in CB2 BM and WM obtained by thermodynamic calculations for the temperature range from 400°C to 1000°C. As it can be seen, the highest phase amount in equilibrium at 625°C is reached by Cr-rich $M_{23}C_6$ carbide. The Laves phase achieves the second highest

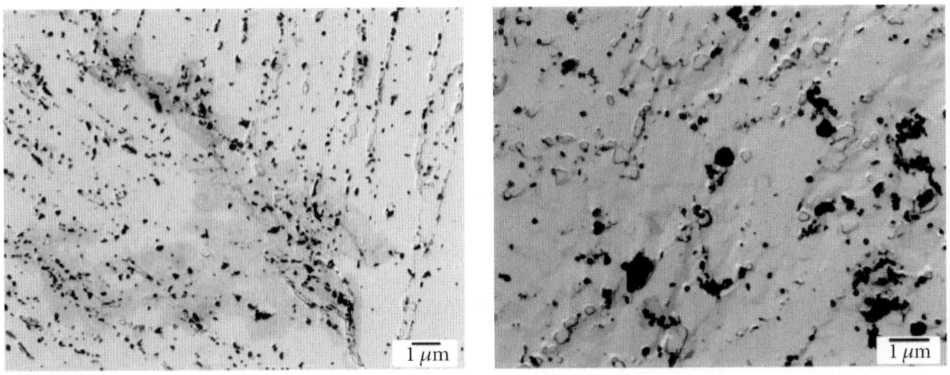

(a) (b)

FIGURE 6: TEM carbon extraction replica visualization of secondary phase precipitates in WM of studied CB2 weldment in PWHT state (a) and after 30 kh ageing at 625°C (b).

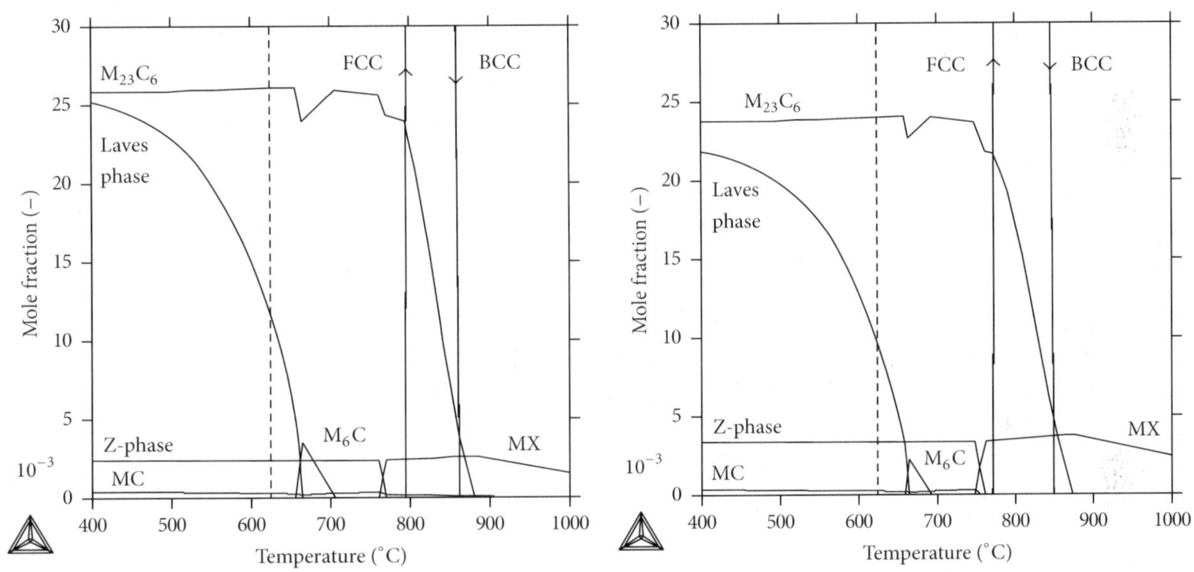

(a) (b)

FIGURE 7: Temperature dependence of mole fractions of equilibrium phases occurring in CB2 steel BM (a) and WM (b).

amount at considered condition. Although the Z-phase and MC carbides exhibit the lowest phase amounts in the CB2 microstructure, their influence on resulting mechanical properties is not insignificant. Detrimental effects of the Z-phase particles coarsening on creep strength are generally known to be directly related to gradual dissolution of fine MX precipitates [19–21] that do not represent the stable phase in CB2 material at 625°C (see Figure 7).

3.2. Mechanical Properties and Fractography

3.2.1. Impact Toughness Test. The results of impact toughness tests are shown in Figure 8. The highest room-temperature (RT) impact toughness (i.e., CVN) value of all the tested weld regions in PWHT state was measured in the WM. The main reasons for the highest WM toughness in PWHT state might be related to its finer initial microstructure (Figure 1(b)), lower boron, and higher nickel content in the used FM compared to the CB2 BM (Table 1). The presence of hard nonmetallic globular inclusions in the WM (Figures 2(c) and 2(d)) might also enhance its toughness by promotion of ductile dimples formation. The use of electrode weaving "temper-bead" TIG welding procedure in present case of CB2 weld preparation could also contribute to the higher WM toughness. In contrast, the overheated HAZ region exhibited the lowest impact toughness as a result of severe welding induced microstructure degradation (Figure 1(c)). By comparing the individual material states with respect to the ageing conditions (PWHT, 625°C/10 kh, 625°C/30 kh), it can be stated that the CVN values measured at RT show their significant and mutually similar decrease for both of the long-term aged states compared to the initial PWHT state (Figure 8). In general, the decrease of RT impact toughness in creep-resistant steel can mainly be attributed to thermal

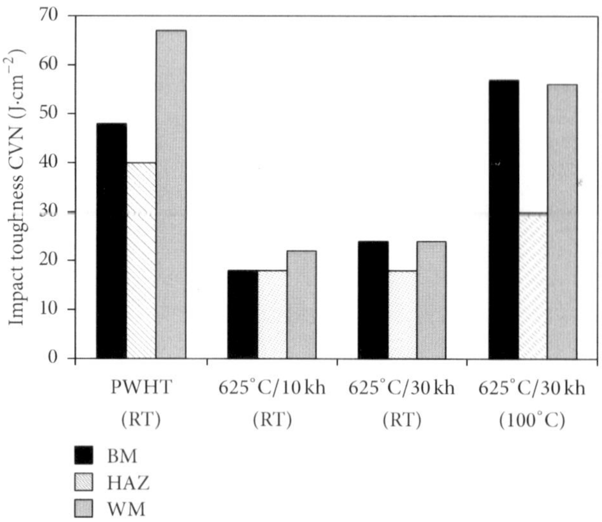

FIGURE 8: Variation of impact toughness values in dependence of CB2 weld region, its material state and testing temperature. CVN values according to [9].

degradation of its microstructure including coarsening of secondary phase precipitates [22]. The observed negligible differences in RT CVN values of the long-term aged states are likely caused by mutually similar level of microstructure degradation of these material states (see Figures 3 and 4). This idea can be supported by the results of other authors [21–23] who also found out that the most remarkable changes in microstructure of ferritic creep-resistant steel proceed during the first several thousand hours of its thermal exposure and afterwards its microstructure and properties undergo only very slow long-term degradation due to high thermal stability of these materials. Summarizing the effects of weld region on resulting impact toughness of all tested material states, it can be concluded that the most significant differences in CVN values were revealed for the PWHT state tested at RT and for the 625°C/30 kh aged state tested at 100°C. Hence subsequent fractographic observations (Figure 9) were focused on the fracture surfaces of broken impact test specimens showing the most pronounced differences in their CVN values in order to reveal substantial differences in fracture mechanisms related to the specific weld regions and their material states. The fracture surfaces of CVN samples in PWHT state after the impact testing at RT with the notch locations in BM and WM are shown in Figures 9(a) and 9(b), respectively. The fracture surfaces exhibit mixed fracture features characterized by the presence of transgranular quasi-cleavage facets as well as ductile dimple areas. In contrast, the fracture mode of the BM sample tested at RT after 30 kh of ageing at 625°C (Figure 9(c)) is almost exclusively formed of transgranular quasi cleavage with coarse facets indicating increased embrittlement of the material by thermal degradation. Moreover, the fracture surface in Figure 9(c) contains also clearly visible secondary cracks and some cleavage reinitiation sites. Similar fracture characteristics are shown in Figure 9(d) which corresponds to the broken CVN sample from the WM region with the same average value of impact toughness (Figure 8). To the contrary, Figure 9(e) represents the SEM fractograph of the long-term aged sample from the WM region after impact testing at 100°C. In this case, the significant fracture features are represented by ductile dimple tearing areas. Such manifestation of fracture behaviour at increased temperature is quite common in ferritic steel and it can generally be related to its thermally enhanced plasticity via increasing mobility of free dislocations in steel matrix. The fracture surface appearance for WM tested at 100°C (Figure 9(e)) indicates that the impact test was carried out above brittle-ductile transition region. Indeed, Buršák et al. [24] reported the T35 value (i.e., transition temperature corresponding to the CVN value equal to 35 J·cm^{-2}) of long-term service exposed Cr-Mo-V creep-resistant steel to be about 52°C. However, in the case of long-term aged HAZ region tested at 100°C (Figure 9(f)) showing notably lower CVN value than that of WM (Figure 8), the fracture surface still exhibits considerable amount of brittle quasi-cleavage features that can be directly related to its increased microstructural heterogeneity enhancing thermal degradation during long-term ageing exposure.

3.2.2. Creep Rupture and Hardness Test. Creep rupture data obtained for studied CB2 weldment creep tested at 625°C in both the PWHT state and the preaged (625°C/30 kh) state and also for the CB2 BM preaged at the same conditions are shown in Figure 10. After the creep tests, the as-crept specimens were longitudinally sectioned and metallographically prepared for further investigations including macroscopic observations, cross-weld hardness measurements, and fractographic analyses.

The observed creep rupture behaviour (Figure 10) generally indicates gradual increase of creep rupture time for all the creep tested CB2 materials (i.e., the preaged CB2 BM as well as the CB2 welded joints in both the PWHT and preaged conditions) with decreasing applied stress. The CB2 BM exhibited the longest creep rupture lives for all the applied stresses, whereas both welded joints exhibited worse creep life performance. According to numerous research findings, for example [25, 26], the most common failure type of welded joints of ferritic creep-resistant steel during long-term service or laboratory creep exposure is related to so-called "type IV failure" in its intercritical HAZ (i.e., the region of BM heated up to the temperature range between Ac_1 and Ac_3 critical transformation temperatures during welding) due to its local microstructure degradation by incomplete ferrite-to-austenite transformation and thus subboundary hardening degradation. In general, the long-term creep exposure characterized by long creep life (i.e., long time to rupture) takes place for a selected material in creep conditions (temperature, applied stress) activating low stress regime (LSR) creep deformation mechanisms causing very slow plastic deformation. However, accelerated creep tests (ACT) in laboratory conditions with relatively short creep rupture lives are commonly performed in high stress regime (HSR). Some studies, for example [27–29], on creep behaviour of 9Cr creep-resistant steel at considered creep temperature indicated transition stress separating the HSR and LSR to be about 100 MPa. Thus, the selected applied

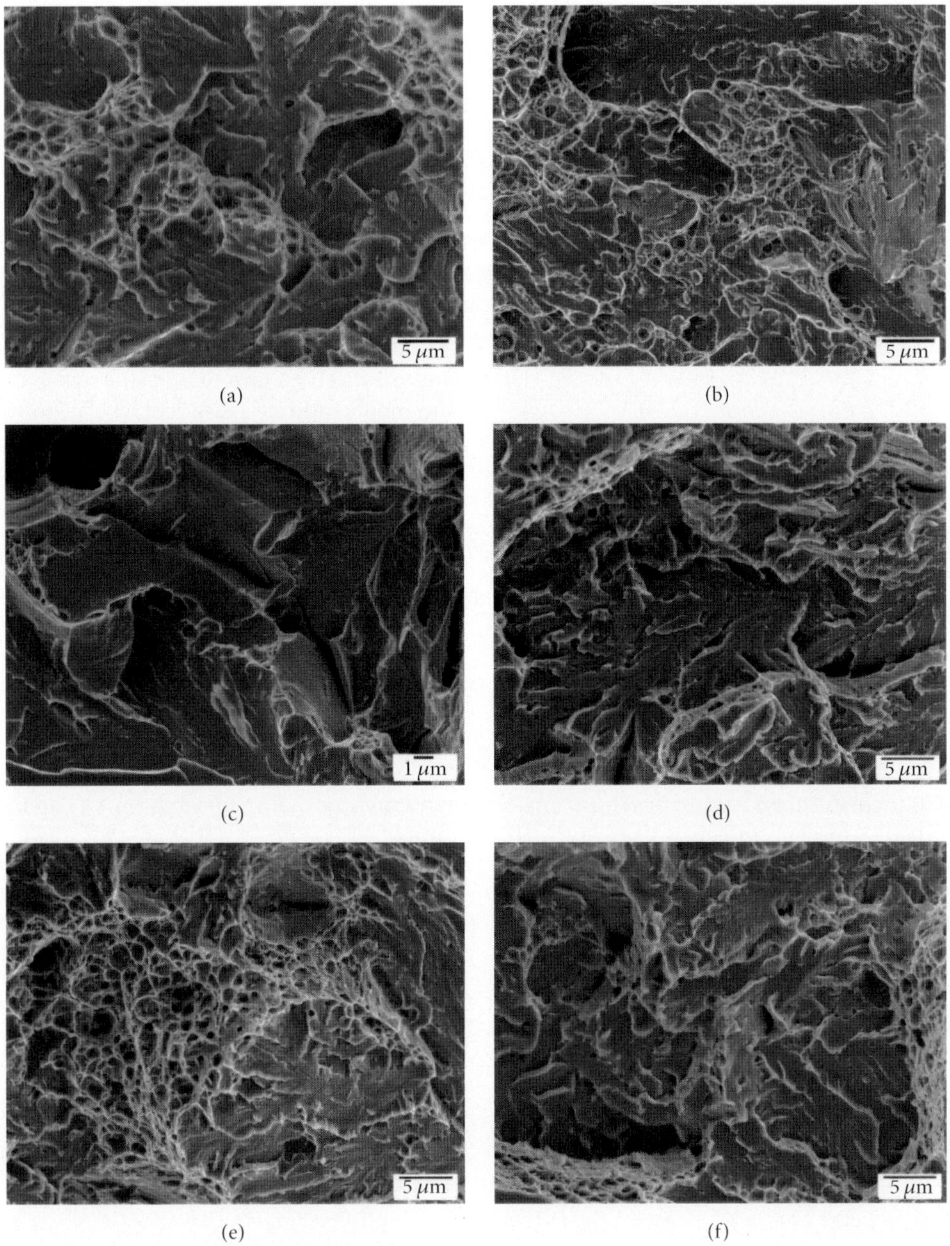

FIGURE 9: Fracture surfaces of CVN samples after impact toughness tests performed for individual weld regions, material states, and testing conditions: BM, PWHT, and RT (a); WM, PWHT, and RT (b); BM, 625°C/30 kh, and RT (c); WM, 625°C/30 kh, and RT (d); WM, 625°C/30 kh, and 100°C (e); HAZ, 625°C/30 kh, and 100°C (f).

stresses for creep testing in present investigation included the stress range around estimated HSR-to-LSR transition. Beside the obtained creep rupture data (Figure 10), the welds failure types occurring in dependence on used creep conditions were identified by macroscopic observations of metallographically prepared (ground, polished, and etched) longitudinal sections of fractured cross-weld creep specimens after the creep rupture tests. Figure 11 clearly shows that at the highest applied stress of 120 MPa, the creep fracture occurred in WM is being referred to as "type I failure" (Figure 11(a)), whereas at both of the lower stresses (80 MPa, 100 MPa) the failure occurrence shifted to FZ at the WM/HAZ interface indicating "type III failure" (Figures 11(b) and 11(c)) according to Schüller's classification [30]. The creep failure in preaged BM is shown in Figure 11(d) for reference. Schuler et al. [31] investigated CB2 welded joints produced using two filler metals with different nickel content aiming for comparison of their creep performance. They indicated failure mode transition from WM at 120 MPa to BM at 100 MPa. In our present study, the failure mode transition occurred in both "only PWHT processed" as well as in "long-term preaged" weldments. It may be likely related not only to the above-discussed stress-dependent deformation mechanism transition but also to several other welding-metallurgical

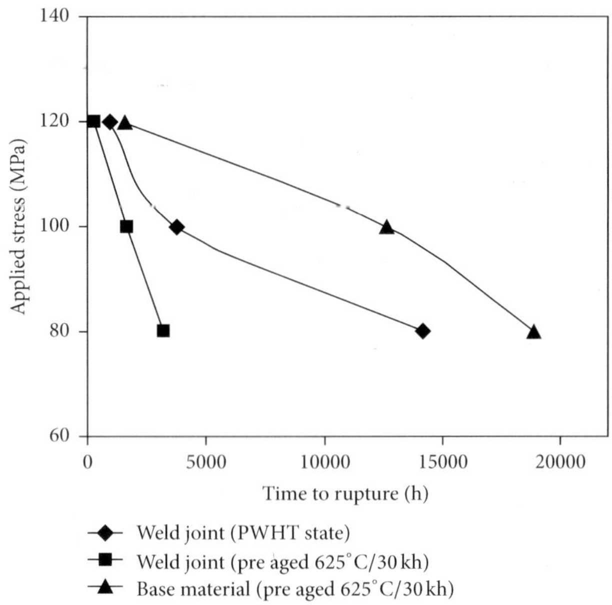

FIGURE 10: Comparison of creep rupture data obtained from creep tests at 625°C for CB2 preaged base material and its welded joint in both PWHT and preaged condition.

factors influencing local chemical and microstructural characteristics of the studied weldment. In general, the "type III failure" mode occurrence in creep conditions has typically been observed in dissimilar weldments exhibiting specific redistribution of their interstitial alloying elements (mainly carbon and nitrogen) during the heat treatment and/or creep exposure. The occurrence of "type III failure" in CB2 welded joint studied in present investigation might also be related to similar redistribution phenomena of boron between BM and WM as well as specific boron alloying effects on the local FZ microstructure (Figure 12). The occurrence of the observed failure mode transition is likely to be related to different propensity of individual weld regions and their microstructures for the localization of creep deformation depending on the applied stress (Figure 11). The differences between the BM and WM chemical compositions may be another important factor influencing the plastic deformation resistance of individual weld regions. The observed rapid softening of higher nickel and lower boron containing WM at the highest applied stress could be related to the higher propensity of nonequilibrium weld microstructure to its degradation by enhanced secondary phase coarsening (compare Figures 5 and 6), dynamic recovery processes, and dislocation-controlled creep deformation localization leading to final high deformation failure (see Figure 11(a)). In contrast, at lower stresses very low deformation failures occurred in fusion zone at the WM/HAZ interface (Figures 11(b) and 11(c)). Unlike the previous case, the occurrence of such extremely brittle creep fractures is likely related to local substructural degradation driven by gradual diffusion-controlled alloying elements redistribution (presumably boron) within the fusion zone between the WM and BM during the weld preparation and subsequent exposure leading to the observed failure (Figure 12). The detailed fracture path and creep cavity analyses of broken creep test specimens clearly indicated quasi-brittle intergranular creep fracture characteristics for creep tests performed at 100 MPa and 80 MPa (see Figures 13 and 14). In contrast, the creep tests of both the CB2 weldment and weld-free BM performed at 120 MPa resulted in mixed intergranular/transgranular shear fractures corresponding well to HSR creep deformation (see Figures 15 and 16).

In order to estimate local mechanical properties variation across the studied weldments after individual creep tests performed at different testing conditions, cross-weld hardness (HV10) profiles were determined with respect to the fusion line position (Figure 17). Performed hardness profile measurements indicated different level of softening or even small hardening effects on mutual fracture counterparts. Such behaviour may indicate the local areas of either suppressed or enhanced precipitation (e.g., see the right portions of Figures 12(a) and 12(b)) promoting combined softening/embrittling effects leading to low deformation creep failures in low stress conditions. Significant scattering of hardness values within creep specimen gauge lengths can likely be explained by notable heterogeneity of CB2 weld HAZ region (Figure 1(c)) compared to CB2 BM (Figure 1(a)). Therefore, only the BM samples in the form of thin foils prepared from both the PWHT and preaged (625°C/30 kh) and creep-exposed (625°C/80 MPa, 18864 h) material states were subjected to substructural TEM observations (Figure 18). The as-crept sample was taken from the area located approximately 5 mm beneath creep fracture surface. By comparison of TEM images in Figure 18 it can be stated that the CB2 BM in PWHT state exhibits highly tempered martensitic lath substructure (Figure 18(a)), whereas its preaged and creep-exposed substructure exhibits notable recovery features and coarsening of secondary phase precipitates (Figure 18(b)). The observed substructural changes correlate well with creep rupture strength degradation and failure mode transition of creep-exposed CB2 welds studied in present investigation.

4. Conclusions

The effects of long-term ageing of 9Cr-1.5Mo-1Co-VNbBN (CB2) steel weldment on its impact toughness, creep rupture behaviour, and cross-weld hardness were studied in relation to microstructure and fracture characteristics. The obtained results can be summarized in the following conclusions:

(i) Performed long-term ageing expositions at 625°C for 10 and 30 thousand hours resulted in significant decrease of room-temperature Charpy V-notch impact toughness (CVN) values in all studied weld regions compared to those of the initial PWHT state. However, the individual weld regions such as WM, HAZ, and BM exhibited different measure of impact toughness degradation corresponding to differing level of their local microstructure degradation. The most deteriorated CVN values of all weld regions in all material states were obtained for overheated HAZ region with the most heterogeneous microstructure containing nonequilibrium delta-ferrite residuals.

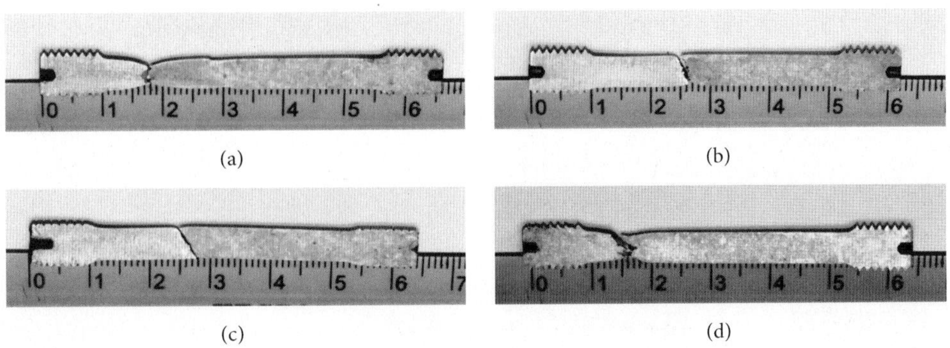

FIGURE 11: Longitudinal sections of creep specimens counterparts indicating the following: "type I failure" in WM after creep test of CB2 welded joint at 625°C/120 MPa (a), "type III failure" in FZ after creep test at 625°C/100 MPa (b), "type III failure" in FZ after creep test at 625°C/80 MPa (c), and failure in CB2 BM of weld-free specimen after creep test at 625°C/120 MPa (d).

FIGURE 12: Light-optical microstructures beneath "type III failure" of CB2 weldment crept at 625°C/80 MPa for 14145 h showing the following: microstructural gradient between WM and FZ (a) and microstructure of FZ in detail (b).

FIGURE 13: Fracture path (a, b) and creep cavity (b, c) analyses of CB2 weldment ruptured by "type III failure" after the creep test at 625°C, 100 MPa.

(ii) The highest room-temperature CVN value in PWHT state was measured in the WM which might be related to its finer initial microstructure, self-tempering effects of temper-bead welding procedure, the lower boron and higher nickel content compared to the BM, and the presence of globular inclusions promoting the formation of ductile dimples during plastic deformation.

(iii) The most significant decreases of room-temperature CVN values corresponding to all individual weld regions were observed after the ageing of studied weldment at 625°C for 10 thousand hours. A variation in CVN values with further ageing up to 30 thousand hours was rather insignificant. This finding was additionally supported by microstructural observations showing the most significant microstructural changes

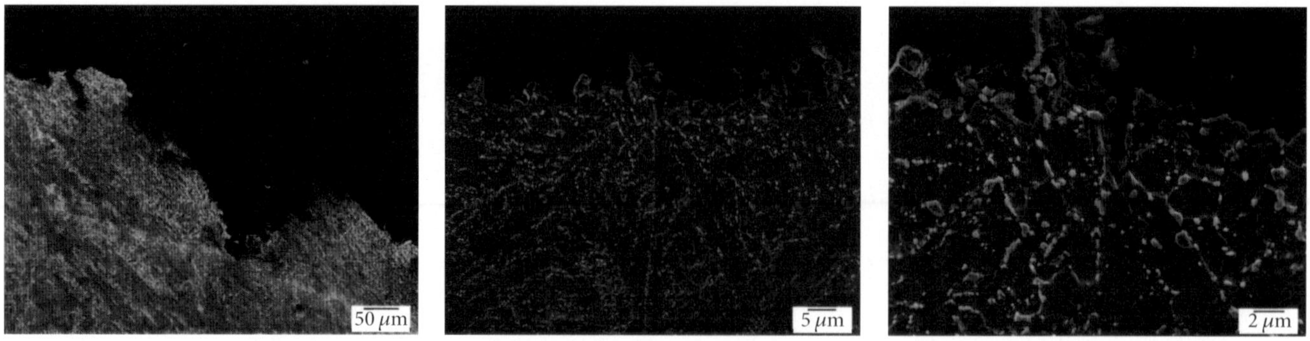

(a) (b) (c)

FIGURE 14: Fracture path (a, b, c) and creep cavity (c) analyses of CB2 weldment ruptured by "type III failure" after the creep test at 625°C, 80 MPa.

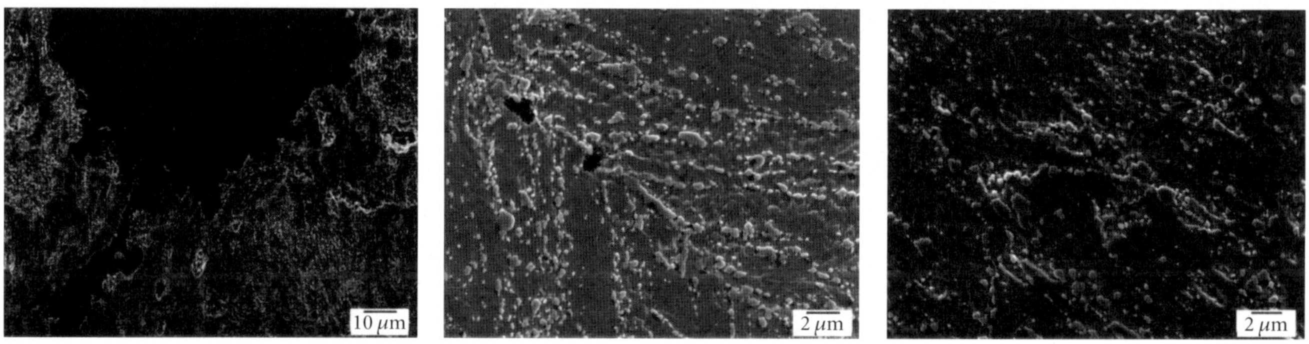

(a) (b) (c)

FIGURE 15: Fracture path (a) and creep cavities in WM (b) and nonruptured necked FZ (c) of CB2 weldment ruptured in WM by "type I failure" after the creep test at 625°C, 120 MPa.

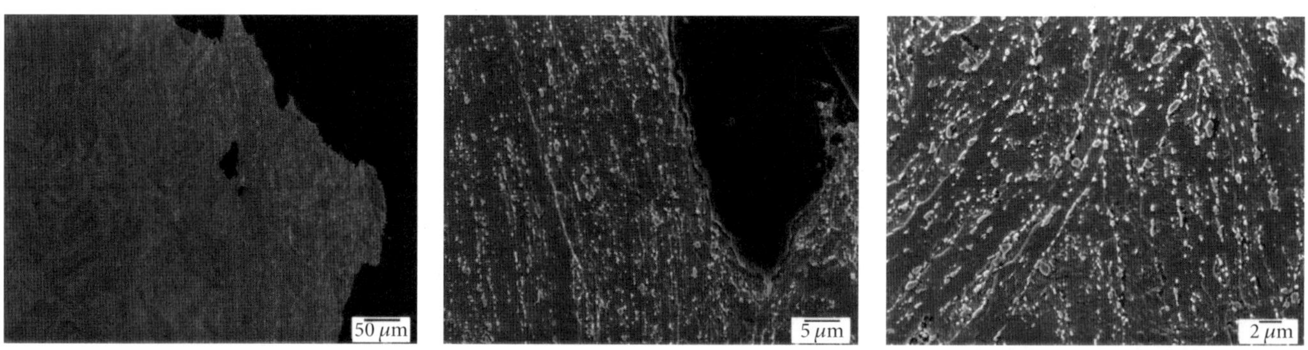

(a) (b) (c)

FIGURE 16: Fracture path (a, b) and creep cavity (c) analyses of CB2 BM specimen ruptured by high deformation shear failure after the creep test at 625°C, 120 MPa.

(matrix recovery and precipitate coarsening) already after the first period of ageing followed by only small microstructural changes with increasing ageing duration.

(iv) According to phase analyses performed on TEM carbon extraction replicas, only two types of the precipitates were revealed in PWHT state, namely, fine intragranular MX carbonitrides (M = V, Nb, Cr; X = C, N) and large intergranular Cr-rich $M_{23}C_6$ carbides. In contrast, during subsequent long-term ageing at 625°C the additional precipitation of newly formed secondary phase particles has been revealed, namely, the $(Fe,Cr)_2Mo$ based Laves phase, Cr(V,Nb)N type Z-phase, and Nb-rich MC carbides. The occurrence of all these secondary phases qualitatively agrees well with performed equilibrium thermodynamic calculations.

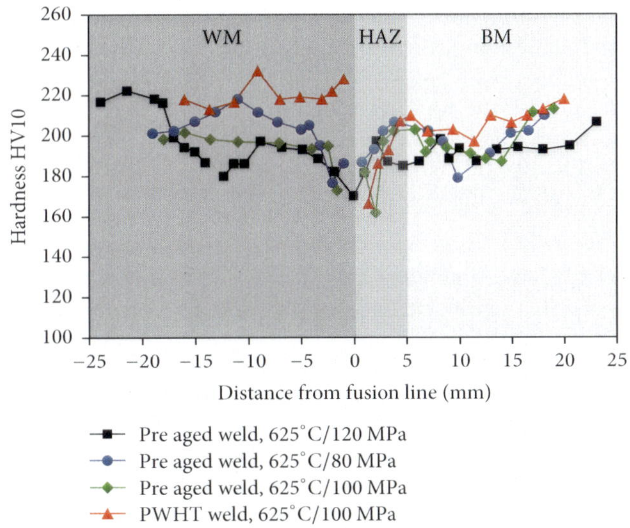

FIGURE 17: Cross-weld hardness profiles of several creep-exposed CB2 weldments.

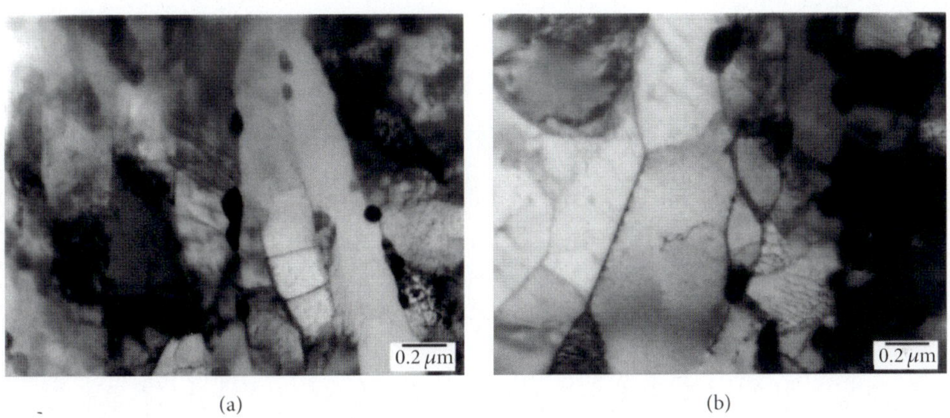

FIGURE 18: Bright field TEM images of CB2 BM substructure in initial PWHT state (a) and after the creep test at 625°C, 80 MPa for 18864 h of the preaged (625°C/30 kh) material (b).

(v) In comparison to the PWHT state, experimentally determined mean chemical composition of individual secondary phases indicated their clear compositional variations in both ageing and calculated equilibrium condition. These changes were related to the increase of chromium content in $M_{23}C_6$ carbides at the expense of other alloying elements. The Z-phase created at the expense of MX particles exhibited notably higher vanadium content in the aged condition compared to thermodynamic prediction. The experimental composition of Nb-rich MC carbides in the aged condition exhibited much lower niobium content compared to equilibrium calculation. In contrast, the chemical composition of Laves phase was found to be very similar to theoretically calculated composition. Thus all these findings can likely be related to different precipitation kinetics and evolution of individual secondary phases.

(vi) The obtained creep rupture data indicated gradual increase of creep rupture time for all the creep tested CB2 materials (i.e., the preaged CB2 BM as well as the CB2 welded joints in both the PWHT and preaged conditions) with decreasing applied stress. The CB2 BM exhibited the longest creep rupture lives for all the applied stresses, whereas both welded joints exhibited worse creep life performance.

(vii) For both the PWHT processed and the preaged welded joints, the failure mode transitions were observed. At the applied stress of 120 MPa, the creep fracture occurred in WM, whereas at lower stresses the failure occurred in FZ at the WM/HAZ interface. The observed failure mode transition may likely be related to several factors such as stress-dependent creep deformation mechanism transition, different propensity of individual weld regions for creep deformation localization, and strain-enhanced substructural coarsening and alloying elements redistribution between BM and WM indicated by performed microstructural and fracture path observations.

(viii) It can be generally stated that short-term creep rupture tests performed at the highest applied stress exhibited high deformation failures, whereas rapid plasticity decrease was observed for the long-term creep tests performed at lower applied stresses. This observation is believed to be directly related to the same aforementioned factors affecting the observed stress-dependent failure mode changes.

(ix) Cross-weld hardness profiles exhibited notable scattering of hardness values due to significant microstructural heterogeneity of studied weldment. However, the hardness profile measurements were revealed to be a suitable tool for the indication of local areas with enhanced creep degradation within creep-exposed weldments. The major fracture mechanisms revealed in present study corresponded well with observed microstructure degradation and were mainly related to transgranular quasi cleavage and intergranular creep cracking in impact and creep testing conditions, respectively.

Competing Interests

The authors declare that they have no competing interests.

Acknowledgments

The present work has been supported by the Slovak National Scientific Grant Agency VEGA under Grant nos. 2/0151/16 and 2/0153/15, International Collaborative Project COST Action 536, and also Bilateral Project SAV-AVCR 15-11. Thanks are also due to Dr. A. Výrostková (IMR SAS retired, Košice, Slovakia) for the TEM replicas analyses, fractography, and helpful discussions.

References

[1] F. Abe, M. Tabuchi, H. Semba et al., "Feasibility of MARBN steel for application to thick section boiler components in USC power plant at 650∘C," in *Proceedings of the 5th International Conference on Advances in Materials Technology for Fossil Power Plants*, pp. 92–106, Marco Island, Fla, USA, October 2007.

[2] S. S. M. Tavares, J. M. Pardal, G. C. Souza et al., "Study of cracks in the weld metal joint of p91 steel of a superheater steam pipe," *Engineering Failure Analysis*, vol. 56, pp. 464–473, 2015.

[3] C. Pandey, A. Giri, and M. M. Mahapatra, "Effect of normalizing temperature on microstructural stability and mechanical properties of creep strength enhanced ferritic P91 steel," *Materials Science and Engineering A*, vol. 657, pp. 173–184, 2016.

[4] M. Cieśla and G. Junak, "The influence of load history on durability of P92 steel used for the construction of energy pipelines," *Archives of Metallurgy and Materials*, vol. 60, no. 3, pp. 1853–1858, 2015.

[5] L. Falat, L. Čiripová, J. Kepič, J. Buršík, and I. Podstranská, "Correlation between microstructure and creep performance of martensitic/austenitic transition weldment in dependence of its post-weld heat treatment," *Engineering Failure Analysis*, vol. 40, pp. 141–152, 2014.

[6] L. Falat, J. Kepič, L. Čiripová, P. Ševc, and I. Dlouhý, "The effects of postweld heat treatment and isothermal aging on T92 steel heat-affected zone mechanical properties of T92/TP316H dissimilar weldments," *Journal of Materials Research*, vol. 31, no. 10, pp. 1532–1543, 2016.

[7] F. Abe, M. Tabuchi, S. Tsukamoto, and T. Shirane, "Microstructure evolution in HAZ and suppression of Type IV fracture in advanced ferritic power plant steels," *International Journal of Pressure Vessels and Piping*, vol. 87, no. 11, pp. 598–604, 2010.

[8] COST Action 536 International Collaborative Project, *Alloy Development for Critical Components of Environmentally Friendly Power Plant*, Project Head: Professor John Hald, Technical University of Denmark, Lyngby, Denmark, 2011.

[9] L. Falat, A. Výrostková, J. Kepič, and L. Čiripová, "The influence of isothermal annealing on degradation of mechanical properties of homogeneous weldment of the 9Cr-Mo steel," *Chemicke Listy*, vol. 106, no. 3, pp. s405–s406, 2012.

[10] A. Výrostková, L. Falat, J. Kepic, P. Brziak, and J. Pecha, "Microstructure and fracture of 9%CR-MO-CO-B steel (CB2) weldment after isothermal ageing," in *Proceedings of the 19th International Conference on Metallurgy and Materials (METAL '10)*, pp. 409–414, Rožnov pod Radhoštěm, Czech Republic, May 2010.

[11] http://www.thermocalc.com/.

[12] A. Kroupa, J. Havránková, M. Coufalová, M. Svoboda, and J. Vřešťál, "Phase diagram in the iron-rich corner of the Fe-Cr-Mo-V-C system below 1000 K," *Journal of Phase Equilibria*, vol. 22, no. 3, pp. 312–323, 2001.

[13] A. Zieliński, G. Golański, and M. Sroka, "Assessment of microstructure stability and mechanical properties of X10CrWMoVNb9-2 (P92) steel after long-term thermal ageing for high-temperature applications," *Kovove Materialy*, vol. 54, no. 1, pp. 61–70, 2016.

[14] V. Sklenička, K. Kuchařová, M. Svobodová, M. Kvapilová, P. Král, and L. Horváth, "Creep properties in similar weld joint of a thick-walled P92 steel pipe," *Materials Characterization*, vol. 119, pp. 1–12, 2016.

[15] D. Meng, F. Lu, H. Cui, Y. Ding, X. Tang, and X. Huo, "Investigation on creep behavior of welded joint of advanced 9%Cr steels," *Journal of Materials Research*, vol. 30, no. 2, pp. 197–205, 2015.

[16] D. Jandová, J. Kasl, and E. Chvostová, "Microstructure of CB2 steel before and after long-term creep tests," *Materials Science Forum*, vol. 782, pp. 311–318, 2014.

[17] C. G. Panait, W. Bendick, A. Fuchsmann, A.-F. Gourgues-Lorenzon, and J. Besson, "Study of the microstructure of the Grade 91 steel after more than 100,000 h of creep exposure at 600∘C," *International Journal of Pressure Vessels and Piping*, vol. 87, no. 6, pp. 326–335, 2010.

[18] R. Agamennone, W. Blum, C. Gupta, and J. K. Chakravartty, "Evolution of microstructure and deformation resistance in creep of tempered martensitic 9–12%Cr–2%W–5%Co steels," *Acta Materialia*, vol. 54, no. 11, pp. 3003–3014, 2006.

[19] L. Cipolla, H. K. Danielsen, P. E. Di Nunzio, D. Venditti, J. Hald, and M. A. J. Somers, "On the role of Nb in Z-phase formation in a 12% Cr steel," *Scripta Materialia*, vol. 63, no. 3, pp. 324–327, 2010.

[20] K. Sawada, H. Kushima, K. Kimura, and M. Tabuchi, "Z-phase formation and its effect on long-term creep strength in 9–12%Cr creep resistant steels," *Transactions of the Indian Institute of Metals*, vol. 63, no. 2-3, pp. 117–122, 2010.

[21] F. Abe, "Heat-to-heat variation in long-term creep strength of some ferritic steels," *International Journal of Pressure Vessels and Piping*, vol. 87, no. 6, pp. 310–318, 2010.

[22] P. Mohyla, Z. Kubon, R. Cep, A. Janasek, and I. Samardzic, "Secondary hardening of low-alloyed creep-resistant steel welds," *Metalurgija*, vol. 53, no. 1, pp. 25–28, 2014.

[23] K. Sawada, H. Kushima, T. Hara, M. Tabuchi, and K. Kimura, "Heat-to-heat variation of creep strength and long-term stability of microstructure in Grade 91 steels," *Materials Science and Engineering A*, vol. 597, pp. 164–170, 2014.

[24] M. Buršák, J. Miche', J. Janek, and M. Vojtko, "The contribution to apraisal of residual life cycle of creep resistant low alloy steels," *Chemicke Listy*, vol. 105, no. 16, pp. s621–s623, 2011.

[25] P. Mayr and H. Cerjak, "The impact of welding on the creep properties of advanced 9–12% Cr steels," *Transactions of the Indian Institute of Metals*, vol. 63, no. 2-3, pp. 131–136, 2010.

[26] F. Abe, M. Tabuchi, and S. Tsukamoto, "Metallurgy of type IV fracture in advanced ferritic power plant steels," *Materials at High Temperatures*, vol. 28, no. 2, pp. 85–94, 2011.

[27] L. Kloc and V. Sklenička, "Transition from power-law to viscous creep behaviour of P-91 type heat-resistant steel," *Materials Science and Engineering A*, vol. 234–236, pp. 962–965, 1997.

[28] L. Falat, V. Homolová, J. Kepič, M. Svoboda, and A. Výrostková, "Microstructure and properties degradation of P/T 91, 92 steels weldments in creep conditions," *Journal of Mining and Metallurgy B: Metallurgy*, vol. 48, no. 3, pp. 461–469, 2012.

[29] B. K. Choudhary, "Tertiary creep behaviour of 9Cr-1Mo ferritic steel," *Materials Science and Engineering A*, vol. 585, pp. 1–9, 2013.

[30] H. J. Schüller, L. Hagn, and A. Woitscheck, "Risse im Schweißnahtbereich von Formstücken aus Heißdampfleitungen-Werkstoffuntersuchungen," *Der Maschinenschaden*, vol. 47, no. 1, pp. 1–13, 1974.

[31] M. Schuler, S. Baumgartner, R. Schnitzer, and N. Enzinger, "Creep investigation and simulation of CB2 joints using similar rutile CB2 flux-cored wire," *Welding in the World*, vol. 58, no. 6, pp. 903–913, 2014.

8

A Bayesian Approach for Estimating the Thinning Corrosion Rate of Steel Heat Exchanger in Hydrodesulfurization Plants

J. C. Velázquez,[1,2] F. Caleyo,[3] R. Cabrera-Sierra,[1] G. Teran,[1] E. Hernandez-Sanchez,[4] S. Capula-Colindres,[1] H. Herrera-Hernández,[5] and C. C. Ortiz-Herrera[1]

[1]*Departamento de Ingeniería Química Industrial, ESIQIE, Instituto Politécnico Nacional, Zacatenco, México City 07738, Mexico*
[2]*Gerencia de Ingeniería y Costos, Pemex Transformación Industrial, Piso 6, Edificio B1, Centro Administrativo Pemex, Marina Nacional 329, Col. Petróleos Mexicanos, México City 11311, Mexico*
[3]*Departamento de Ingeniería Metalúrgica, ESIQIE, Instituto Politécnico Nacional, Zacatenco, México City 07738, Mexico*
[4]*Instituto Politécnico Nacional, UPIBI, Avenida Acueducto s/n Barrio La Laguna Ticomán, México City 07340, Mexico*
[5]*Universidad Autónoma del Estado de México, IIN-Lab. de Electroquímica Aplicada y Corrosión de Materiales Industriales, Blvd. Universitario s/n, Atizapán de Zaragoza, Estado de México 54500, Mexico*

Correspondence should be addressed to J. C. Velázquez; jcva8008@yahoo.com.mx

Academic Editor: Marco Cannas

Fuel consumption has been increasing in recent years, especially that of diesel and jet fuel. For this reason, the necessity to build new plants to reduce their sulfur content has arisen. Sometimes, just revamping existing plants is feasible, but determining which pieces of equipment are in the appropriate condition to be reused is also necessary. In order to select the equipment, it is essential to have information about the wall thickness of vessels. Sometimes, the information is limited; consequently, the application of advanced statistical techniques is needed. The Bayesian Data Analysis (BDA) used in this study has the goal of determining a more accurate, unobserved thinning rate distribution for existing heat exchangers, taking into consideration all the information available about the thinning rate of the heat exchangers that cool down the effluent of the hydrotreating reactors in Mexican oil refineries. The information obtained from BDA was compared with existing shell wall thickness obtaining favorable results.

1. Introduction

Middle distillates (e.g., jet fuel, diesel, or kerosene) are widely used as fuel in motor vehicles (e.g., cars, buses, and trucks), in airplanes, and in some industrial machines (e.g., locomotives, ships, and farm equipment) [1]. Diesel engines are about 40% more fuel-efficient than comparable gasoline engines [2]. Nevertheless, they suffer from associated contaminants, such as NO_x and SO_x emissions, that are dangerous to human health. Sulfur, a natural part of crude oil from which diesel fuel is derived, is one of the key causes of particulates, or soot, in diesel [3]. In Mexico, the environmental regulation [4] has established a maximum limit of 15 wppm in total sulfur content for middle distillates. To produce middle distillates with low sulfur content, Pemex (the Mexican state-owned petroleum company) decided to launch the ULSD project (Proyecto DUBA in Spanish) in all Mexican oil refineries. This project consists of building new units and revamping others. To decide which units are adequate for revamping, it is necessary to estimate the cost. This estimate has to include the new equipment being added because of the modernization of the process and because of the expiration of the vessels. In this sense, having the capacity to accurately estimate the vessels' remaining life is important. In the middle distillates' hydrotreating units, the heat exchangers that preheat the feed to the reactor using the effluent stream from the same reactor are the vessels that undergo the fastest deterioration. In the hydrodesulfurization process, these heat exchangers can suffer deterioration from atmospheric corrosion (externally, when the coating cannot completely protect the external surface) and by ammonium bisulfide corrosion (internally, when ammonium salts are formed) [5–7].

The hydrodesulfurization and hydrodenitrogenation are carried out through the following reactions [8–14]:

$$\text{R-SH} + H_2 \longrightarrow \text{R-H} + H_2S_{(g)} \quad (1)$$

$$\text{R-N} + 2H_2 \longrightarrow \text{R-H} + NH_{3(g)} \quad (2)$$

In addition to hydrogen sulfide and ammonia, water vapor, in the form of CO_2 and CO, is also produced [8–14]:

$$2R'\text{-}R''\text{-COOH} + 2H_2 \longrightarrow 2R'\text{-H} + 2R''\text{-H} + H_2O_{(g)} + CO_{2(g)} + CO_{(g)} \quad (3)$$

The environment also induces HCl production if the middle distillate contains organic salts [8–14]:

$$\text{R-Cl} + H_{2(g)} \longrightarrow \text{R-H} + HCl_{(g)} \quad (4)$$

According to the chemical reactions described above, the gaseous stream that flows out of the reactor contained H_2S, NH_3, and HCl, which can precipitate acid salts such as NH_4Cl and NH_4HS, which cause corrosion in heat exchangers.

According to the paper published in 2016 by NACE titled "International Measures of Prevention, Application, and Economics of Corrosion Technologies Study" [15], the global cost of corrosion is estimated to be 3.4% of the global GDP. This study also indicates that the cost of corrosion tends to be higher in developing countries than in developed countries (Table 2-2 in Reference [15]) perhaps due to fewer specialists in this area and fewer resources focused on the prediction, monitoring, prevention, and maintenance of these structures.

In order to monitor the corrosion damage in vessels and pipes, ultrasonic inspection systems are used. These systems help measure the wall thickness in plates and pipes in the downstream oil industry sector. The devices used for the ultrasonic monitoring (UT) are sensitive to errors, such as time of flight, variation of speed of sound (test block of similar material), material grade, material heat treatment, and operational temperature of the equipment or pipe [16]. These errors in the digital thickness gauge reading, in addition to other human errors, can affect the accuracy and the repeatability of the results. Measurement errors are random, additive, independent, and identically distributed [17, 18]. While considering that the bias of the measurement is very close to zero and assuming sizing uncertainty of the magnitude of what is measured, it means that the dimensions of the corrosion defect are measured with an error independent of their values; measurement error could be described using a normal distribution with a mean of zero and variance (σ_{ME}). Mathematically, it can be written as follows:

$$f_{ME}(\mu_{ME}, \sigma_{ME}) \sim N(0, \sigma_{ME}). \quad (5)$$

Quantifying these kinds of errors shall be important if one wants to estimate the remaining life of the pipes and vessels more accurately and reliably. For example, the wall thickness sometimes seems to be increasing over time, something which is completely absurd, but this is the result of the inclusion of errors in the measurement process. Besides the measurement errors, another problem that presents itself in the examination of the vessels is related to the sampling inspection. Usually, it is not possible to monitor, or, at least, to measure, the wall thickness for the entire vessel, so, consequently, only some points of the vessels are measured. The selected points are established by the owner or by a mutual agreement between the owner and the inspector. Bear in mind that sampling inspection is a concept used in quality control and, according to ISO-2859 "Sampling Procedures for inspection by attributes" [19], sampling inspection requires the producer to submit lots, products, or raw materials at a quality which is at or better than a mutually agreed level. Applying this concept to mechanical integrity, it means that the company that makes the inspection must execute this task with a certain degree of confidence (in agreement with the owner). However, for the owner, the problem is determining the correct degree of confidence in the inspection to get the best technical-economical benefit. In summary, getting a very high confidence level in the inspection could be very expensive and very difficult because, in addition to having the correct device with high resolution and a very highly qualified inspector, it is sometimes also necessary to shut down a plant to perform a good inspection. Usually, in order to avoid shutting down a process unit, the owner decides to make a sampling inspection of a vessel. It means that partially inspecting a vessel can be enough to estimate, with some confidence, its remaining lifespan. In order to establish which part of the vessel needs to be monitored, it is necessary to determine the points most susceptible to corrosion. The API-510 Standard "Pressure Vessel Inspection Code: In-Service Inspection, Rating, Repair and Alteration" [20] named the points to be measured in vessels "Condition Monitoring Locations (CMLs)." In short, the point is that, with limited data, it is possible to estimate the corrosion rate of the equipment. Bayes' theorem is a useful tool for estimating the corrosion rate when it is not possible to collect a large amount of data with its variability. For this reason, in this study, a methodology is presented to estimate the corrosion rate in shell and tube heat exchangers that preheat the feed of the reactor in a hydrotreating unit using Bayes' rule, and the information collected in some HDS plants was distributed in all Mexican oil refineries.

2. Theoretical Framework

2.1. Bayes' Theorem and the Application in Predictions. The classic methods of estimation are based on information that gives a random sample. These methods read into probabilities like relative frequencies [21]. Another estimation method is based on the Bayesian methodology. The purpose of this method comes from the Bayes' rule [21–23]:

$$\pi(\boldsymbol{\theta} \mid \boldsymbol{x}) = \frac{f(\mathbf{x} \mid \boldsymbol{\theta})\pi(\boldsymbol{\theta})}{g(\mathbf{x})}, \quad (6)$$

where

(i) $\pi(\boldsymbol{\theta} \mid \mathbf{x})$ denotes the posterior probability of parameters $\boldsymbol{\theta}$ after taking into account the observed data $\boldsymbol{x}$.

(ii) $\mathbf{x}$ corresponds to the vector of the new data observed.

(iii) $\pi(\boldsymbol{\theta})$is the prior probability distribution. It is the probability distribution of $\boldsymbol{\theta}$ parameters before x, and it is described by the vector of hyperparameters $\boldsymbol{\alpha}$.

(iv) $\boldsymbol{\alpha}$ is the vector of hyperparameters, mathematically expressed as $\boldsymbol{\theta} \sim p(\boldsymbol{\theta} \mid \boldsymbol{\alpha})$.

(v) $f(\mathbf{x} \mid \boldsymbol{\theta})$ is the probability of observing $\mathbf{x}$, given $\boldsymbol{\theta}$. This is the likelihood function.

(vi) $g(\mathbf{x})$ is the marginal distribution of $\mathbf{x}$. This is a normalization factor that symbolizes the evidence and represents the probability that the data follows the chosen model under marginalization over all the parameter values, mathematically denoted as follows: $g(\mathbf{x}) = \int_{\boldsymbol{\theta}} p(\mathbf{x} \mid \boldsymbol{\theta}) p(\boldsymbol{\theta} \mid \boldsymbol{\alpha}) d\boldsymbol{\theta}$.

This posterior distribution $\pi(\boldsymbol{\theta} \mid \mathbf{x})$ is a conditional probability conditioned on randomly observed data. Therefore, it is considered a random variable because the true value of $\boldsymbol{\theta}$ is uncertain. Nonetheless, this posterior distribution can rarely be used directly in engineering applications. In order to apply this distribution, it is necessary to obtain the posterior predictive distribution, which denotes the new unobserved set of data points $\tilde{x}$ given a set of existing observations $(\mathbf{x})$ and the hyperparameters $\boldsymbol{\alpha}$; thus, it reflects how the new data behave. If the data can be expected to have a distribution $M(x)$, the posterior predictive distribution is represented by the following expression [23, 24]:

$$\pi_p(\tilde{x} \mid \mathbf{x}; \boldsymbol{\alpha}) = \int_{\boldsymbol{\theta}} M(x \mid \boldsymbol{\theta}) \pi(\boldsymbol{\theta} \mid \mathbf{x}). \quad (7)$$

3. Bayes' Theorem on the Estimation of the Remaining Life in Heat Exchangers

After the explanation of the mathematical background described in Section 2, the narrative of its application is necessary. The first step in applying the Bayesian inference was to determine the prior distribution that denotes the beliefs about the variable before some evidence is taken into account. To propose this prior distribution, the wall thickness loss rate (WTLR) data was obtained from the low alloy steel shells of heat exchangers present in all the middle distillate units in Pemex refineries. For this criterion, it was taken into account that the characteristics of the middle distillates are roughly the same. Likewise, the shells of all heat exchangers studied are made from low alloy steel (Material Specification ASTM SA 387.GR-5 CL-1). All these vessels are also exposed to atmospheric corrosion inside the refineries, which means that the shells of the heat exchangers studied are exposed to both external corrosion and internal corrosion (usually ammonium bisulfide corrosion). The wall thickness was monitored by ultrasonic techniques in order to estimate the thinning. The wall thickness loss rate (WTLR) data were acquired from 38 shells of heat exchangers with at least 30 observation points for each one. For each shell analyzed, the thinning rate was fitted to a Generalized Extreme Value Distribution (GEVD); mathematical expression (8) describes the cumulative distribution function for GEVD [25]:

$$\mathrm{F}(x; \mathrm{a}, \mathrm{b}, \xi) = \mathrm{Exp}\left\{-\left[1 + \xi\left(\frac{x-a}{b}\right)\right]^{-(1/\xi)}\right\}, \quad (8)$$

where x is the random variable and is the wall thickness loss rate and a, b, and ξ are the location parameter, scale parameter, and shape parameter, respectively. This probability function has already been applied successfully in corrosion topics [26–33]. In order to perform a statistical analysis, the availability of data are necessary, but this information is often too scarce for an estimation of the vessels' remaining life. For example, in the case of pipelines, field studies with data regarding low-carbon steel pipeline corrosion in soil were published in 2010 by Velázquez et al. [32]. Velázquez et al. in 2010 [33], in another study, showed a field survey that helped analyze the distribution of the number of corrosion defects and their size (depth and length) in oil and gas pipelines [17]. For the sake of contributing to the corrosion field information accessible in the literature, Table 1 shows details of the wall thickness thinning rate data obtained for the 38 heat exchangers studied after fitting each one to a GEV distribution. These details describe the location, scale, and shape parameters of the fitted GEV distribution; the number of observations or number of thinning rates obtained from each heat exchanger; and the p value returned by the Kolmogorov–Smirnov test [34]. This p value is an indicator of the acceptance degree for the GEV distribution that represents the observed data [34]. It means that, in this case, GEV distribution that fits the observed data with low probability of error can be accepted if confidence levels are higher than 0.05 (p value higher or equal than 0.05) [34]. For the sake of schematization, Figure 1 illustrates the thinning rate data from heat exchanger number 36 (entry 36) and its fitting.

A few reports on corrosion studies of carbon steel immersed in ammonium hydrosulfide solutions had been reported in the literature [35]. In the former, the electrochemical behavior of the carbon steel immersed in 0.1 M $(NH_4)_2S$ had been characterized using linear polarization and impedance spectroscopy techniques. Based on these findings, the corrosion current density and corrosion rate of the steel can be estimated [35] using the following equations [36]:

$$j_{\mathrm{corr}} = \frac{\beta_{\mathrm{a}} \times \beta_{\mathrm{c}}}{2.303 \times (\beta_{\mathrm{a}} + \beta_{\mathrm{c}})} \times \frac{1}{R_{\mathrm{p}}},$$
$$\mathrm{CR} = \frac{k \times j_{\mathrm{corr}} \times \mathrm{EW}}{\rho}, \quad (9)$$

where j_{corr} is the corrosion current density, $\mu A/cm^2$; β_{a} and β_{c} are the slopes anodic and cathodic Tafel reactions when plotted on base 10 logarithmic, V/decade; R_{p} is the true polarization resistance, $\Omega \cdot cm^2$; k is a constant value depending of the units, 3.27×10^{-3} mm·g/μA·cm·yr; EW is the equivalent weight, 27.92; ρ is the density of the carbon steel, 7.86 g/cm^3.

Based on the linear polarization plots (Figure 1) reported in the literature [37], β_{a} and β_{c} parameters were determined

Table 1: Generalized Extreme Value Distribution parameters after fitting wall thickness thinning rate data, number of observations, and p value returned by K–S test from thirty-eight low alloy steel shell heat exchangers used in middle distillate units to preheat the feed with the effluent from the reactor.

Entry	a Location parameter (mm/year)	b Scale parameter (mm/year)	ξ Shape parameter	Number of observations	p value returned by K–S test
1	0.0584	0.0432	0.333	38	0.67
2	0.0488	0.0292	0.287	39	0.51
3	0.2112	0.1399	0.321	38	0.32
4	0.1622	0.1213	0.124	35	0.69
5	0.4468	0.2578	−0.095	33	0.34
6	0.2911	0.2311	0.136	32	0.41
7	0.4706	0.2294	−0.535	40	0.22
8	0.4745	0.2408	−0.028	41	0.21
9	0.0536	0.0424	0.441	37	0.40
10	0.2285	0.1781	0.168	39	0.53
11	0.2465	0.1781	0.136	38	0.47
12	0.3800	0.2697	−0.129	35	0.32
13	0.0758	0.0667	0.425	34	0.78
14	0.2535	0.2032	0.262	33	0.24
15	0.0999	0.0785	0.189	31	0.54
16	0.0755	0.0537	0.246	33	0.36
17	0.1008	0.0770	0.179	40	0.12
18	0.0463	0.0273	0.131	43	0.09
19	0.0512	0.0350	0.383	37	0.18
20	0.2265	0.1788	0.158	38	0.43
21	0.1094	0.0785	0.234	38	0.58
22	0.1178	0.1018	0.226	37	0.32
23	0.1046	0.1349	0.549	40	0.18
24	0.2379	0.1529	−0.013	41	0.11
25	0.1237	0.1195	0.131	33	0.72
26	0.4305	0.5093	0.243	31	0.81
27	0.1014	0.1018	0.139	40	0.11
28	0.2219	0.2295	0.410	37	0.23
29	0.0658	0.0524	0.388	35	0.69
30	0.1264	0.1271	0.345	37	0.51
31	0.2693	0.0422	0.133	38	0.43
32	0.3160	0.0625	0.184	38	0.32
33	0.1671	0.0760	0.162	32	0.28
34	0.1692	0.0994	0.062	31	0.88
35	0.1740	0.0501	0.199	40	0.11
36	0.1316	0.0346	0.386	47	0.17
37	0.1301	0.0225	0.504	31	0.91
38	0.0674	0.0319	0.341	32	0.29

at 0 h of immersion being 0.1 and 0.12 V, respectively. Unlike these parameters, the resistance used for calculating the corrosion rate [37] should be related with the rate-controlling step in the oxidation of the steel evaluated by fitting the EIS diagrams using an equivalent circuit and Boukamp program. It is noteworthy the oxidation of the steel involves the formation of iron sulfides depending on the immersion time. According to the EIS analysis is suggested a charge transfer resistance and the presence of different diffusional process through the sulfide film, e.g., iron ions and atomic hydrogen [37], as well as the bisulfide ions favoring the growth of the sulfide film. In order to compare the corrosion rates determined in the field only is considered the R_2 contribution (2273 $\Omega{\cdot}cm^2$) at the beginning of the immersion (0 h) [37].

One of the great challenges in Bayesian inference is to find an adequate prior probability distribution. This distribution is obtained from past information and expresses the beliefs about the probability of the studied event before some evidence is taken into account. In this study, the prior distribution is proposed to be the GEV distribution, because in all the cases studied, it was fitted properly, and because, when all the data were fitted to a GEVD, it was also fitted with a low probability of rejection (Figure 2).

The probability distribution of each parameter is obtained from the distribution from all the GEV distribution parameters contained in Table 1. Figure 3 shows the results of fitting the histogram of these parameters. Analyzing these results, according to Figure 3, it is possible to state that the location parameter could be represented by a Generalized Pareto distribution scale parameter and the shape parameter could be represented by GEV distribution. The following equation represents the Generalized Pareto distribution that was already used in corrosion modeling [38]:

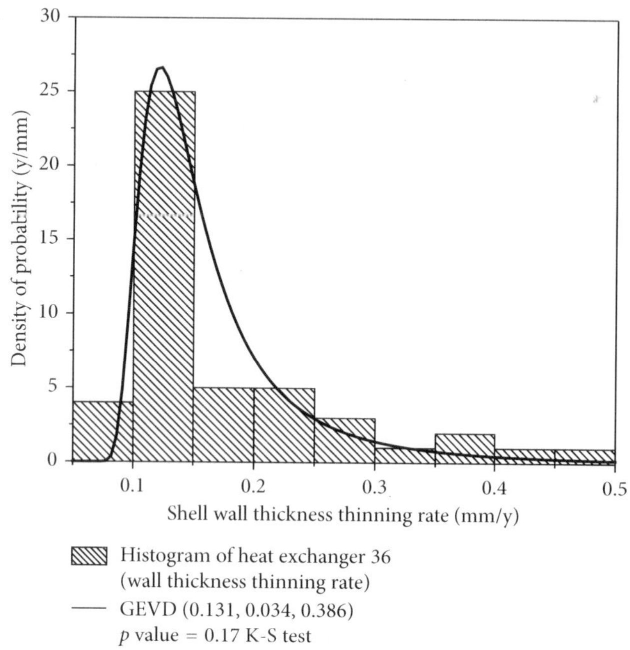

Figure 1: Wall thickness thinning rate of heat exchanger histogram and its fitting to GEVD (all the Probability Density Function (PDF) values plotted in this paper were adjusted in order for the distribution to "look like" the histogram. To do this, it is necessary to convert the PDF to the histogram's size using the bins' width and the number of bins. Reference [44] explains this process in detail).

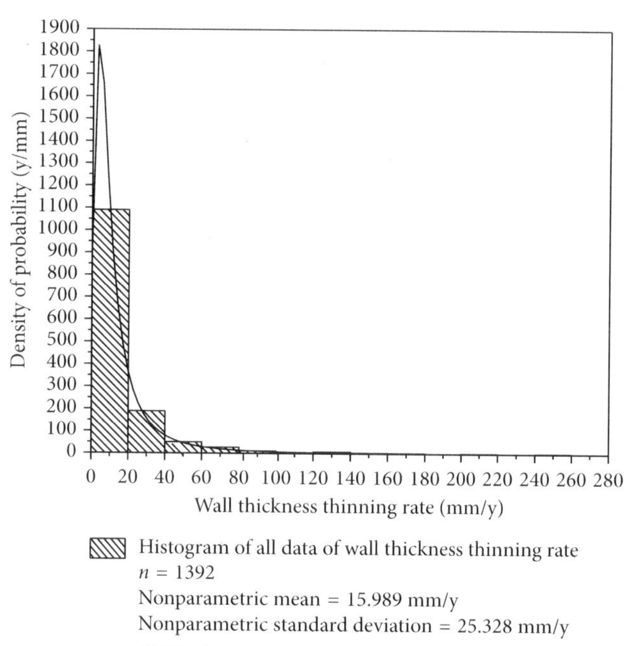

Figure 2: Wall Thickness Thinning Rate in all data for all heat exchanger analyzed (all the Probability Density Function (PDF) values plotted in this paper were adjusted in order for the distribution to "look like" the histogram. To do this, it is necessary to convert the PDF to the histogram's size using the bins' width and the number of bins. Reference [44] explains this process in detail).

$$g(x) = \frac{1}{\beta}\left(1 + \xi\frac{x-\alpha}{\beta}\right)^{-1-(1/\xi)}, \quad (10)$$

where α is the location parameter, β is the scale parameter, and ξ represents the shape parameter for Generalized Pareto distribution.

All these aforementioned parameters are called hyper-parameters because they describe the parameters of the prior distribution ($\boldsymbol{\theta} \sim p(\boldsymbol{\theta} \mid \boldsymbol{\alpha})$). In all cases, the theoretical distribution was chosen with high certainty (p value ≥ 0.05 in K–S test). Once the prior distribution is specified and the distributions of the parameters are also specified, the question about how to estimate the marginal distribution of $\mathbf{x}$ ($g(\mathbf{x})$ in Equation (2)) emerged. In this study, a numerical approximation is applied via a grid (Chapter 6 in Reference [24]), where a fine grid of the vector of parameters $\boldsymbol{\theta}$ is created. With this method, it is feasible to estimate the posterior distribution by defining the prior distribution through a grid of parameter values. In this method, it is not necessary to do any analytical integration; in other words, the denominator of Bayes' rule is converted in a sum over many discrete parameter values in lieu of an integral. Mathematical expression (11) summarizes how this numerical estimation could be done [18]:

$$\pi(\boldsymbol{\theta} \mid \mathbf{D}) = \frac{f(\mathbf{D} \mid \boldsymbol{\theta})\pi(\boldsymbol{\theta})}{\sum_{\boldsymbol{\theta}} f(\mathbf{D} \mid \boldsymbol{\theta})\pi(\boldsymbol{\theta})}. \quad (11)$$

The sum in the denominator correlates with the finite number of discrete values of $\boldsymbol{\theta}$, and $\mathbf{D}$ is the vector of observed data value.

The posterior distribution obtained is applied in order to get the posterior predictive distribution (Expression (3)), so that it will represent the probability of any possible value of the studied random variable. Applying the grid method, the predictive distribution can be obtained using Expression (12):

$$\pi_p^*(\tilde{x} \mid \mathbf{x}; \boldsymbol{\alpha}) = \sum_{\boldsymbol{\theta}} M^*(x \mid \boldsymbol{\theta})\pi^*(\boldsymbol{\theta} \mid \mathbf{x})\Delta\boldsymbol{\theta}, \quad (12)$$

where π_p^*, M^*, and π^* are the probability mass function (probability of a discrete random variable) counterparts of the corresponding PDF file. In this study, $M(x)$ is represented by the Generalized Extreme Value Distribution (Expression (4)) because it is expected that the new unobserved data follow this distribution. This expectation is supported by the fact that in all heat exchangers analyzed in this study, the GEVD is considered an acceptable distribution that represents the metal loss thinning rate. The same situation happens when all the data from all heat exchangers studied are used.

4. Applications and Results

For the sake of estimating the remaining life in heat exchangers using the wall thickness thinning rate obtained by the predictive distribution at 80% and 90% of probability values, some real exemplifications were carried out. Two

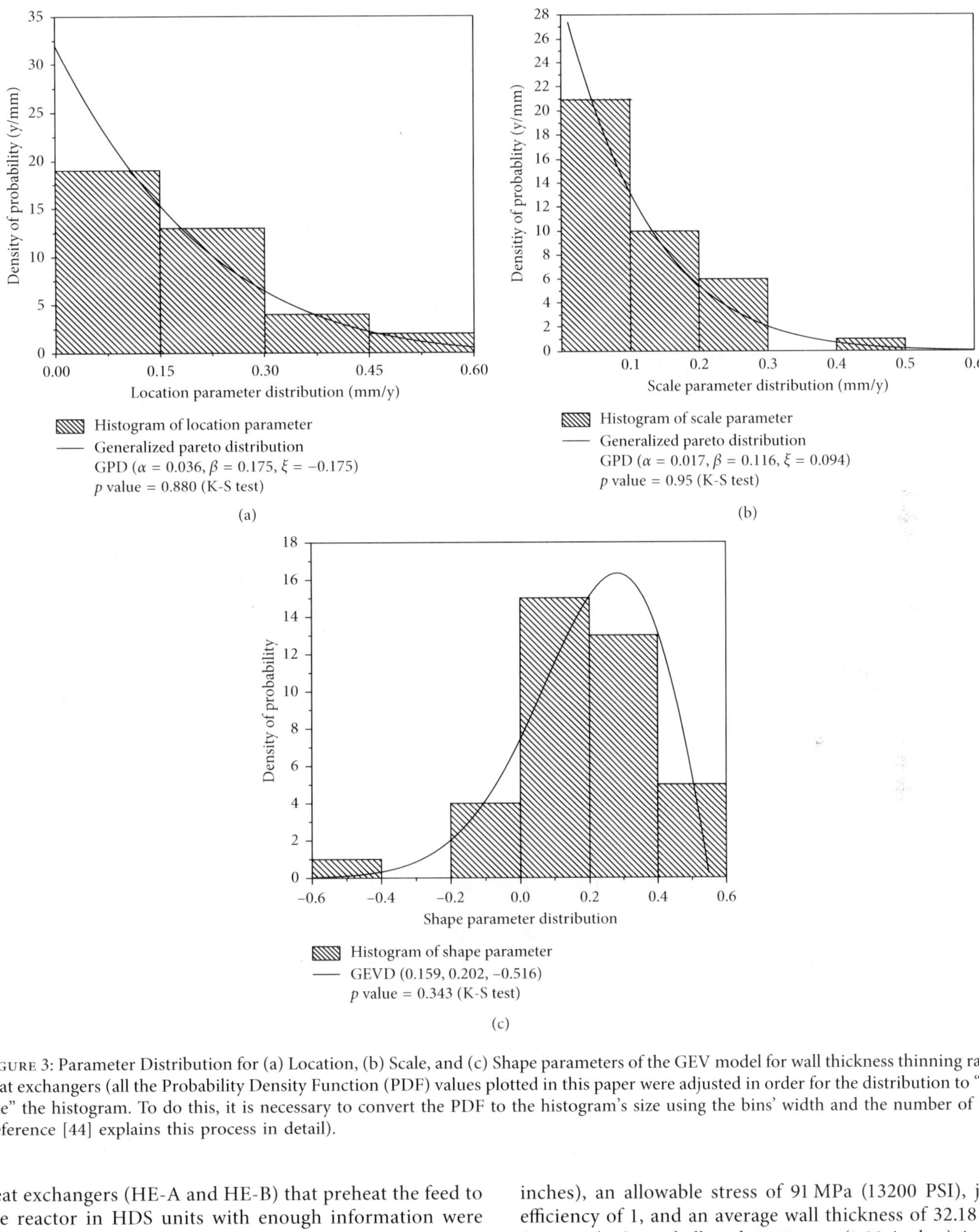

FIGURE 3: Parameter Distribution for (a) Location, (b) Scale, and (c) Shape parameters of the GEV model for wall thickness thinning rate in heat exchangers (all the Probability Density Function (PDF) values plotted in this paper were adjusted in order for the distribution to "look like" the histogram. To do this, it is necessary to convert the PDF to the histogram's size using the bins' width and the number of bins. Reference [44] explains this process in detail).

heat exchangers (HE-A and HE-B) that preheat the feed to the reactor in HDS units with enough information were used in the study in order to compare the results when it is considered that the thinning rate is constant over time (a practical common criteria in the industry when there is not enough information). The shells of those heat exchangers (HE-A and HE-B) are also made of low alloy steel (ASTM SA 387 GR-5 CL-1), with a diameter (D) of 889 mm (35 inches), an allowable stress of 91 MPa (13200 PSI), joint efficiency of 1, and an average wall thickness of 32.18 mm (1.26 inches) in shell and 27.71 mm (1.09 inches) in the elliptical heads. Expressions (13) and (14) allow determining the Maximum Allowable Operating Pressure (MAOP) [39, 40] for a safety operation in heat exchangers in the shell or cylinder (Expression (7)) and in the heads (Expression (8)) [41]:

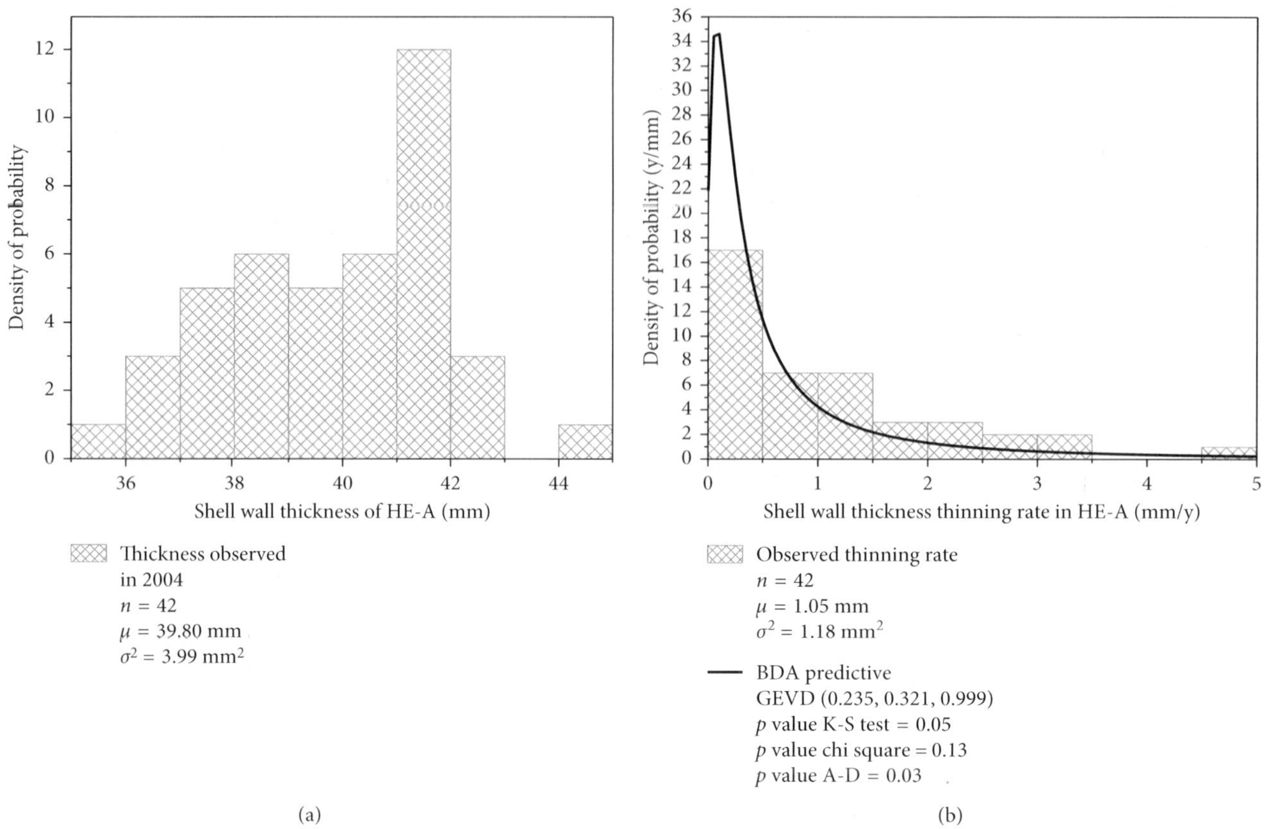

FIGURE 4: Wall thickness in HE-A in 2004 (a) and wall thickness thinning rate in HE-A in 2004 (b) (all the Probability Density Function (PDF) values plotted in this paper were adjusted in order for the distribution to "look like" the histogram. To do this, it is necessary to convert the PDF to the histogram's size using the bins' width and the number of bins. Reference [44] explains this process in detail).

$$\text{MAOP} = \frac{SEt}{R + 0.6t}, \quad (13)$$

$$\text{MAOP} = \frac{2SEt}{KD + 0.2t}, \quad (14)$$

In Equations (13) and (14), S represents the allowable stress, E is the joint efficiency, t is the wall thickness, R is the radius, D is the inside diameter, and $K = 0.66$ that is obtained from Table 1-4-1 of ASME Section VIII Div. 1 [36]. From all these variables, the wall thickness is the only variable that tends to change (by diminishing) over time. The allowable stress is influenced by the operational temperature of the vessel, but, in this study, it is considered constant, because the purpose is to study the effects of corrosion.

In order to begin applying the Bayesian approach proposed in this paper, it is necessary to have the wall thickness from a specific time and the thinning rate (corrosion rate). In this context, Figures 4 and 5 represent the wall thickness histogram and their corresponding thinning rate for HE-A and HE-B in a given time. In Figures 4(b) and 5(b), they also include distribution obtained after applying the Bayesian Data Analysis (BDA) from the thinning rate observed and their corresponding fitted parameter, evaluating the fitting by Kolmogorov–Smirnov test, chi-squared test, and Anderson–Darling test. In the two cases, at least two of the tests were obtained with enough acceptance.

In order to verify the reliability of the method proposed in this study, it is necessary to compare the results obtained with the wall thickness values observed later. For example, in the case of Heat Exchanger A (HE-A), the following inspection campaign was carried out in 2009. This campaign did not have the same number of Condition Monitoring Locations (CMLs) that were obtained in 2004. This situation happens often when a vessel is inspected. In this case, there were only 35 CMLs inspected, as opposed to the 42 CMLs used in 2004 inspection. With this information, it would be necessary to estimate the wall thickness in the future. In the oil industry, often the future wall thickness is calculated considering that the thinning rate is constant over time ($d = (t_2 - t_1)/\tau$), that is, the wall thickness difference divided by the interval time of the inspections. Figure 6 illustrates the shell wall thickness in HE-A in 2009.

In order to apply the Bayesian approach developed in this study, the shell wall thickness was estimated using the 80% and 90% ($\widehat{x}_{80\%}$ and $\widehat{x}_{90\%}$) probability values of the thinning rate distribution obtained. These values are shown in Table 2 for HE-A and HE-B.

For the sake of illustration, Figures 7 and 8 show the predictive shell wall thickness histograms using both 80% and 90% probability values of the predicted thinning rate ($\widehat{x}_{80\%}$ and $\widehat{x}_{90\%}$) and the comparison with the observed wall thickness.

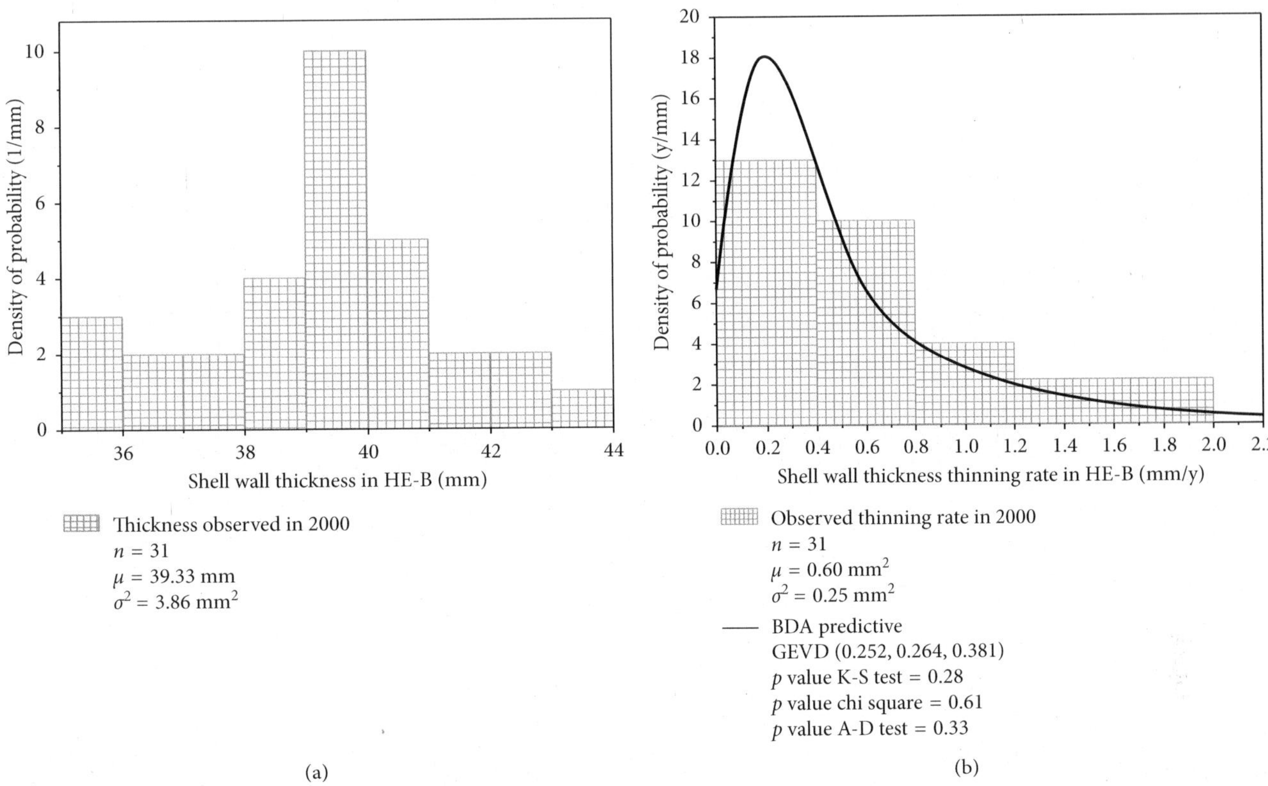

FIGURE 5: Wall thickness in HE-B in 2000 (a) and wall thickness thinning rate in HE-B in 2000 (b) (all the Probability Density Function (PDF) values plotted in this paper were adjusted in order for the distribution to "look like" the histogram. To do this, it is necessary to convert the PDF to the histogram's size using the bins' width and the number of bins. Reference [44] explains this process in detail).

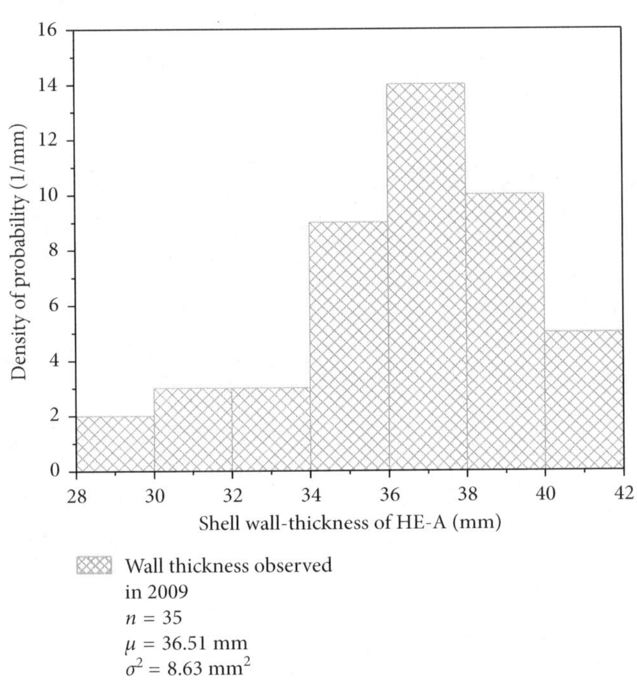

FIGURE 6: Wall Thickness observed for HE-A in 2009.

Using the results displayed in Figures 7 and 8, the same exercise can be repeated for longer periods of time for each heat exchanger. In addition to estimating the wall thickness

TABLE 2: GEVD parameters of the distributions obtained after the Bayesian approach in heat exchangers HE-A and HE-B.

Heat exchanger	GEVD parameters after applying the Bayesian approach			$\hat{x}_{80\%}$ (mm/y)	$\hat{x}_{90\%}$ (mm/y)
	a (mm/y)	b (mm/y)	ξ		
HE-A	0.235	0.321	0.999	1.351	2.956
HE-B	0.252	0.264	0.381	0.786	1.192

in the future with all the information available for a specific time, the methodology developed in this study was also applied to a sample with a smaller number of observations. This was done in order to determine if performing a random sampling of a reduced number of observations has a significant impact on the estimation of future wall thickness. The mean and variance of the error distribution percentage are used as a reference to establish the convenience of this methodology. The results of all these exercises are summarized in Tables 3 and 4.

In the previous tables, it is possible to observe that when the number of observations is reduced, the proposed method in this paper can still be implemented with satisfactory results. For example, the mean and variance of the error percentage are quite close to those obtained with all the observations available (Figures 7 and 8). This fact confirms that the purpose of developing a reliable statistical method

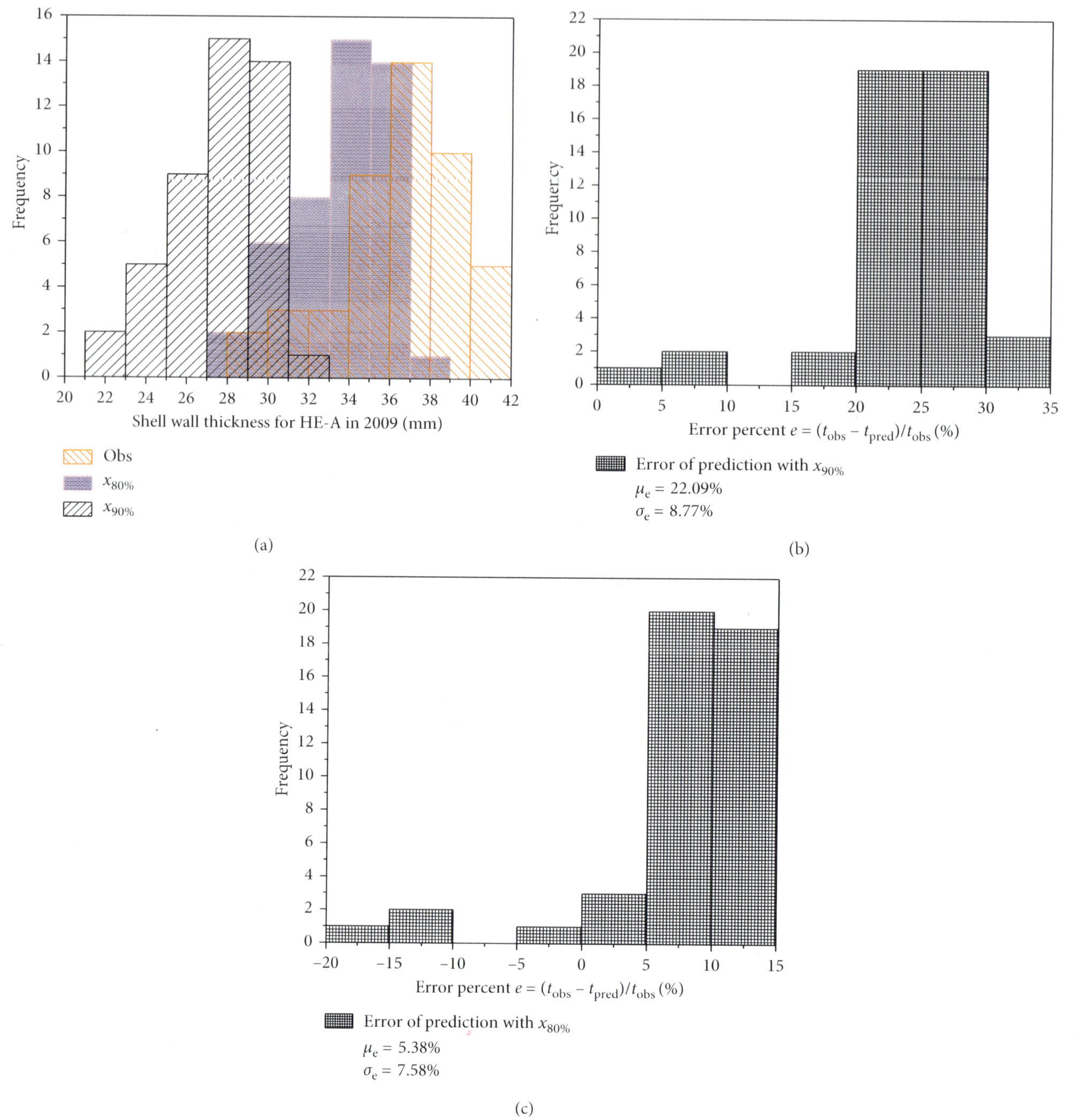

FIGURE 7: Histogram of the predicted wall thickness and the observations in HE-A (a) with the prediction error histogram using $\widehat{x}_{90\%}$ (b) and $\widehat{x}_{80\%}$ (c).

that can estimate the thinning rate with a relatively small quantity of information (at least twenty observation points) has been accomplished.

In Figures 7(a) and 8(a), one can notice that a greater thinning rate represented by the percentile 90 ($\widehat{x}_{90\%}$) motivates an overestimation of the damage causing a thinner wall thickness to be estimated. The bins that represent the wall thickness estimated with the percentile 90 are, on average, less than those represented by the bins that represent the wall thickness with the percentile 80 ($\widehat{x}_{80\%}$), which means that the remaining life is lower when the percentile 90 is used. In the oil and gas industry, it is common to use the corrosion rate average to predict the vessel's remaining life; however, the greater corrosion rates are the real threat to the pipes' and vessels' hermeticity. For that reason, using the percentiles 80 or 90 sounds more reasonable for the estimation of real exchangers than the average, which is close in value to the percentile 50. In our case, the best results were obtained using the percentile 80 of the thinning rate, and the mean values of the error (μ_e) in both cases (HE-A and HE-B) were closer to zero, even when the number of samples was reduced from 25 to 20 (Tables 3 and 4). The GEVD

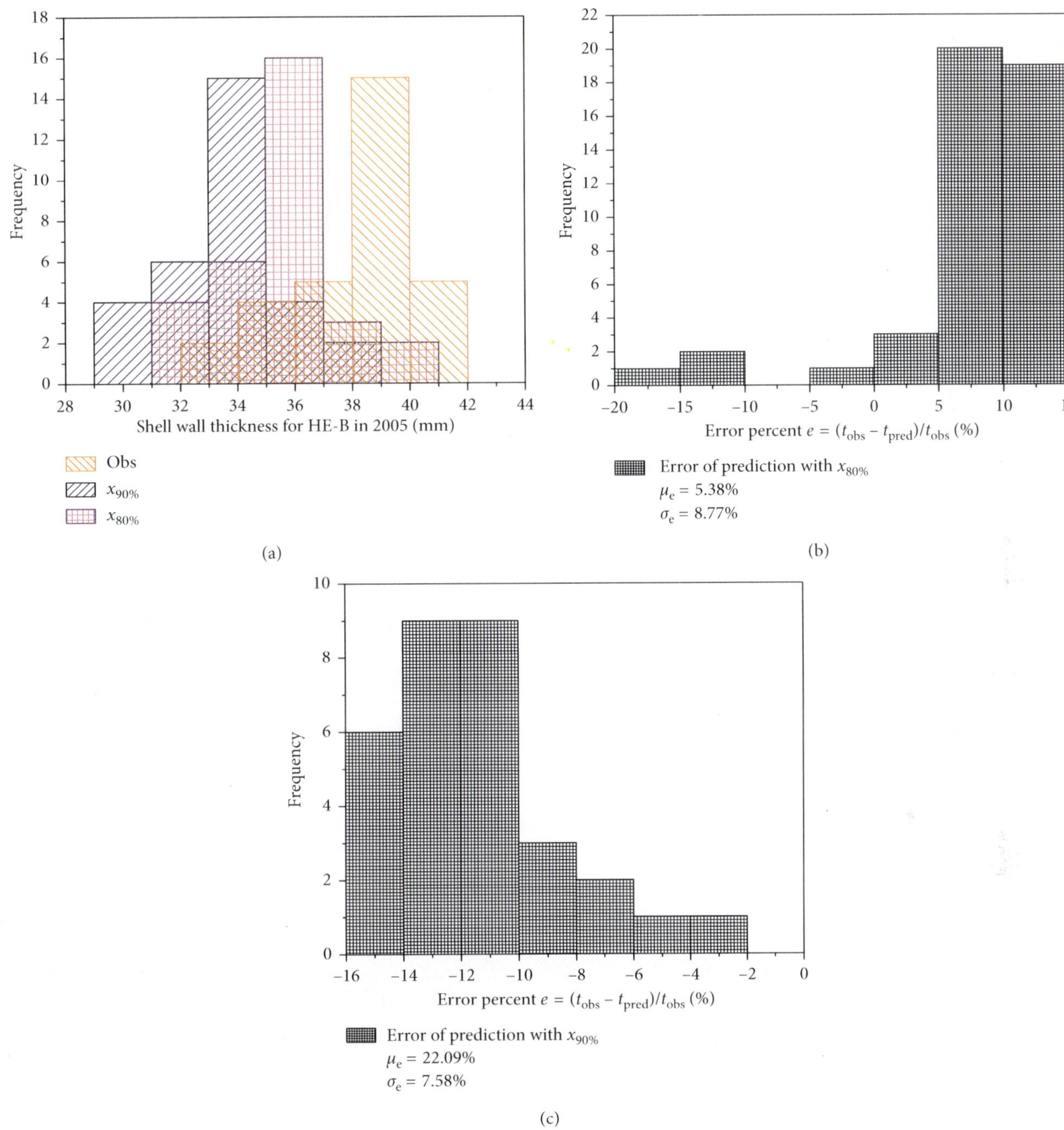

FIGURE 8: Histogram of the predicted wall thickness and the observed in HE-B (a) with the prediction error histogram using $\widehat{x}_{90\%}$ (b) and $\widehat{x}_{80\%}$ (c).

parameters obtained by the Bayesian approach ((a, b, ξ) in Tables 3 and 4 can be useful for estimating the remaining life of other heat exchangers that preheat a similar hydrocarbon mix.

Even though the oil and gas industry has developed several techniques to manage the materials' deterioration (in our case, thinning corrosion) such as the RBI (Risk-Based Inspection) methodology [42], which involved concepts of probability of failure, consequence of failure and risk of process, it does not take into consideration the sensitivity of the number of Condition Monitoring Locations (CML) that are available in a piece of equipment. This factor is important because sometimes it is not possible to have the same number of CMLs or to monitor and record the same CML for different periods of time. A simple way to average the last corrosion rate is not always the correct tool because the corrosion rate is not constant over long periods of time as explained in the literature [29, 32, 43]. From a practical point of view, it is feasible to consider the corrosion rate constant only for close periods of time, but the corrosion rate must be quite representative of the phenomena. If one takes a greater thinning rate (e.g., thinning rate at the percentile 90 ($\widehat{\boldsymbol{x}}_{90\%}$)),

Table 3: GEV distribution parameters for HE-A with different numbers of sampling units ($n = 25$ and $n = 20$) in 2009 and its correspondence error in comparison with the observed data.

	2009 $n = 25$								2009 $n = 20$							
Vessel	a (mm/y)	b (mm/y)	ξ	$\hat{x}_{90\%}$ (mm/y)		$\hat{x}_{80\%}$ (mm/y)		p value (K–S test)	a (mm/y)	b (mm/y)	ξ	$\hat{x}_{90\%}$ (mm/y)		$\hat{x}_{80\%}$ (mm/y)		p value (K–S test)
HE-A				2.363		1.190						2.19		1.01		
	0.24	0.31	0.85	μ_e (%)	σ_e (%)	μ_e (%)	σ_e (%)	0.07	0.18	0.25	0.97	μ_e (%)	σ_e (%)	μ_e (%)	σ_e (%)	0.05
				19.19	7.84	5.83	8.81					21.28	7.98	6.89	8.97	

Table 4: GEV distribution parameters for HE-B with different numbers of sampling units ($n = 25$ and $n = 20$) in 2005 and its correspondence error in comparison with the observed data.

	2005 $n = 25$								2005 $n = 20$							
Vessel	a (mm/y)	b (mm/y)	ξ	$\hat{x}_{90\%}$ (mm/y)		$\hat{x}_{80\%}$ (mm/y)		p value (K–S test)	a (mm/y)	b (mm/y)	ξ	$\hat{x}_{90\%}$ (mm/y)		$\hat{x}_{80\%}$ (mm/y)		p value (K–S test)
HE-B				1.12		0.76		0.27	0.21	0.22	0.50	1.15		0.717		
	0.26	0.25	0.35	μ_e (%)	σ_e (%)	μ_e (%)	σ_e (%)					μ_e (%)	σ_e (%)	μ_e (%)	σ_e (%)	0.06
				10.98	2.91	6.13	2.99					11.33	2.90	6.54	3.00	

it will be necessary to replace equipment that is in a safe condition, provoking unnecessary spending. On the contrary, if a lower thinning rate is used, the possibility of having a risky event because of unsafe conditions increases. In this sense, this paper provides a methodology that can be used in order to manage the vessels' integrity by taking into account a small number of available CMLs in the inspection history.

5. Conclusions Remark

(i) In hydrotreating units, the thinning rate observed in the shell heat exchangers could be fitted to a Generalized Extreme Value distribution with high confidence and could be used as a prior distribution. Figures 1 and 2 in this paper illustrate this affirmation.

(ii) Pareto Distribution and Generalized Extreme Value Distribution can be used to represent the hyperparameters of the prior distribution selected in this study.

(iii) It is feasible to use a Bayesian approach with a grid technique in order to estimate the posterior distribution of the thinning rate in heat exchangers.

(iv) The predictive distribution obtained satisfactorily represents the unobserved thinning rate distribution in heat exchangers, even when little information is available.

(v) The application of a Bayesian approach proposed in this paper helps answer the key question about the amount of data required to find reliable estimations of the thinning rate data in the heat exchangers that preheat the feeding to a hydrotreating reactor.

Conflicts of Interest

The authors declare that they have no conflicts of interest in this study.

Acknowledgments

The authors are grateful to CONACYT and ESIQIE-IPN for the economic support provided for this research. The authors are also grateful to PEMEX for the information provided.

References

[1] G. W. Mushrush, E. J. Beal, D. R. Hardya, and J. Hughes, "Nitrogen compound distribution in middle distillate fuels derived from petroleum, oil shale, and tar sand sources," *Fuel Processing Technology*, vol. 61, no. 3, pp. 197–210, 1999.

[2] M. Jeftić, G. T. Reader, and M. Zheng, "Impacts of low temperature combustion and diesel post injection on the in-cylinder production of hydrogen in a lean-burn compression ignition engine," *International Journal of Hydrogen Energy*, vol. 42, no. 2, pp. 1276–1286, 2016.

[3] A. Stanislaus, A. Marafi, and M. S. Rana, *Catalysis Today*, vol. 153, no. 1-2, pp. 1–68, 2010.

[4] Norma Oficial Mexicana NOM-086-SEMARNAT-SENER-SCFI-2005, "Especificaciones de los Combustibles Fósiles Para la Protección Ambiental", http://www.profepa.gob.mx/innovaportal/file/1278/1/nom-086-semarnat-sener-scfi-2005.pdf, 2017.

[5] API Publication 571, *Damage Mechanisms Affecting Fixed Equipment in the Refining Industry*, American Petroleum Institute, Washington, DC, USA, 2nd edition, 2011.

[6] L. Garverick, *Corrosion in the Petrochemical Industry*, ASM International, Russell Township, OH, USA, 3rd edition, 1994.

[7] S. Papavinasam, *Corrosion Control in the Oil and Gas Industry*, Gulf Professional Publishing, Amsterdam, Netherlands, 2013.

[8] R. J. Horvath, M. S. Cayard, and R. D. Kane, "Prediction and assessment of ammonium bisulfide corrosion under refinery sour water service conditions," NACE Paper No. 06576, NACE International, Houston, TX, USA, 2006.

[9] M. S. Cayard, W. G. Giesbrecht, R. J. Horvath, R. D. Kane, and V. V. Lagad, "Prediction Of Ammonium Bisulfide Corrosion And Validation With Refinery Plant Experience," Corrosion NACE, Paper No. 06577, NACE International, Houston, TX, USA, 2006.

[10] P. Alvisi and V. Freitas Cunha Lins, "Acid salt corrosion in a hydrotreatment plant of a petroleum refinery," *Engineering Failure Analysis*, vol. 15, no. 8, pp. 1035–1041, 2008.

[11] L. Sun, M. Zhu, G. Ou, H. Jin, W. Kai, and K. Wang, "Corrosion investigation of the inlet section of REAC pipes in the refinery," *Engineering Failure Analysis*, vol. 66, pp. 468–478, 2016.

[12] M. Zhu, L. Sun, G. Ou, K. Wang, K. Wang, and Y. Sun, "Erosion corrosion failure analysis of the elbow in sour water stripper overhead condensing reflux system," *Engineering Failure Analysis*, vol. 62, pp. 93–102, 2016.

[13] R. Murata, J. Benaquisto, and C. Storey, "A methodology for identifying and addressing dead-legs and corrosion issues in a Process Hazard Analysis (PHA)," *Journal of Loss Prevention in the Process Industries*, vol. 35, pp. 387–392, 2015.

[14] R. Cabrera-Sierra, E. Sosa, M. T. Oropeza, and I. González, "Electrochemical study on carbon steel corrosion process in alkaline sour media," *Electrochimica Acta*, vol. 47, no. 13-14, pp. 2149–2158, 2002.

[15] G. Koch, J. Varney, N. Thompson, O. Moghissi, M. Gould, and J. Payer, *International Measures of Prevention, Application, and Economics of Corrosion Technologies Study*, NACE International, Houston TX, USA, 2016.

[16] P. Hammond, *On Resolution, Accuracy and Calibration of Digital Ultrasonic Thickness Gauges*, NDT.net. 1997, http://www.ndt.net/article/wt1097/hammond/hammond.htm.

[17] A. Valor, F. Caleyo, L. Alfonso, J. Vidal, and J. M. Hallen, "Statistical Analysis of Pitting Corrosion Field Data and Their Use for Realistic Reliability Estimations in Non-Piggable Pipeline Systems," *Corrosion*, vol. 70, pp. 1090–1100, 2014.

[18] F. Caleyo, A. Valor, L. Alfonso, J. Vidal, E. Perez-Baruch, and J. M. Hallen, "Bayesian analysis of external corrosion data of non-piggable underground pipelines," *Corrosion Science*, vol. 90, pp. 33–45, 2015.

[19] ISO 2859-1, *Sampling Procedures for Inspection by Attributes - Part 1: Sampling Schemes Indexed by Acceptance Quality Limit (AQL) for Lot-by-lot Inspection*, International Organization for Standardization, Geneva, Switzerland, 1999.

[20] API 510 American Petroleum Institute, *Pressure Vessel Inspection Code*, API Publishing Services, Washington, DC, USA, 2006.

[21] R. E. Walpole, R. H. Myers, S. L. Myers, and K. E. Ye, *Probability and Statistics for Engineers and Scientists*, Prentice-Hall, Boston, USA, 9th edition, 2011.

[22] F. M. Dekking, C. Kraaikamp, H. P. Lopuhaa, and L. E. Meester, *A Modern Introduction to Probability and Statistics: Understanding Why and How*, Springer-Verlag, London, UK, 2005.

[23] T. Leonard and J. S. J. Hsu, *Bayesian Methods: An Analysis for Statisticians and Interdisciplinary Researchers*, Cambridge University Press, Cambridge, UK, 1999.

[24] J. Kruschke, *Doing Bayesian Data Analysis: a Tutorial Introduction with R*, Academic Press, Burlington MA, USA, 2011.

[25] S. G. Coles, *An Introduction to Statistical Modeling of Extreme Values*, Springer, London, UK, 2001.

[26] J. C. Velazquez, A. Valor, F. Caleyo et al., "Pitting corrosion models improve integrity management, reliability," *Oil & Gas Journal*, vol. 107, no. 28, pp. 56–62, 2009.

[27] F. Caleyo, A. Valor, V. Venegas, J. H. Espina-Hernandez, J. C. Velazquez, and J. M. Hallen, "Pipeline Integrity—1: accurate corrosion modeling improves reliability estimations," *Oil & Gas Journal*, vol. 110, pp. 122–129, 2012.

[28] M. Kowaka, *Introduction to Life Prediction of Industrial Plant Materials: Application of the Extreme, Value Statistical Method for Corrosion Analysis*, Allerton Press, New York, NY, USA, 1994.

[29] J. C. Velázquez, J. A. M. Van Der Weide, E. Hernández-Sanchez, and H. Herrera-Hernández, "Statistical modelling of pitting corrosion: extrapolation of the maximum pit depth-growth," *International Journal of Electrochemical Science*, vol. 9, pp. 4129–4143, 2014.

[30] L. A. Alfonso, F. Caleyo, J. M. Hallen, J. Escamilla-Davish, and J. H. Espina-Hernandez, "Application of extreme value statistics to the prediction of maximum pit depth in non-piggable, buried pipelines," in *Proceedings of 7th International Pipeline Conference*, New York, NY, USA, September-October 2008.

[31] L. A. Alfonso, F. Caleyo, J. M. Hallen, and J. Araujo, "On the applicability of extreme value statistics to the prediction of maximum pit depth in non-piggable, buried pipelines," in *Proceedings of 8th International Pipeline Conference*, New York, NY, USA, September-October 2010.

[32] J. C. Velázquez, F. Caleyo, A. Valor, and J. M. Hallen, "Predictive model for pitting corrosion in buried oil and gas pipelines," *Corrosion*, vol. 65, no. 5, pp. 332–342, 2009.

[33] J. C. Velázquez, F. Caleyo, A. Valor, and J. M. Hallen, "Technical note: field study—pitting corrosion of underground pipelines related to local soil and pipe characteristics," *Corrosion*, vol. 66, no. 1, pp. 016001–016001-5, 2010.

[34] Engineering Statistics Handbook, Kolmogorov-Smirnov Goodness-of-Fit Test, 2016, http://www.itl.nist.gov/div898/handbook/eda/section3/eda35g.htm.

[35] Z. A. Foroulis, "Role of solution pH on wet H2S cracking in hydrocarbon production," *Corrosion Prevention and Control*, vol. 40, no. 4, pp. 84–89, 1993.

[36] ASTM G102-89, *Calculation of Corrosion Rates and Related Information from Electrochemical Measurements, Annual Book of ASTM Standards*, American Society for Testing and Materials, Philadephia, PA, USA, 2004.

[37] R. Cabrera-Sierra, I. García, E. Sosa, T. Oropeza, and I. González, "Electrochemical behavior of carbon steel in alkaline sour environments measured by electrochemical impedance spectroscopy," *Electrochimica Acta*, vol. 46, pp. 487–497, 2000.

[38] D. Rivas, F. Caleyo, A. Valor, and J. M. Hallen, "Extreme value analysis applied to pitting corrosion experiments in low carbon steel: Comparison of block maxima and peak over threshold approaches," *Corrosion Science*, vol. 50, no. 11, pp. 3193–3204, 2008.

[39] ASME boiler & pressure vessel Code. Section VIII, division 1, 2010.

[40] ASME boiler & pressure vessel Code. Section VIII, division 2, 2010.
[41] API 579-1/ASME FFS-1, *Fitness-for-Service, Jointly Published by the,* American Petroleum Institute and the American Society for Mechanical Engineers, New York, NY, USA, 2007.
[42] M. K. Chang, R. R. Rong-Chang, C. M. Shu, and K. N. Lin, "Application of risk based inspection in refinery and processing piping," *Journal of Loss Prevention in the Process Industries*, vol. 18, no. 4-6, pp. 397–402, 2005.
[43] J. C. Velázquez, J. C. Cruz-Ramirez, A. Valor, V. Venegas, F. Caleyo, and J. M. Hallen, "Modeling localized corrosion of pipeline steels in oilfield produced water environments," *Engineering Failure Analysis*, vol. 79, pp. 216–231, 2017.
[44] https://math.stackexchange.com/questions/497878/how-to-convert-a-histogram-to-a-pdf, 2018.

9

Effect of Hot Deformation Temperature on Texture Formation Regularity of CGO Silicon Steel under High Deformation Rate

Zhi-chao Li,[1] Ning Dang,[2] and Zhen-li Mi[2]

[1]*Collaborative Innovation Center of Steel Technology, University of Science and Technology, Beijing 100083, China*
[2]*Institute of Engineering Technology, University of Science and Technology, Beijing 100083, China*

Correspondence should be addressed to Zhi-chao Li; lizhichao1225@163.com

Academic Editor: Jingwei Zhao

The influence of hot deformation temperature on microstructure and texture of CGO silicon steel under the condition of a high deformation rate (100 s^{-1}) was studied by SEM and EBSD techniques. The results indicate that the typical microstructures at room temperature consist of ferrite and pearlite under different hot deformation temperatures. The higher deformation temperature is beneficial to obtain a more uniform recrystallization microstructure and lower pearlite content. Cubic texture {100}⟨001⟩ and rotated cubic texture {100}⟨011⟩ are dominant texture components in the tested steels, and {111}⟨112⟩ texture in γ fiber is also strong, the intensity of which is higher than that of {111}⟨110⟩ texture. Goss texture {110}⟨001⟩ is weak. With the rising of the hot deformation temperature, {100}⟨011⟩ texture decreases firstly and then increases at 1100°C. When the hot deformation temperature raises from 800°C to 900°C, {111}⟨112⟩ texture shows an increasing trend, while {111}⟨110⟩ texture content is stable. When the temperature further increases to 1100°C and 1150°C, {111}⟨112⟩ and {111}⟨110⟩ textures are slightly weakened.

1. Introduction

Grain-oriented silicon steel is an important soft magnetic material which is widely applied in electronic and electric power industry and mainly used for the manufacture of transformers and so forth [1, 2]. The metallurgical factors which affect the magnetic properties include the steel composition, microstructure, crystallographic texture, grain size, and internal stress, of which microstructure and texture are important influencing factors [3]. Due to the obvious inheritance of the textures [4] during manufacturing, the hot rolling textures have significant effect [5] on the texture evolution during the following process of grain-oriented silicon steel. The research on the origin and evolution of textures is beneficial for the optimization of texture control in the production processes. The deformation rate applied to grain-oriented silicon steel in actual hot rolling process is relatively high. Despite numerous researches on the effect of the hot rolling temperature on the microstructure and texture of grain-oriented silicon steel at present, the research under the condition of a high strain rate is not much. Liu et al. [6] studied hot rolling texture distribution at different temperatures using a thermal simulator under low deformation rate condition. In this study, the texture evolution during hot rolling process was simulated by uniaxial compression. The results show that gamma textures which have strong inheritance play an important role in the subsequent process and improve the concentration of textures after secondary recrystallization. Zhang et al. [7] studied hot rolling textures of 3.2% Si grain-oriented silicon steel and the inheritance of textures between hot rolling and subsequent processes, this research focused on microstructure, and texture evolution of the static recrystallization occurred in the process of annealing after the deformation of CGO silicon steel. The results show that there is no obvious heterogeneity along the thickness direction after primary recrystallization, and {111}⟨110⟩ texture is the strongest texture component, and there is a certain intensity of the Goss texture. The texture formation mechanism is different between dynamic recrystallization and static recrystallization. The texture is definitely different between hot compression and hot rolling which is used industrially, and Zhichao et al. [8, 9] have done some researches on the formation and evolution of textures in hot rolling process, but the thermal compression process is not involved in the same deformation conditions, and this process has significant impact on the texture evolution during hot deformation. The rough rolling of CGO steel usually has a high rolling speed which is closed to 100 s^{-1}, so that the texture evolution during this process is difficult to analyze, and limited work is available in the literature on this impact. It is therefore important to summarize the texture evolution law under the large deformation rate with different temperatures. This paper investigate the effect of hot deformation temperature on the microstructure and texture of grain-oriented silicon steel under the condition of a high deformation rate, which provides reference to the texture control during the manufacturing process of grain-oriented silicon steel.

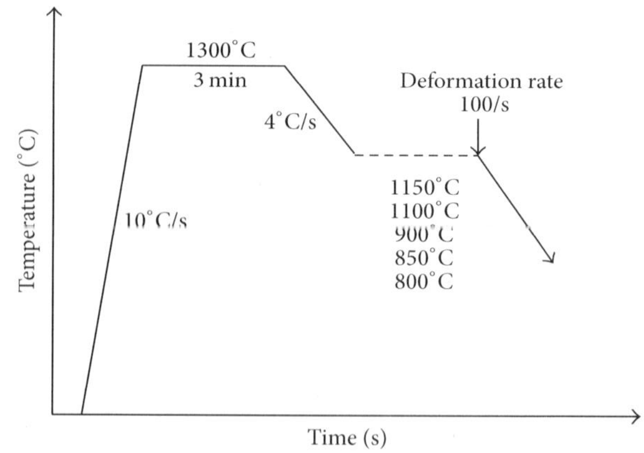

FIGURE 1: Processing route of thermal simulation experiment.

2. Experimental Materials and Methods

Experimental materials are cut from an ingot which has a chemical composition of 0.065% C, 2.95% Si, 0.15% Mn, and 0.027% S and processed into cylinder specimens. The experimental process is illustrated in Figure 1. The specimens were heated to 1300°C at 10°C/s, held for 3 min, and then cooled, respectively, to 800, 850, 900, 1100, and 1150°C at 4°C/s and conducted uniaxial compression with the same deformation rate of 100 s^{-1}. Test samples were cut along the axis of compression specimen and their microstructures in center layer were analyzed by field emission scanning electron microscopy (SEM) after mechanical polishing and etching. The orientation distributions were measured by electron back-scattered diffraction (EBSD) attached to Zeiss ultra 55 SEM. The experimental data were analyzed by Channel 5 software and the contents of typical textures were obtained.

3. Results and Discussion

3.1. Microstructures of Experimental Steels. The phase diagrams of the experimental steels are calculated using Thermo-Calc software. Figure 2(a) presents the effect of carbon content on the microstructure of the steel. Figure 2(b) presents the dependence of the relative qualities of austenite, ferrite, cementite, and so forth on the temperature. It can be seen from Figure 2 that when the C content is in a range of 0.04–0.78%, the high temperature microstructure at 1200°C is single-phase austenite. During the cooling process, the proeutectoid ferrite occurs firstly, and two-phase range of austenite and ferrite exists. With the formation of ferrite, the concentration of carbon in ferrite is very low and reduces with the decreasing of temperature, and C atoms diffuse to the surrounding austenite. This process is accompanied with the decreasing of temperature and the formation of ferrite continuously. At the five experimental temperatures between 800 and 1100°C, deformation specimens are at the "ferrite + austenite" two-phase range, the austenite has the highest relative content at 1180°C, and the content will reduce with the decreasing of temperature.

Figures 3(a)–3(e) present the microstructures of the experimental steels at different deformation temperatures under a high deformation rate. Figure 3(f) illustrates the average grain size of different specimens. Typical microstructures consist of ferrite and pearlite, and the microstructure morphologies of the samples show significant differences under various deformation temperatures. The recrystallization grains of the specimens deformed at 850 and 900°C are coarse, and their grain size distributions are inhomogeneous. At the deformation temperatures of 1100°C and 1150°C, the recrystallization degrees of part grains are low and the content of pearlite is higher. With the increases of deformation temperature, the recrystallization grain size decreases gradually and the grain size distribution is more uniform. It can be obtained from Figure 2(b) that, with the increasing of deformation temperature, the austenite proportion becomes larger and the supersaturation of C in austenite decreases. The deformed austenite recrystallizes sufficiently, leading to fine and uniform size of deformed austenite. At the same time, crystal distortion is small and is not conducive for the diffusion of the C atoms from inside of grains to grain boundaries where pearlite nucleates, which lead to the difficulty of alternative growth of pearlite lamellae. The pearlite layer is thinner after high temperature deformation and pearlite area is smaller. In addition, due to the increasing of deformed austenite, pearlite distribution becomes more uniform after phase transformation. During 800~900°C, recrystallization degree is weak, distortion of deformed austenite is large, and it is conductive to the diffusion of C atoms. Coarse pearlite is formed after phase transformation and distributed intensively because of the decreasing of deformed austenite. During rolling in two-phase range of ferrite and austenite, the degree of recrystallization is lower at low temperature. At a higher deformation temperature, ferrite and austenite are compressed and form dislocation-substructure inside of the grains [9]. The higher temperature provides driving force for the grain growth, which leads to the recrystallization and the growth of ferrite grains in a wide range. When the deformation temperature is low, austenite grains are elongated without recrystallization and dislocation-substructures accumulated within grains, which become nucleation sites of phase transformation. The elongated austenite grain boundaries restrict the growth of newly formed microstructures.

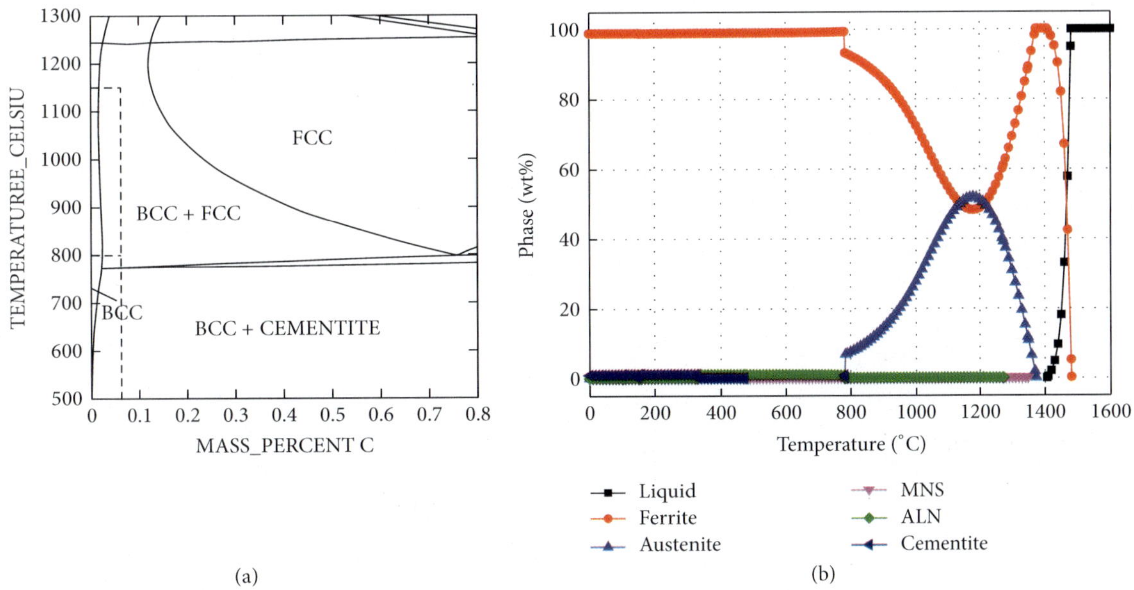

FIGURE 2: Thermodynamic calculated phase diagram. (a) Fe-C phase diagram; (b) the relative amount of different phases change with temperature.

So it can be observed from Figure 3(b) that the grain bands did not grow up sufficiently between the gaps of larger size grains. During the cooling, ferrite and cementite will generate inside of austenite and the newly formed ferrite has small size because of the limitation of austenite grain boundaries. At lower deformation temperature, there are long strip recovery microstructures which remain in the matrix, such as Figures 3(b) and 3(c).

3.2. Textures of Experimental Steels. The initial textures before high temperature deformation are distributed randomly because the experimental materials are cut from forged CGO silicon steel. The orientation image maps of the experimental steels with high deformation rate at different deformation temperatures were shown in Figure 4 and the area fractions of different textures were listed in Table 1 and Figure 5. Various colors represent grains with different orientation, Goss {110}⟨001⟩ oriented grains is red, {111}⟨121⟩ oriented grains is yellow, {111}⟨100⟩ oriented grains is blue, cubic oriented grains is brown, rotated cube {100}⟨011⟩ oriented grains is green, brass-type {110}⟨112⟩ oriented grains is purple, and copper-type {112}⟨111⟩ oriented grains is blue.

Cubic texture {100}⟨001⟩, rotated cube texture {100}⟨011⟩, gamma fiber textures including {111}⟨112⟩ and {111}⟨110⟩ textures, Goss texture {110}⟨001⟩, copper-type texture {112}⟨111⟩, and brass-type {110}⟨112⟩ texture are important texture components in grain-oriented silicon steel and their existence in hot rolled plate will influence the subsequent texture development [11–13].

As can be seen in Table 1, in general, cube texture {100}⟨001⟩ and rotated cube texture {100}⟨011⟩ are dominant texture components at different hot deformation temperatures. In addition, the contents of {111}⟨112⟩ in γ fiber and copper-type {112}⟨111⟩ texture are also higher. Goss texture {110}⟨001⟩ is weak and its content is low. The contents of texture components will change at various deformation temperatures. For instance, with the rising of hot deformation temperature, the content of rotated cube texture {100}⟨011⟩ decreases first and then increases further. The content of {111}⟨112⟩ texture and copper-type texture {112}⟨111⟩ increase first and then reduce.

TABLE 1: Texture contents of samples under different hot deformation temperatures in tested steels (%).

Hot deforming temperatures/°C	800	850	900	1100	1150
{100}⟨001⟩	19.4	7.8	13.4	13.2	17.9
{100}⟨011⟩	17.2	15.4	4.8	11.0	22.9
{110}⟨001⟩	2.2	2.0	8.6	2.9	1.3
{110}⟨112⟩	4.8	5.3	5.2	7.7	2.9
{111}⟨110⟩	9.5	6.5	9.9	7.5	6.7
{111}⟨112⟩	8.1	8.5	11.2	8.0	7.5
{112}⟨111⟩	8.3	16.6	8.5	5.9	7.5

From Table 1 and Figure 6 it can be seen that the content of Goss texture {110}⟨001⟩ is low at most of the hot deformation temperatures. Goss texture of hot rolled plate originates from the broken columnar crystal with ⟨100⟩ direction in casting billet [14–16]. During the hot deformation process, Goss texture formed by shear strain will transit to other orientations under high deformation rate because of the migration of the high angle grain boundaries. Goss oriented grains will transform to brass-type texture {110}⟨112⟩ and copper-type {112}⟨111⟩ texture because of the orientation conversion relationship. The intensity of copper-type texture {112}⟨111⟩ is higher than the Goss texture {110}⟨001⟩ which can be seen in Table 1. During the cold rolling, Goss

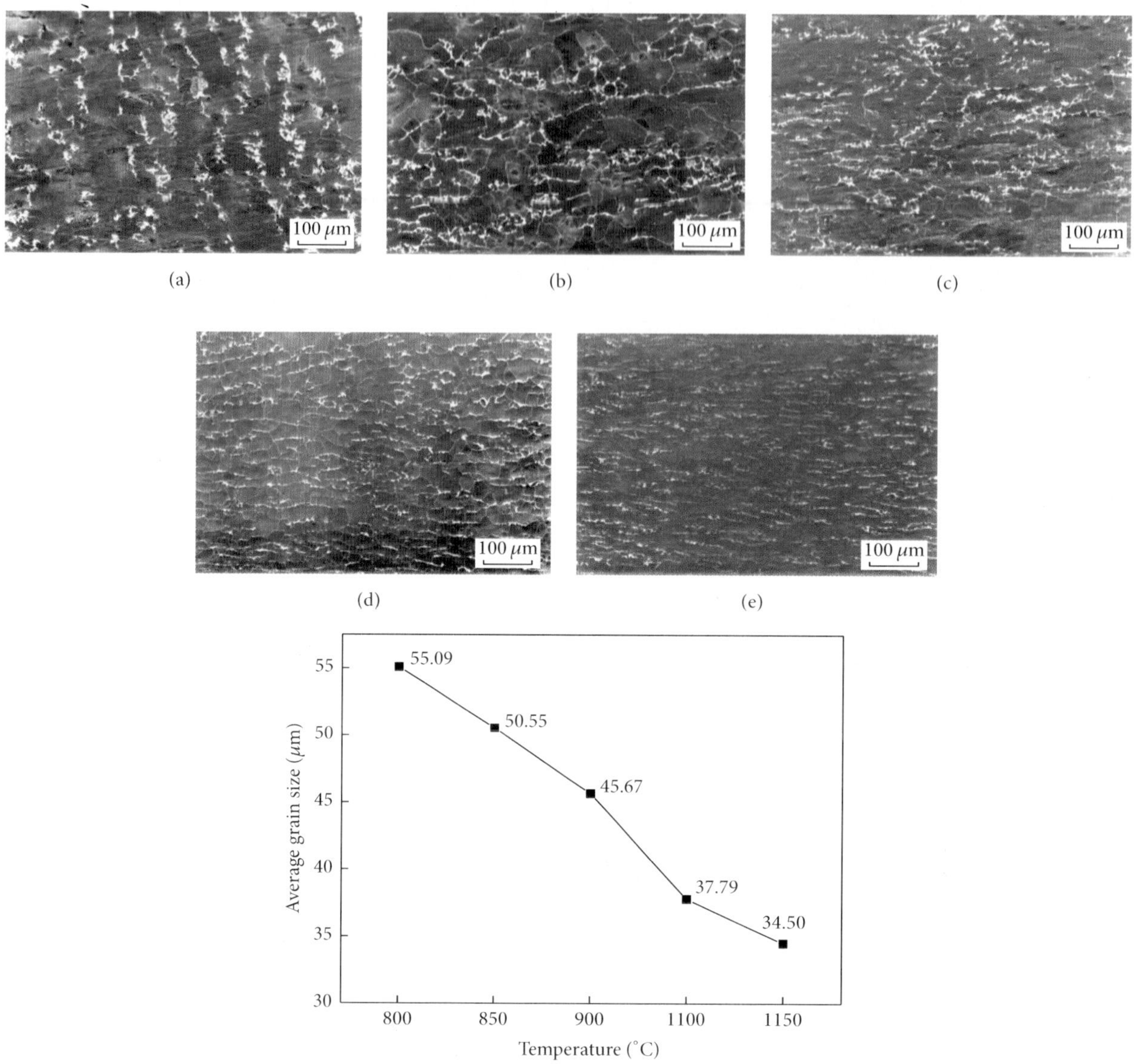

FIGURE 3: Microstructure of samples under different hot deformation temperatures. (a) 800°C; (b) 850°C; (c) 900°C; (d) 1100°C; (e) 1150°C; (f) average grain size variation.

oriented grains descend from the broken columnar grains with ⟨100⟩ direction of casting billet and are prone to transit to {223}⟨110⟩ texture along certain path. However, it is difficult to obtain the final {223}⟨110⟩ texture because of the restriction of certain conditions. In brief, the content of Goss texture {110}⟨001⟩ is low in the steels which experienced thermal deformation.

Rotated cube texture {100}⟨011⟩ is the common component in hot rolled plate. From Table 1 and Figure 5 it can be seen that, with the rising of deformation temperature, the content of {100}⟨011⟩ texture decreases first and then increases. The texture characteristics of single crystal under different rolling reductions indicate that when the rolling reduction was greater than 60%, {211}⟨011⟩ and {100}⟨011⟩ texture components are enhanced because of {211}⟨111⟩ slip system during thermal deformation [17, 18], which explains the fact that the content of rotated cube texture {100}⟨110⟩ is high at all the deformation temperature. The rotated cube texture {100}⟨110⟩ is difficult to recrystallize [19] and should be eliminated in the subsequent process. The typical texture during austenite recrystallization is cubic texture. All steels with different deformation temperature therefore have a high proportion of {100}⟨001⟩ cubic texture. However, the cubic texture is an unfavorable component which is formed by the initial columnar crystal in grain-oriented silicon steel. Researches [10] indicated that, in the subsequent process, the repeated rolling and annealing can weaken the heredity of {100} texture and decrease the {100}⟨001⟩ cubic texture.

As can be seen in Figure 4, for the grains with {111} crystal plane, when the hot deformation temperature increased from 800°C to 900°C, {111}⟨112⟩ texture shows an increasing trend, while {111}⟨110⟩ texture is relatively stable. When

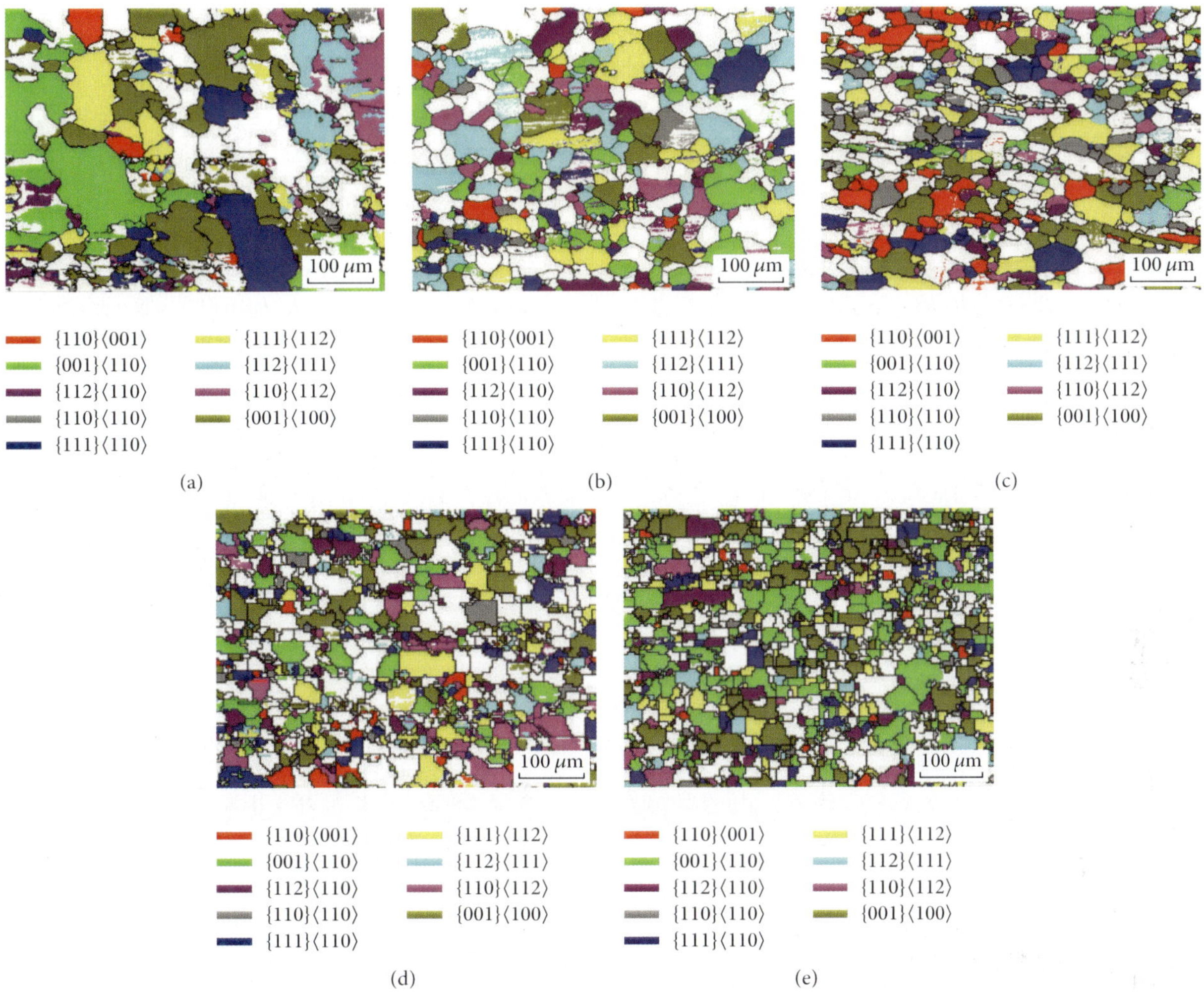

FIGURE 4: EBSD analysis of samples under different hot deformation temperatures. (a) 800°C; (b) 850°C; (c) 900°C; (d) 1100°C; (e) 1150°C.

the hot deformation temperature was increased to 1100°C and even 1150°C further, {111}⟨112⟩ texture and {111}⟨110⟩ texture are slightly weakened. {111} plane textures are the typical recrystallization texture, whose proportion in experimental steels is closely related to the recrystallization degree. {111} plane textures provide nucleation sites for the abnormal growth of Goss grains during secondary recrystallization process [1, 12], which are important growth environment for Goss oriented grains [20, 21]. In the recrystallization after deformation, the stored energies are different in various texture components, which are {001}⟨110⟩, {112}⟨110⟩, {111}⟨110⟩, and {111}⟨112⟩ in an order from strong to weak. {111}⟨110⟩ and {111}⟨112⟩ oriented grains can nucleate and grow preferentially due to higher stored energies. From Figure 4 it can be seen that {111}⟨110⟩ and {111}⟨112⟩ oriented grains present a higher degree of recrystallization. Because it is easier for {111}⟨112⟩ texture to form shear bands compared with {111}⟨110⟩ texture [22], {111}⟨112⟩ oriented grains are more likely to grow up. From Table 1 it can be seen that the intensity of {111}⟨112⟩ texture is stronger than {111}⟨100⟩ texture in multiple thermal deformation temperatures.

During the hot compression with high strain rate of CGO silicon steel, scattered α fiber texture merged into {100}⟨011⟩ texture and cube {100}⟨011⟩ texture was formed because of strong rotation of grains. Under high stain rate, cubic texture {100}⟨001⟩ keeps stable and takes a great proportion. Copper texture {112}⟨111⟩ and brass texture {110}⟨112⟩ are typical shear textures. Brass texture {110}⟨112⟩ is derived from shear layer which forms at the subsurface of the plate during hot rolling and is closely related to Goss texture {110}⟨001⟩. The formation of brass texture {110}⟨112⟩ is due to the inhibition of grain boundaries. The rotation of Goss oriented grains along TD direction is impeded so that the grains rotate along ND direction instead.

4. Conclusions

(1) The typical microstructures at room temperature of CGO silicon steel after hot deformation are ferrite and pearlite, and there are some differences in grain size distributions and morphologies. A high hot deformation temperature is benefit to obtain a more uniform recrystallized microstructure and lower pearlite

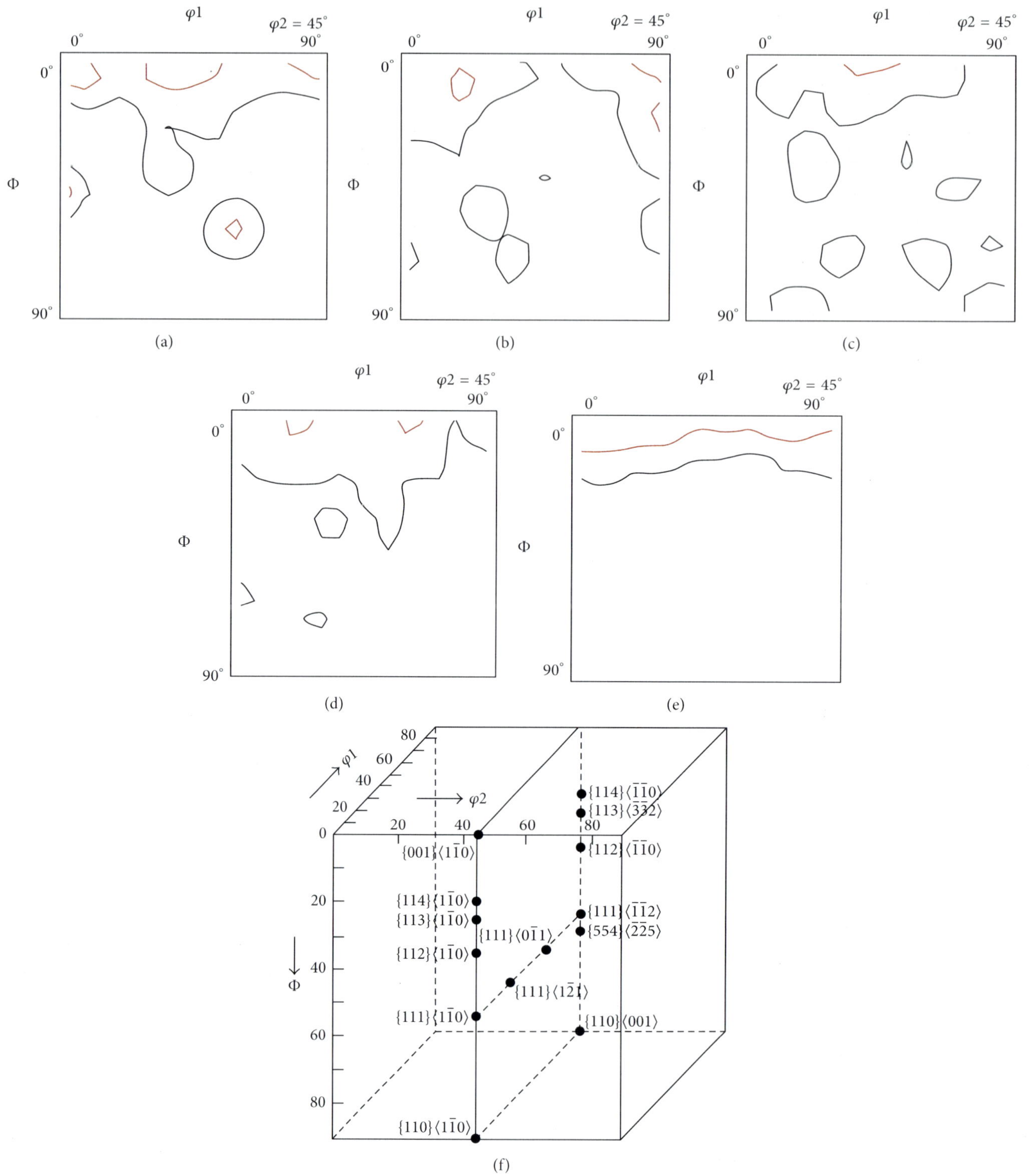

FIGURE 5: ODF analysis of samples under different hot deformation temperatures. (a) 800°C; (b) 850°C; (c) 900°C; (d) 1100°C; (e) 1150°C; (f) distribution of main orientations in Euler space for silicon steel [10].

content. A low deformation temperature leads to higher pearlite content and partial recrystallization.

(2) The cubic texture {100}⟨001⟩ and rotated cubic texture {100}⟨011⟩ are dominant texture components in the tested steels after hot deformation. The {111}⟨112⟩ texture in γ fiber is also strong, and its strength is higher than that of {111}⟨110⟩ texture, whereas the content of Goss texture {110}⟨001⟩ is very low.

(3) With the rising of deformation temperature, the content of the rotated cube {100}⟨011⟩ texture decreases

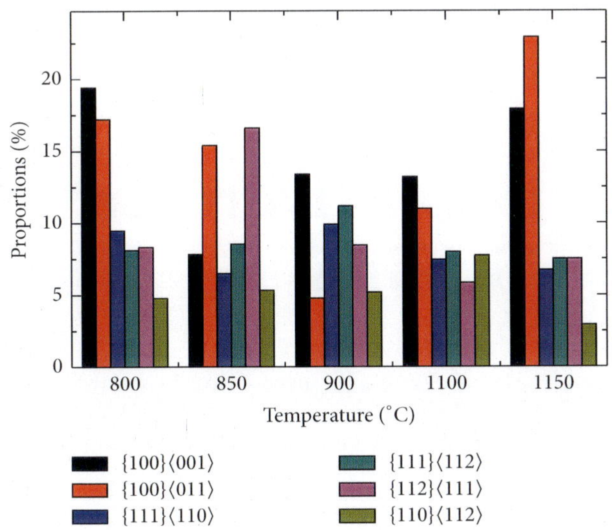

FIGURE 6: Texture contents of samples under different hot deformation temperatures in tested steel.

first and then increases when the deformation temperature further achieved to 1100°C; when the deformation temperature raises from 800°C to 900°C, {111}⟨112⟩ texture shows an increasing trend, and {111}⟨110⟩ texture is more stable. When the hot deformation temperature continues to increase to 1100°C and 1150°C, {111}⟨112⟩ texture and {111}⟨110⟩ texture are slightly weakened.

Competing Interests

The authors declare that they have no competing interests.

Acknowledgments

This works is supported by Fundamental Research Funds for the Central Universities (TP-A3), no. FRF-TP-15-063A3.

References

[1] J. Y. Park, K. S. Han, J. S. Woo, S. K. Chang, N. Rajmohan, and J. A. Szpunar, "Influence of primary annealing condition on texture development in grain oriented electrical steels," *Acta Materialia*, vol. 50, no. 7, pp. 1825–1834, 2002.

[2] Y. Hayakawa and M. Kurosawa, "Orientation relationship between primary and secondary recrystallized texture in electrical steel," *Acta Materialia*, vol. 50, no. 18, pp. 4527–4534, 2002.

[3] T. Yonamine and F. J. G. Landgraf, "Correlation between magnetic properties and crystallographic texture of silicon steel," *Journal of Magnetism and Magnetic Materials*, vol. 272–276, supplement 1, pp. E565–E566, 2004.

[4] C. Gheorghies and A. Doniga, "Evolution of texture in grain oriented silicon steels," *Journal of Iron and Steel Research International*, vol. 16, no. 4, pp. 78–83, 2009.

[5] L. Cheng, N. Zhang, P. Yang et al., "Retaining {100} texture from initial columnar grains in electrical steels," *Scripta Materialia*, vol. 67, no. 11, pp. 899–902, 2012.

[6] L.-L. Liu, Y.-L. Feng, and B. Hou, "Microstructure and texture in hot rolling strip of low temperature oriented silicon steel," *Hot Working Technology*, vol. 43, no. 12, pp. 99–102, 2014.

[7] Z.-Z. Zhang, W. Ren, and J.-F. Fang, "Texture evolution of primary recrystallization and deformation of grain-oriented electrical steel with low temperature slab reheating," *Iron and Steel*, vol. 46, no. 8, pp. 83–88, 2011.

[8] L. I. Zhichao, S. U. N. Qiang, M. Zhenli, and L. Ruimin, "Lin Ruimin. Effect of cooling rate on annealing textures in hot rolled 1. 35%Si non-oriented electrical steel," *Transactions of Materials and Heat Treatment*, vol. 37, no. 2, pp. 104–109, 2016.

[9] Q. Sun, L. Zhichao, M. Zhenli, and l. Ruimin, "Effect of chemical composition on recrystallized microstructure and texture of hot rolled non-oriented electrical steel," *Transactions of Materials and Heat Treatment*, vol. 37, no. 4, pp. 82–87, 2016.

[10] S. Wang, *Influence of Rolling Direction on the Texture Evolution of Grain-Oriented Silicon Steel*, University of Science and Technology, Liaoning, Anshan, China, 2013.

[11] N. Bernier, E. Leunis, C. Furtado, T. Van De Putte, and G. Ban, "EBSD study of angular deviations from the Goss component in grain-oriented electrical steels," *Micron*, vol. 54-55, pp. 43–51, 2013.

[12] Y.-Y. Shao, P. Yang, Y.-J. Fu, and W.-M. Mao, "Texture evolution of columnar grains in electrical steel during hot rolling," *Journal of Iron and Steel Research International*, vol. 20, no. 10, pp. 99–106, 2013.

[13] L. Xie, P. Yang, N. Zhang et al., "Formation of {100} textured columnar grain structure in a non-oriented electrical steel by phase transformation," *Journal of Magnetism and Magnetic Materials*, vol. 356, pp. 1–4, 2014.

[14] Y.-Y. Shao, P. Yang, and W.-M. Mao, "Evolution of orientations and their influence on precipitation during hot compression of columnar-grained electrical steel," *Journal of Materials Engineering*, vol. 10, pp. 75–81, 2014.

[15] D. Dorner, S. Zaefferer, L. Lahn, and D. Raabe, "Overview of microstructure and microtexture development in grain-oriented silicon steel," *Journal of Magnetism and Magnetic Materials*, vol. 304, no. 2, pp. 183–186, 2006.

[16] U. Yoshiyuki, M. Masato, F. Masahiro, K. Takeshi, F. Hiroyasu, and M. Kenichi, "Recent development of low-loss grain-oriented silicon steel," *Journal of Magnetism and Magnetic Materials*, vol. 254-255, pp. 307–314, 2003.

[17] N. A. Castro, M. F. de Campos, and F. J. G. Landgraf, "Effect of deformation and annealing on the microstructure and magnetic properties of grain-oriented electrical steels," *Journal of Magnetism and Magnetic Materials*, vol. 304, no. 2, pp. e617–e619, 2006.

[18] L. Fangfang, *Study on Hi-B Steel Recrystallization During the Hot Deformation Process by Thermal Simulation*, Wuhan University of Science and Technology, 2015.

[19] Z. Shifeng, *Study on Microstructure and Texture During Hot Rolling and Normalizing of Twin-Roll Cast*, Northeastern University, 2009.

[20] Z.-L. Jin, H.-P. Ren, W. Li, and Y.-M. Li, "Effect of thermomechanical treatment history upon annealing microstructure and texture of a cold rolled nonoriented silicon steel," *Transactions of Materials and Heat Treatment*, vol. 33, no. 2, pp. 122–126, 2012.

[21] Y. B. Xu, Y. X. Zhang, Y. Wang et al., "Evolution of cube texture in strip-cast non-oriented silicon steels," *Scripta Materialia*, vol. 87, pp. 17–20, 2014.

10

Electrochemical Determination of Hydrogen Entry to HSLA Steel during Pickling

Jari Aromaa, Antero Pehkonen, Sönke Schmachtel, Istvan Galfi, and Olof Forsén

Department of Chemical Engineering and Metallurgy, School of Chemical Technology, Aalto University, P.O. Box 16200, Aalto 00076, Finland

Correspondence should be addressed to Jari Aromaa; jari.aromaa@aalto.fi

Academic Editor: Jun Liu

Pickling with hydrochloric acid is a standard method to clean steel surfaces before hot-dip galvanizing. When normal low strength steels are pickled, hydrogen formed in pickling reactions does not have any significant harmful effect on the mechanical properties of steel. However, in pickling of steels with higher strength, the penetration of hydrogen into the steel may cause severe damages. The effect of pickling of high-strength low-alloy (HSLA) steels was investigated using a cell construction based on the Devanathan-Stachurski method with modified anodic surface treatment and hydrogen production using acid. The penetration and the permeability of hydrogen were measured using an electrochemical cell with hydrochloric acid on the one side of the steel sample and a solution of NaOH on the other side. No protective coating, for example, palladium on the anodic side of the sample, is needed. The penetration rate of hydrogen into the steel and exit rate from the steel were lower for higher strength steel.

1. Introduction

The number and the applications of new high-strength low-alloy (HSLA) steels with enhanced formability are constantly increasing. It is generally accepted that the transition from mild steel to HSLA steel occurs at yield strength of about 300 MPa. These steels have the potential to introduce weight savings while improving performance. Depending on design, the higher strength can evolve into better fatigue and crash performance, while maintaining or even reducing required material thickness [1]. However, HSLA steels may also give rise to some risks. For example, under certain circumstances, hydrogen introduced into steel during its manufacture, subsequent component fabrication, or in service may result in brittle failures at applied stresses far below the yield strength or the nominal design stress for the materials [2].

In this paper, we have studied hydrogen intrusion to HSLA steels during acid pickling to remove surface films before hot-dip galvanizing. The electrolytically produced hydrogen that penetrates into metal lattice can originate from corrosion, pickling, or electroplating [3]. In the pre-treatment of the hot-dip galvanizing process, hydrogen may be absorbed in the steel during the pickling stage through contact with the hydrochloric acid which oxidizes the metal and forms adsorbed hydrogen that partly enters the metal lattice. After hot-dip galvanizing, the zinc layer prevents or delays the exit of the hydrogen from the steel [4]. Steels with tensile strength at the level of 800–1000 MPa or higher are considered to be the most susceptible ones to hydrogen embrittlement [5, 6]. The majority of steels have generally tensile strength in the range of 200–450 MPa, and they are apparently not subject to hydrogen embrittlement problems. Special consideration must be given to high strength steels, if they are required to be pickled [3].

Inhibitors are used to prevent pitting of steel surface in case of overpickling. They also minimize the dissolution of iron, decrease the acid consumption, and lower the formation of harmful fumes. Normally, inhibitors provide good protection against hydrogen absorption during acid cleaning. It is known that inhibitors such as benzotriazole obstruct or decrease the diffusion of atomic hydrogen into the steel. In addition, some nitrogen bearing organic compounds such as imidazoline have been found to be highly effective inhibitors of hydrogen penetration into low carbon steels [2, 7].

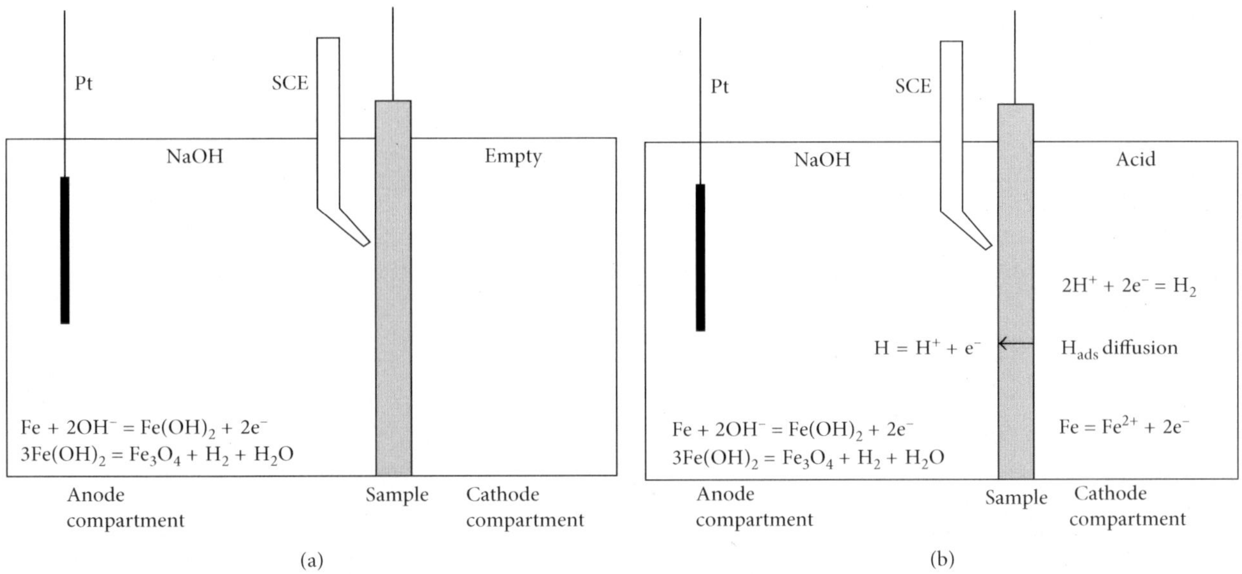

FIGURE 1: Schematic setup of the test cell with anodic and cathodic reactions at passivation and permeation test stages. Pt is platinum counterelectrode, and SCE is saturated calomel reference electrode. (a) During passivation. (b) During testing.

The Devanathan-Stachurski method is the most frequently utilized technique to measure the hydrogen permeation in metals. Hydrogen is produced on the cathodic side of the double cell using cathodic polarisation. Part of this hydrogen diffuses through the sample, and on the anodic side exiting hydrogen atoms are oxidized by applying an anodic current. Thickness of steel samples in hydrogen permeation experiment is typically 0.3–1.0 mm. Galvanostatic current to produce hydrogen has varied from 0.05 to 100 mA/cm^2 [8–10]. The aim of this investigation was to measure the permeation of hydrogen through HSLA steels using the Devanathan-Stachurski method with modified anodic surface treatment and hydrogen production using acid. The permeation is commonly studied using external polarisation to produce the hydrogen, but we used the pickling acid as the hydrogen source. The samples had no Pd coating on the exit surface, but the exit surface was passivated potentiostatically using NaOH before permeation measurements. Passivation was done to produce a steady background current on the steel. A schematic diagram on the reactions and phenomena during the measurements is shown in Figure 1. In the anodic compartment, steel is first dissolved and passivated at a constant potential. The cathodic compartment is empty during the passivation stage. During the permeation tests in the cathodic compartment, steel is corroded by acid at the free corrosion potential, and part of the produced hydrogen enters the steel diffusing to the exit surface in the anode compartment. When hydrogen diffuses through the steel, hydrogen evolution begins on the anodic surface and this is detected as current increase.

Usually the anodic side of the steel sample is coated with palladium to ensure a uniform surface and to minimize the anodic dissolution of the metal during permeation current measurements. Palladium film favours the kinetics of hydrogen oxidation. Without Pd layer, the oxidation of hydrogen atoms could be incomplete due to the presence of an oxide film that could build up on a metal sample when anodically polarised in an alkaline medium. The oxide film acts as a barrier against hydrogen release. According to Manolatos et al. [11], without a palladium coating on the exit side of the sample the surface phenomena are not controlling and the passive layer on the exit side changes with time preventing stabilization of the hydrogen concentration on the exit side, and therefore stationary conditions cannot be obtained. This is the reason why most of the investigations have been carried out using palladium coating. The variation in hydrogen concentration is observed as the nonsteady state of the charging current curve. However, it is possible to obtain reproducible results by controlling the parameters related to the formation of the passive layer [11]. Our approach was to use long passivation times and subtract the steady passive current from hydrogen oxidation transient currents.

The charging of steel with hydrogen is usually done electrolytically or in hydrogen-containing atmosphere (cf. [12]), and the gaseous charging and electrolytic charging are considered equivalent. Identical setup with freely corroding charging side has been used, for example, in [13] to study the effect of pickling inhibitors for hydrochloric acid and in [14, 15] to study corrosion of pipeline steels. In [13], the effect of inhibitors on hydrogen permeation was calculated by comparing permeation current with and without inhibitor. In [14], the permeation current was used to evaluate the effect of additives and corrosion films on hydrogen absorption. In [15], the permeation current increased when the H_2S concentration in hydrogen entry environment increased. These examples indicate that use of freely corroding surface can be used to charge steel with hydrogen. In our work, the permeation of hydrogen was investigated using 1–4 mm thick steel samples, and several analysis methods were applied for estimating the diffusion coefficients.

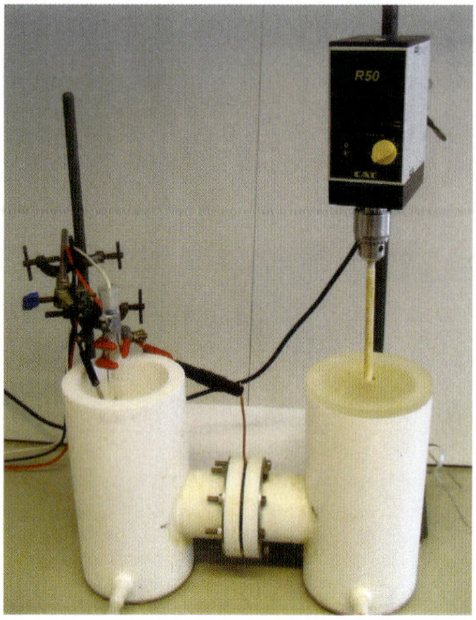

Figure 2: Test cell for permeation experiments, with compartment with pickling solution on the right side and 0.2 M NaOH on the left side.

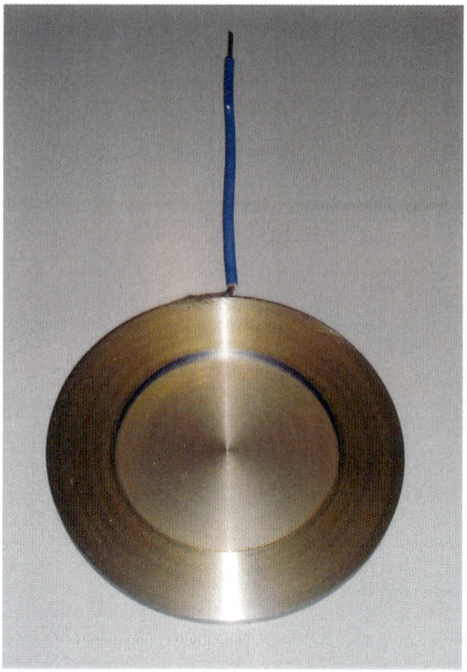

Figure 3: Test sample after turning. The middle part is for hydrogen permeation and the machined side was in the hydrogen evolving side.

Table 1: Chemical compositions of steels used in experiments.

Steel grade	C_{max}	Si_{max}	Mn_{max}	P_{max}	S_{max}	Al_{min}	CEV_{max}
S355	0.12	0.50	1.50	0.025	0.020	0.015	0.39
S500	0.12	0.50	1.70	0.025	0.015	0.015	0.43
S700	0.12	0.60	2.10	0.025	0.015	0.015	0.41
S900	0.20	0.60	2.20	0.025	0.010	0.015	0.57
S235	0.17	—	1.40	0.035	0.035	—	0.35

2. Experimental

A double-compartment test cell was constructed to measure the permeation of hydrogen through steel (Figure 2). Material of this double cell was Teflon, and the steel sample was installed between the two separate cells. The electrochemical tests were done using Autolab PGSTAT30 Potentiostat with Autolab Software 4.9.

Test materials were high-strength low-alloy (HSLA) steels with yield strength of 355, 500, 700, and 900 MPa. Samples for experiments were cut from rectangular hollow sections 50 mm wide with 6 mm material thickness. The material in these hollow sections is hardened, hot-rolled steel, and corresponding grade carbon steel S235 was used as a reference material. The tensile strength of this steel is 235 MPa. Chemical compositions of test materials are shown in Table 1. CEV (carbon equivalent for estimation of steel weldability) values calculated using $CEV = C + Mn/6 + (Cr + Mo + V)/5 + (Ni + Cu)/15$ were taken for small material thicknesses used in this study. The analyses and material properties were supplied by the manufacturer and referred to general product qualities.

The surface area of the sample that was in contact with solution was 19.6 cm^2. The thickness of the sample area exposed to the solutions was reduced by turning (Figure 3). The thickness of exposed sample area varied from 1 to 4 mm.

After thickness adjustment, the anodic surface of the sample was ground with emery paper (grade 400) followed by washing in citric acid just before electrochemical experiments. No coating was used on the anodic side of the sample.

The anodic potential in potentiostatic permeation experiments was determined by potentiodynamic polarisation measurements for carbon steel S235 and S700 in 0.2 M NaOH solution at room temperature. Anodic polarisation curve from the open circuit was measured with a potential sweep rate 100 mV/min.

Hydrochloric acid diluted to 16.5% (5.4 M) was used in the permeation measurements to produce hydrogen on steel surface. The acid was used both without dissolved iron and with Fe^{2+} concentrations 14 g/L (0.25 M) and 45 g/L (0.8 M) to simulate a used pickling bath. Iron was added as $FeCl_2{\cdot}4H_2O$. Temperature of the solution was 20°C. The sample was installed between the two compartments of the cell, and 1 L of 0.2 M NaOH solution was added into the anodic side of the test cell. Thereafter, sample was polarised to a potential of 0 mV versus SCE to form a passive layer on the surface, and the current was measured up to steady state situation (about 15 h). The passive film in alkaline solution is usually magnetite [16, 17]. The passivation procedure causes initially an increase in the current due to the corrosion of the steel when the oxide layer is formed, and this current decreases with time as the passive film reaches a steady state. In the passive state, steel dissolves, but at a low and constant rate. When the passive film had reached a steady state, hydrochloric acid (1 L) was introduced in the cathodic compartment. When hydrogen started to diffuse to the exit side a current increase was seen. When the current reached

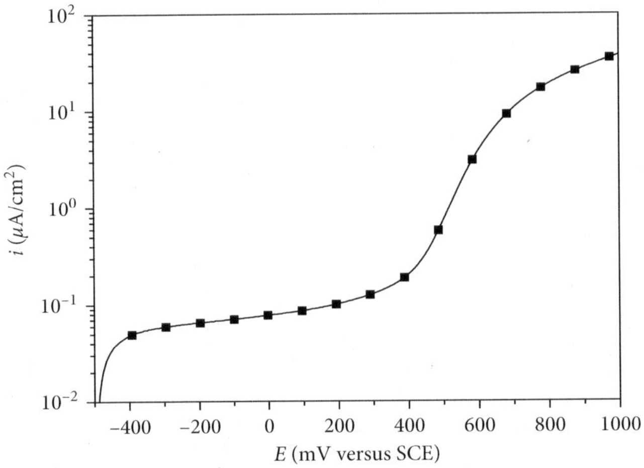

FIGURE 4: The polarisation curve of carbon steel S253 in 0.2 M NaOH at 20°C.

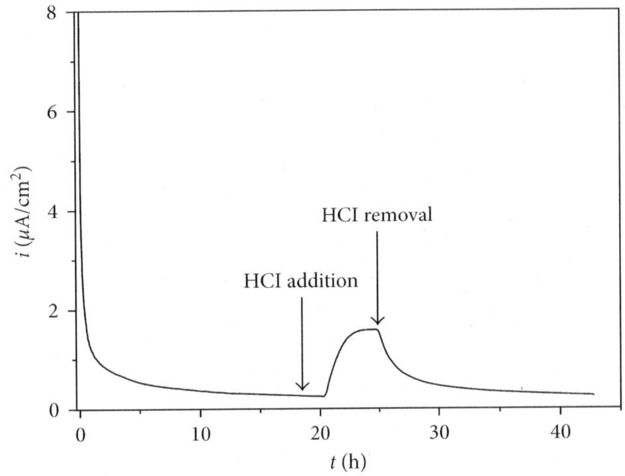

FIGURE 5: A typical current density versus time curve measurement in passivation and permeation experiment in 16.5% HCl for steel S355.

a steady maximum level, the acid was removed through the bottom valve, and after some time, a decrease of the anodic current was observed. During the whole measurement, the anodic side of the sample was kept at constant potential. In the beginning of the test, the current was due to anodic oxidation and passivation of steel. Oxidation of hydrogen that had diffused to exit side was seen as current increase. When hydrogen was stopped by removing the acid, current started to decrease and reached finally the same passive current level as before the introduction of acid. This background current is subtracted from the current measured during the permeation measurements. Thus, the effect of only hydrogen diffusion on the current could be observed.

To increase the uptake of hydrogen into the metal, recombination poisons such as arsenic, hydrogen sulphide, and thiourea can be used [18]. These elements block adsorption sites on electrode surface diminishing the coverage by adsorbed hydrogen. The adsorbed H is an intermediate resulting from discharge of H^+, and its recombination produces H_2 in the cathodic hydrogen evolution. No tests were done with hydrogen poisons as they are not used in pickling baths. The use of recombination poisons was not found to be necessary as detectable amount of hydrogen was generated from the concentrated hydrogen chloride solutions used in this work.

3. Results and Discussion

The open circuit potential of carbon steel S253 was −490 mV versus SCE and a wide passive range up to the potential of +400 mV versus SCE was found in 0.2 M NaOH solution (Figure 4). Based on these results, the potential of 0 mV versus SCE was chosen for the permeation measurements. The selected potential is on the passive range of tested steels, and it is higher than the equilibrium potential of H^+/H_2 reaction so oxidation of diffused hydrogen was guaranteed. The current density at this potential is less than 0.1 μA/cm^2. The dissolution rate of steel is at very low level, approximately 1 μm/year.

A typical current density versus time curve in an experiment on steel S355 is shown in Figure 5. The current is related only to the reaction rates on the steel surface in the anodic compartment containing NaOH. A high current density is observed at the beginning of the experiment, when the steel surface is still in active state, and much lower current density of about 0.3 μA/cm^2 was reached when the steel passivates. Hydrochloric acid was added on the cathodic side of the cell after 18 h when the anodic surface had reached the passive state. After a delay of over two hours, an increase in the current was seen in the anodic compartment. This current increase is due to hydrogen that had diffused through the steel and was oxidized on the anodic side. A steady maximum was observed after about 5 h from the acid addition indicating steady flux of hydrogen through the steel sample. At this point, the hydrochloric acid was removed from the cathodic side of the cell. The anodic current density started to decrease as no more hydrogen was produced on the acid side, and hydrogen charged into the steel was consumed. At the end of the test, the current was approximately same passive current as before acid addition.

Results for hydrogen permeation experiments for high strength steel are shown in Figure 6. The curves show only the part after adding HCl in the cathodic compartment. The hydrogen penetration time through the sample is similar for all steels with similar thickness, as shown in Figure 6. This is seen as an increase in the current density after HCl addition. The maximum current density is lowest for the steel S900 and highest for steel S500. The grade of the steel does not have direct correlation with the maximum current. The maximum current density gives the maximum penetration rate of hydrogen through the steel. Theoretically, integration of current versus time curve gives the amount of hydrogen penetrated through the sample.

Introduction of iron into the pickling acid has only a minor effect on the current versus time curve (Figure 7). As an example, the influence of ferrous ion concentrations 0 g/L, 14 g/L, and 45 g/L on hydrogen entry for S700 steel is

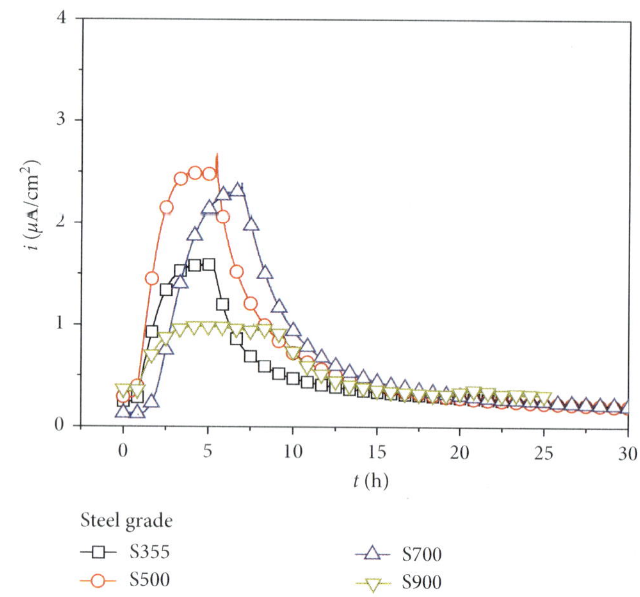

FIGURE 6: Permeation of hydrogen in 16.5% HCl for different steel grades with sample thickness 1 mm.

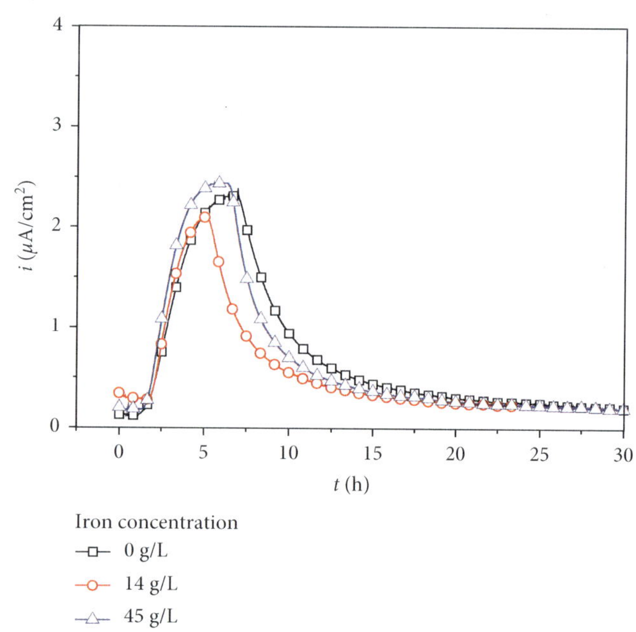

FIGURE 7: The influence of iron content of the HCl on the permeation in steel S700.

shown in Figure 7. Dissolved iron had only a minor effect on the permeation of hydrogen in the studied steels.

The hydrogen diffusion coefficient was calculated by using four different methods given in [19, 20]:

(1) The time lag method

(2) Breakthrough time

(3) Half-rise time

(4) Analysis of the beginning decay transient

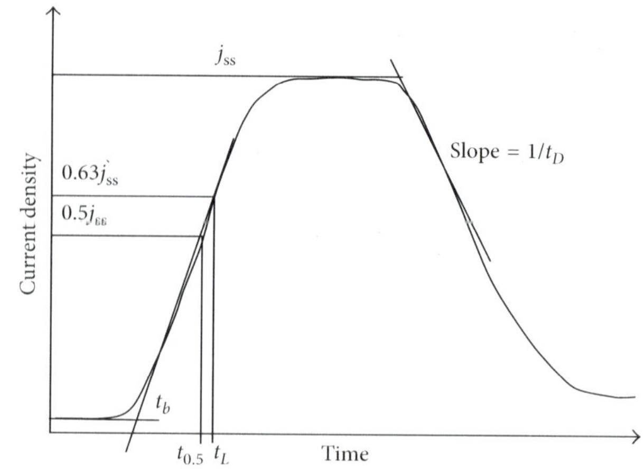

FIGURE 8: Determination of characteristic times for calculation of diffusion coefficients: time lag t_D, breakthrough time t_b, half-rise time $t_{0.5}$, and decay time constant t_D.

The analysis methods are based on the solution of the diffusion equation with constant concentration boundary conditions. The solution for the charging transient, derived by separation of variables, is shown in the following equation:

$$I = nFAD\frac{c_1}{L}\left[1 + 2\sum_{n=1}^{\infty}(-1)^n \exp\left(\frac{-Dn^2\pi^2 t}{L^2}\right)\right], \quad (1)$$

and for the decay transient (both ends at zero concentration and linear initial condition), the following solution is obtained:

$$I = nFAD\frac{c_1}{L}2\sum_{n=1}^{\infty}(-1)^{n+1} \exp\left(\frac{-Dn^2\pi^2 t}{L^2}\right). \quad (2)$$

Figure 8 shows the principles for determination of characteristic times for calculation of diffusion coefficients.

In the time lag method, analysis is done by integrating the rising curve to calculate the quantity of hydrogen, which has permeated through the sample as a function of time. An extrapolation of the plot of quantity against time gives the time lag t_L which is related to the diffusion constant by

$$D_L = \frac{1}{6}\cdot\frac{L^2}{\pi^2\cdot t_L}. \quad (3)$$

The t_L in (3) can be obtained by determining the time at which the rate of permeation is 0.63 times the steady-state value described by current density j_{ss} [19].

In the breakthrough time method, the time t_b is obtained by drawing tangent in the rising part of the curve. The tangent crosses the time axis at point t_b [19]:

$$D_B = 0.5\cdot\frac{L^2}{\pi^2\cdot t_b} \quad (4)$$

In the half-rise time method, the characteristic time $t_{0.5}$ corresponds to the current density that is 0.5 times the current density at steady-state situation j_{ss}. Diffusion coefficient can then be calculated as follows [19]:

TABLE 2: Average values of diffusion coefficients (cm^2/s) for test materials calculated using the four methods.

Method/steel	355 MPa	500 MPa	700 MPa	900 MPa
$D_{0.5}$	$1.82{\cdot}10^{-7}$	$1.6{\cdot}10^{-7}$	$1.61{\cdot}10^{-7}$	$2.5{\cdot}10^{-7}$
D_L	$1.16{\cdot}10^{-7}$	$8.78{\cdot}10^{-8}$	$3.2{\cdot}10^{-8}$	$2.54{\cdot}10^{-8}$
D_B	$1.96{\cdot}10^{-7}$	$1.49{\cdot}10^{-7}$	$1.73{\cdot}10^{-7}$	$1.93{\cdot}10^{-7}$
D_D	$1.76{\cdot}10^{-7}$	$1.81{\cdot}10^{-7}$	$9.76{\cdot}10^{-8}$	$1.26{\cdot}10^{-7}$

$$D_{0.5} = 0.138 \cdot \frac{L^2}{t_{0.5}}. \quad (5)$$

In the decay transient analysis, a characteristic time constant t_D is determined using the slope of current decay after the steady state. The obtained current versus time is corrected for offsets and divided by its maximum value at the steady state. Taking the logarithm of the obtained normalized current, one obtains for the beginning decay transient a linear section from which the slope is the inverse decay time constant [20]:

$$D_D = \frac{1}{t_D}\frac{L^2}{\pi^2}. \quad (6)$$

When the sample thickness increases, the diffusion coefficient value in replicate tests seems to become more reproducible. At sample thicknesses of 1 mm, the diffusion coefficient values were from 10^{-9} to 10^{-6} cm^2/s. When the sample thickness was 1.5–2 mm, the range was from 10^{-8} to 10^{-6} cm^2/s and with sample thicknesses of 4 mm from 10^{-7} to 10^{-6} cm^2/s. This effect was clearer with 500 MPa and 700 MPa samples. This effect agrees with the finding of Charca et al. [21] by using Armco iron. It is generally easier to attain solubility limits with thinner metallic samples than thicker ones, which contain more material defects and hence trapping sites.

The diffusion coefficient was not dependent on the analysis method. The coefficient decreased slightly with increasing strength. The results presented in Table 2 show that both charging and discharging transients have similar diffusion coefficients. This is important for the argumentation, whether deep traps are significantly involved and filled during the charging.

Filling of traps would lead to an internal accumulation of hydrogen, and this hydrogen cannot be detected in the charging transient anymore. Since deep traps are preferentially filled before the hydrogen can diffuse on, the penetration of the hydrogen is delayed, which would show up in the analysis as changed diffusion coefficients. For the decay transient, when all the deep traps were already filled, the diffusion of the hydrogen would proceed normally. Thus, whenever there would be deep traps present, this should be visible as a significant difference in the diffusion coefficient determination methods for charging (3–5) compared to the discharge method (6). That the diffusion coefficients obtained from the decay transient are similar to the other method shows, thus, that an involvement of traps for the tested steels can be neglected. That iron-based samples do not have a significant number of traps has been already mentioned in [22].

To evaluate a concrete problem case, the here obtained diffusion coefficients can be used to simulate the penetration of hydrogen during pickling and the release of it during the time between pickling and hot dip galvanizing. This, however, is very much depending on the dimensions of the steel product and on the degree of deformation within that product. All this we leave for future investigations.

4. Conclusions

The purpose of this paper was to study the intrusion of hydrogen during pickling of steel. Hydrogen permeation can be caused by pickling acid, such as HCl, instead of cathodic hydrogen production that is mainly used in investigations found in the literature.

No protective coating, for example, palladium on the anodic side of the sample, is required. Permeation measurement needs to be carried out after passivation of steel, and this low background current is to be subtracted from the measured current before the calculation of the diffusion coefficient.

The permeation rate of hydrogen into the steel and desorption rate from the steel are slightly lower for higher strength steel. This means that if these steels are plated soon after pickling, for example in hot-dip galvanizing, hydrogen has no time to leave steel and hydrogen-induced failures may occur.

Iron dissolved in pickling acid has no essential effect on the permeation of hydrogen.

Conflicts of Interest

The authors declare that they have no conflicts of interest.

References

[1] D. Schaeffler, "Introduction to advanced high-strength steels–Part I," *Stamping Journal*, vol. 16, pp. 22–28, 2004.

[2] T. P. Groeneveld, E. E. Fletcher, and A. R. Elsea, "A study of hydrogen embrittlement of various alloys, National Aeronautics and Space Administration Report,"Annual Summary Report, p. 40, NASA, Washington, DC, USA, 1966, https://ntrs.nasa.gov/archive/nasa/casi.ntrs.nasa.gov/19660025404.pdf.

[3] AZoM, *Galvanised Steel-Embrittlement Due to Hot Dip Galvanising*, AZoM, Sydney, Australia, 2002, https://www.azom.com/article.aspx?ArticleID=1407.

[4] A. El Hajjami, M. Gigandet, M. De Petris-Wery et al., "Hydrogen permeation inhibition by zinc–nickel alloy plating on steel XC68," *Applied Surface Science*, vol. 255, no. 5, pp. 1654–1660, 2008.

[5] E. Hörnlund, J. K. T. Fossen, S. Hauger, C. Haugen, T. Havn, and T. Hemmingsen, "Hydrogen diffusivities and concentrations in 520 M carbon steel under cathodic protection in 0.5 M NaCl and the effect of added sulphite, dithionite, thiosulphate, and sulphide," *International Journal of Electrochemical Science*, vol. 2, pp. 82–92, 2007.

[6] J. Ćwiek and K. Nikiforov, "Hydrogen degradation of high-strength weldable steels in seawater," *Materials Science*, vol. 40, no. 6, pp. 831–835, 2004.

[7] N. Amokrane, C. Gabrielli, G. Maurin, and L. Mirkova, "Effect of organic additives on hydrogen permeation into an iron

membrane studied by frequency analysis techniques," *Electrochimica Acta*, vol. 53, no. 4, pp. 1962–1978, 2007.

[8] J. O'M Bockris and P. K. Subramanyan, "Hydrogen embrittlement and hydrogen traps," *Journal of the Electrochemical Society*, vol. 118, no. 7, pp. 1114–1119, 1971.

[9] R.-H. Sang, S. Pyun Il, and R. A. Oriani, "Hydrogen permeation through the passivation film on iron by time-lag method," *Journal of the Electrochemical Society*, vol. 137, no. 6, pp. 1703–1706, 1990.

[10] M. A. V. Devanathan and Z. Stachurski, "The adsorption and diffusion of electrolytic hydrogen in palladium," *Proceedings of the Royal Society London, Series A Mathematical and Physical Sciences*, vol. 270, no. 1240, pp. 90–102, 1962.

[11] P. Manolatos, M. Jerome, C. Duret-Thuali, and J. Le Coze, "The electrochemical permeation of hydrogen in steels without palladium coating. Part I: interpretation difficulties," *Corrosion Science*, vol. 37, no. 11, pp. 1773–1783, 1995.

[12] Q. Liu and A. Atrens, "A critical review of the influence of hydrogen on the mechanical properties of medium-strength steels," *Corrosion Reviews*, vol. 31, no. 3–6, pp. 85–103, 2013.

[13] B. Ramesh Babu and R. Holze, "Corrosion and hydrogen permeation inhibition for mild steel in HCl by isomers of organic compounds," *British Corrosion Journal*, vol. 35, no. 3, pp. 204–209, 2000.

[14] Y. Cheng, L. Yang, and F. King, "Analysis of hydrogen permeation through pipeline steel in near-neutral pH SCC environments," in *Proceedings of 2000 International Pipeline Conference*, vol. 2, pp. 1479–1485, Alberta, Canada, October 2000.

[15] J. Yang, L. Zhang, L. Xu, and M. Lu, "Influence of H_2S and CO_2 corrosion scales on hydrogen permeation in X65 steel," *Corrosion*, vol. 08412, p. 8, 2008.

[16] R. O. Rihan and S. Nesic, "Erosion–corrosion of mild steel in hot caustic. Part I: NaOH solution," *Corrosion Science*, vol. 48, no. 9, pp. 2633–2659, 2006.

[17] T. D. Burleigh, P. Schmuki, and S. Virtanen, "Properties of the nanoporous anodic oxide electrochemically grown on steel in hot 50% NaOH," *Journal of the Electrochemical Society*, vol. 156, pp. C45–C53, 2009.

[18] S. Y. Qian, B. E. Conway, and G. Jerkiewicz, "Kinetic rationalization of catalyst poison effects on cathodic H sorption into metals: relation of enhancement and inhibition to H coverage," *Journal of the Chemical Society, Faraday Transactions*, vol. 94, no. 19, pp. 2945–2954, 1998.

[19] B. S. Chaudhari and T. P. Radhakrishnan, "Boundary conditions during the electropermeation of hydrogen through pure iron," *Corrosion Science*, vol. 30, no. 12, pp. 1219–1234, 1990.

[20] J. O'M Bockris and M. A. V. Devanathan, "The determination of the coverage on nickel and steel during electrolytic hydrogen evolution," Tech. Rep. 4, Contract No. 551(22) NR036-028, Office of Naval Research, p. 41, Arlington, VL, USA, 1961, http://www.dtic.mil/docs/citations/AD0254474.

[21] S. Charca, O. Uwakweh, B. Shafiq, and V. Agarwala, "Characterization of hydrogen permeation in armco-fe during cathodic polarization in aqueous electrolytic media," *Journal of Materials Engineering and Performance*, vol. 17, no. 1, pp. 127–133, 2008.

[22] C. Gabrielli, G. Maurin, L. Mirkova, H. Perrot, and B. Tribollet, "Transfer function analysis of hydrogen permeation through a metallic membrane in a Devanathan cell. I. Theory," *Journal of Electroanalytical Chemistry*, vol. 590, no. 1, pp. 1–14, 2006.

11

Experimental Study on Basic Mechanical Properties of Steel Fiber-Reinforced Siliceous Wet Shotcrete

Xu Yan, Li-min Liu, Jin-peng Zhang, Yan-hui Li, and He Wang

College of Mining and Safety Engineering, Shandong University of Science and Technology, Qingdao 266590, China

Correspondence should be addressed to Xu Yan; yanxuhhu@163.com

Academic Editor: Paolo Andrea Carraro

In order to solve the problem of low strength and easy cracking of shotcrete in permanent support of tunnel single shell lining, the effects of steel fiber-reinforced siliceous on mechanical properties of wet shotcrete were investigated by the tests of compressive strength, splitting tensile strength, flexural strength, shear strength, and bending toughness. The strength and the cracking resistance mechanism of steel fiber-reinforced siliceous wet shotcrete were analyzed by the bending toughness evaluation method. The results show that (1) the steel fiber-reinforced siliceous can improve splitting tensile, flexural, and shear capacity of the shotcrete, and the maximum growth rates were 77.42%, 72.73%, and 98.31%. The steel fiber plays a major role, and silica fume plays a subsidiarity role. (2) The effect of different types and contents of steel fibers on the compressive strength of wet shotcrete are not obvious. (3) The strengthening effect of shear undaform steel fiber on concrete is obviously better than that of shear flat steel fiber. (4) The flexural toughness of wet shotcrete with the silica content of 10% and the shear steel fiber of 60 kg/m^3 is the best.

1. Introduction

With the rapid reduction of traditional resources, the scarcity of resources has become a key factor in restricting economic development. At present, as a raw material widely used in the field of engineering infrastructure, concrete has a huge demand and consumption. So the research on high-efficient and environmental-friendly concrete to replace traditional consumption resources is the key to the development of modern concrete. Steel fiber-reinforced siliceous wet shotcrete is a kind of new composite material, which is mixed with steel fiber, silica fume, and high-performance admixtures in common concrete, and the wet method is used in the spray technology. In recent years, shotcrete and cast-in-situ concrete have been widely used in tunnel support field. Cast-in-situ concrete needs three processes including supporting-formwork, pouring, and demoulding, so the operation is complicated and the cost is high. But shotcrete can be used in high pressure to carry out long-distance transportation, and the supporting efficiency is high. In addition, shotcrete technology can be divided into spray drying and wet spurt. The wet spurt technology has been widely used in the process of tunnel construction, because of its low spring rate and less dust.

The permanent support technique of tunnel single shell lining is mainly divided into three steps. The first step is that waterproof concrete is sprayed in time, after the roadway is excavated. The second step is that cable anchor and steel frame support are carried out according to the surrounding rock classification. The third step is that one or more layers of reference concrete are sprayed. The reference concrete belongs to brittle material, which has many problems such as insufficient mechanical strength index, high hydration heat, and serious self shrinkage. Moreover, the complex force of the shotcrete lining structure in the permanent support of single lining of tunnel, which causes the shotcrete prone to cracking, water inrush, and mud sudden accident [1–3]. Therefore, studying on the mix design method [4] and the mechanical properties of shotcrete [5, 6] has become a key in permanent support of tunnel single shell lining. In the past years, schol1ars had done a lot of research on shotcrete at home and abroad. For example, Zhang et al. [7] had studied the effect of silica fume and fly ash on the performance of shotcrete through laboratory tests. The result of the tests concluded that the optimum ratio of new shotcrete can effectively reduce the amount of cement, the cracking, the rebound, and dust. Liu et al. [8] analyzed the shrinkage performance of silica fume concrete through the orthogonal test. The research result indicated that the ratio of water to cement, sand rate, and silica fume addition significantly affected the shrinkage performance of silica fume concrete. Ardalan et al. [9] and Fallah and Nematzadeh [10] found that the strength and durability of shotcrete can be improved by mixing silica fume. Wang et al. [11] and Su et al. [12] found that the mixing of steel fiber can effectively improve the mechanical properties of shotcrete. Naqvi et al. [13] explained the seismic behavior principle of steel fiber shotcrete. Nguyen et al. [14] studied on the impact resistance of steel fiber shotcrete. Nehdi et al. [15] and Hoang and Fehling [16] studied on the effect of steel fiber with different lengths and contents on mechanical properties of shotcrete. Marar et al. [17] discussed the effect of steel fiber with different aspect ratios and contents on shear strength of shotcrete. Yoo et al. [18] analyzed the effect of deformed steel fiber on strength and bending toughness of shotcrete. Nattaj and Nematzadeh [19] obtained the effect of steel fiber with different contents on mechanical properties of silica fume shotcrete. Ghavidel et al. [20] tested on the stability of shotcrete during the steel fiber was pulled out of the shotcrete matrix. Abbass et al. [21] obtained the evaluation method of steel fiber-reinforced concrete with different strength. Zhang et al. [22] had studied the flexural toughness of steel fiber-reinforced wet concrete.

The above scholars had studied the mechanical properties of steel fiber or silica fume shotcrete, but little research had been done on the mechanical properties of steel fiber-reinforced siliceous wet shotcrete. Therefore, 10 groups of contrast tests were designed in this paper. The effects of steel fiber-reinforced siliceous on mechanical properties of wet shotcrete were investigated by the test of compressive strength, splitting tensile strength, flexural strength, shear strength, and bending toughness. Finally, the optimum mix ratio of wet shotcrete was obtained. The new steel fiber-reinforced siliceous wet shotcrete makes full use of industrial waste to replace nonrenewable raw materials, which has green environmental properties. It also significantly enhances the mechanical strength index, antiseepage, and anticracking ability. Moreover, it not only improves the supporting effect of tunnel engineering, but also avoids the occurrence of water bursting and mud outburst.

2. Experimental

2.1. Raw Material. The shotcrete of this test adopted No. 42.5 Portland cement. The fine aggregate was sand, and the fineness modulus was 2.64. The well-graded coarse aggregate was crushed stone, and the particle size was 5–10 mm. Silica fume was produced by Ark New Materials (Shandong) Co., Ltd. The performance parameters of silica fume are shown in Table 1. Steel fiber was produced by Huaxing(Yueyang)Co., Ltd. It was divided into two types: shear undaform and shear flat. Its specific parameters are shown in Table 2. The admixture adopted a low alkaline liquid setting accelerator of FZ-100C. Photographs of steel fiber and silica fume are shown in Figure 1.

2.2. Mix Proportion Design of Wet Shotcrete. The specification GB 50086-2015 of bolt shotcrete support technology was used to carry out the mix proportion design of wet shotcrete. The dust, resilience, and pumping capacity of wet shotcrete were also considered. The apparent density of shotcrete was 2400 kg/m^3. The ratio of cement to sand was 1 : 2.4. The water-cement ratio was 0.55. The content of gravel was 598 kg/m^3. The silica fume was 10% cement quality. The content of accelerator was 1.2 kg/m^3, and accelerator was 5% cement quality. The maximum content of steel fiber was 60 kg/m^3. There were 10 groups of contrast tests, and the specific mix proportion of each group is shown in Table 3.

2.3. Experimental Method. The experiment was divided into ten groups, with 15 specimens in each group. The 6 cube specimens of 100 mm × 100 mm × 100 mm were used to carry out the tests of compressive and splitting tensile strength. The 3 cube specimens of 100 mm × 100 mm × 300 mm were used to carry out the test of shear strength. The 6 rectangular specimens of 100 mm × 100 mm × 400 mm were used to carry out the tests of flexural and bending toughness strength. This test adapted forced action mixer of 60 L. The time of entry template was 4 minutes later, and the time of dismantling template was 24 h later. The vibrating time was 60 s. The specimens shall be carried out for 28 days at standard curing condition. The electronic universal testing machine CSS44100 of 300 kN was used by this test, and the method of steel fiber concrete JG/T 472–2015 was used to carry out the continuous loading on shotcrete specimens. The test of bending toughness adopted the loading method of four-point bending, and the loading rate was 0.2 mm/min. When the discreteness was lower, the wet shotcrete intensity of this group was defined as the average value of the intensities of three specimens. When it was higher, the wet shotcrete intensity of a group was defined as the median value of the intensities of three specimens if the difference between the maximum or minimum value and the median value was larger than 15 percent of the median value, if not, this experiment was invalid. The electronic universal testing machine is shown in Figure 2.

3. Results and Discussion

3.1. Combined Effect of Silica Fume and Steel Fiber. Table 4 shows the test results of compression, splitting tensile, flexural, shear, and first-crack strength.

Comparison of group 2 and group 1 shows that when silica fume was mixed only, the compressive strength, splitting tensile strength, flexural strength, shear strength, and initial splitting strengths of silica fume concrete were improved by 15.94%, 7.53%, 37.97%, 50.56%, and 18.92%, respectively, compared with the reference concrete. This showed that silica fume had a good enhancement effect. The silica fume of high pozzolanic activity and water did not

TABLE 1: Silica fume performance parameter.

Average particle size (μm)	Specific surface area ($m^2 \cdot kg^{-1}$)	Fineness (%)	Density (g/cm^3)
0.1	15000	0.05	2.3

TABLE 2: Steel fiber performance parameter.

Steel fiber type	Length (mm)	Equivalent diameter (mm)	Aspect ratio	Tensile strength
Shear undaform	30	0.3	100	Greater than 1150 MPa
Shear flat	25	0.3	83	Greater than 1150 MPa

(a) (b) (c)

FIGURE 1: (a) Shear undaform steel fiber. (b) Shear flat steel fiber. (c) Silica fume.

TABLE 3: Mix proportion of wet shotcrete.

Group	Water-cement ratio	Accelerator (kg/m^3)	Gravel (kg/m^3)	Silica fume	Steel fiber shape	Steel fiber content (kg/m^3)
1	0.55	1.2	598	Nothing	Nothing	0
2	0.55	1.2	598	10% cement quality	Nothing	0
3	0.55	1.2	598	Nothing	Shear undaform	50
4	0.55	1.2	598	Nothing	Shear flat	50
5	0.55	1.2	598	10% cement quality	Shear undaform	50
6	0.55	1.2	598	10% cement quality	Shear flat	50
7	0.55	1.2	598	10% cement quality	Shear undaform	60
8	0.55	1.2	598	10% cement quality	Shear flat	60
9	0.55	1.2	598	10% cement quality	Shear undaform	40
10	0.55	1.2	598	10% cement quality	Shear flat	40

involve in the hydration reaction but silica fume and hydration product $Ca(OH)_2$ of cement involved in the secondary hydration reaction to produce C-S-H gel, thus improving viscosity of the mixture. In addition, the microscopic pore structure of hardened cement paste can be filled with silica fume with microaggregate effect and secondary hydration products. Not only it can improve the microscopic pore structure but also increase the density and strength of silica fume concrete.

Comparison of group 3 and group 1 shows that when shear undaform steel fiber was mixed only, the compressive strength, splitting tensile strength, flexural strength, shear strength, and initial splitting strength of steel fiber concrete were improved by 18.12%, 33.33%, 39.57%, 78.09%, and 28.38%, respectively, compared with the reference concrete. It concluded that the crack resistance of steel fiber was better than that of silica fume. Moreover, the splitting tensile strength, flexural strength, and shear strength of the steel fiber concrete were obviously improved. Due to gravity, steel fibers and solid particulate matter took place subsidence during the hardening process of steel fiber-reinforced concrete, but the distributed steel fiber hindered the subsidence of the solid particles, and it played a reinforcing role, which increased the density of the steel fiber concrete and reduced the occurrence of the original microcracks. In addition, the bridging action of the steel fiber across the cracks prevented the coalescence of microcracks, and it reduced the depth and number of cracks, thus played a role in crack resistance.

Comparison of group 5 and group 1 shows that when shear undaform steel fiber of 50 kg/m^3 and silica fume were mixed together, the compressive strength, splitting tensile

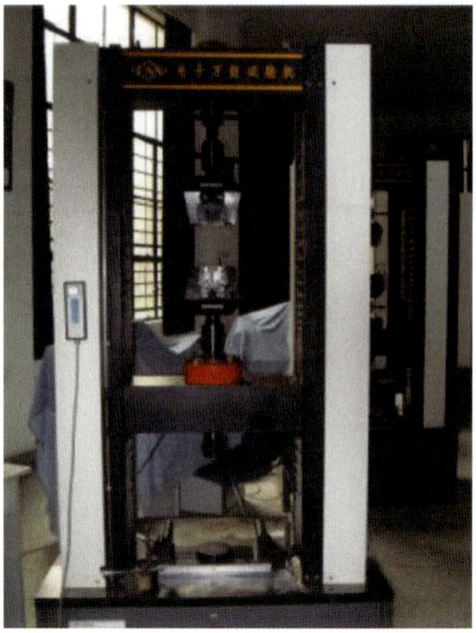

FIGURE 2: Electronic universal testing machine.

TABLE 4: Test results of compression, splitting tensile, flexural, shear, and first-crack strength.

Group	Compressive strength (MPa)	Splitting tensile strength (MPa)	Bending strength (MPa)	Shear strength (MPa)	First-crack strength (MPa)
1	13.8	0.93	1.87	1.78	1.48
2	16	1	2.58	2.68	1.76
3	16.3	1.24	2.61	3.17	1.9
4	16.9	1.1	2.68	3.08	1.88
5	17.1	1.62	3	3.3	1.97
6	17	1.48	2.89	3.13	1.95
7	17.8	1.65	3.23	3.53	2.03
8	17.4	1.61	3.11	3.44	1.98
9	16.8	1.52	2.7	2.83	1.95
10	16.6	1.31	2.59	2.73	1.93

strength, flexural strength, shear strength, and initial splitting strength of steel fiber-reinforced siliceous shotcrete were improved by 23.91%, 74.19%, 60.43%, 85.39%, and 33.11%, respectively, compared with the reference concrete. It found that the edge of the reference concrete was broken and dropped seriously, and there was an explosive spalling phenomenon during the test. Finally, brittle failure took place. The steel fiber-reinforced siliceous shotcrete was relatively complete during the test, and there was no shedding dregs phenomenon. Finally, ductile failure took place. The C-S-H gel produced by secondary hydration can increase the cohesion force between the matrix and the steel fiber. So, it not only decreased segregation and bleeding of aggregate but also reduced the shrinkage of hardening cement paste.

By comparing the data, it is concluded that the compressive strength of steel fiber-reinforced siliceous shotcrete was not very different from the former two kinds of concrete. When shotcrete was subjected to axial pressure, the lateral expansion deformation of the shotcrete was restrained. Because of the cohesive force between the steel fiber and the matrix, the compressive strength can only be indirectly improved. Therefore, the effect of compression was not obvious.

But the splitting tensile strength, flexural strength, shear resistance, and initial crack strength of steel fiber-reinforced siliceous shotcrete had obvious effect on enhancement and crack resistance. However, the steel fiber and the matrix beared the load together, which hindered the extension and expansion of crack in the initial loading process. When the ultimate limit load of the matrix was reached, the steel fibers of across or near the crack tip played a bridging action, which forced the crack to change the direction of development or destroy the steel fiber. When the ultimate limit load was reached, pulling-out and shearing phenomena took place during the test. The pulling out, debonding, and destruction of steel fibers with high modulus of elasticity consumed most of the fracture energy, so it can greatly increase the splitting tensile strength, flexural strength, and shear strength of steel fiber-reinforced siliceous shotcrete.

Finally, the macrocracks of steel fiber-reinforced siliceous shotcrete were expanding and coalescence. Ductile failure took place because of the short bearing capacity of steel fiber and matrix together. A comparison of compressive and shear failure of steel fiber-reinforced siliceous shotcrete is shown in Figures 3 and 4.

3.2. Influence of Steel Fiber Content and Type. Comparison of groups 1 and 5–10 in Table 4 shows that when the content of shear undaform steel fiber was increased from 40 kg/m^3 to 50 kg/m^3 and 60 kg/m^3, the growth rate of compressive strength and initial crack strength of steel fiber-reinforced siliceous shotcrete was little changed. For compressive strength, the content of steel fiber can not change the lateral expansion limited deformation and then indirectly increased the compressive strength. For the initial crack strength, the steel fiber and the matrix beared the load together before the initial crack. The matrix was unloaded but the load was carried by the steel fiber after the initial crack. Finally, it reached the limit load. The initial crack strength of steel fiber-reinforced siliceous shotcrete was mainly controlled by the matrix, while the limited strength was mainly controlled by steel fiber. Therefore, the content of steel fiber had little influence on the initial crack strength of steel fiber-reinforced siliceous shotcrete.

When the content of steel fiber was increased from 40 kg/m^3 to 50 kg/m^3, the growth rate of splitting tensile strength, flexural strength, and shear strength of steel fiber-reinforced siliceous shotcrete was very greatly changed; but from 50 kg/m^3 to 60 kg/m^3, the growth rate was gradually decreased. With the increasing content of steel fiber and nonmixing uniformity of shotcrete, which reduced the slump and workability of shotcrete. So it gradually increased porosity and microcracks. Some steel fibers appeared to the agglomerates by observing the cross section of steel fiber-reinforced siliceous shotcrete.

Comparison of group 5 and group 6, group 7 and group 8, and group 9 and group 10 shows that the shape of the steel fiber had little effect on the compressive strength, but the splitting tensile strength, flexural strength, shear strength, and initial crack strength of mixing the shear undaform steel fiber were about 10% higher than that of mixing the shear flat steel fiber. Because the special section of shear undaform steel fiber had anchoring effect, and the bonding area was large between shear undaform steel fiber and matrix, which increased the cohesion force. So pulling out, tensile failure, and shearing of undaform steel fiber required higher energy. The steel fiber-reinforced siliceous shotcrete section is shown in Figure 5.

3.3. Analysis and Evaluation of Bending Toughness of Wet Shotcrete. According to the JG/T 472-2015 test method of steel fiber concrete, the bending toughness index and the variation coefficients of load-carrying capability were calculated. The bending toughness of wet shotcrete was evaluated by comparing two parameters with two parameters of the ideal elastoplastic material. The bending toughness index was expressed by I_5, I_{10}, and I_{30}, and the variation coefficients of load-carrying capability was expressed by $\xi_{m,n,5}$, $\xi_{m,n,10}$, and $\xi_{m,n,30}$. The bending toughness test results are shown in Table 5, and the load deflection curves of different types of wet shotcrete are shown in Figure 6.

As is shown in Table 5, when silica fume was mixed only, wet shotcrete took place brittle failure rapidly after reaching the initial crack strength. So silica fume can not improve the bending toughness of wet shotcrete. When shear undaform steel fiber of 50 kg/m^3 was mixed only, the I_5, I_{10}, and I_{30} of wet shotcrete was 4.86, 9.26, and 24.91, which was significantly higher than that of reference concrete. When shear undaform steel fiber of 50 kg/m^3 and silica fume were added together, the I_5, I_{10}, and I_{30} of wet shotcrete were increased by 0.35, 0.47, and 1.96, respectively, compared with steel fiber concrete. Although the bending toughness index had been improved to some extent, it had not changed greatly. When the content of shear undaform steel fiber was increased from 40 kg/m^3 to 50 kg/m^3, the bending toughness index had changed greatly. But from 50 kg/m^3 to 60 kg/m^3, its growth rate was decreased obviously.

As is shown in Figure 6, the peak load of the seventh group of shotcrete was maximum. At first, the bending deformation ability was not as good as that of the ninth group, and the ninth group was declined faster. But the bending deformation ability of the seventh group was more than the ninth group beyond a certain deflection. When the content of steel fiber exceeded a certain limit, the improvement effect of bending toughness was weakened. The bending toughness index of mixing the shear undaform steel fiber was slightly larger than that of mixing the shear flat steel fiber. The change trend of the variation coefficients of load-carrying capability was similar to the bending toughness index, and the change mechanism of bending toughness was similar to splitting tensile strength. The content of silicon fume was cement quality of 10% and shear undaform steel fiber was 60 kg/m^3, and the bending toughness of wet shotcrete was the best by comprehensive analysis.

4. Conclusions

(1) The mixing of steel fiber and silica fume can significantly improve the splitting tensile strength, flexural strength, shear strength, and bending toughness of wet shotcrete. The steel fiber played a major role, and silica fume played a subsidiarity role. When the content of shear undaform steel fiber was 60 kg/m^3, the maximum growth rate of the shear strength was 98.31%.

(2) The effect of different types and contents of steel fibers on the compressive strength of wet shotcrete was not obvious. When shear undaform steel fiber and silica fume were mixed together, the maximum growth rate of the compressive strength was 28.99%.

(3) When steel fiber and silica fume were mixed together, the splitting tensile strength, flexural strength, shear strength, and initial crack strength of mixing the shear undaform steel fiber were about 10% higher than that of mixing the shear flat steel fiber.

(a) (b)

Figure 3: Compression failure true thickness. (a) Reference concrete. (b) Steel fiber-silica fume shotcrete.

(a) (b)

Figure 4: Shear failure true thickness. (a) Reference concrete. (b) Steel fiber-silica fume shotcrete.

(a) (b)

Figure 5: Section of steel fiber-reinforced silica fume concrete.

(4) When silica fume was mixed only, it can not improve the bending toughness of wet shotcrete. When steel fiber and silica fume were mixed together, the bending toughness index had a little improved. The seventh group of wet shotcrete was the optimum mixture ratio; moreover, the bending toughness

TABLE 5: Test results of bending toughness.

Group	Cracking load (N)	Cracking deflection (0.01 mm)	I_5	I_{10}	I_{30}	$\xi_{m,n,5}$	$\xi_{m,n,10}$	$\xi_{m,n,30}$
1	4944	8.7	1	1	1	0	0	0
2	5868	9.5	1	1	1	0	0	0
3	6349	10.3	4.86	9.26	24.91	0.93	0.84	0.65
4	6258	9.9	4.63	9.07	24.38	0.82	0.79	0.61
5	6506	10.4	5.21	9.73	26.87	1.11	0.94	0.78
6	6489	9.9	4.92	9.49	26.09	0.96	0.89	0.73
7	6756	10.7	5.34	9.92	27.53	1.17	0.98	0.83
8	6606	10.5	5.13	9.67	26.89	1.07	0.93	0.79
9	6497	10.2	4.98	9.45	25.65	0.99	0.88	0.7
10	6443	9.7	4.75	9.31	25.28	0.88	0.85	0.67

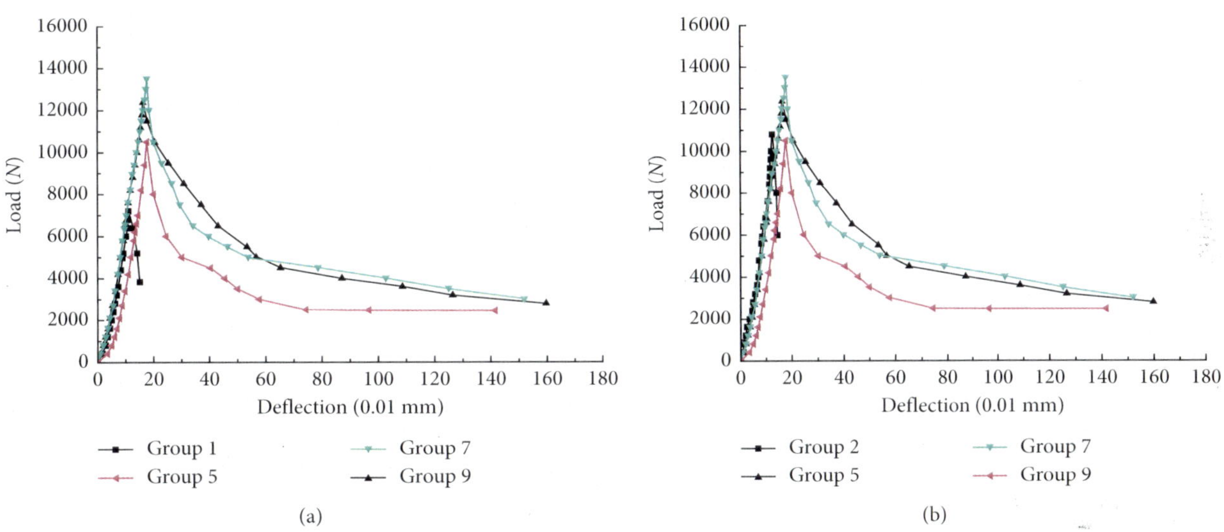

FIGURE 6: Shotcrete load-deflection curve. (a) Comparison of shear undaform steel fiber concrete and reference concrete. (b) Comparison of shear undaform steel fiber concrete and silica fume concrete.

index and the variation coefficients of load-carrying capability were closest to the corresponding parameters of ideal elastic-plastic material.

(5) The ordinary steel fiber-reinforced wet shotcrete has been initially applied in the tunnel of Brunel Thames in Britain and the Riva tunnel in Turkey, but the steel fiber-reinforced siliceous wet shotcrete could more effectively enhance the antiseepage and anticracking ability of tunnels, underground stations, and mine construction. Moreover, it could also avoid the water inrush and mud outburst disaster.

Conflicts of Interest

The authors declare that they have no conflicts of interest.

Acknowledgments

This work was fully supported by the Key Research and Development Plan of Shandong Province of China (2018GSF116006).

References

[1] M. Rostami and K. Behfarnia, "The effect of silica fume on durability of alkali activated slag concrete," *Construction and Building Materials*, vol. 134, pp. 262–268, 2017.

[2] A. J. Daniel, S. Sivakamasundari, and A. Nishanth, "Study on partial replacement of silica fume based geopolymer concrete beam behavior under torsion," *Procedia Engineering*, vol. 173, pp. 732–739, 2017.

[3] K. Alireza, M. R. Elias, H. Payam, and T. Hamidreza, "Mechanical performance of self-compacting concrete reinforced with steel fibers," *Construction and Building Materials*, vol. 51, pp. 179–186, 2014.

[4] J. Yu, S. M. Lei, X. W. Liang et al., "A design method of mix proportion for a new type of hybrid steel fiber reinforced self compacting concrete," *Journal of Building Materials*, vol. 20, no. 4, pp. 611–615, 2017.
[5] W. F. Cheng, X. P. Lin, Y. N. Ding et al., "Study on engineering application of silica fume additives and steel fiber shotcrete," *Journal of Rock Mechanics and Engineering*, vol. 30, no. 11, pp. 2311 2329, 2011.
[6] X. R. Liu, Y. H. Zhu, X. H. Li et al., "Experimental research on single-layer tunnel lining of steel fiber shotcrete," *Geotechnical Mechanics*, vol. 30, no. 8, pp. 2319–2323, 2009, in Chinese.
[7] L. C. Zhang, S. C. Li, S. Li et al., "Influence of silica fume and fly ash on Shotcrete performance," *Journal of Shandong University*, vol. 46, no. 5, pp. 103–109, 2016, in Chinese.
[8] Y. Liu, R. Ma, J. Yang et al., "Shrinkage properties of silica fume concrete," *Journal of Guangxi University*, vol. 40, no. 4, pp. 909–914, 2015, in Chinese.
[9] R. B. Ardalan, A. Joshaghani, and R. D. Hooton, "Workability retention and compressive strength of self-compacting concrete incorporating pumice powder and silica fume," *Construction and Building Materials*, vol. 134, pp. 116–122, 2017.
[10] S. Fallah and M. Nematzadeh, "Mechanical properties and durability of high-strength concrete containing macro-polymeric and polypropylene fibers with nano-silica and silica fume," *Construction and Building Materials*, vol. 132, pp. 170–187, 2017.
[11] L. C. Wang, P. Q. Jiang, Y. Q. Liang et al., "Experimental study on dynamic mechanical properties of steel fiber concrete under biaxial compression," *Journal of Building Materials*, vol. 20, no. 1, pp. 106–111, 2017.
[12] C. D. Su, Y. Li, Z. Q. Xiong et al., "Splitting strength and deformation characteristics of steel fiber reinforced high performance concrete," *Journal of Building Materials*, vol. 17, no. 4, pp. 587–591, 2014.
[13] S. Naqvi, K. Mahmoud, and E. El-Salakawy, "Effect of axial load and steel fibers on the seismic behavior of lap-spliced glass fiber reinforced polymer-reinforced concrete rectangular columns," *Engineering Structures*, vol. 134, pp. 376–389, 2017.
[14] T. Nguyen, T. T. Nguyen, and W. Pansuk, "Experimental study of the punching shear behavior of high performance steel fiber reinforced concrete slabs considering casting directions," *Engineering Structures*, vol. 131, pp. 564–573, 2017.
[15] M. L. Nehdi, S. Abbas, and A. M. Soliman, "Exploratory study of ultra-high performance fiber reinforced concrete tunnel lining segments with varying steel fiber lengths and dosages," *Engineering Structures*, vol. 101, pp. 733–742, 2015.
[16] A. L. Hoang and E. Fehling, "Influence of steel fiber content and aspect ratio on the uniaxial tensile and compressive behavior of ultra high performance concrete," *Construction and Building Materials*, vol. 153, pp. 790–2806, 2017.
[17] K. Marar, Ö. Eren, and H. Roughani, "The influence of amount and aspect ratio of fibers on shear behaviourof steel fiber reinforced concrete," *Journal of Civil Engineering*, vol. 21, no. 4, pp. 1393–1399, 2017.
[18] D. Y. Yoo, S. W. Kim, and J. J. Park, "Comparative flexural behavior of ultra-high-performance concrete reinforced with hybrid straight steel fibers," *Construction and Building Materials*, vol. 132, pp. 219–2229, 2017.
[19] F. H. Nattaj and M. Nematzadeh, "The effect of forta-ferro and steel fibers on mechanical properties of high-strength concrete with and without silica fume and nano-silica," *Construction and Building Materials*, vol. 137, pp. 557–572, 2017.
[20] R. Ghavidel, R. Madandoust, and M. M. Ranjbar, "Reliability of pull-off test for steel fiber reinforcedself -compacting concrete," *Measurement*, vol. 73, pp. 628–639, 2015.
[21] W. Abbass, M. I. Khan, and S. Mourad, "Evaluation of mechanical properties of steel fiber reinforced concrete with different strengths of concrete," *Construction and Building Materials*, vol. 168, pp. 556–569, 2018.
[22] J. P. Zhang, L. M. Liu, Z. D. Zhu, and J. Z. Cao, "Flexural fracture toughness and first-crack strength tests of steel fiber-silica fume concrete and its engineering applications," *Strength of Materials*, vol. 50, no. 1, pp. 166–175, 2018.

12

Experimental and Theoretical Study of Sandwich Panels with Steel Facesheets and GFRP Core

Hai Fang,[1] Huiyuan Shi,[2] Yue Wang,[1] Yujun Qi,[1] and Weiqing Liu[3]

[1] *College of Civil Engineering, Nanjing Tech University, Nanjing, China*
[2] *School of Civil Engineering, Southeast University, Nanjing, China*
[3] *Advanced Engineering Composites Research Center, Nanjing Tech University, Nanjing, China*

Correspondence should be addressed to Weiqing Liu; wqliu@njtech.edu.cn

Academic Editor: Gianfranco Palumbo

This study presented a new form of composite sandwich panels, with steel plates as facesheets and bonded glass fiber-reinforced polymer (GFRP) pultruded hollow square tubes as core. In this novel panel, GFRP and steel were optimally combined to obtain high bending stiffness, strength, and good ductility. Four-point bending test was implemented to analyze the distribution of the stress, strain, mid-span deflection, and the ultimate failure mode. A section transformation method was used to evaluate the stress and the mid-span deflection of the sandwich panels. The theoretical values, experimental results, and FEM simulation values are compared and appeared to be in good agreement. The influence of thickness of steel facesheet on mid-span deflection and stress was simulated. The results showed that the mid-span deflection and stress decreased and the decent speed was getting smaller as the thickness of steel facesheet increases. A most effective thickness of steel facesheet was advised.

1. Introduction

In recent years, sandwich structures have been widely applied. A sandwich panel is a typical structural form with good structural performance and excellent material advantages, such as higher strength, better performance under shock absorption, and also a lower density compared to traditional monolayer materials [1–5]. Worldwide, study of metal sandwich panel structures has been ongoing since the 1950s and the related studies can be found in the publications of Allen [6]. Via experimentation, Davies [7] proposed the corresponding deflection calculation formulas on metal-faced sandwich panels, and Frostig and Baruch [8] studied the mechanical behavior of sandwich beams when the upper and lower facesheets were identical or different. Finnish scholar Frank et al. [9] evaluated the fatigue strength of laser-stake-welded, web-core, steel sandwich panels. A 3D-panel bending response was analyzed using the finite element method. The simulated values were compared with the experimental values and showed good agreement. Cambridge University [10] developed a type of sandwich beam comprising a Y-frame and corrugated cores that were manufactured by assembling and brazing prefolded AISI type 304 stainless steel sheets together and conducted three-point bending experiments to study its extreme loading capacity and failure mode; infinite element software was also used to simulate the number of core materials, the thickness of the core material, and the angle of the corrugation, and to propose an effective design method. Swedish scholars Kolsters and Zenkert [11] performed a compression test on the laser-welded, steel sandwich panels and determined the lateral pressure failure mode, the lateral pressure ultimate bearing capacity, and the corresponding strain curve for this type of sandwich panel. Based on panels' buckling theory, those authors also predicted the ultimate bearing capacity under the lateral pressure state for sandwich panels, and their calculated values were shown to be in good agreement with the experimental results. Swiss Keller [12] and his team developed a new type of hybrid bridge girder and conducted a series of studies that included static tests, fatigue tests, and theoretical calculations. They also used lightweight wood as the core material and GFRP as facesheet to make sandwich bridge decks. Huo et al. [13] found that the facesheet thickness affected the displacement ductility of GFFW decks. Seo et al. [14] presented the concept of material hybridization increasing elastic modulus of GFRP bars by using steel. Özes and Neşer [15] evaluated the adhesive bonding performance and the results reveal that the surface roughness of steel has a significant effect on the bonding performance of steel to FRP combinations. Metal sheets are used at the outer surfaces to maximize rigidity while inner lightweight cores are adhesively bonded with them to keep the whole structure together; this is inconsistent with the concept developed by Mamalis et al. [16]. The ASTM [17] formulated the buckling experiment methods and design specifications for multilayer sandwich panels.

Currently, there are still plenty of shortcomings to use the steel plate and sandwich steel plates used in bridge girders and ships, such as too many welding joints needed, easy corrosion, and difficultly maintained. It is rarely suitable to applications in complex stress states. Fiber reinforced polymer composite materials are gaining popularity as the materials of choice to replace deteriorated concrete bridge decks; two different FRP bridge decks were modeled for their dynamic responses by Prachasaree et al. [18]. In this paper, a novel composite sandwich panel is presented, consisting of steel slab as the top and bottom facesheet and middle hollow GFRP square tubes as core. The GFRP tubes are molded together with the steel facesheet during the manufacturing process. This type of composite sandwich panel has several advantages over traditional slabs: (1) Welding work can be avoided in process of manufacture and the amount of welding seam is reduced, (2) the panel has better antifire and antiageing performance as the GFRP tubes are protected by the steel facesheet, (3) the steel slab located on the surface is easy to maintain, and (4) the hollow GFRP square tube core can reduce the work required to reserve a variety of pipes and lines. It is mainly proposed for applications in ships, bridges, and buildings as its light weight, high strength, high stiffness, and good wear resistance. Use this kind of sandwich panels as bridge decks can avoid the disadvantages of low stiffness and easily split which the integral composite bridge decks have. Besides, applying this kind of sandwich panel in steel box girders can obtain smaller negative moment region due to its relatively high rigidity. In order to explore the flexural behavior of this steel-GFRP composite sandwich panel, a series of experimental studies was conducted under four-point bending. The ultimate bending capacity and mid-span deflection were calculated by the transformed section method. An FEM analytical model was developed to predict the ultimate bending capacity as well as the mid-span deflection of the panels.

2. Experimental Program

2.1. Specimen Preparation. The specimens created in this study were composed of steel facesheet, glass fiber reinforced plastic (GFRP) composite core, which was pultruded as hollow square tubs. Use the structural adhesive to bond the GFRP tubes together and the core was sharped. Apply structural adhesive on the surface of steel slab, paste the tube core onto it, place another piece of slab onto the tube core, and then apply pressure. The shaped specimen is shown in Figure 1.

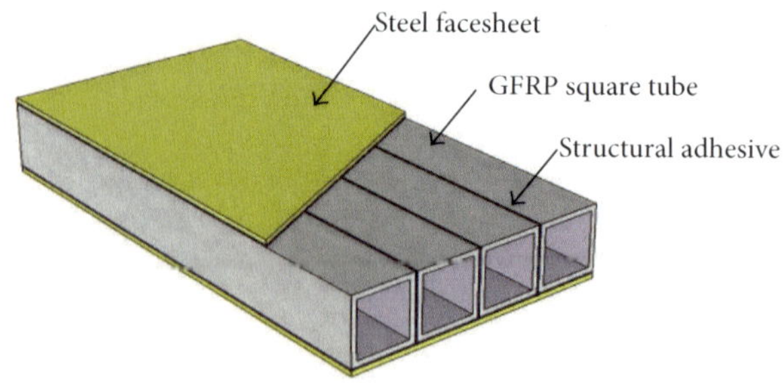

FIGURE 1: Schematic of the sandwich panel.

TABLE 1: Description of specimens of steel-GFRP sandwich panels.

Specimen	Specimen
S(0)-GFRP	GFRP hollow tube (without facesheet)
S(4)-GFRP	Steel facesheet (thickness: 4 mm) + GFRP hollow tube core
S(6)-GFRP	Steel facesheet (thickness: 6 mm) + GFRP hollow tube core
S(8)-GFRP	Steel facesheet (thickness: 8 mm) + GFRP hollow tube core
S(10)-GFRP	Steel facesheet (thickness: 10 mm) + GFRP hollow tube core
F(6)-GFRP	GFRP facesheet (thickness: 6 mm) + GFRP hollow tube core

Note: in the specimens' names, "S" represents a steel facesheet, and the figure following it represents the thickness of steel facesheet; "F" represents GFRP facesheet, and "GFRP" represents the composite tube core.

TABLE 2: Size of specimens (unit: mm).

Specimens	Length	Width	Thickness of facesheet	Height of core
S(0)-GFRP	1800	300	—	75
S(4)-GFRP	1800	300	4	75
S(6)-GFRP	1800	300	6	75
S(8)-GFRP	1800	300	8	75
S(10)-GFRP	1800	300	10	75
F(6)-GFRP	1800	300	6	75

One GFRP hollow specimen, four sandwich specimens with steel facesheet and GFRP core, and one specimen with GFRP facesheet and GFRP core were prepared, as shown in Table 1. The heights of the core materials were all 75 mm, and the widths were all 300 mm. The thicknesses of steel facesheets were 4, 6, 8, and 10 mm, and that of GFPR facesheet was 6 mm, with identical width of 300 mm. The sizes of the specimens are shown in Table 2.

2.2. Material Properties. The tensile strength and Young's modulus of GFRP were determined as per ASTM D3039/D 3039M-14 [19]. The strength and modulus and Poisson's ratio of the steel slab, GFRP slab, and GFRP core of this innovative composite sandwich panel are given in Table 3, respectively. It was concluded, therefore, that the sequence of steel and GFRP

TABLE 3: The material properties of GFRP and steel.

Beam component	E_1 (Pa)	E_2 (Pa)	E_3 (Pa)	ν_{12}	ν_{13}	ν_{23}	G_{12} (Pa)	G_{13} (Pa)	G_{23} (Pa)
GFRP facesheet	1.36E10	2.50E9	2.50E9	0.15	0.15	0.15	—	—	—
GFRP core	2.42E10	9.37E9	9.37E9	0.18	0.18	0.18	3.63E8	2.75E8	2.75E8
Steel facesheet	1.95E11	1.95E11	1.95E11	0.31	0.31	0.31	—	—	—

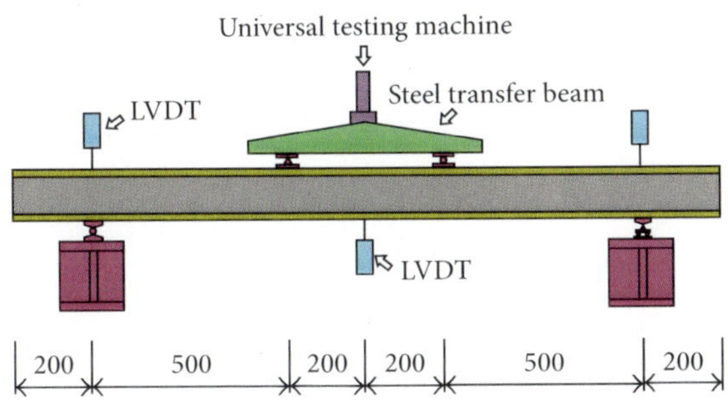

FIGURE 2: Schematic illustration of the test.

FIGURE 3: Flexural test setup.

might give optimized structural stiffness and load carrying capacity of the resulting sandwich configuration.

2.3. Experimental Setup and Loading Procedure. A 4-point static bending test was performed on the innovative sandwich panels with different thicknesses of steel or GFRP facesheet in accordance with the ASTM C393-00 standard [17]. The experiments mainly evaluated the flexural stiffness and ultimate flexural strength of sandwich panels with different configurations. The experiments in this study used a universal testing machine to apply static loads based on the experimental method to determine the bending performance of the sandwich structure. The schematic illustration and loading setup used in the experiments are shown in Figures 2 and 3, respectively.

Specimens in the bending tests were named according to the two parts, for example, S(4)-GFRP and F(6)-GFRP, where the first part with S (or F) represents that the material of facesheet was steel (or GFRP), the number represents the thickness of the layer in mm, and the second part GFRP represents the material of the core.

The loading was applied 250 mm from the mid-span through a spreader panel using a 200 kN universal testing machine with a loading rate of 1 kN/min. Before each test, the loading pins were set to almost touch the upper surface of the composite sandwich specimen. The applied load, displacement, and strains were recorded using a data logger, linear variable differential transducer (LVDT, made by Mitutoyo Corporation, with a precision of 25 micrometers), and strain gauges (MM CEA-06-1250UW-350). Failure mechanisms were also monitored and recorded.

3. Experimental Results and Discussions

3.1. Failure Behavior. Experimental investigation showed that the sandwich panels with different facesheets exhibited different failure behaviors. These different failure modes are shown in Figures 4–9. These test results showed that, due to the support of the steel or GFRP facesheet, the panels showed better stiffness and bearing abilities. No collapse of the facesheet occurred. Meanwhile, local failure of the GFRP tube core was avoided in the loading point region.

Specimen SF had a low stiffness as it did not have an upper or lower facesheet. As the load increases, the deflection quickly increased as well. When the load was increased to 94 kN, the GFRP tube suddenly produced a ripping sound, and cracks were noted on the web beneath the loading points. With continued loading, cracks appeared on the web between the loading points, as shown in Figure 4(a), and a significant cracking sound from GFRP was heard, and cracks also appeared on the facesheet between the loading points, as shown in Figure 4(b); these cracks spread to both sides of the specimen, as shown in Figure 4(c). Keeping loading, the load remained stable while deflection increased rapidly.

The upper and lower facesheets of specimen S(4)-GFRP were 4 mm steel plates. As the load increased, the deflection increased gradually without other significant phenomena. When the load had increased to 104 kN, a "snapping" sound was heard. After the load was increased to 128 kN, the steel plate of the upper facesheet between the loading points peeled off and delaminated from the composite square tube, as shown in Figure 5(a). Keeping loading, a continuous tearing sound was heard, and tearing failure appeared where the web between the loading points and the upper and lower facesheets intersected, as shown in Figure 5(b).

The thickness of the upper and lower steel facesheets of specimen S(6)-GFRP was 6 mm. No significant phenomenon was noted during the initial loading stage. When the load increased to 132 kN, "snapping" sounds were heard. As the load increased, sounds could be heard continuously. When a sudden loud sound was heard, the web peeled off and delaminated from the lower facesheet, as shown in Figure 6(a). The

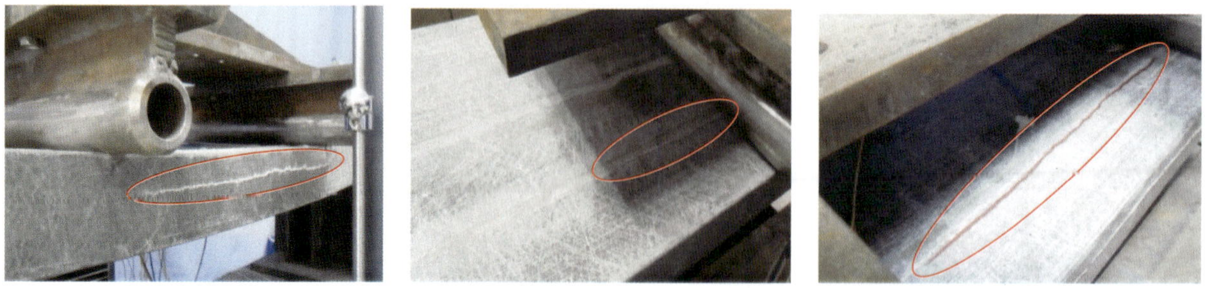

(a) (b) (c)

FIGURE 4: Failure of specimen S(0)-GFRP: (a) horizontal cracks appeared on the web, (b) cracks appeared on the facesheet, and (c) cracks spreading.

(a) (b)

FIGURE 5: Failure of specimen S(4)-GFRP: (a) peeling off and delaminating and (b) tearing failure of the web where it intersected the facesheet.

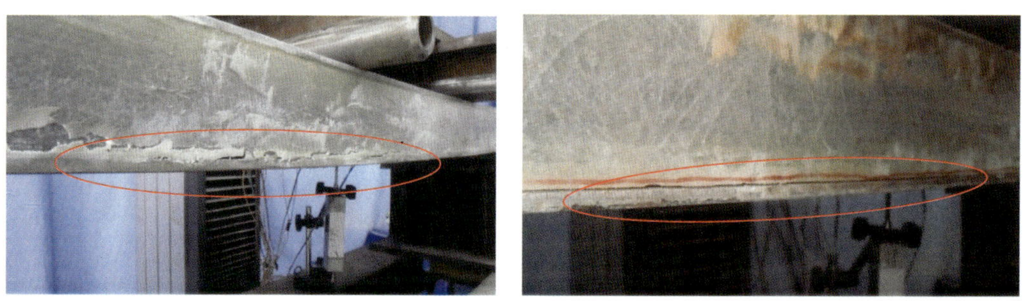

(a) (b)

FIGURE 6: Failure of specimen S(6)-GFRP: (a) web peeling off from the lower facesheet and (b) web on the other side peeling off from the lower facesheet.

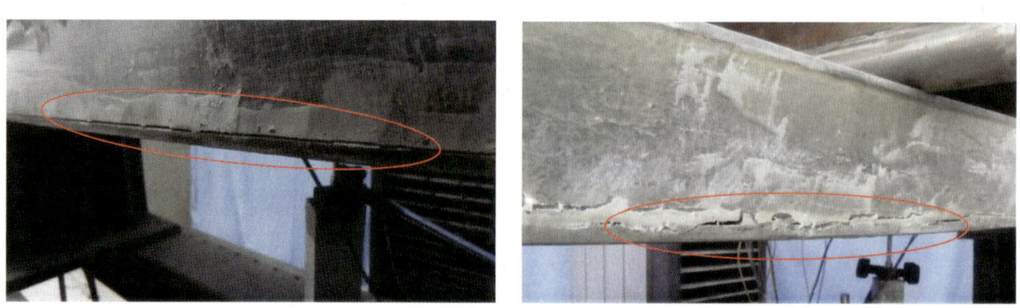

(a) (b)

FIGURE 7: Failure of specimen S(8)-GFRP: (a) web peeling off from the lower facesheet and (b) web on the other side delaminating from the lower facesheet.

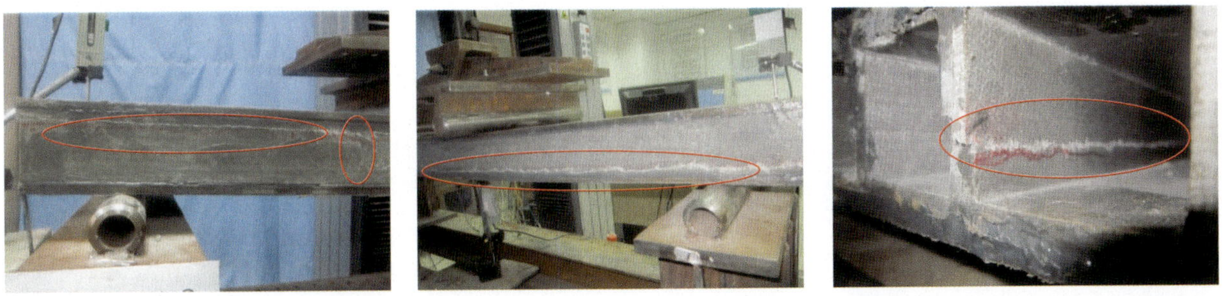

(a) (b) (c)

FIGURE 8: Failure of specimen S(10)-GFRP: (a) spreading of cracks, (b) cracks appearing where the web and the lower facesheet intersected, and (c) breaking of specimen rib.

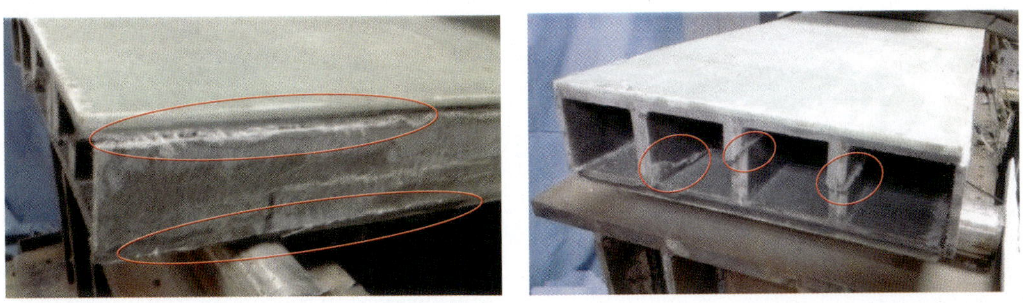

(a) (b)

FIGURE 9: Failure of specimen F(6)-GFRP: (a) cracks spreading horizontally and (b) rib fracture.

web on the other side also peeled off from the lower facesheet. Keeping loading, this phenomenon became more significant, as shown in Figure 6(b), and the specimen failed.

The thickness of the upper and lower steel facesheets of specimen S(8)-GFRP was 8 mm. Because the steel plate was thicker compared to the previous tests, this specimen had a high rigidity. Small sounds could be heard during the loading process. However, there was no significant or abnormal phenomenon that was visible. When the load increased to 148 kN, the specimen produced a small snapping sound. With continuous loading, small sounds could be heard, but no visible cracks appeared. When the load was 152 kN, a loud tearing sound was heard, and the load decreased to 85 kN rapidly at the same time. The specimen had peeled off from the lower facesheet, as shown in Figure 7(a). Keeping loading, the web on the other side of the specimen also delaminated from the lower facesheet, as shown in Figure 7(b).

The thickness of the upper and lower steel facesheets of specimen S(10)-GFRP was 10 mm As the load increased, the displacement grew slowly. When the load was 189 kN, the specimen produced a small sound. With increasing load, continuous "snapping" sounds could be heard. When the load was increased to 204 kN, a significant tearing sound was heard. At this time, cracks appeared where the web and the upper facesheet intersected, and as the load increased, the cracks expanded along the fiber direction, as shown in Figure 8(a). At this time, horizontal cracks also appeared where the web on the other side and the facesheets intersected, as shown in Figure 8(b). When the load was increased to 212 kN, a loud sound was heard, and the load decreased to 76 kN; at this time, the middle rib of the specimen broke, as shown in Figure 8(c), which indicated that the specimen had failed.

The upper and lower layers of specimen F(6)-GFRP were 6 mm GFRP plates. As the load increased, small sounds could be heard coming from the specimen. No abnormal phenomenon was found using naked-eye observation. When the load was increased to 180 kN, cracking sounds were heard but no abnormal phenomenon was visible. When the load was increased to 187 kN, another significant cracking sound could be heard; cracks were then found where the web and the upper facesheet intersected, as shown in Figure 9(a). Keeping loading, cracks also appeared at interface of the web and the lower facesheet and spread along the horizontal direction. When the load was increased to 191 kN, a big sound was heard, and the middle rib of the specimen was found to have broken, as shown in Figure 9(b); loading was stopped at this time.

3.2. Effect of Steel Thickness on Panel Flexural Properties. Analyzing the load-deflection curve in Figure 10, it is shown that the specimens exhibited good agreement before failure. As the thickness of the facesheets increased, the mid-span deflection decreased gradually, and the declining scope became more evident as the facesheet thickness increased. When the facesheet thickness is increased from 4 to 10 mm in 2 mm steps, the mid-span displacements are reduced

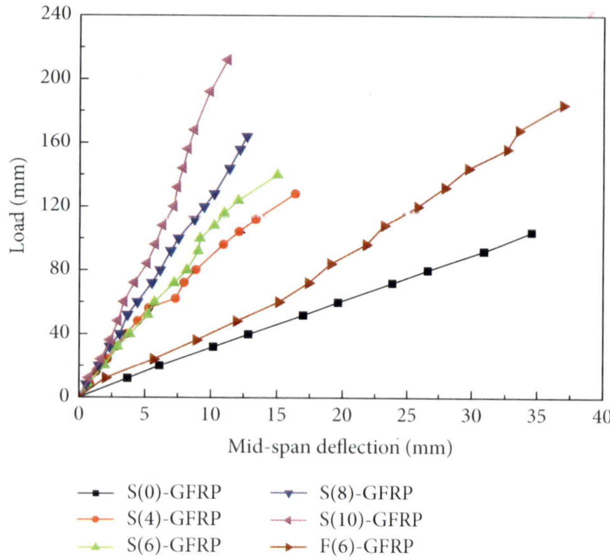

FIGURE 10: Load-deflection curves.

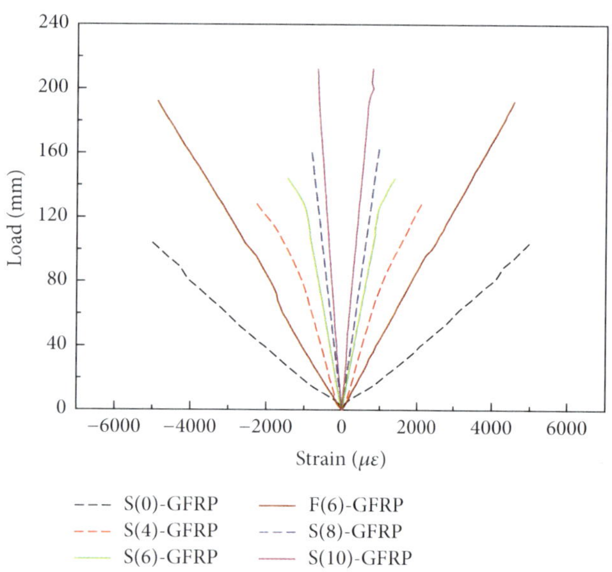

FIGURE 11: Mid-span load-strain curves.

by 8.4%, 15.4%, and 11.5%, respectively. The ultimate load of the specimens also increased when the thickness of facesheets increased. The specimen F(6)-GFRP had greater stiffness than S(0)-GFRP, but its stiffness was much smaller compared with the specimen with steel facesheet. Adhesive failure occurred to the facesheet and the sandwich core for specimens S(4)-GFRP, S(6)-GFRP, and S(8)-GFRP. This type of failure happened suddenly, which should be avoided as much as possible in real designs.

As shown in Figure 11, the sandwich panel without a facesheet exhibited poor ductility and the curve was almost linear. With the thickness of the steel facesheets increasing, the mid-span strain decreased. It showed that the upper and lower steel facesheets' strains were evenly distributed when

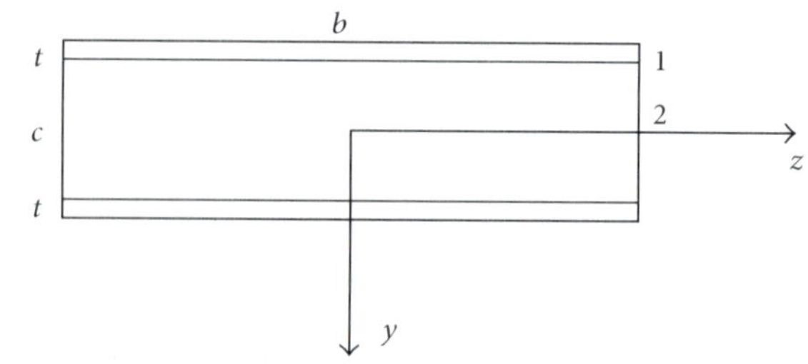

FIGURE 12: Section of the sandwich panel.

loads were moderate, and the stiffness of the system was fully used. Overall, the strains at the upper and lower symmetrical positions of the neutral axis were similar. Besides, the elastic behavior of the strain distribution was observed at the early stage of loading in all specimens.

4. Theoretical Analysis

This study used sandwich beam theory and transformed section theory combined with the material mechanics method to deduce the formula for the stress and mid-span deflection of the sandwich panel. Steel and GFRP materials are both considered ideal elastomers. The transition between different materials is continuous, and the slippage between interfaces is neglected.

4.1. Transformed Section Theory. The sandwich panels that consisted of steel and composite materials can be transformed into the section of the same material. For the steel facesheet-GFRP core section unit, it is assumed that the area is A_f, the elastic modulus is E_f, and there is a strain ε_f under the action of stress σ_f. Then, based on the equivalent condition that the strain is the same, and the total force is constant, the steel facesheet unit can be used to describe the elastic modulus E_f, the stress σ_f, and the core's equivalent converted section area A_{cf}. The diagram of the panel section is shown in Figure 12, where t and c describe the thicknesses of the steel plate and the GFRP core, respectively; and the number 1 indicates the steel face, and the number 2 indicates the GFRP core.

Taking the steel facesheet as an example, as the strain is the same, it is assumed that the coefficient α_{E_f} is the ratio of the core's elastic modulus E_c to the steel plate's elastic modulus E_f; thus

$$\frac{\sigma_f}{\sigma_c} = \frac{E_f}{E_c} = \frac{1}{\alpha_{E_f}}. \quad (1)$$

Based on the condition that the total force is constant, it can be derived that

$$A_{cf} = \frac{A_f}{\alpha_{E_f}}. \quad (2)$$

During the section's conversion process, it is required that the section centroid remains unchanged before and after conversion; therefore, the thickness of the steel plate must

FIGURE 13: Conversion section.

remain unchanged, and thus, the area conversion is equal to the width conversion (i.e., b is converted to b_{cf}):

$$b_{cf} = \frac{b}{\alpha_{E_f}}. \tag{3}$$

The converted width b_{cf} can now be derived. Then, based on the converted section (Figure 12), the inertia moment I_{cf} can be calculated for the converted GFRP section, as shown in Figure 13.

4.2. Stress Analysis. On the basis of the distribution of the strain as shown in Figure 11, the plane section assumption is applicable for the early stage of loading. The full section remains plane until the specimen reaches the limit condition.

Combining each layer as a whole, the strain can be expressed as

$$\varepsilon_x = \frac{y}{\rho}, \tag{4}$$

where ρ is the radius of deflection curve and y is the distance from any point to the neutral. The stress of the steel facesheet and GFRP core can be expressed as

$$\sigma_f = \frac{y}{\rho} E_f, \qquad \sigma_c = \frac{y}{\rho} E_c, \tag{5}$$

where E_f and E_c are the elastic modulus of the facesheet and core, respectively. The bending moment within the beam's cross section is derived to be

$$M = 2\int_{A_f} y\sigma_f \, dA_f + \int_{A_f} y\sigma_c \, dA_c. \tag{6}$$

Substitute (4) and (5) into (6) and simplify

$$M = \frac{1}{\rho}\left(2E_f I_f + E_c I_c\right), \tag{7}$$

where the integral areas $A_f = bt$ and $A_c = bc$ and t and c are the thicknesses of the steel plate and the GFRP core, respectively; b is the beam width; I_f and I_c are the moment of inertia of the facesheet and core, respectively.

Substitute (7) into (5), and then the stresses for the steel facesheet and the core material should, respectively, be

$$\sigma_c = \frac{MyE_c}{2E_f I_f + E_c I_c} = \frac{My}{2\left(E_f/E_c\right) I_f + I_c}, \tag{8}$$

$$\sigma_f = \frac{MyE_f}{2E_f I_f + E_c I_c} = \frac{My}{2I_f + \left(E_c/E_f\right) I_c}. \tag{9}$$

Simplify the formula, and then

$$\sigma_f = \frac{My}{I_{cf}}, \qquad \sigma_c = \frac{My\alpha_{E_f}}{I_{cf}}, \tag{10}$$

where I_{cf} is the equivalent cross-sectional moment of inertia.

To solve the interlayer shearing force, the differential equation of the equilibrium of the beam unit is first solved:

$$\frac{\partial \sigma_x}{\partial x} + \frac{\partial \tau_{yx}}{\partial y} + f_x = 0, \qquad \frac{\partial \sigma_y}{\partial y} + \frac{\partial \tau_{yx}}{\partial x} + f_y = 0. \tag{11}$$

If the shearing force between the facesheet and core is to be solved, σ_f should be incorporated into the differential equation of equilibrium to derive the integral form:

$$\int_y^{c/2+t} \frac{\partial \sigma_f}{\partial x} dy + \int_{\tau_{xy}}^{0} \frac{\partial \tau_{xy}}{\partial y} dy = 0, \qquad \tau_{xy} = \frac{P}{2I_{czf}} \cdot \frac{(c/2+t)^2 - y^2}{2}. \tag{12}$$

The shearing force of the facesheet and core is solved as

$$\tau_f = \frac{VS_f}{I_{cf} t_{eqf}}, \qquad \tau_c = \frac{VS_c}{I_{cf} t_w}, \tag{13}$$

where S_f and S_c are the static moments of the area of transformed section below the calculating point to the neutral axis.

The bending stress is calculated based on (10), and the bending interlayer shearing force is calculated based on (13). Using specimen S(8)-GFRP as an example, an external load of 60 kN is used for analysis in this study because this theory is valid only under the linear elastic range, and plasticity is often displayed in the failure margin of the sandwich beam. Based on the calculations, the distribution of the bending stress and shearing force of the sandwich beam section is shown in Figure 14. Table 4 compares the predicted maximum tensile and compressive stress with experimental results which were relative with strain of different components under a load of 60 kN. The stresses at the edge of tensile and compressive region of the sandwich panels can be calculated while the maximum tensile and compressive stress can be derived from homologous strain value measured in the experiment. The result of calculations shows an approximate agreement with the experimental values. The relative error between experimental and analytical results for specimens with steel facesheet and GFRP core is not more than 7.83%. The maximum relative error is 10.64%, which occurred in specimen S(0)-GFRP.

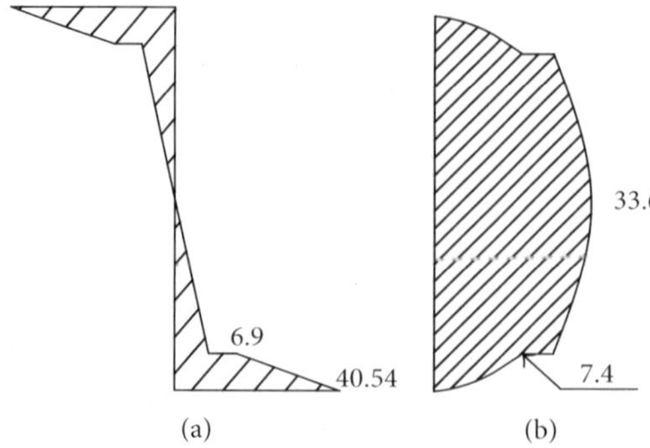

FIGURE 14: Sectional stress distribution at mid-span for specimen S(8)-GFRP: (a) normal stress and (b) shear stress (unit in MPa).

TABLE 4: Comparisons of maximum tensile and compressive stress between experimental and analytical results.

Specimen	Experimental strain ($\mu\varepsilon$)		Experimental stress (MPa)		Predicted stress (MPa)		Relative error (%)	
	Tensile	Compressive	Tensile	Compressive	Tensile	Compressive	Tensile	Compressive
S(0)-GFRP	2998	2985	40.7728	40.596	36.85	36.85	10.64	10.16
F(6)-GFRP	1462	1448	35.3804	35.0416	32.26	32.26	9.67	8.62
S(4)-GFRP	345	351	67.275	68.445	63.73	63.73	5.56	7.39
S(6)-GFRP	261	275	50.895	53.625	49.73	49.73	2.34	7.83
S(8)-GFRP	211	217	41.145	42.315	40.54	40.54	1.49	4.37
S(10)-GFRP	167	171	32.565	33.345	32.11	32.11	1.41	3.84

Note: the relative error is calculated based on the following formula: (results from the predicted stress − experimental stress)/experimental stress.

4.3. Deflection Analysis. When calculating the deflection of the panels, the effects of shear deformation should be considered because the shear modulus of the hollow core is small. Using first-order shear deformation theory, the mid-span deflection of the panels is equal to

$$\Delta = \frac{ap}{48D}\left(3l^2 - 4a^2\right) + \frac{pl}{6AG}, \tag{14}$$

where a is the distance from the loading point to the abutment; l is the distance between the beam abutments; p is the load; and D is the flexural rigidity, which can be derived by the equivalent inertia moment I_{cf} using the transformed section method multiplied by the elastic modulus E of the GFRP core; AG is the shear stiffness.

Based on the shear displacement theory of a classical sandwich panel, and considering the real conditions of specimens in this study, as shown in Figure 15, the shear displacement can be calculated based on the geometric relation [20].

As seen in Figure 15, the straight line ACB becomes the curve $A'CB'$ due to shear displacement, and the point p generates the corresponding displacement u. The distance BB' can be calculated by $z = c/2$; thus

$$BB' = \frac{Q}{GD}\left(\frac{E_f tdc}{4} + \frac{E_c c^3}{24}\right). \tag{15}$$

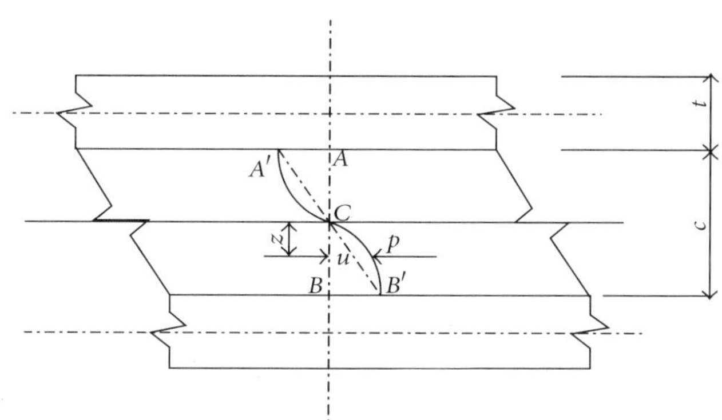

FIGURE 15: Shear displacement of sandwich panel.

It is assumed that the core is replaced by an antiplane core material whose shear modulus is G' but elastic modulus is D. The horizontal displacement of BB' should be expressed as

$$BB' = \frac{Q}{G'D}\left(\frac{E_f tdc}{4}\right). \tag{16}$$

The equivalent shear modulus G' of the core can be derived as [6]

$$G' = \frac{G}{1 + E_c/6E_f \cdot c^2/t\,(c+t)}. \tag{17}$$

TABLE 5: Comparisons of flexural capacity between experimental and FEM results.

Serial number of specimen	Ultimate flexural capacity (kN)			Ultimate displacement (mm)		
	Finite element results	Experimental value	Relative error (%)	Finite element results	Experimental value	Relative error (%)
S(0)-GFRP	91	104	−12.5	17.82	19.87	−10.3
F(6)-GFRP	169	184	−8.1	14.1	15.13	−6.8
S(4)-GFRP	114	128	−10.9	6.53	7.13	−8.5
S(6)-GFRP	129	140	−7.8	5.4	5.72	−5.6
S(8)-GFRP	171	164	4.3	4.27	4.41	−3.2
S(10)-GFRP	197	212	−7.1	3.02	3.5	−13.7

Note: the relative error is calculated based on the following formula: (results from the finite element method − experimental results)/experimental values.

The equivalent shear stiffness of the sandwich panel could be approximated as

$$AG = \frac{bG' d^2}{c}, \tag{18}$$

where d is the sum of the thickness of the core and the thickness of a single facesheet; c is the thickness of core; t is the thickness of the facesheet; b is the width of the sandwich beam; A is the area of the shear section; and G is the shear modulus. To verify the validity of the deflection calculation method, comparison between experimental and analytical results for specimens S(6)-GFRP and S(8)-GFRP under load of 60 kN is shown in Figures 17 and 18.

5. FEM Simulation

5.1. Finite Element Modeling and Validation. The finite element model was further developed to predict the mechanical performance of the sandwich by using ANSYS™ commercial program. The models of sandwich panel with different facesheet thickness were modeled using 3D linear finite element analysis. Four-node SHELL 181 elements were used to simulate the skins and eight-node SOLID 185 elements were used to simulate the core. In total, 8100 SHELL 181 elements were used, as the upper and lower steel facesheet were divided into 4050 elements, respectively. At the same time, 16740 SOLID 185 elements were utilised for modeling the GFRP core. Each material had a lamina property with a longitudinal elastic modulus of direction (E_x), a transverse elastic modulus (E_y), an elastic modulus in the depth direction (E_z), shear modulus of rigidity (G_{xz}, G_{yz}, G_{xy}), and Poisson ratio ($\nu_{xy}, \nu_{yz}, \nu_{zx}$). The applied load was imposed as four points bending in accordance with the experimental setup. The solution was obtained through a series of load increments. The simulated results were then compared with the experimental and theoretical values. Figure 16 shows the finite element model of the steel facesheet-GFRP core sandwich panel.

Based on Table 5, the FE and experimental results showed good agreement. Due to the deficiencies of the specimens in the experiment and other uncertain factors, the values of the mid-span deflection obtained experimentally were smaller than the theoretically calculated values, which were conservative; this is plausible for structure designs. Under the action of the load applied in the experiments, the materials were all in their elastic-deformation region, and the developing tendencies of the load-displacement curve for the three specimens were found to generally be in agreement. These results suggest that the FE model could be used as a design approach to evaluate the bending performance of sandwich panels.

Under the same load (60 kN), using specimens S(6)-GFRP and S(8)-GFRP as examples to verify the precision of the transformed section method, the elastic stage of each specimen was investigated. A comparison of the experimental, theoretical, and simulated values of the mid-span displacement is shown in Figures 17 and 18. The comparisons of numerical and analysis results to the experimental data show that the method is rather precise and is a useful auxiliary means for research work, design, and test.

5.2. Parametric Analysis. The finite element simulation of the steel facesheet composite core, sandwich panel structure's four-point bending stress exhibits good precision. Therefore, an ANSYS infinite element model is used to simulate more parameters of the bending performance of the structure to mitigate the lack of additional specimens, which is unsuitable for expanded analyses.

The validated modeling approach was then used to investigate the bending behavior of the sandwich beams with various thicknesses of steel facesheet. It is assumed that the total height of the steel-faced, composite, sandwich panel structure is 95 mm; the facesheet thickness for modeling is set as 0, 1.5, 3, 4.5, 6, 7.5, 9, 10.5, and 12 mm, and the load remains equal to 60 kN. The mid-span deflection and stress changed as the facesheet thickness increased, as shown in Figures 19 and 20. Both the mid-span deflection and stress decreased with the facesheet thickness increasing. The mid-span deflection of specimen decreased by 5.5, 3.7, 3.1, 0.7, 1.3, 0.4, 0.6, and 0.2 mm when the thickness of the facesheet increased by 1.5 mm from 0 to 12 mm; at the same time, the mid-span stress of decreased by 28.8, 17.9, 16.9, 6.6, 6.5, 4.1, 5.8, and 1.6 mm. When the thickness of the facesheet changes from 0 to 4.5 mm, the speed at which the sandwich beam's

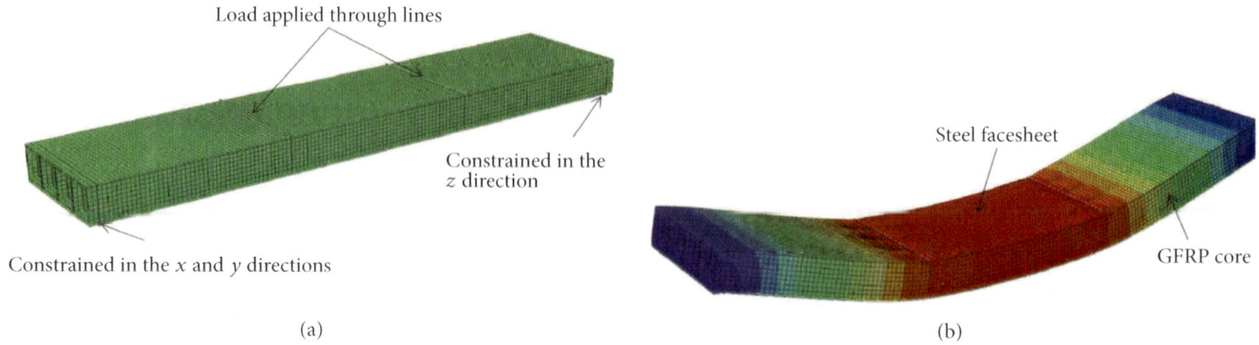

FIGURE 16: Finite element model for the sandwich panel specimen: (a) meshing and loading setup and (b) deformed shape.

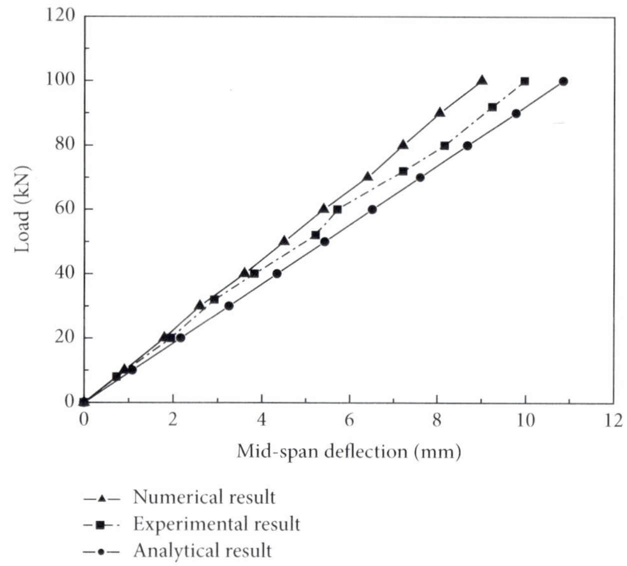

FIGURE 17: Contrast of the mid-span deflections of specimen S(6)-GFRP.

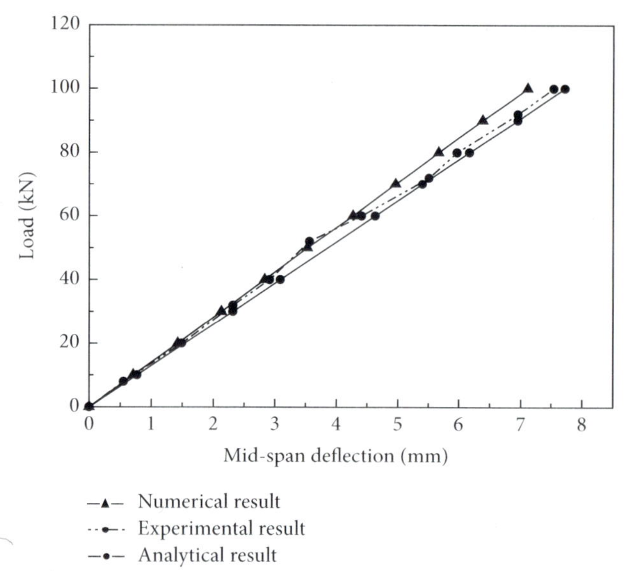

FIGURE 18: Contrast of the mid-span deflections of specimen S(8)-GFRP.

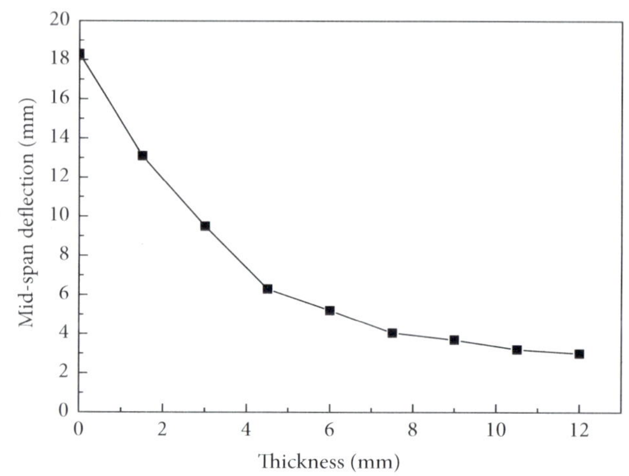

FIGURE 19: Effect of thickness of steel facesheet on mid-span deflection.

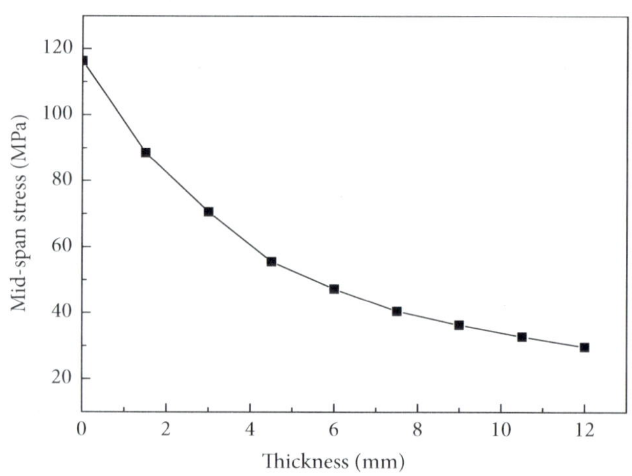

FIGURE 20: Effect of thickness of steel facesheet on mid-span stress.

mid-span deflection decreases is large; conversely, when the thickness of the facesheet is increased above 4.5 mm, the rate of decrease is less significant. This is also true for the normal stress of the mid-span panel. Therefore, for a sandwich beam with a total height of 95 mm, it is most effective to use a 4.5 mm thick facesheet.

6. Conclusions

An innovative steel-GFRP sandwich panel was introduced in this study. Experiments were conducted to investigate the effects of the steel and GFRP facesheets on the overall structural performance in bending. FE modeling was used to parametrically analyze the material efficiency and to obtain an optimal solution for the thickness of steel facesheet. From this work, the following conclusions can be drawn:

(1) The adhesion between steel facesheet and GFRP core showed satisfactory mechanical performance. The use of steel facesheet is to protect the GFRP tube core and provide good flexural stiffness.

(2) Increases of the thickness of the steel facesheet could significantly increase the flexural stiffness and ultimate load of the sandwich beams. The flexural rigidity is apparently improved and the mid-span displacement decreases as the facesheet thickness increases. When the facesheet thickness is increased from 4 to 10 mm in 2 mm steps, the mid-span displacements are reduced by 8.4%, 15.4%, and 11.5%, respectively.

(3) The transformed section method is used for the theoretical calculation of the sandwich panel, and the method of material mechanics is used to deduce the computational formula for the mid-span deflection; the theoretical and experimental values are shown to be in agreement.

(4) The FE modeling approach based on material properties and failure modes obtained from experimental investigations could evaluate the bending performance of steel-GFRP sandwich panels. Results produced by the infinite element simulation and the theoretical value are nearly the same, and the infinite element result is marginally smaller than the experimental value, which is caused by the deficiency of the specimens and other uncertain factors. A 4.5 mm thick facesheet is appropriate for a sandwich beam with a total height of 95 mm.

Competing Interests

The authors declare that they have no competing interests.

Acknowledgments

The research described here was supported by the National Natural Science Foundation of China (Grant nos. 51238003 and 51578285).

References

[1] S. W. Davey, G. M. V. Erp, and R. Marsh, "Fibre composite bridge decks—an alternative approach," *Composites Part A: Applied Science and Manufacturing*, vol. 32, no. 9, pp. 1339–1343, 2001.

[2] H. Hu, S. Belouettar, M. Potier-Ferry, and A. Makradi, "A novel finite element for global and local buckling analysis of sandwich beams," *Composite Structures*, vol. 90, no. 3, pp. 270–278, 2009.

[3] A. F. Daly and J. R. Cuninghame, "Performance of a fibre-reinforced polymer bridge deck under dynamic wheel loading," *Composites Part A: Applied Science and Manufacturing*, vol. 37, no. 8, pp. 1180–1188, 2006.

[4] T. Keller and A.-X. Zhou, "Fatigue behavior of adhesively bonded joints composed of pultruded GFRP adherends for civil infrastructure applications," *Composites Part A: Applied Science and Manufacturing*, vol. 37, no. 8, pp. 1119–1130, 2006.

[5] M.-F. He and W.-B. Hu, "A study on composite honeycomb sandwich panel structure," *Materials and Design*, vol. 29, no. 3, pp. 709–713, 2008.

[6] H. G. Allen, *Analysis and Design of Structural Sandwich Panel*, Pergamon Press, New York, NY, USA, 1969.

[7] J. M. Davies, *Lightweight Sandwich Construction*, University of Manchester, Manchester, UK, 2001.

[8] Y. Frostig and M. Baruch, "Bending of sandwich beams with transversely flexible core," *AIAA Journal*, vol. 28, no. 3, pp. 523–531, 1990.

[9] D. Frank, J. Romanoff, and H. Remes, "Fatigue strength assessment of laser stake-welded web-core steel sandwich panels," *Fatigue & Fracture of Engineering Materials & Structures*, vol. 36, no. 8, pp. 724–737, 2013.

[10] V. Rubino, V. S. Deshpande, and N. A. Fleck, "The three-point bending of Y-frame and corrugated core sandwich beams," *International Journal of Mechanical Sciences*, vol. 52, no. 3, pp. 485–494, 2010.

[11] H. Kolsters and D. Zenkert, "Buckling of laser-welded sandwich panels: ultimate strength and experiments," *Proceedings of the Institution of Mechanical Engineers, Part M: Journal of Engineering for the Maritime Environment*, vol. 224, no. 1, pp. 29–45, 2010.

[12] T. Keller, J. Rothe, J. de Castro, and M. Osei-Antwi, "GFRP-balsa sandwich bridge deck: concept, design, and experimental validation," *Journal of Composites for Construction*, vol. 18, no. 2, pp. 301–310, 2013.

[13] R. Huo, W. Liu, L. Wan, Y. Fang, and L. Wang, "Experimental study on sandwich bridge decks with GFRP face sheets and a foam-web core loaded under two-way bending," *Advances in Materials Science and Engineering*, vol. 2015, Article ID 434721, 12 pages, 2015.

[14] D.-W. Seo, K.-T. Park, Y.-J. You, and S.-Y. Lee, "Experimental investigation for tensile performance of GFRP-steel hybridized rebar," *Advances in Materials Science and Engineering*, vol. 2016, Article ID 9401427, 12 pages, 2016.

[15] Ç. Özes and N. Neşer, "Experimental study on steel to FRP bonded lap joints in marine applications," *Advances in Materials Science and Engineering*, vol. 2015, Article ID 164208, 6 pages, 2015.

[16] A. G. Mamalis, K. N. Spentzas, N. G. Pantelelis, D. E. Manolakos, and M. B. Ioannidis, "A new hybrid concept for sandwich structures," *Composite Structures*, vol. 83, no. 4, pp. 335–340, 2008.

[17] ASTM International, "Standard test method for flexural properties of sandwich constructions," ASTM C393-2000, ASTM International, West Conshohocken, Pa, USA, 2000.

[18] W. Prachasaree, A. Sangkaew, S. Limkatanyu, and H. V. S. GangaRao, "Parametric study on dynamic response of fiber reinforced polymer composite bridges," *International Journal of Polymer Science*, vol. 2015, Article ID 565301, 13 pages, 2015.

13

Shot Peening Effects on Subsurface Layer Properties and Fatigue Performance of Case-Hardened 18CrNiMo7-6 Steel

H. S. Ho, D. L. Li, E. L. Zhang, and P. H. Niu

School of Mechanical Engineering and Henan Key Engineering Laboratory of Anti-Fatigue Manufacturing Technology, Zhengzhou University, Science Road 100, Zhengzhou 450001, China

Correspondence should be addressed to E. L. Zhang; erliang.zhang@zzu.edu.cn

Academic Editor: Zhiping Luo

The present study is conducted with a dual-aim: firstly, to examine the effect of several single shot peening conditions on the subsurface layer properties and fatigue performance of the case-hardened 18CrNiMo7-6 steel, and secondly, to propose an optimized peening condition for improved fatigue performance. By carrying out the subsurface integrity analysis and fatigue testing, the underlying relationships among the peening process, subsurface layer property and fatigue performance are investigated, the way peening conditions affect the fatigue life and its associated scatter for the case-hardened 18CrNiMo7-6 steel is quantitatively assessed. The in-depth study shows that dual peening can be an optimized solution, for it is able to produce a subsurface layer with enhanced properties and eventually gain a significant improvement in fatigue performance.

1. Introduction

High-strength steels are massively used in the aeronautical and automotive applications for their high-loading capacity. With the increasing demand for high reliability products, the dual properties of high strength and high ductility are expected for such steel materials to give a longer service lifetime of engineering machinery. This motivates the design of a new class of material, referred to as gradient nanostructured (GNS) high-strength steel [1]. As fatigue failures generally occur on the material surface and propagate into the interior, a structural architecture comprising a GNS subsurface region and a coarse-grained interior are considered optimal for enhanced fatigue behavior. Loosely speaking, this is because fatigue crack initiation would be suppressed by the strong work-hardened subsurface layer of a fine-grained structure, and fatigue crack propagation would be arrested by the in-plane compressive residual stress produced by the misfit strain between the bulk and the subsurface material [2]. The GNS subsurface layer can be formed through surface strengthening treatments by means of mechanical methods such as shot peening (SP), which entails impacting a surface with small, round, and hard particles with force sufficiently high to create localized plastic deformation for the formation of a deep work-hardened and compressed subsurface layer in metals and alloys [3, 4]. In the recent several years, there have been many new types of peening techniques such as laser shock peening, ultrasonic shot peening, water cavitation peening, and severe shot peening which uses more intense parameters compared to conventional shot peening [5–8]. Despite the development of these new variants, conventional shot peening method remains one of the most important means in the industrial production due to its high flexibility and productivity, low cost, and environmental friendliness. For these reasons, shot peening process is of particular interest in the present work.

Mechanical methods can be combined with other surface strengthening treatments such as thermochemical methods for further improvement in fatigue strength and fatigue life of steel components [9–13]. As the thermochemical method is mostly benefited by increasing the microhardness and the mechanical method is mostly benefited by generating the compressive residual stress, a positive synergistic effect can therefore be anticipated if these two methods are combined. It is highlighted that compared with the single mechanical method, the combination of mechanical and thermochemical methods can be more advantageous in terms of not only the microhardness and residual stress distributions, but

Table 1: Chemical composition of 18CrNiMo7-6 steel.

Element	C	S	Si	Mn	P	Cr	Ni	Mo	Fe
Composition (Wt.%)	0.15–0.21	≤0.035	≤0.40	0.50–0.90	≤0.025	1.50–1.80	1.40–1.70	0.25–0.35	Balance

Table 2: Shot peening parameters for case-hardened 18CrNiMo7-6 steel.

Variant	Shot diameter (mm)	Coverage (%)	Almen intensity (mmA)
SP0	—	—	—
SP1	0.3	200	0.25
SP2	0.6	200	0.25
SP3	0.6	200	0.45

also the overall mechanical properties, and the hybrid process can finally obtain a remarkable improvement of fatigue limit for surface treated low-alloy steel regardless of the sequence how nitriding and severe shot peening are combined [10]. However, the hybrid process cannot be always advantageous to the application of steels. For instance, it is shown in [11] that when both carbonitrided gear and carbonitrided plus peened gear made of AISI 4130 steel are subjected to contact fatigue testing, fatigue damages, which are ought to be in different modes, are found to occur after about the same testing time for both types of methods. This could most probably be a consequence of the use of inappropriate shot peening condition; the extreme shot bombardment on the surface can cause roughness increment and flaw occurrence and therefore outweigh the beneficial effect brought by the hybrid process. Therefore, efforts are made in the present work to address the issue questioning the appropriate mechanical process condition that should be selected posterior to thermochemical method in order to maximize its beneficial effect on the fatigue life improvement of structural metallic materials.

Thus, the selection of an optimal mechanical process condition which enhances the fatigue performance of case-hardened steels is the main concern of the present work, with the ultimate aim to understand from an experimental point of view how the mechanical process parameters affect the heat-treated surface and subsurface layer properties, which in turn influence the fatigue performance. To tackle this issue, high-strength 18CrNiMo7-6 steel, widely employed in high-speed and heavy-duty gear applications, is selected as the target material. The carburizing and shot peening are respectively selected as thermochemical and mechanical methods since they are the versatile surface strengthening treatments commonly used in the automotive industry. In the present work, a few single shot peening conditions with different Almen intensities and shot diameters, which are recognized as ones of the measurable key parameters of shot peening process, are carried out on 18CrNiMo7-6 steel posterior to the carburizing process. By carrying out the surface and subsurface layer property analysis and fatigue testing, this work is aimed at establishing qualitative interrelationships between peening process conditions, surface and subsurface layer properties, and fatigue performance. Subsequently, based on the obtained qualitative relationships, we propose an optimized dual peening condition that can produce superior surface and subsurface layer properties characterized by work hardening, compressive residual stress, microstructural alteration, and surface roughness, leading to not only longer service lifetime but also smaller dispersion degree of fatigue life data. The proposed protocol is demonstrated and validated on industrial gears, which can be valuable for providing a design guideline for engineering industries.

2. Experiment Setup

2.1. Material. In the present study, the target material is 18CrNiMo7-6 steel and the supplier-given raw material chemical composition is summarized in Table 1. All the specimens are carburized in a controllable gas carburizing furnace, then subsequently quenched in oil to room temperature, and finally followed by tempering and cooling in air. The thermochemical conditions are utilized in agreement to gear heat treatment standard ANSI/AGMA B89-2004 [14].

2.2. Shot Peening Treatment. The shot peening treatment is performed with the air blast shot peening machine. The specimens are treated by shot peening with different intensities and shot diameters. The intensities of peening are measured by the arc height of Almen specimens, varying from 0.25 mmA to 0.45 mmA. Cast steel balls with diameters of 0.3 mm and 0.6 mm are selected as shot blasting media in accordance to ISO 26910-1 [15]. The coverage of shot peening is 200% for all peening processes, which is obtained by doubling the exposure time needed to achieve full coverage (100%). Note that the roughness increases gradually with the increasing degree of coverage until reaching full coverage and remains constant thereafter, and thus, to ensure that the specimens can benefit fully the advantages of the applied surface treatment methods in terms of subsurface layer properties, it is of prime importance to keep constant the level of the generated surface roughness [16]. The different shot peening configurations of immediate interest, which are determined based on the previous laboratory testing, are presented in Table 2.

2.3. Measurement Methods. The specimens under test are gear teeth prepared using wire-cut electrical discharge machining, the cut surface area of the specimen being located far away from the surface area of interest. The cross-section of the specimen is first mechanically polished using standard metallographic techniques, then chemically etched by 8% Nital.

The cross-section microstructure of the specimen is subsequently investigated by optical microscopy (OM) and scanning electron microscopy (SEM) using Zeiss Auriga FIB-SEM at room temperature.

The microhardness is measured along the depth of the cross-section of the specimen using a Vickers microhardness tester with a load of 200 g and a holding time of 10 s. Considering the material's heterogeneities and measurement errors, three microhardness measurements are acquired at each depth, whose average values are used in this work.

The residual stress is measured by the X-ray stress analyzer (LXRD, Proto, Canada) using Cr $K\alpha$ radiation and martensitic {211} crystalline plane. The voltage and current are, respectively, 30 kV and 25 mA. The measured interference peaks are evaluated by the $\sin 2\psi$ method with the diffraction angle (2θ) varying from -45° and 45°. The in-depth profiles of the residual stresses are determined by performing first iterative electrolytic removal of thin surface layers followed by X-ray in-depth measurements.

In addition to the analysis of the cross-section area performed on the specimen, the surface state of the specimen is examined by Veeco Wyko NPFLEX optical profiler. The measured surface topography is characterized by amplitude surface roughness parameters in agreement with ISO 25178-2012 [17], which are frequently used to measure the vertical characteristics of the surface deviation [18]. Taking into account the spatial surface roughness variations, the presented surface roughness data are the values averaged over three measurement zones on each specimen.

2.4. Fatigue Testing. Following the setup and procedure for conducting the SAE single-tooth bending fatigue test (cf. Figure 1) using a servohydraulic fatigue testing machine [19], the applied cyclic fatigue load is 0.82 of the yield stress of the bulk material, the stress ratio $R = 0.01$, and the nominal cyclic frequency is 89 Hz at room temperature. Each test series includes six gear tooth specimens. All the specimens are tested until occurrence of fracture, else the tests are terminated at 3×10^6 cycles.

3. Results and Discussions

3.1. Microstructure. In Figure 2, the cross-sectional surface by SEM is presented for carburized specimen, where three distinct regions with different microstructures can be recognized. As can be seen in Figure 2(a), a very thin white carbon oxide layer of a few microns is formed at the top surface; the pores, carbide particles, and microcracks are presented within the oxidized layer, as exhibited in the inserted figure of Figure 2(a). Beneath the oxidized layer is known as the so-called case-hardened zone, which consists mainly of martensite microstructure with a very small quantity of retained austenite and finely dispersed undissolved spherical carbide particles. In the innermost zone, the substrate without any evidence of microstructural change is observed (cf. Figure 2(b)).

Though carburizing technique has generally a beneficial effect on the fatigue behavior of specimens, the oxidized layer and the remained amount of retained austenite can be detrimental to high-cycle fatigue, as the presence of the retained austenite can decrease the elastic limit and the yield strength [20]. For this reason, shot peening, a cold working process consisting of bombarding a ductile surface with numerous hard particles at velocity sufficiently high to create localized plastic deformation, is often performed posterior to carburizing. Considering the high brittleness of oxidized layer, the impingement of hard shot particles on the outermost surface layer could sometimes cause the formation of highly damaged surface area covered with plenty of microcracks if inappropriate shot peening condition is selected.

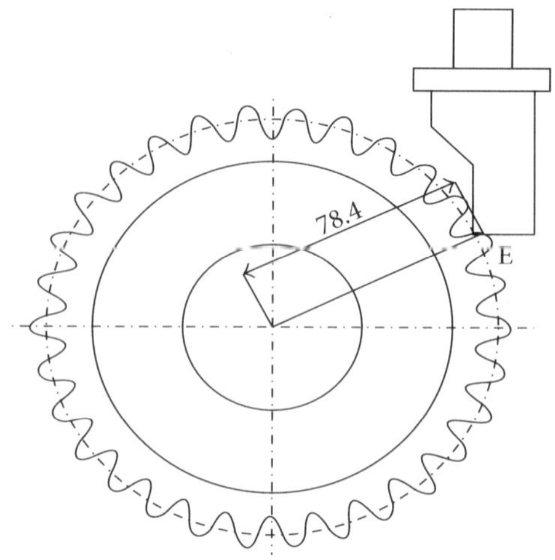

FIGURE 1: Schematic diagram of the bending fatigue testing of the cylindrical gear.

Figure 3 presents SEM micrographs of the cross-section of carburized 18CrNiMo7-6 steel after shot peening treatment. From Figure 3(a), it can be found that shot peening performed after carburizing can not only suppress the porous structure and eliminate the carbide particles at the outer oxidized layer, but also modify the microstructure of the near surface region of case-hardened 18CrNiMo7-6 steel; the plastic deformation-induced grain refinement and strain-induced phase transformation are the principal microstructural changes that occur during shot peening treatment, which are well recognized to be a consequence of the accumulated localized plastic strain caused by the repeated impingement by shot particles on the specimen surface. This shows that shot peening allows not only improving the material response, but also modifying the microhardness and (residual) stress fields as discussed later; the results are similar to [4, 21].

Shot peening is a process which projects high-velocity shot particles by the action of an external force. So, when the external force reaches the critical stress required for dislocation movement, dislocation slip is generated, and well-defined slip band structure is therefore formed. On the same slip plane, when the amount of dislocations impeded at the grain boundaries increases, the pile-ups of dislocations are formed [22]. Consequently, the phenomenon of stress concentration can easily occur in the vicinity of grain boundaries due to the heavy accumulation of the dislocations pile-ups against the boundaries. When a critical stress value is reached, the grain boundaries are destroyed and a high-density nanoscale deformation twins is formed. The intensive interactions between twin boundaries and dislocations can then lead to subsequent transformation of

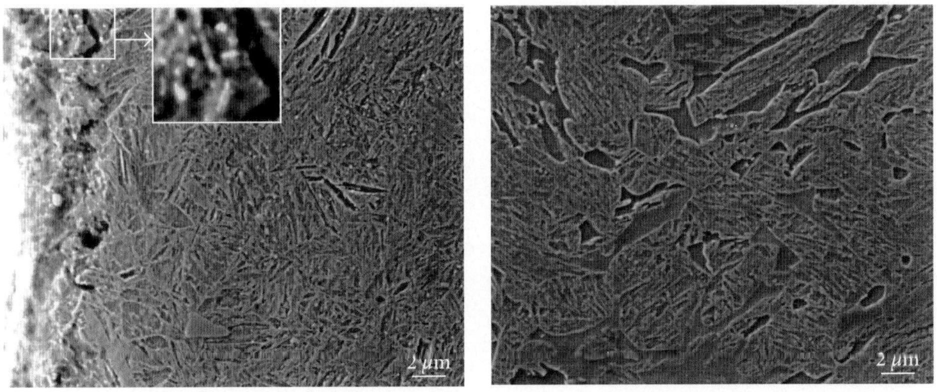

(a) (b)

Figure 2: Cross-sectional SEM of carburized SP0 specimen; the microstructure of (a) top surface region and (b) bulk material, with the inserted image in (a) showing an enlarged representation of pores, carbide particles, and microcracks.

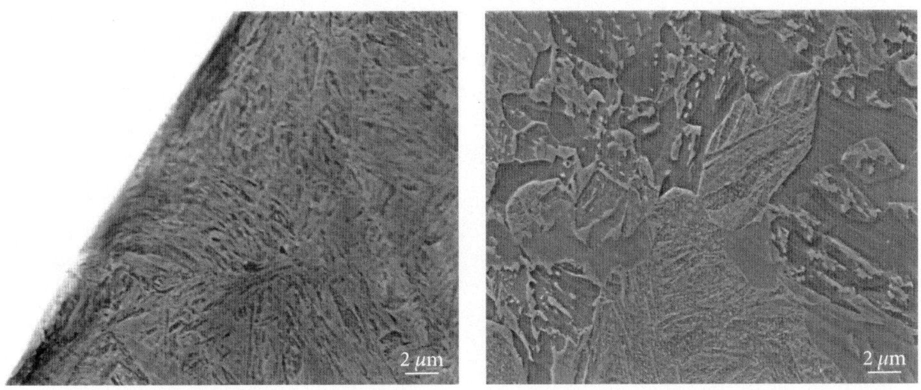

(a) (b)

Figure 3: Cross-sectional SEM of carburized and peened SP3 specimen; the microstructure of (a) top surface region and (b) bulk material.

two-dimensional twin lamellae structures into three-dimensional nanograins with their boundary characteristics lying in the form of low angle and high angle. Herein, there is a considerable fraction of low-angle grain boundaries and a high density of dislocations resided in the nonequilibrium high-angle grain boundaries [23]. In addition to plastic deformation-induced grain refinement, strain-induced phase transformation is another mechanism by which nanosurface forms during shot peening. The strain-induced nucleation of martensite transformation can take place more dominantly at higher-energy grain boundaries due to their atomic structures and energies. Besides, the strain localization bands and deformation twins that accumulate high strain energy storage could also promote martensitic transformation. All these are elaborated here to show that the principal microstructural changes induced by shot peening consist of plastic deformation-induced grain refinement and strain-induced phase transformation.

In general, the microstructure is deemed to have a strong correlation with the microhardness, where the former's modification can be reflected by the change of the latter [22]. Due to the reduction of the content of retained austenite caused by martensite transformation and austenite grain refinement led by the shot peening process, a microhardness improvement at the top surface region of carburized and peened specimens is clearly present, as is detailed in the next section.

3.2. Hardness. Conventionally, the microhardness value is used to characterize the work-hardened properties. In some cases, the size of the microhardness marks impressed along the depth direction of the subsurface layer can also be considered as an option used to illustrate very roughly the variation of the microhardness along the depth direction. Thereby, it is presented in Figure 4 the distributions of the in-depth microhardness of all specimens. For carburized-only SP0 specimens, the maximum microhardness appears on the surface (denoted as maximum surface microhardness), then gradually decreases with increasing depth and finally reaches a stabilized value. It is well accepted that this is mainly due to the declination of the carbon content from the surface to the center. Meanwhile, when shot peening is performed on case-hardened steels, regardless of shot peening condition, a similar trend of the microhardness distribution is observed. This can generally be attributed to the peening-induced reduction of plastic deformation from the outer surface to the center due to the lower grain refinement, martensite transformation, and dislocation density when approaching the center.

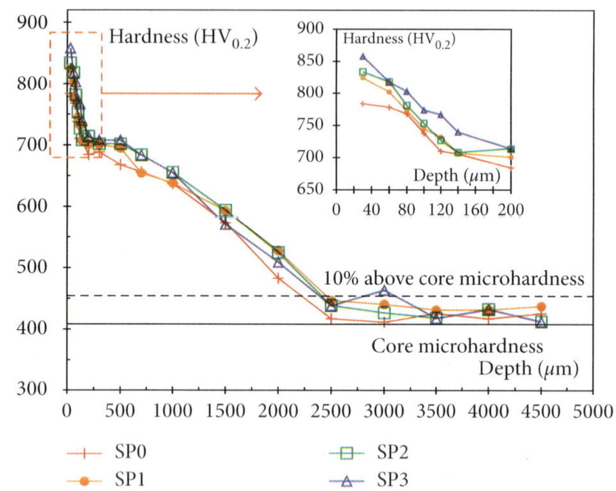

Figure 4: Microhardness distribution curves of all surface treated specimens.

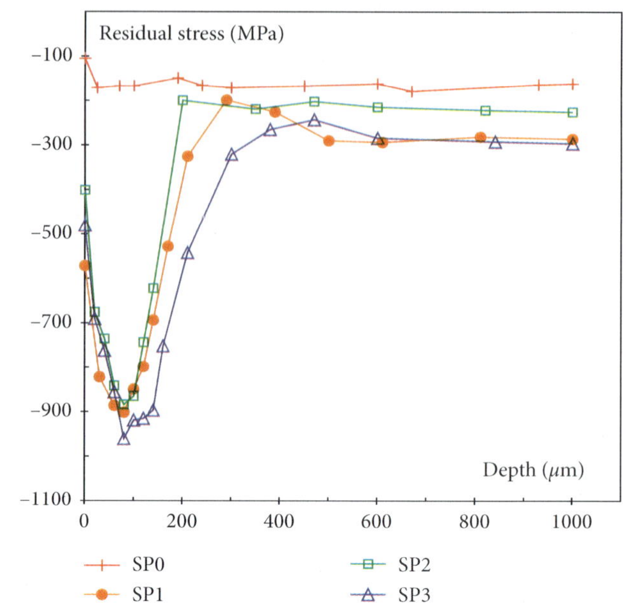

Figure 5: Residual stress distributions of all surface treated specimens.

Case depth after thermochemical treatments is very often considered as a matter of convention. When fatigue performance is oriented, a microhardness value of 10% above the core microhardness is generally used to characterize the case depth [13]. The case depth is approximated to be ~2250 μm, as exhibited in Figure 4. It is shown that the case depth is remained unchanged after shot peening is carried out posterior to carburizing.

In comparison to the single carburizing process, the hybrid process of carburizing and shot peening has caused an improvement in surface microhardness, in particular maximum surface microhardness, from the top surface up to a limited depth estimated to ~140 μm, as clearly seen in the inserted image of Figure 4. Compared to the maximum surface microhardness of carburized-only SP0 specimens (784 $HV_{0.2}$), that of carburized and peened are estimated to increase by 9.3% for SP3 specimens (857 $HV_{0.2}$), by 6.3% for SP2 specimens (833 $HV_{0.2}$), and by 5.2% for SP1 specimens (825 $HV_{0.2}$). The improvement in surface microhardness can be attributed to the peening-induced microstructural changes in the hardened surface region due to the reduction of the content of retained austenite caused by martensite transformation and austenite grain refinement. Refer to Figure 3(a) for a very dense structure consisting mainly of the martensitic state formed on the hardened top surface region after shot peening.

Regarding the two peening parameters (shot diameter and Almen intensity), it can be seen that the increase in the maximum surface microhardness caused by the increase of shot diameter is quite comparable to that resulted by the increase of Almen intensity. Indeed, the values of the maximum surface microhardness are found to increase by ~2.9% with the increment of the Almen intensity and by ~1.0% with the increase of the shot diameter, which shows that the maximum surface microhardness increment is restricted to some extent. This finding infers that the plastic deformation of 18CrNiMo7-6 steel subjected to the combined processing of carburizing and shot peening has reached a critical condition. Indeed, when the hardened region near the surface is treated with shot peening, a very small plastic deformation is produced. This is because the hardened surface region can no further accumulate localized plastic strain during the peening process due to the presence of a very high dislocation density and carbon content induced by the prior carburizing process, resulting in a slight increase in the maximum surface microhardness. The results hereby obtained show that the thermochemical surface treatment is advantageous in terms of microhardness distribution compared to mechanical surface treatment, which is consistent with the results obtained in [10].

3.3. Residual Stress. In Figure 5, the in-depth distributions of the residual stress for all the surface treated specimens are depicted. In the range of testing, all the specimens are found to have similar trend of obvious compressive residual stress distribution; the top surface compressive residual stress first increases to a maximum (denoted as maximum compressive residual stress), then decreases gradually along the depth direction of the hardened layer and remains constant thereafter.

From Figure 5, both the top surface compressive residual stress and maximum compressive residual stress are exhibited to be much more important for the hybrid processing of carburizing and shot peening than the individual carburizing process. This observation shows that shot peening process is highly beneficial in forming compressive residual stress as compared with the carburizing process which is previously seen to be responsible for increasing microhardness. The compressive residual stress is well recognized to be related to the plastic deformation, due to the creation of an elevated level of the plasticity unevenness; it can subsequently give rise to compressive residual stress in the steel [24]. In general, two types of plastic deformations can be distinguished [25]: (i) plastic deformation due to surface hammering which tends to localize the peak on the surface and (ii) plastic deformation induced by Hertzian pressure which is responsible for the subsurface peak. Yet, in the case of shot peening performed posterior to carburizing, a third process heavily linked with the content of retained austenite is considered as predominant especially in generating compressive residual stress; that is, the intensity of the local stress field produced by the impacts of shot peening can trigger the diffusionless phase transformation from austenite to martensite [26]. As a result of the constraint exerted by the surrounding material on the volume expansion, the compressive residual stress would then become deeper [21]. Consequently, the peak of the residual stress would then be seen moving to the region where the Hertzian pressure is the highest.

Regardless of the shot peening process parameter, the values of the compressive residual stress are shown to be lower at the top surface than at the peak position, and the maximum compressive residual stresses are found to be located in the subsurface region at the depth of ~90 μm. The former is most likely a consequence of the counteracting effects by tensile residual stress and high surface roughness produced by prior machining processes. And the latter suggests that the steel material, which undergoes plastic deformation, has most probably reached a critical plastic deformation limit, and thus the continuous impingement of the shot particles does not further deepen the compressed subsurface region.

In Figure 5, the values of the compressive residual stress observed at the top surface and peak position increase with Almen intensity. This is because the increase in Almen intensity, which accelerates the shot velocity, produces more kinetic energy and thereby induces a greater amount of plastic deformation on the hardened surface region. However, with an increase of the shot diameter, the top surface and maximum compressive residual stresses are on the decrease. One of the reasons for this phenomenon is that when larger diameter shots are projected on the surface of specimens with a small-scale finite kinetic energy, a larger impact area is created. Thereby, it can result in a reduction of equivalent plastic strain and consequently the compressive residual stress is lowered. Besides, from Figure 5, it can be seen at the top surface region that the effect of shot diameter is more significant than that of Almen intensity on the compressive residual stress; the decrease of shot diameter induces an increase of compressive residual stress by 42.0% while the increase of Almen intensity causes an increase of compressive residual stress by only 19.7%. It can also be observed from Figure 5 that the values of the compressive residual stress at the peak position (~90 μm) appear to be less severely affected by shot diameter rather than Almen intensity; the increase of Almen intensity leads to an increase in compressive residual stress by 8.5% while the compressive residual stress grows by only 1.8% with the decrease of shot diameter.

3.4. Roughness. In what follows, the roughness characterization results are presented for both the individual carburizing and the hybrid process of carburizing and shot peening. The surface roughness parameter S_a, which denotes the arithmetic mean height and is the counterpart of the parameter arithmetical mean deviation of the roughness profile (R_a), is a commonly used roughness parameter for characterization of measured surface textures. Thus, S_a is used in what follows to discuss the surface states of treated specimens. In Table 3, the values of S_a are presented. For the specimen SP0 with the S_a value of ~2.51 μm, the surface of carburized specimens can be considered as rather rough. In this instance, the roughness can be ascribed to the texture left by machining (cf. Figure 6(a)) but also to the formation of the oxide layer produced during carburizing process (cf. Figure 2(a)).

Table 3: Values of the surface roughness parameter S_a of all surface treated specimens.

Variant	SP0	SP1	SP2	SP3
S_a (μm)	2.51	1.97	2.69	1.89

It can be seen from Table 3 that the magnitude of roughness due to shot peening induced at the carburized specimen surface depends on applied shot peening conditions. The surface roughness of SP1 specimens (with $S_a \approx 1.97\,\mu$m, cf. Figure 6(b)) is exhibited to decrease by 21.5% compared to that of SP0 specimens (with $S_a \approx 2.51\,\mu$m, cf. Figure 6(a)). This can be attributed to the fact that when Almen intensity is low, the impact of shot peening using smaller shot blasting particles is equivalent to the effect of surface polishing, which is however unable to induce sufficient plastic deformation that allows modifying completely the surface texture left by heat treatments and machining operations (cf. Figure 6(b)). In contrast, if shot peening process is performed using larger shot blasting particles, the surface roughness is increased. So the roughness of SP2 specimens (with $S_a \approx 2.69\,\mu$m, cf. Figure 6(c)) is found to be 36.5% higher than that of SP1 specimens. This is because the use of larger shot media can produce a higher level of surface plastic deformation that allows not only to enhancing the magnitude of the surface roughness increment, but also to obtaining a more evenly distributed texture, as illustrated in Figure 6(c).

With increasing Almen intensity from 0.25 mmA to 0.45 mmA, the surface state of the specimens is improved, and the surface roughness is thus reduced. For example, the S_a value of SP3 specimens ($\approx$1.89 μm, cf. Figure 6(d)) is approximated to be 29.7% less than that of SP2 specimens. This can be explained by the fact that the hard heat-treated surface region has been subjected to the so-called "intense" shot peening process condition. It is worth noting that using a high Almen intensity at double coverage can be interpreted as if every portion of the hard surface region has been treated with very high impact energy shot blasting particles for at least two times. Thereby, when the hard surface region is intensely shot peened, only a small amount of plastic deformation is induced as a consequence of the attainment of a critical limit of the plastic deformation, and as a result, the surface roughness is difficultly being altered. Due to the continuous impingement of the shot particles on the specimen surface, the surface state is thereby improved with lowered surface roughness (cf. Figure 6(d)). This observation is found to be in accordance with the results exhibited in the literature [9, 10, 27].

3.5. Fatigue Test Data. The fatigue tests are carried out to examine the fatigue performance of all the surface treated

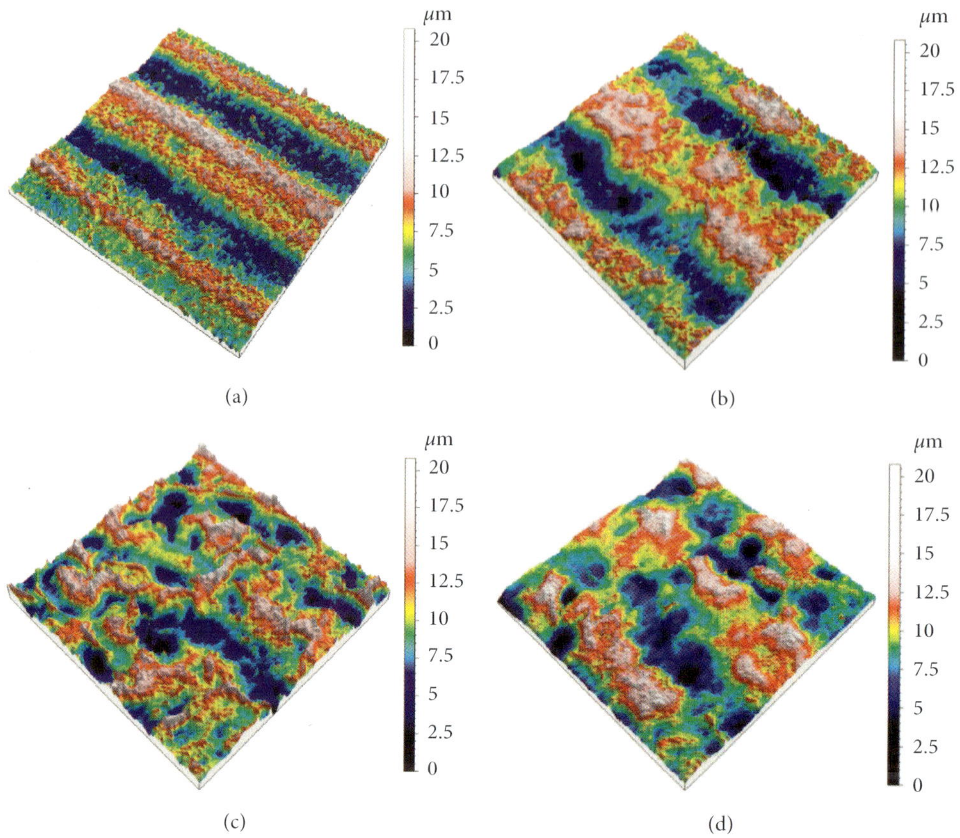

FIGURE 6: Surface topographies at the root surfaces of (a) SP0, (b) SP1, (c) SP2, and (d) SP3 specimens.

specimens. In Figure 7, the fatigue lives (N_f) of these surface treated specimens are depicted. N_f for case-hardened specimens is determined in the range of $14.8–57.9 \times 10^3$ cycles. It is noticeable that the shot peening process can considerably improve the fatigue lives of case-hardened specimens, which is attributed to the microstructure refinement, retained austenite reduction, microhardness amelioration, compressive residual stress generation, and surface roughness alteration induced by shot peening processes. These improved attributes are revealed to exhibit beneficial effects on fatigue strength and bending fatigue life [28]. It is worth noting that the trend of fatigue life improvement is highly dependent on the surface and subsurface layer properties which are controlled by shot peening conditions.

It is clearly shown in Figure 7 that N_f for SP2 specimens is the lowest, and it is improved by increasing Almen intensity, as exhibited by SP3 specimens. This significant improvement is solely a consequence of the presence of deeper compressed work-hardened subsurface layer with ameliorated surface roughness in SP3 specimens. Meanwhile, SP1 specimens are found to have the highest N_f despite the moderate work hardening and compressive residual stress. The remarkable increase in N_f for SP1 specimens can be related to the fact that the magnitude of the surface roughness is the lowest among all peened specimens (cf. Table 3). Nevertheless, the dispersion of the fatigue life of SP1 specimens is found to be rather large; the longest lifetime is estimated to be

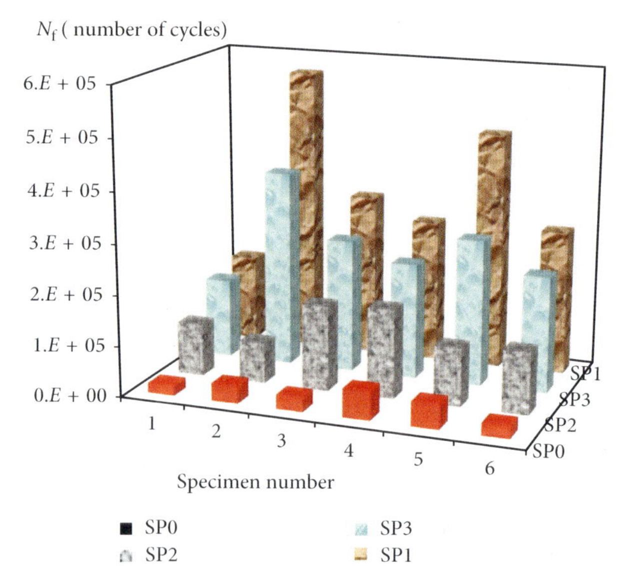

FIGURE 7: Bending fatigue lives of surface treated specimens.

about 69% more than the shortest one. However, the variation scale of the fatigue data obtained for SP1 specimens can be considered to be rather similar to the level of the fatigue data reported on shot peening [29, 30].

The presence of the scatter in the fatigue life can originate from the peening induced average surface and subsurface layer properties. Two important uncertain sources are noted here.

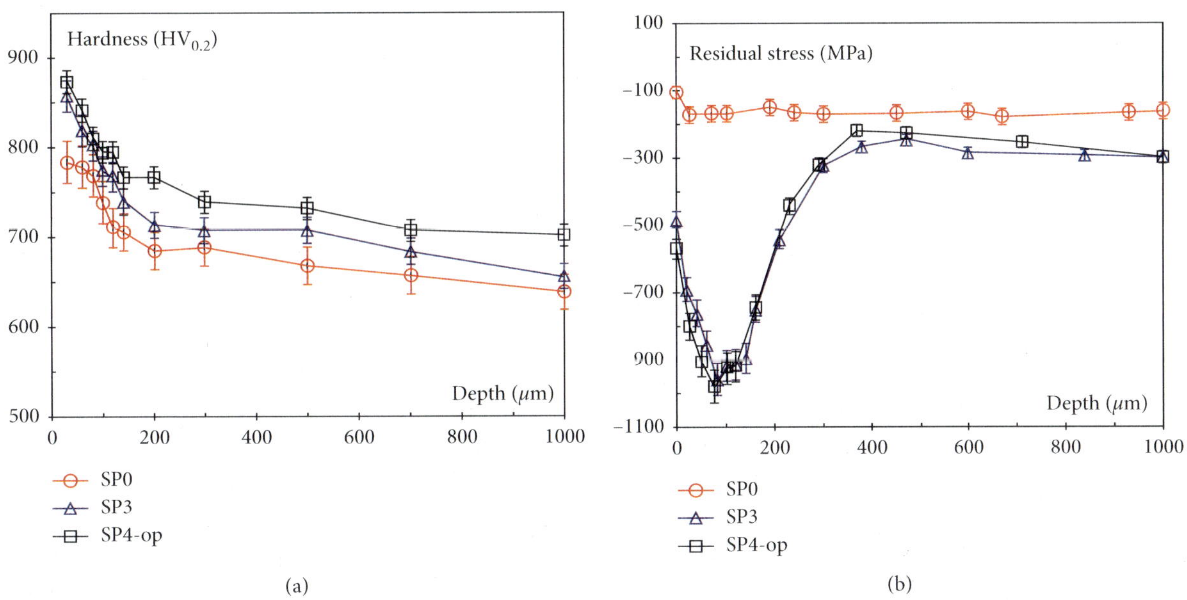

FIGURE 8: In-depth (a) microhardness and (b) residual stress distributions of the surface treated specimens.

One is the nonuniformity of the subsurface layer property, in particular, the uneven spatial distribution of the compressive residual stress. It is recently shown by [31] that if shot peening causes the unevenness of the residual stress distribution in a very small region of the surface, this can neutralize the beneficial effect of the compressive residual stress and hence degrade the fatigue resistance. The other is the fatigue strength endowed by the surface and subsurface layer properties, which is unable to completely cause a shift in fatigue crack initiation site from a surface to subsurface location and therefore cause the simultaneous occurrence of both surface and subsurface fatigue crack initiation in a fatigue testing series. This certainly intensifies the dispersion of the fatigue life.

3.6. Determination of Optimal Shot Peening Process

3.6.1. Optimal Shot Peening Conditions. In the previous section, it is investigated that the peening induced surface and subsurface layer properties can shift the fatigue life; it is of particular importance to optimize the shot peening conditions with the purpose to maximize their beneficial effect for enhanced fatigue performance. The testing results indicate that on one hand, SP1 specimens, possessing a moderate subsurface layer properties with reduced surface roughness, exhibit a considerable improvement in fatigue life which is however associated with a rather large scatter, and on the other hand, SP3 specimens, owing to its superior subsurface layer properties with rather low surface roughness, have a comparable estimate of the fatigue life with a moderate scatter. These investigations motivate us to propose an optimal shot peening conditions by grouping different peening conditions. The proposed dual peening process consists of two steps, as indicated by the name. The first peening process is aimed at generating a deep compressed work-hardened subsurface region by using a large media with very high kinetic energy, that is, with SP3

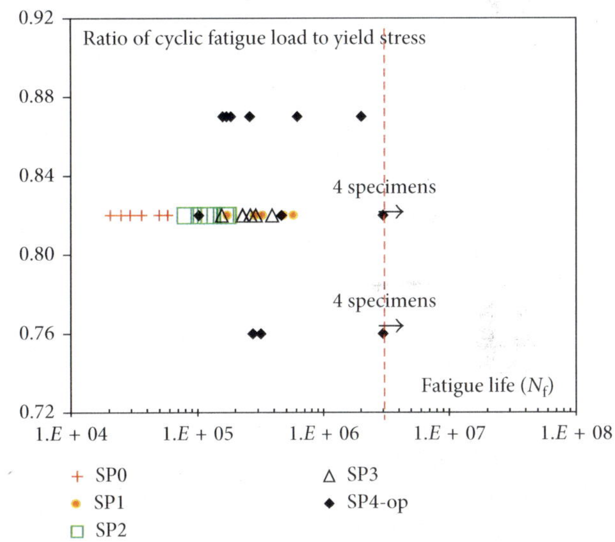

FIGURE 9: The ratio of applied cyclic load to yield stress versus fatigue life data.

process, and the repeening process is then performed using a smaller media with rather low kinetic energy, that is, with SP1 process, in order to improve the surface state (e.g., reduce the roughness and enhance its uniformity).

3.6.2. Validity of the Proposed Solution. In the following, the specimens subjected to the proposed optimal shot peening condition are denoted as SP4-op specimens. The experimental results show that the maximum microhardness of 874 $HV_{0.2}$ appears on the surface (cf. Figure 8), the values of top surface and maximum compressive residual stresses are, respectively, 569 MPa and 980 MPa, and the estimated value of S_a is equal to ~1.86 μm. Compared to the previous single peening method, the dual peening method by using first

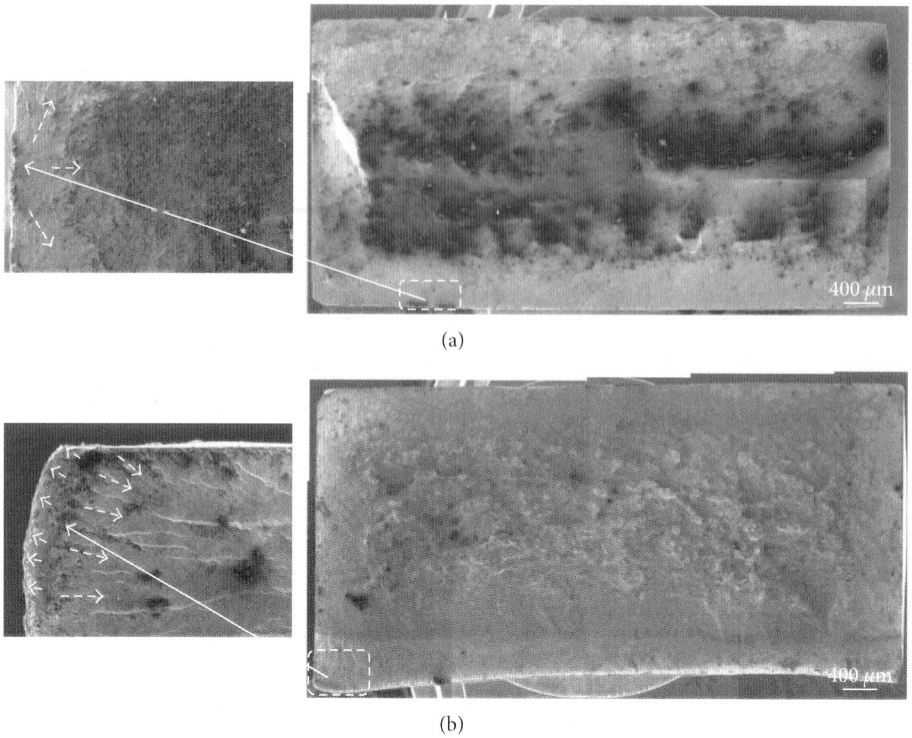

FIGURE 10: SEM micrographs of the surface treated specimens obtained at the ratio of applied cyclic load to yield stress of (a) 0.87 and (b) 0.82.

a large media with very high kinetic energy followed by a smaller media with rather low kinetic energy can indeed be more advantageous in producing improved quality of the surface and subsurface layer of 18CrNiMo7-6 steel. It is emphasized that the repeening process in the dual peening method has double benefits: firstly, inducing a more regular surface state by smoothing sharp corners and somehow closing the microcracks generated during the first shot peening process; secondly, generating a more homogeneous surface residual stress and surface microhardness [29, 32].

To validate the proposed shot peening process condition for fatigue life enhancement, the fatigue test is carried out for SP4-op specimens under three different stress levels; the ratio of the applied cyclic fatigue load to yield stress is 0.76, 0.82, and 0.87. The fatigue data presented in Figure 9 show that for the first two cyclic load levels, the bending fatigue life of gear teeth is greatly improved, that is, not only the fatigue life is increased marvelously (with $N_f \geq 3 \times 10^6$ cycles), but also the degree of dispersion in the fatigue-life data is also minimized. This affirms the capability of the dual peening process in delaying the fatigue crack initiation phase by enhancing the surface and subsurface layer properties of 18CrNiMo7-6 steel. However, when the ratio of applied cyclic fatigue load to yield stress is increased up to 0.87, the fatigue performance of SP4-op specimens is drastically degraded.

The aforementioned observations are understood as follows. Under the high cyclic fatigue load, fatigue cracks are most likely to initiate from the surface (cf. Figure 10(a)), in contrast to the case of lower fatigue loads, where fatigue cracks are found to be prone to originate from the interior of the subsurface layer region (cf. Figure 10(b)). When the load level is low, the peak stress of the applied fatigue load does not exceed the increased fatigue strength at the surface; however, at some depth below the surface, the applied stress can be larger than the local fatigue strength. Using the concept of the local fatigue strength applied for surface treated material [10], the intersection of the applied stress level with the local fatigue strength curve gives the site of crack initiation; subsurface crack initiation does occur. In the case of high load being applied, crack initiation happens at the surface because the peak stress at the free surface exceeds the increased fatigue strength.

4. Conclusion

In this paper, the effect of different shot peening conditions on the surface and subsurface layer properties and the fatigue performance of case-hardened 18CrNiMO7-6 steel has been investigated. The analysis of the surface and subsurface integrity reveals that the subsurface region of the case-hardened 18CrNiMO7-6 steel can no further accumulate enormous localized plastic deformation during shot peening process as a consequence of the attainment of a critical limit of the plastic deformation for 18CrNiMo7-6 steel due to the presence of a high dislocation density and carbon content induced by the prior carburizing process. Based on the quantitative assessment of subsurface layer properties and fatigue performance, the interrelationships between peening process, subsurface layer properties, and fatigue performance are qualitatively established; shot peening application with lower intensity and smaller shots causing moderate subsurface layer properties results in an increase in both fatigue life and its associated scatter, while shot peening application with higher intensity and larger shots producing superior subsurface layer properties exhibits comparable estimate of fatigue life with decreased scatter in fatigue data. The experimental study suggests that a two-step shot peening process, where a large media with very high kinetic energy is used in the first step and a smaller media with rather low kinetic energy in the second step, can be considered as an effective way to enhance the surface and subsurface layer characteristics and the fatigue performance; the optimality of the proposed dual peening process is experimentally validated and discussed.

Conflicts of Interest

The authors declare that they have no conflicts of interest.

Acknowledgments

The authors are grateful to the National Natural Science Foundation of China–International Cooperation and Exchange Program (Grant no. 51650110502) and Henan Provincial Research Bureau (Grant nos. 15A460030 and 17A460025) for their financial support.

References

[1] K. Lu, "Making stronger materials ductile with gradients," *Science*, vol. 345, no. 6203, pp. 1455-1456, 2014.

[2] T. Hanlon, E. D. Tabachnikova, and S. Suresh, "Fatigue behavior of nanocrystalline metals and alloys," *International Journal of Fatigue*, vol. 27, no. 10–12, pp. 1147–1158, 2005.

[3] Y. K. Gao and X. R. Wu, "Experimental investigation and fatigue life prediction for 7475-T7351 aluminum alloy with and without shot peening-induced residual stresses," *Acta Materialia*, vol. 59, no. 9, pp. 3737–3747, 2011.

[4] P. Fu, K. Zhan, and C. H. Jiang, "Micro-structure and surface layer properties of 18CrNiMo7-6 steel after multistep shot peening," *Materials and Design*, vol. 51, pp. 309–314, 2013.

[5] S. Huang, Y. Zhu, W. Gui et al., "Effects of laser shock processing on fatigue crack growth in Ti-17 titanium alloy," *Journal of Materials Engineering and Performance*, vol. 25, no. 2, pp. 813–821, 2017.

[6] V. Singh and M. Marya, "Surface modification of oilfield alloys by ultrasonic impact peening: UNS N07718, N07716, G41400 and S17400," *Journal of Materials Engineering and Performance*, vol. 25, no. 1, pp. 338–347, 2016.

[7] S. J. Kim, K. Y. Hyun, and S. K. Jang, "Effects of water cavitation peening on electrochemical characteristic by using micro-droplet cell of Al-Mg alloy," *Current Applied Physics*, vol. 12, pp. S24–S30, 2012.

[8] L. Trsko, M. Guagliano, O. Bokuvka, F. Novy, M. Jambor, and Z. Florkova, "Influence of severe shot peening on the surface state and ultra-high-cycle fatigue behavior of an AW 7075 aluminum alloy," *Journal of Materials Engineering and Performance*, vol. 26, no. 6, pp. 2784–2797, 2017.

[9] Y. Lv, L. Q. Lei, and L. N. Sun, "Influence of different combined severe shot peening and laser surface melting treatments on the fatigue performance of 20CrMnTi steel gear," *Materials Science and Engineering A*, vol. 658, pp. 77–85, 2016.

[10] S. M. Hassani-Gangaraj, A. Moridi, M. Guagliano, A. Ghidini, and M. Boniardi, "The effect of nitriding, severe shot peening and their combination on the fatigue behavior and microstructure of a low-alloy steel," *International Journal of Fatigue*, vol. 62, pp. 67–76, 2014.

[11] A. C. Batista, A. M. Dias, J. L. Lebrun, J. C. Le Flour, and G. Inglebert, "Contact fatigue of automotive gears: evolution and effects of residual stresses introduced by surface treatments," *Fatigue and Fracture of Engineering Materials and Structures*, vol. 23, no. 3, pp. 217–228, 2000.

[12] P. G. Ranaware and M. J. Rathod, "Combined effect of shot peening, subcritical austenitic nitriding, and cryo-treatment on surface modification of AISI 4140 steel," *Materials and Manufacturing Processes*, vol. 32, no. 4, pp. 349–354, 2017.

[13] F. Ashrafizadeh, "Influence of plasma and gas nitriding on fatigue resistance of plain carbon (Ck45) steel," *Surface and Coatings Technology*, vol. 174-175, pp. 1196–1200, 2003.

[14] ANSI/AGMA B89, *Gear Material and Heat Treatment Manual*, American Gear Manufacturers Association, Alexandria, VI, USA, 2004.

[15] ISO 26910-1, *Springs–Shot Peening–Part 1: General Procedures*, International Organization for Standardization, Geneva, Switzerland, 2009.

[16] V. Llaneza and F. J. Belzunce, "Study of the effects produced by shot peening on the surface of quenched and tempered steel: roughness, residual stress and work hardening," *Applied Surface Science*, vol. 356, pp. 475–485, 2015.

[17] ISO 25178-2, *Geometrical Product Specifications (GPS)–Surface Texture: Areal–Part 2: Terms, Definitions and Surface Texture Parameters*, International Organization for Standardization, Geneva, Switzerland, 2012.

[18] E. S. Gadelmawla, M. M. Koura, T. M. A. Maksoud, I. M. Elewa, and H. H. Soliman, "Roughness parameters," *Journal of Materials Processing Technology*, vol. 123, no. 1, pp. 133–145, 2002.

[19] R. W. Buenneke, M. B. Slane, C. R. Dunham, M. P. Semenek, M. M. Shea, and J. E. Tripp, "Gear single tooth bending fatigue test," in *Proceedings of SAE Technical Paper 821042*, Milwaukee, WI, USA, 1982.

[20] M. Zaccone and G. Krauss, "Elastic limit and microplastic response of hardened steels," *Metallurgical and Materials Transaction A*, vol. 24, no. 10, pp. 2263–2277, 1993.

[21] M. Benedetti, V. Fontanari, B. R. Hohn, and T. Tobie, "Influence of shot peening on bending tooth fatigue limit of case hardened gears," *International Journal of Fatigue*, vol. 24, no. 11, pp. 1127–1136, 2002.

[22] S. M. Hassani-Gangaraj, K. S. Cho, H. J. L. Voight, M. Gugliano, and C. A. Schuh, "Experimental assessment and simulation of surface nanocrystallization by severe shot peening," *Acta Materialia*, vol. 97, pp. 105–115, 2015.

[23] X. H. An, Q. Y. Lin, S. D. Wu, and Z. F. Zhang, "Improved fatigue strengths of nanocrystalline Cu and Cu-Al alloys," *Materials Research Letters*, vol. 3, no. 3, pp. 135–141, 2015.

[24] K. Dalaei, B. Karlsson, and L. E. Svensson, "Stability of shot peening induced residual stresses and their influence on fatigue lifetime," *Materials Science and Engineering A*, vol. 528, no. 3, pp. 1008–1015, 2011.

[25] M. Kobayashi, T. Matsui, and Y. Murakami, "Mechanism of creation of compressive residual stress by shot peening," *International Journal of Fatigue*, vol. 20, no. 5, pp. 351–357, 1998.

[26] R. E. Reed-Hill and R. Abbaschian, *Physical Metallurgy Principles*, Publishing Company, Boston, MA, USA, 1992.

[27] O. Unal and R. Varol, "Almen intensity effect on microstructure and mechanical properties of low carbon steel subjected to severe shot peening," *Applied Surface Science*, vol. 290, pp. 40–47, 2014.

[28] N. Kawagoishi, T. Nagano, M. Moriyama, and E. Kondo, "Improvement of fatigue strength of maraging steel by shot peening," *Materials and Manufacturing Processes*, vol. 24, no. 12, pp. 1431–1435, 2009.
[29] R. R. Rego, J. O. Gomes, and A. M. Barros, "The influence on gear surface properties using shot peening with a bimodal media size distribution," *Journal of Materials Processing Technology*, vol. 213, no. 12, pp. 2152–2162, 2013.
[30] S. Bagherifard, I. Fernandez-Pariente, R. Ghelichi, and M. Gugliano, "Effect of severe shot peening on microstructure and fatigue strength of case iron," *International Journal of Fatigue*, vol. 65, pp. 64–70, 2014.
[31] S. M. H. Gangaraj and G. H. Farrahi, "Side effects of shot peening on fatigue crack initiation life," *International Journal of Engineering, Transactions A: Basics*, vol. 24, pp. 275–280, 2011.
[32] K. Zhan, C. H. Jiang, and V. Ji, "Uniformity of residual stress distribution on the surface of S30432 austenitic stainless steel by different shot peening processes," *Materials Letters*, vol. 99, pp. 61–64, 2013.

Durability of Steel Fibres Reinforcement Concrete Beams in Chloride Environment Combined with Inhibitor

AbdelMonem Masmoudi[1] and Jamel Bouaziz[2]

[1]*Department of Civil Engineering, LA2MP Laboratory, ENIS, Université de Sfax, Sfax, Tunisia*
[2]*Department of Industrial Chemistry, National School of Engineers of Sfax, Sfax, Tunisia*

Correspondence should be addressed to AbdelMonem Masmoudi; abdelmonem.masmoudi@enis.rnu.tn

Academic Editor: Velu Saraswathy

This paper presented the effect of the combination of an inhibitor and steel fibre reinforced concrete (SFRC) for concrete structures in chloride environments. Twelve beams were cast and tested to study their flexural behavior. The morphology of steel surfaces using the inhibitor after observing the scanning electron microscope showed a low layer of corrosion products. The steel surface immersed in the inhibitor free solution was seen to have been subject to chloride ions attacks as shown in this study. The interest to the field of the present study is the relatively higher durability of the performance when using the inhibitor. Crack width and crack spacing for beams under the same load showed that the use of SFRC with the inhibitor for concrete structures in chloride environments must have transferred tension across cracks that led to reducing crack spacing without any chloride ions attack.

1. Introduction

When the concrete is exposed to seawater or deicing salts, it tends to cause localised breakdown of the passive film, a phenomenon termed pitting corrosion. This can result in serious local loss of the bars' cross section in the affected regions while the surrounding regions remain virtually unaffected, if sufficient water and oxygen are available at the reinforcement surface.

According to a study carried out by Beeby [1] and Tuutti [2], results showed that macrocracks play an important role in the transport of aggressive substances. If engineers would significantly improve the life span of concrete structures they must be controlling these cracks. As a way to ensure the durability of the reinforced concrete structures, current regulations define the maximum allowed crack widths, based on exposure conditions. Even though the effect of cracks on the initiation of corrosion has been dealt with by several authors, the effect of cracks on durability, for example, is still debated. The only consensus amongst researchers is that if the cracks exceed a certain size, they will have a negative impact on durability. Berrocal et al. [3] and Otieno et al. [4] showed that the application of steel fibre reinforced concrete (SFRC) is still limited to industrial floors, slabs, and pavements, because of the random distribution and orientation of the fibres, which reduce the mechanical efficiency when compared to conventional reinforcing bars.

Since the first studies on SFRC in the early 1960s, a significant amount of research has been carried out to achieve a deeper understanding of the mechanical properties of the material. The important result is that cracks can be modified by correct reinforced concrete members in direct tension. Another study done by Abrishami and Mitchell [5] focused on the influence of steel fibres on reinforcement concrete beams indicated that the transverse cracks were more closely and smaller compared with specimens without steel fibres. This leads to the fact that adding fibres could effectively control splitting cracks. In a similar study concluded on the SFRC uses by Bischoff [6], it was showed that SFRC is efficient in transferring tension across cracks, consequently reducing crack spacing and increasing tension stiffening. The influence of adding different types of steel fibres on chloride penetration has been studied by Mangat and Gurusamy [7] and discussed elsewhere by Buratti et al. [8]. They found that fibres had an insignificant effect in sound concrete. In addition, when the structures are exposed to marine

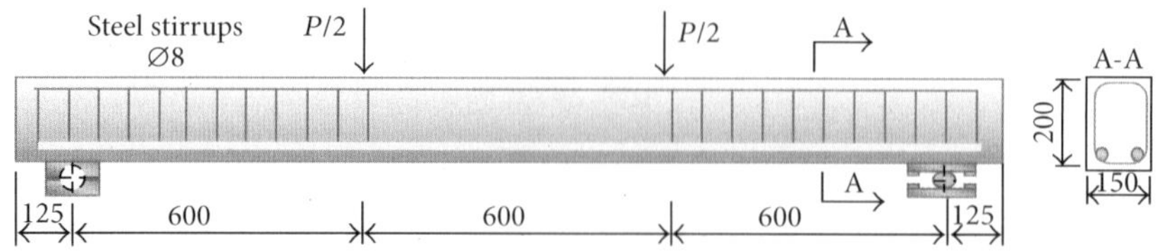

FIGURE 1: Geometric details of the tested beams.

environments, they do not exhibit any effects to the sound and also elements remain uncracked.

The addition of fibres for cracked concrete had a marginal effect for cracks of less than 0.2 mm. This addition became important for cracks of more than 0.5 mm. A critical crack width has been estimated to values ranging between 0.10 and 0.25 mm, which showed that hybrid fibres would be very close to the concrete surface. Some researchers have directed their investigations towards the mechanical characterization when fibres are exposed to extreme environments [9]. This study makes in evidence the fact that the fibres are influenced by external agents, and, consequently, a higher degradation would be expected Danso et al. [10]. Another study carried by Hongfang [11] used a simulated impressed current cathodic protection system with 3% NaCl solution without inhibitor. Experimental results showed that, during anodic polarization, the epoxy was corroded. Kobayakawa et al. [12] investigated the influence of superficial fibres accompanied by substantial stain rust appearing at the concrete surface. Results showed that reduction of damage in the fibres is the consequence of reducing the water cement ratio of the concrete mix as well as limiting the region where fibres are prone to suffering severe corrosion to depths as small as 0.2 mm [13, 14].

The critical chloride content is generally accepted to be in the range of 0.4–1.0% Cl^- (by weight of cement) for conventional reinforced concrete structures. The critical chloride content, or chloride threshold value, represents a basic concept used today by most of the current service life models. Mangat and Gurusamy [15] showed that fibres embedded in concrete remained free from corrosion for chloride concentrations up to 1.7% Cl^-. This is in agreement with the results of Janotka et al. [16] and Ganesan et al. [17] who found that the necessary concentration of chloride to initiate corrosion in steel fibres was at least 3 times higher compared to conventional reinforcing steel.

On the other hand, the experimental investigation carried out in solutions simulating the concrete pore electrolyte by Elsener et al. [18] revealed that the amino alcohol tested compounds are efficient against steel corrosion. Moreover, their use as mortar admixtures is not detrimental to the physicomechanical properties. Triethanolamine (TEA) use at a convenient concentration equal to 0.5 mL/50 mL guaranteed steel corrosion inhibition in mortar contaminated by sodium chloride NaCl at 0.5 mole/L. The electrochemical impedance spectroscopy technique EIS associated to periodic chloride concentration measurements proved the effect of TEA in delaying the corrosion process initiation. Furthermore, its capacity of diffusion through mortar cover was noted. Consequently, TEA can be exploited through both preventive and curative modes of use [19–21].

This paper aimed at determining the viability of using steel bars immersed in a chloride solution combined with inhibitor to improve the relatively better durability performance.

2. Experimental Test Program

2.1. Beams Specifications. The beams were cast into two series with the same target concrete strength (30 MPa). For each series, a total of 8 concrete beams of 150 × 200 × 2050 mm (Figure 1) and with a clear cover equal to 20 mm were with two different amounts of steel bars (two diameters 10 mm and two diameters 12 mm) placed at the tension side (bottom). In this study, groups of beams were tested before and after being immersed for 28 days in two Cl^- solutions (S_1 and S_2). The two chloride solutions' concentrations are equal to 5.1% Cl^- (three times the minimum value likely to cause corrosion for the steel fibres which is around 1.7% Cl^-). The solution S_2 having the same chloride concentration (5.1% Cl^-) is combined with triethanolamine (TEA) type inhibitor. Changes in capacity flexure behavior values are often used as indicators of the durability performance. Additionally, for comparison purposes, 2 beams were reinforced with 2Ø10 steel bars and 2 beams were reinforced with 2Ø10 steel bars and steel fibre reinforcement thus having a similar stiffness as the other beams that were immersed for 28 days in water (L_0 equal zero % Cl^-).

The beams were designated as Sx-Myyz, where S denotes the chloride solution (S_1, S_2), respectively, combined with inhibitor and not combined. For comparison purposes, a reference solution S_0 without chloride nor inhibitor was chosen; M is the material of reinforcement (S = Stell, SSF = steel with steel fibres) and yy is the diameter of the bar (10, 12 mm); additionally, *z* differentiates between the twin beams. The geometry and designation of the beams are shown in Figure 1 and Table 1.

2.2. Material Properties

2.2.1. Concrete. Concrete with a target 30 MPa was used for both beam series, "S_1" and "S_2." The concrete was provided by a local ready-mix supplier with a maximum aggregate size of 12 mm. The proportions of the concrete mix are summarized in Table 2. For both series, the cylindrical specimens were kept in the same exposure conditions of temperature and humidity until testing. From these cylinders the compressive strength f_c, compressive modulus of elasticity E_c, and tensile

Table 1: Beams designation.

Beam designation	Rebar material	Ø bar (mm)	Reinforcement ratio ρ (%)	Concentrations of Cl^- (%)
S-beams				
S_1-S_{10a}	Steel	10	1.0	5.1
S_1-S_{10b}	Steel	10	1.0	5.1
S_2-S_{12a}	Steel	12	1.4	5.1 + TEA
S_2-S_{12b}	Steel	12	1.4	5.1 + TEA
S_1-SSF_{10a}	Steel + steel fibres	10	1.0	5.1
S_1-SSF_{10b}	Steel + steel fibres	10	1.0	5.1
S_2-SSF_{12a}	Steel + steel fibres	12	1.4	5.1 + TEA
S_2-SSF_{12b}	Steel + steel fibres	12	1.4	5.1 + TEA
S_o-S_{10a}	Steel	10	1.0	0
S_o-S_{10b}	Steel	10	1.0	0
S_o-SSF_{10a}	Steel + steel fibres	10	1.0	0
S_o-SSF_{10b}	Steel + steel fibres	10	1.0	0

Table 2: Composition of concrete.

Component	Units	Concrete mix
Water	Kg/m^3	175
Cement (CEMI 32.5 N)	Kg/m^3	350
w/c ratio		0.51
Fine aggregate	Kg/m^3	900
Coarse aggregate	Kg/m^3	820
Plasticizer BV 40	Cement weight (%)	1.2
Target compressive strength	MPa	30

strength f_{ct} were evaluated and the test results are represented in Table 3.

2.2.2. Rebars. The steel bars used in these experiments were of HA 400 type, with specified design modulus of elasticity of 200 GPa and characteristic yield and tensile strengths of 400 MPa and 420 MPa, respectively.

2.2.3. Steel Fibres. The steel fibres used in these experiments were with specified design modulus of elasticity of 210 GPa and characteristic yield and tensile strengths of 1000 MPa and 1150 MPa, respectively.

2.2.4. Triethanolamine. The inhibitor (TEA) used in conditioning beams has a convenient concentration equal to 0.5 mL/50 mL of water.

2.3. Test Setup and Instrumentation. All beam specimens were tested under a static four-point load test. A servo-controlled hydraulic machine with a capacity of 300 kN was used to apply the load to the test beam (Figure 2). The load was applied in displacement control mode at a displacement rate of 0.7 mm/min, and all data were collected by a data acquisition system. The test was stopped every 10–20 kN to register the evolution of cracks, and crack widths were measured with an optical micrometer with an accuracy of 0.05 mm.

Figure 2: Flexure test with data acquisition system.

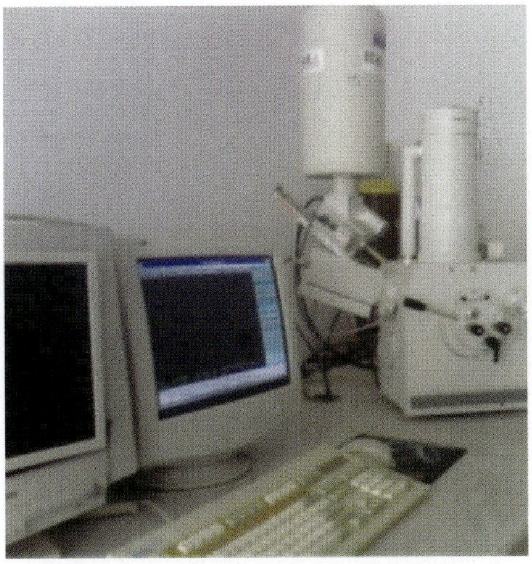

Figure 3: Philips XL 30 SEM.

Scanning electron microscope (SEM) observations and image analyses were performed to observe the specimens microstructure before and after immersion in the chloride environment. All specimens observed in the SEM were first cut, polished, and coated with a thin gold layer. After the coating of surfaces, microstructural observations were performed on a Philips XL 30 SEM (Figure 3).

TABLE 3: Experimental values of f_c, E_c, and f_{ct}.

Compressive strength (f_c)		Compressive modulus of elasticity (E_c)		Tensile strength (f_{ct})	
Age (days)	Mean (MPa)	Age (days)	Mean (MPa)	Age (days)	Mean (MPa)
31	*32.1*	*31*	*34125*	*31*	*2.8*

(a) (b)

FIGURE 4: SEM images of steel bars imbedded in chloride. (a) Using an inhibitor; (b) without any inhibitor.

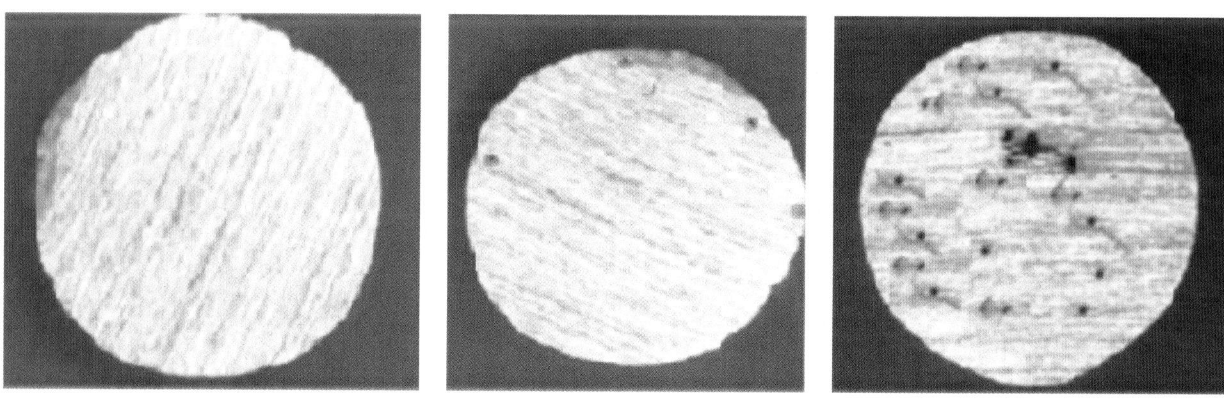

(a) (b) (c)

FIGURE 5: Micrographs of steel surfaces immersed for 30 days in different environments (a) S_o, (b) S_1, and (c) S_2.

3. Test Results and Discussion

3.1. Microstructure Evaluations. Figure 4 shows the morphology of steel surfaces after observing the scanning electron microscope (SEM) at the end of immersion in solutions S_1 and S_2. The steel surface immersed in the solution using inhibitor (Figure 4(a)) has a low layer of corrosion products, while the steel surface immersed in the solution without inhibitor (Figure 4(b)) clearly shows the attack of steel surface by chloride ions. Chloride ions are responsible for corrosion, resulting in the decrease of the steel bar section.

The microstructure comparisons of the two SEM bars revealed that bars immersed in inhibitor solutions had relatively better durability performance, which is in agreement with the results of Avci [19], revealing that the inhibitor used in conditioning with a convenient concentration equal to 0.5 mL/50 mL of water is efficient against steel corrosion. This highlights the protective effect of the inhibitor (TEA).

A comparison of the micrographs steel surfaces immersed in three different solutions levels S_o, S_1, and S_2, showed clearly that there is no attack of the steel surfaces immersed in solution S_o (Figure 5(a)). No significant attack was observed for the steel surfaces immersed in the solution S_1. However the steel surfaces in the solution S_2 show considerable attack by chloride ions. It can be concluded that the necessary concentration of chloride to initiate corrosion in SFRC combined with an inhibitor was at least 4 times higher compared to the conventional reinforcing steel.

Concrete mortar samples cut from specimens with and without TEA (0.5 mL/50 mL) and having been treated for 28 days in water have been observed in SEM (Figure 6). The comparison of two figures (Figures 6(a) and 6(b)) reveals that

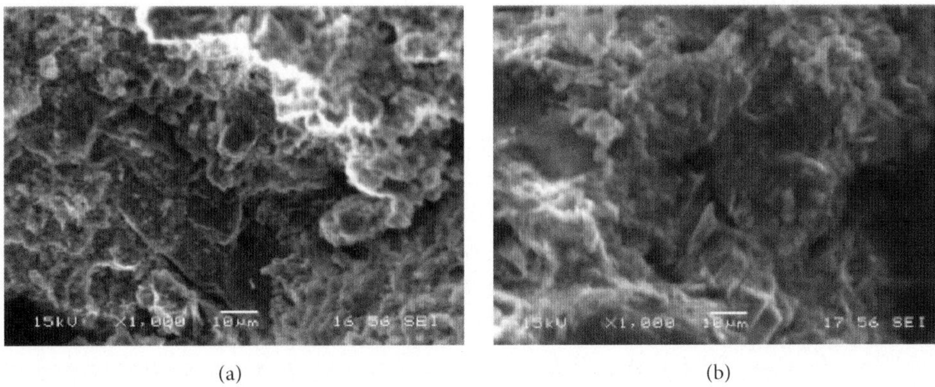

Figure 6: SEM micrographs mortar sample. (a) Without an inhibitor; (b) with any inhibitor.

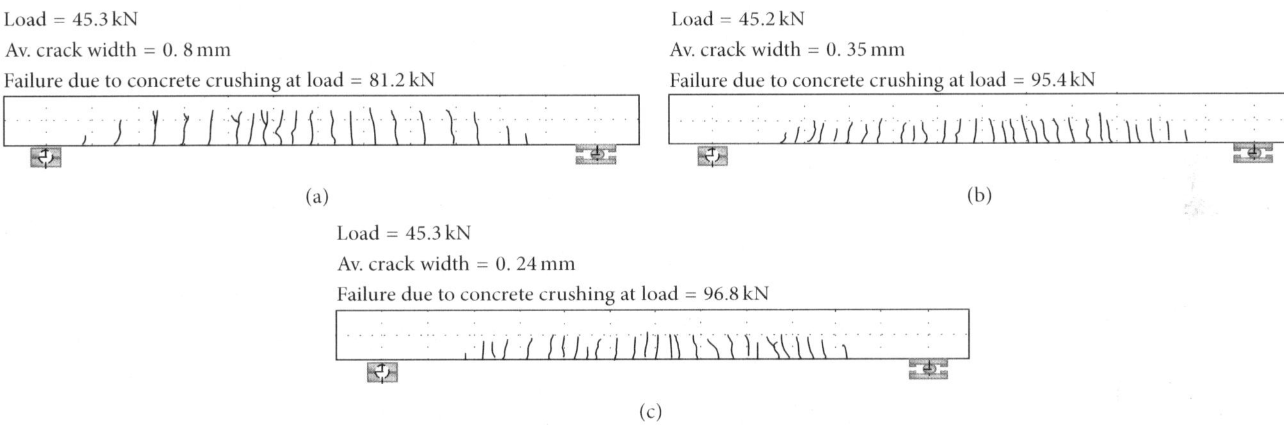

Figure 7: Crack pattern for beams under same load. (a) Without SFRC, (b) with SFRC, and (c) with SFRC and an inhibitor.

the morphology of concrete mortar was not modified by the incorporation of the TEA inhibitor.

3.2. Results on Cracking. In Figure 7, the typical experimental crack widths are represented as a function of the average crack width at same load step for two specimens with and without steel fibre reinforced concrete (SFRC). As depicted, during the crack formation phase, cracks form at random positions and are mostly vertical and start at the central zone. As the load increases, additional inclined cracks appear due to the effect of shear forces on the principal tensile stresses, and those that were vertical at the shear span also start to incline because of the combined flexural-shear effects in the zones with larger bending moment.

Under the same load, crack pattern for beams showed that the use of SFRC should transfer tension across cracks which led to reducing crack spacing (0.35 mm compared to 0.8 mm) and increased tension stiffening (Figure 7(b)). The transverse cracks were smaller and more closely spaced compared to beams without steel fibres. Finally, no more cracks appeared and the existing ones widened. The experimental minimum, average, and maximum crack spacing were measured, at the same load steps, in the pure bending central zone at the height of the reinforcement. The crack spacing ranged between 30 mm and 190 mm for beams without SFRC. However crack spacing ranged between 14 mm and 85 mm for beams with SFRC. This may lead us to conclude that the fibres provide good distribution of cracks, resulting in a reduction of the crack opening. The load level P at which cracking stabilized (P_{sta}) ranged between 12% and 43.1% of the ultimate load Pu, with a mean value of 25.1% Pu and a standard deviation of 9.3% Pu for beams without SFRC. This load level P rose for beams with SFRC, and P_{sta} ranged between 19% and 63.1% of the ultimate load Pu, with a mean value of 41.05% Pu and a standard deviation of 9.7% Pu. For the same load, the average crack width for beams with the inhibitor is less that those conditioning in solution without inhibitor. The inhibitor changes the behavior and protects structure in chloride environments. The interest to the field of using the SFRC combined with an inhibitor in RC beams in chloride environments has the positive impact of reducing the crack width and therefore increasing the durability of the structure (Figure 7(b)).

4. Conclusion

Based on the experimental study and microstructure observations, the following conclusions can be drawn:

(i) The microstructure for the two SEM bars revealed that bars immersed in a chloride solution combined with an inhibitor had relatively better durability performance.

(ii) The necessary concentration of chloride to initiate corrosion in SFRC combined with an inhibitor is at least 3 times higher compared to conventional reinforcing steel.

(iii) Steel fibres combined with an inhibitor could be used in reinforced concrete structures exposed to chloride environments improving their overall durability performance

(iv) The use of SFRC should transfer tension across cracks leading to reduced crack spacing and the inhibitor changes the behavior and protects structure in chloride environments.

(v) The average crack width for beams with an inhibitor is less that conditioning in solution without any inhibitor.

Competing Interests

The authors declare that there is no conflict of interests.

Acknowledgments

The authors would like to thank the manufacturer of el Fouled at Bizerte for providing the steel bars and the Faculty of Sciences of Sfax for using Philips XL 30 SEM. The research was conducted in National Engineering School of Sfax.

References

[1] A. W. Beeby, "Cracking: what are crack width limits for?" *Concrete*, vol. 12, no. 7, pp. 31–33, 1978.

[2] K. Tuutti, "Corrosion of steel in concrete," CBI Report 4:82, The Swedish Cement and Concrete Research Institute, Stockholm, Sweden, 1982.

[3] C. Berrocal, K. Lundgren, and I. Lofgren, "Influence of steel fibers on corrosion of reinforcement in concrete in chloride environments: a review Fibre concrete," in *Proceedings of the 7th International Conference (FIBRE CONCRETE '13)*, Prague, Czech Republic, September 2013.

[4] M. B. Otieno, M. G. Alexander, and H.-D. Beushausen, "Corrosion in cracked and uncracked concrete—influence of crack width, concrete quality and crack reopening," *Magazine of Concrete Research*, vol. 62, no. 6, pp. 393–404, 2010.

[5] H. H. Abrishami and D. Mitchell, "Influence of steel fibers on tension stiffening," *ACI Structural Journal*, vol. 94, no. 6, pp. 769–776, 1997.

[6] P. H. Bischoff, "Tension stiffening and cracking of steel fiber-reinforced concrete," *Journal of Materials in Civil Engineering*, vol. 15, no. 2, pp. 174–182, 2003.

[7] P. S. Mangat and K. Gurusamy, "Chloride diffusion in steel fibre reinforced marine concrete," *Cement and Concrete Research*, vol. 17, no. 3, pp. 385–396, 1987.

[8] N. Buratti, C. Mazzotti, and M. Savoia, *Long-Term Behavior of Cracked SFRC Elements Exposed to Chloride Solutions*, ACI Special Publication, 2011.

[9] J.-L. Granju and S. U. Balouch, "Corrosion of steel fibre reinforced concrete from the cracks," *Cement and Concrete Research*, vol. 35, no. 3, pp. 572–577, 2005.

[10] H. Danso, D. Brett Martinsona, M. Ali, and J. B. Williams, "Physical, mechanical and durability properties of soil building blocks reinforced with natural fibres," *Construction and Building Materials*, vol. 101, part 1, pp. 797–809, 2015.

[11] S. Hongfang, "Corrosion behavior of carbon reinforced polymer anode in simulated impressed current cathodic protection system with 3%NaCl solution," *Construction and Building Materials*, vol. 112, pp. 539–546, 2016.

[12] A. Kobayakawa, D. Homma, H. Mihashi, and K. Shimozawa, "Corrosion durability of fiber reinforced cementitious composites," in *Proceedings of the 2nd RILEM Workshop on Concrete Durability and Service Life Planning*, pp. 311–318, 2009.

[13] K. Kosa and A. E. Naaman, "Corrosion of steel fiber reinforced concrete," *ACI Materials Journal*, vol. 87, no. 1, pp. 27–37, 1990.

[14] S. U. Balouch, J. P. Forth, and J.-L. Granju, "Surface corrosion of steel fibre reinforced concrete," *Cement and Concrete Research*, vol. 40, no. 3, pp. 410–414, 2010.

[15] P. S. Mangat and K. Gurusamy, "Permissible crack widths in steel fibre reinforced marine concrete," *Materials and Structures*, vol. 20, no. 5, pp. 338–347, 1987.

[16] I. Janotka, L'. Krajčí, K. Komloš, and D. Frt'alová, "Chloride corrosion of steel fibre reinforcement in cement mortar," *The International Journal of Cement Composites and Lightweight Concrete*, vol. 11, no. 4, pp. 221–228, 1989.

[17] N. Ganesan, R. Abraham, and S. Deepa Raj, "Durability characteristics of steel fibre reinforced geopolymer concrete," *Construction and Building Materials*, vol. 93, pp. 471–476, 2015.

[18] B. Elsener, M. Büchler, F. Stalder, and H. Böhni, "Migrating corrosion inhibitor blend for reinforced concrete: part 1—prevention of corrosion," *Corrosion*, vol. 55, no. 12, pp. 1155–1163, 1999.

[19] G. Avci, "Inhibitor effect of N,N'-methylenediacrylamide on corrosion behavior of mild steel in 0.5 M HCl," *Materials Chemistry and Physics*, vol. 112, no. 1, pp. 234–238, 2008.

[20] V. Afroughsabet and T. Ozbakkaloglu, "Mechanical and durability properties of high-strength concrete containing steel and polypropylene fibers," *Construction and Building Materials*, vol. 94, pp. 73–82, 2015.

[21] B. Elsener, *Corrosion Inhibitors for Steel in Concrete. State of the Art Report*, EFC Series no. 33, IOM Communications, Institute of Materials, London, UK, 2001.

15

Compressive Behavior and Mechanical Characteristics and their Application to Stress-Strain Relationship of Steel Fiber-Reinforced Reactive Powder Concrete

Baek-Il Bae,[1] Hyun-Ki Choi,[2] Bong-Seop Lee,[2] and Chang-Hoon Bang[2]

[1]*Research Institute of Industrial Science, Hanyang University, 17 Haengdang-Dong, Seongdong-Gu, Seoul 04763, Republic of Korea*
[2]*Department of Fire and Disaster Prevention Engineering, Kyungnam University, Gyeongsangnam-do 51767, Republic of Korea*

Correspondence should be addressed to Hyun-Ki Choi; chk7796@kyungnam.ac.kr

Academic Editor: Juan J. Del Coz Díaz

Although mechanical properties of concrete under uniaxial compression are important to design concrete structure, current design codes or other empirical equations have clear limitation on the prediction of mechanical properties. Various types of fiber-reinforced reactive powder concrete matrix were tested for making more usable and accurate estimation equations for mechanical properties for ultra high strength concrete. Investigated matrix has compressive strength ranged from 30 MPa to 200 MPa. Ultra high strength concrete was made by means of reactive powder concrete. Preventing brittle failure of this type of matrix, steel fibers were used. The volume fraction of steel fiber ranged from 0 to 2%. From the test results, steel fibers significantly increase the ductility, strength and stiffness of ultra high strength matrix. They are quantified with previously conducted researches about material properties of concrete under uniaxial loading. Applicability of estimation equations for mechanical properties of concrete was evaluated with test results of this study. From the evaluation, regression analysis was carried out, and new estimation equations were proposed. And these proposed equations were applied into stress-strain relation which was developed by previous research. Ascending part, which was affected by proposed equations of this study directly, well fitted into experimental results.

1. Introduction

The compressive strength of concrete is an important parameter in the design of reinforced concrete structures according to current design criteria [1]. In recent years, performance-based designs have been increasingly used, with accompanying increases in the diversity of types and strengths of concrete and reinforcement used. As a result, prediction of the compressive strength and other mechanical properties of concrete have become important to use various types of materials, because these parameters control the stress-strain behavior of concrete. The concrete stress-strain models developed in previous studies [2–11] and used extensively are based on coefficients determined from experiments. These stress-strain relations cannot be used without additional investigation because many coefficients for stress-strain relation of concrete are determined by limited number of experiments. The reason for this is that there are limits to the strength ranges to which such models are applicable, depending on the conditions of the tests conducted to develop these models. Therefore, an equation for use in estimating the mechanical properties of ultrahigh strength fiber-reinforced concrete was derived. These equations can serve as a basis for describing the stress-strain relationships for such materials, even beyond the limits of the currently used empirical formulas or code provisions.

Usually, normal strength concrete members are designed using rectangular stress block parameters under flexure. Current design codes provide the rectangular stress block parameters for simplified design methodology. However, these stress blocks are semiempirical parameters. They are determined by tests of reinforced concrete columns and they have apparent limitations. Rectangular stress block can be used because the shape of stress-strain relation of concrete is similar to the trapezoid. However, shape of stress-strain relationship of concrete changed into triangle as increase of

Table 1: Mix proportions.

ID	w/b	Weight (kg/m^3)							f_{ck}
		Cement	Water	Silica fume	Sand	Filler	Steel fiber**	Super-plasticizer	[MPa]
30-0 and f series*	0.43†	344	172	—	635	1180††	0, 37, 74, 147	0.0	30
80-0 and f series*	0.30	780	255	60	1097	114	0, 37, 74, 147	0.5	80
100-0 and f series	0.25	809	222	80	1052	162	0, 37, 74, 147	1	100
150-0 and f series	0.20	820	190	112	918	186	0, 37, 74, 147	1.04	150
200-0 and f series	0.17	830	176	207	912	246	0, 37, 74, 147	1.08	200

*f series means fiber-reinforced mix.
**0 means no fiber, 37 for 0.5% volume fraction of fiber contents, 74 means 1% volume fraction of fiber contents, and 147 means 2% volume fraction of fiber contents.
†Water-cement ratio was used for normal strength concrete.
††Coarse aggregates were used for normal strength concrete.

compressive strength of concrete. For this reason, rectangular stress block parameters depend on the compressive strength of concrete. For the compressive strength of concrete, higher than 76 MPa, rectangular stress block parameters α_1 and β_1 of ACI318 [1] are limited by 0.85 and 0.65, respectively. However, in this case, relation between ultimate strain of concrete and peak stress should be checked again. Because high strength concrete failed with brittle manner usually, they cannot experience the slight and gradual decrease of compressive stress. It may cause the unexpected failure under flexure, especially for ultrahigh strength concrete member.

Making brittle and unexpected failure of ultrahigh strength concrete matrix under compression more ductile, steel fiber can be included in the matrix. Inclusion of steel fiber can change explosive failure of ultrahigh strength concrete and provide more tensile strength and deformability [12]. So steel fiber can usually be used for ultrahigh strength concrete matrix.

Ultrahigh performance concrete usually has much higher compressive strength and tensile strength than normal strength concrete [13, 14]. Shape of stress distribution under compression and effect of tensile strength of concrete shall be considered in design of section. Many related design guidelines were suggested the design methodologies for ultra high performance concrete flexural member but their safety should be investigated and more easy way to design the section shall be found. Therefore, in this study, various types of stress block and distribution combinations were evaluated with experimental result and previous research results.

2. Compression Testing of Ultrahigh Strength Steel Fiber-Reinforced Concrete

In this study, compression tests of cylindrical concrete specimens with compressive strengths in the range of 30–200 MPa were carried out to derive equations that can be safely used to estimate the mechanical properties of concrete even beyond the strength limits given in the design criteria. The test results were analyzed to determine the stress-strain relationship of concrete over a range of design concrete strengths and compare the relationship to others described by existing equations. The important mechanical properties related to the stress-strain relationship of concrete under compressive loading that were considered in this study were the elastic modulus, the stress-to-matrix strength ratio, and the strain at the maximum stress.

2.1. Experiment Design. According to the previous researches on mechanical properties of concrete and fiber-reinforced concrete [2–11], the mechanical properties that determine the stress-strain relationship of fiber-reinforced concrete depend on the proportion of fibers contents and matrix strength of concrete. Therefore, the compressive strength of the concrete matrix and the contents of steel fiber are important parameters in this study. The compressive strength range considered was 30–200 MPa. Because it is hard to make compressive strength exceeding 100 MPa, we use reactive powder concrete (RPC) as steel fiber-reinforced concrete matrix.

Steel fibers were mixed into concrete batches at volumetric ratios of 0.5% to 2% to ensure both improvement of the structural performance of the concrete and workability of the concrete after addition of the fibers. Five cylindrical specimens ($\varphi 100 \times 200$ mm in size) were produced for each mix. Loading was performed using KS F 2405 [15]. The strain rate of the specimen was measured using a compressor meter. Table 1 lists the mixes used in the testing.

2.2. Experimental Results for the Mechanical Properties of Steel Fiber-Reinforced Reactive Powder Concrete. Figure 1 shows typical stress-strain relation of steel fiber-reinforced powder concrete cylinder specimens. According to Figure 1, steel fiber can increase ductility and strength of brittle matrix. Therefore, the most important investigation shall be the increase rate of ductility and strength of matrix according to the inclusion of steel fiber. So we listed the test results for the elastic modulus, the maximum stress, and the strain at the maximum stress which show the ductility and strength of material in Table 2.

The values shown are the average values for the five specimens from each mix. The test results indicate that steel fiber significantly affects the properties related to the ductility of material. Figure 2 shows the trends in the parameters of elastic modulus and strain at peak strength with respect to the steel fiber content.

The mixes with design strengths of 30, 80, 100, 150, and 200 MPa that were reinforced with fibers at a volume

TABLE 2: Test results (mean value of test specimens).

ID	V_f [%]	E_c [MPa]	f_{co} [MPa]	ε_{co} [‰]
C30-0	0	26,756	32.54	2.65
C30-f0.5	0.5	27,659	33.32	2.98
C30-f1.0	1.0	28,751	34.72	3.32
C30-f2.0	2.0	31,221	34.98	3.43
C80-0	0	32,970	80.79	3.16
C80-f0.5	0.5	33,597	82.60	3.55
C80-f1.0	1.0	34,097	85.17	3.64
C80-f2.0	2.0	34,768	89.01	4.04
C100-f0	0	36,233	104.86	3.39
C100-f0.5	0.5	37,376	107.39	3.71
C100-f1.0	1.0	38,732	111.93	3.95
C100-f2.0	2.0	38,099	116.92	4.09
C150-0	0	42,023	149.40	3.97
C150-f0.5	0.5	41,203	154.96	4.47
C150-f1.0	1.0	42,365	159.60	4.77
C150-f2.0	2.0	43,222	162.40	4.79
C200-0	0	45,512	198.21	4.87
C200-f0.5	0.5	45,019	202.70	4.97
C200-f1.0	1.0	46,734	210.40	5.26
C200-f2.0	2.0	47,515	216.52	5.39

V_f: volume fraction of steel fiber (%), E_c: elastic modulus (MPa), f_{co}: compressive strength of concrete (tested value, MPa), ε_{co}: strain corresponding peak stress.

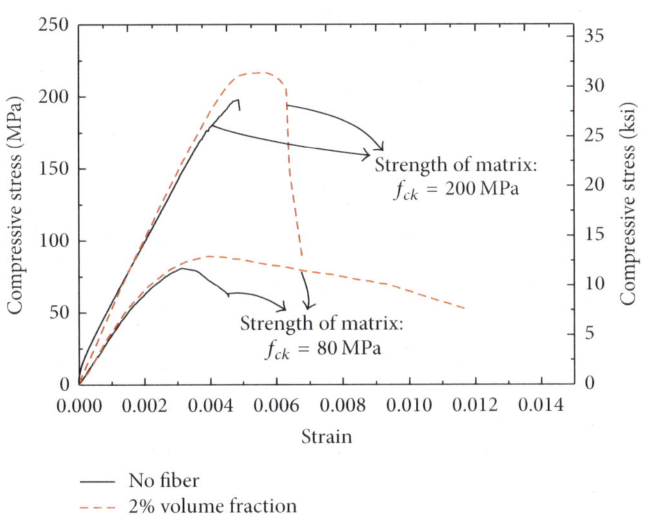

FIGURE 1: Typical uniaxial compressive stress-strain relation of steel fiber-reinforced powder reactive concrete.

of 2% exhibited compressive strength increases of 7.5%, 10.2%, 11.5%, 8.7%, and 9.2%, respectively. The strains at the maximum stress increased by 29.4%, 27.8%, 23.5%, 20.6%, and 10.6%, respectively, with 2% fiber volume fraction. The increases of strain at peak stress are significant in comparison to the increases in strength. Significant increase of strain at peak stress would be caused by the confinement effect of steel fiber to the formation of cracks along the specimen axis, as shown in Figure 3, the failure aspects of fiber-reinforced specimen.

This reinforcing effect on the strain at peak stress was shown in all the cases of compressive strength of matrix. However, the ductility increase of the concrete specimens with strengths greater than 100 MPa was smaller than the increase of the ductility of the 30 and 80 MPa concrete. That is, at higher compressive strengths, the effect of reinforcing effect of steel fibers on improving the ductility decreases.

3. Estimation of the Mechanical Properties of Steel Fiber-Reinforced Reactive Powder Concrete

We collected concrete material test results from a number of studies [30–44] and analyzed them together with the test results obtained in this study to estimate the important parameters needed to derive a stress-strain relationship that can be easily applied over a wider range of concrete strengths than the range encompassed by previous studies. The data collected included 295 results for compressive strength increase, 134 results for the strain at the maximum stress, and 1486 results for the elastic modulus. In this study, the reinforcing effect of the fiber was characterized using a reinforcing index (RI) associated with the aspect ratio and other physical properties of the fiber. The value of RI for a particular type of fiber is determined using (1). The aspect ratio of the fibers, the fiber end shape, and the amount of fiber reinforcement are important parameters in this equation. Consider

$$\mathrm{RI} = \frac{V_f L_f d_f}{D_f}, \qquad (1)$$

where RI is reinforcing index, V_f is volume fraction of steel fiber, L_f is length of steel fiber, and D_f is diameter of steel fiber.

The term d_f in (1) is a correction factor for the end shape of the fiber. In this study, we used correction factors of 1, 2.0, and 1.5 for straight fibers, hooked fibers, and crimped fibers, respectively, to correct for the effect of fiber shape, as suggested in the literature [45].

In order to compare the proposed estimation methods and find the important parameters on mechanical characteristics on steel fiber-reinforced reactive powder concrete we also collect the empirical equations about mechanical characteristic estimation equations for concrete under uniaxial compression.

Table 3 shows an equation for estimating the strain at the maximum stress as a function of the reinforcing effect of the steel fibers and the compressive strength of the concrete. Table 4 shows an equation for estimating the elastic modulus. When examining equations about compressive mechanical characteristics on concrete and fiber-reinforced concrete, the most important parameter is compressive strength of concrete and (1), reinforcing index. Most of the equations consider the effect of steel fiber as additional value independent of compressive strength of concrete. However, as

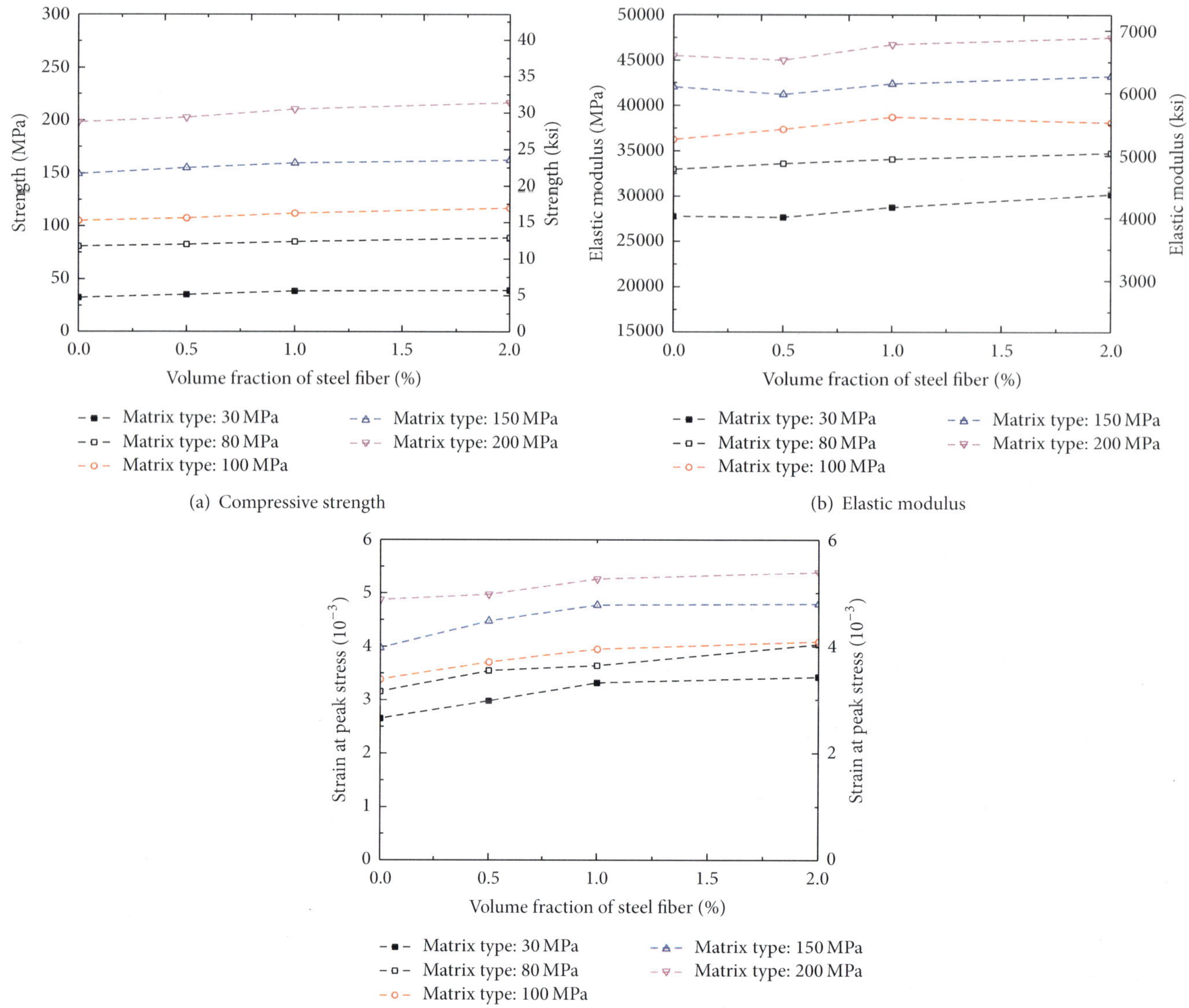

(a) Compressive strength

(b) Elastic modulus

(c) Strain at peak stress

FIGURE 2: Mechanical characteristics of fiber-reinforced reactive powder concrete under uniaxial compression.

FIGURE 3: Failure of specimens under uniaxial compression.

shown in test results, increase rate of mechanical properties of steel fiber-reinforced reactive powder concrete is affected by compressive strength of concrete. In this section we quantify the increase rate of mechanical properties of concrete considering combination of compressive strength of matrix and steel fiber contents.

3.1. Mechanical Properties of Reactive Powder Concrete Matrix. For fiber-reinforced concrete, the elastic modulus and the strain at the maximum stress are often determined from equations that are based on test results for nonreinforced mixes. The accuracy of such estimates is significantly affected by the strength ranges of the specimens used to derive the equations. Elastic modulus and ε_{co} values derived from the results of material testing conducted in previous studies [30–44] and in the present study are shown in Figures 4 and 5, respectively. The associated equations are also shown. Tables 5 and 6 list the statistical parameters of the equations.

As shown in Table 5 and Figure 4, the current code provisions show the highest accuracy in the estimation of the

TABLE 3: Prediction equations for strain at peak stress (previous researches).

Researcher	Equation	Limitation [MPa]
Collins et al. [6]	$\varepsilon_{co} = \dfrac{f_{ck}}{E_c}\dfrac{n}{n-1}, \quad n = 0.8 + \dfrac{f_{ck}}{17}, \ E_c = 3320 f_{ck}^{0.5} + 6900$	$f_{ck} \leq 100$
Wee et al. [7]	$\varepsilon_{co} = \dfrac{180\left(f_{ck}\right)^{0.25}}{10^6}$	$f_{ck} \leq 120$
Ros [16]	$\varepsilon_{co} = \dfrac{\left(0.0546 + 0.003713 f_{ck}\right)}{100}$	$f_{ck} \leq 43$
Fafitis and Shah [17]	$\varepsilon_{co} = 1.491 \times 10^{-5} f_{ck} + 0.00195$	$f_{ck} \leq 66$
De Nicolo et al. [18]	$\varepsilon_{co} = 0.00076 + \left[\left(0.626 f_{ck} - 4.33\right) \times 10^{-7}\right]^{0.5}$	$f_{ck} \leq 90$
CEB fip 228 [9]	$\varepsilon_{co} = \dfrac{0.7\left(f_{ck}\right)^{0.31}}{1000}$	$f_{ck} < 100$
EC2 [19]	$\varepsilon_{co} = \dfrac{\left(2.0 + 0.085\left(f_{ck} - 50\right)^{0.53}\right)}{1000}$, for $f_{ck} \geq 50$ MPa; otherwise $\varepsilon_{co} = 0.002$	$f_{ck} \leq 90$
Soroushian and Lee [20]	$\varepsilon_{cof} = 0.007\text{RI} + 0.0021, \quad \text{RI} = \dfrac{V_f L_f}{D_f}$	—
Nataraja et al. [21]	$\varepsilon_{cof} = 0.002 + 0.006(\text{RI}), \quad \text{RI} = \dfrac{W_f L_f}{D_f}$	$f_{ck} \leq 50$
Dhakal et al. [22]	$\varepsilon_{cof} = 0.002 + 2000\alpha_c V_f^2, \quad \alpha_f = 1.0$ (flat), $\alpha_f = 2.4$ (hooked)	$f_{ck} \leq 40$

E_c: secant elastic modulus of concrete at $0.45 f_{ck}$ (MPa), ε_{co}: strain corresponding to peak stress (normal concrete), f_{ck}: compressive strength of concrete (MPa), RI: reinforcing index, D_f: diameter of steel fiber (mm), L_f: length of steel fiber (mm), V_f: volume fraction of steel fiber, W_f: weight fraction of steel fiber.

TABLE 4: Prediction equations for modulus of elasticity (previous researches).

Researcher	Equation	Limitation
KCI2007 [23]	$E_c = 8{,}500\sqrt[3]{f_{cu}}, \quad E_c = 0.077 w_c^{1.5}\sqrt[3]{f_{cu}}, \quad f_{cu} = f_{ck} + 8$	Did not specify but $\sqrt{f_{ck}}$ cannot exceed 8.4 MPa
KCI2012 [24]	$E_c = 8{,}500\sqrt[3]{f_{cu}}, \quad E_c = 0.077 w_c^{1.5}\sqrt[3]{f_{cu}}, \quad f_{cu} = f_{ck} + \Delta f$, where $\Delta f = 4$ when $f_{ck} \leq 40$ MPa and $\Delta f = 6$ when $f_{ck} > 60$ MPa. Interpolate between 40 MPa and 60 MPa	
ACI318-11 [1]	$E_c = 4{,}700\sqrt{f_{ck}}$ (normal weight concrete), $\quad E_c = w_c^{1.5} 0.043\sqrt{f_{ck}}$	
CEB-fip 228 [9]	$E_{ci} = E_{co}\left[\dfrac{\left(f_{ck} + \Delta f\right)}{f_{cmo}}\right]^{1/3}, \quad \Delta f = 8, \ f_{cmo} = 10, \ E_{co} = 22\text{ GPa}$	$f_{ck} \leq 90$
Martinez et al. [25]	$E_c = 3{,}320\sqrt{f_{ck}} + 6{,}900$	21 MPa $< f_{ck} <$ 83 MPa
Cook [26]	$E_c = 3.385 \times 10^{-5} w_c^{2.55}\left(f_{ck}\right)^{0.315}$	—
Ahmad and Shah [27]	$E_c = 3.385 \times 10^{-5} w_c^{2.55}\left(f_{ck}\right)^{0.325}$	$f_{ck} <$ 84 MPa
Graybeal [11]	$E_c = 3{,}840\sqrt{f_{ck}}$	$f_{ck} <$ 200 MPa
Gao et al. [28]	$E_{cf} = E_c(1 + 0.173(\text{RI}))$	70 MPa $< f_{ck} <$ 85 MPa
Padmarajaiah [29]	$E_{cf} = E_c + 2440(\text{RI})$	$f_{ck} <$ 69 MPa

E_c: secant elastic modulus of concrete at $0.45 f_{ck}$ (MPa), w_c: unit weight of concrete (kg/m^3), f_{ck}: compressive strength of concrete (MPa), RI: reinforcing index, $\text{RI} = V_f L_f / D_f$.

elastic modulus. As shown in statistical values for KCI2012, current Korean design code provision, a standard deviation is not the lowest among other estimation methods.

However, mean value, coefficient of variation, and IAE (integrated absolute error) values for KCI2012 have lowest value among other prediction methods. In general, the estimation equation for the elastic modulus tends to underestimate the modulus of elasticity for normal strength concrete and overestimate the modulus of elasticity for high strength concrete. However, highest accuracy of KCI2012 occurs because the estimation equation was developed to provide relatively safe estimates for both normal strength and high strength concrete, using different value of Δf according to the compressive strength of concrete. Therefore, even for an ultrahigh strength concrete, there should be no significant problems associated with using the estimation equation presented in the current design standards to estimate the elastic modulus of the concrete.

Because ε_{co}, the strain at the maximum stress, is an important boundary condition in defining the stress-strain relationship and important design parameter, we also compared the results of previous studies and the results of this study with respect to this parameter. In previous studies on the stress-strain relationship of concrete, the typical value used for the strain at the maximum stress is 0.002. However, Figure 5 confirms that the strain at the maximum stress tends to increase with the compressive strength of the concrete. It was confirmed that equations from previous studies for estimation of ε_{co}, which take the form of an exponential function of the compressive strength, underestimate the

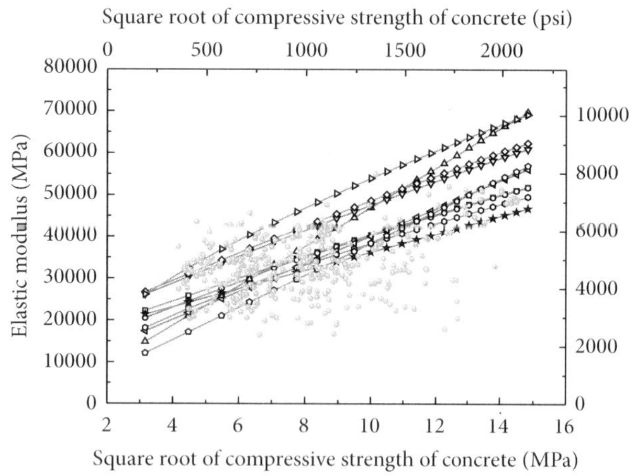

FIGURE 4: Elastic modulus: test results and prediction equations.

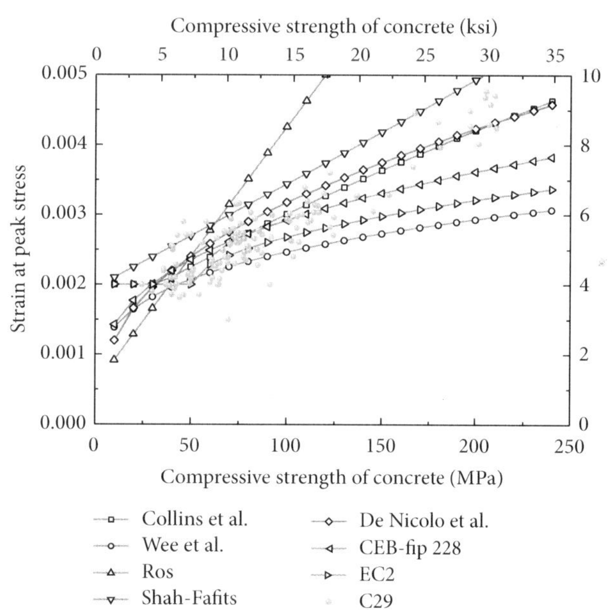

FIGURE 5: Strain at peak stress: test results and prediction equations.

increase in ε_{co} with increasing compressive strength. Table 6 confirms that the accuracy of these equations increases as the compressive strength range considered increases. Given that the equation proposed by Collins et al. [6] has the lowest IAE and average value, Collins et al. [6] equation for ε_{co} was concluded to be suitable for use with high strength mixes. Collins et al. [6] equation, shown in Table 1, reflects the effect of the elastic modulus. Because our analysis confirmed that the equation presented in the current design standards yields the highest accuracy in the estimation of the elastic modulus, this equation was used with Collins et al. [6] equation to

TABLE 5: Statistics on the prediction methods for modulus of elasticity.

Researcher	Mean	SD	CV	IAE
KCI2007 [23]	0.99	0.12	0.12	8.52
KCI2012 [24]	1.01	0.12	0.12	8.42
ACI318-11 [1]	0.93	0.13	0.14	13.72
CEB-fip 228 [9]	0.83	0.10	0.12	20.65
Martinez et al. [25]	1.05	0.13	0.12	9.45
Cook [26]	0.75	0.09	0.12	32.44
Ahmad and Shah [27]	1.07	0.12	0.12	9.83
Graybeal [11]	1.14	0.16	0.14	13.54

SD: standard deviation, CV: coefficient of variation, IAE: integrated absolute error (%).

TABLE 6: Statistics on the prediction methods for strain at peak stress under uniaxial compression.

Researcher	Mean	SD	CV	IAE (%)
Collins et al. [6]	0.99	0.12	0.12	8.74
Wee et al. [7]	1.16	0.17	0.15	16.37
Ros [16]	0.78	0.19	0.25	42.04
Fafitis and Shah [17]	0.84	0.09	0.11	19.49
De Nicolo et al. [18]	0.93	0.11	0.12	11.58
CEB-fip 228 [9]	0.99	0.13	0.13	10.56
EC2 [19]	1.09	0.15	0.14	12.52

SD: standard deviation, CV: coefficient of variation, IAE: integrated absolute error (%).

estimate ε_{co}. As a result, the IAE decreased by 8.67%, and the average of the ratios of experimental values to estimated values was found to be 1.02, indicating that the estimated values were on the safe side.

3.2. Effect of Steel Fiber Contents on Compressive Strength Increase. Figure 6 shows test results from previous studies [30–44] and from this study on the effect of the steel fiber contents on the increase in concrete compressive. The increase of maximum stress of test results was expressed as the ratio of the compressive strength of fiber-reinforced specimens to the compressive strength of non-fiber-reinforced specimens. The average compressive strength increase due to fiber reinforcement was 9.7%. However, for specimens of the same design compressive strength, the compressive strength tended to increase linearly with increasing proportions of fiber. Examination of the results classified by the design compressive strength of the concrete indicated that the increase achieved in compressive strength with increasing fiber content tended to decrease with increasing design compressive strength. However, it is difficult to confirm that a clear correlation exists because of the high degree of variation in the test results. In particular, the correlation between compressive strength increase and fiber content tended to decrease with increasing design compressive strength. The results of our study suggest that the increases achieved in compressive strength with increasing fiber content were comparable over the range of

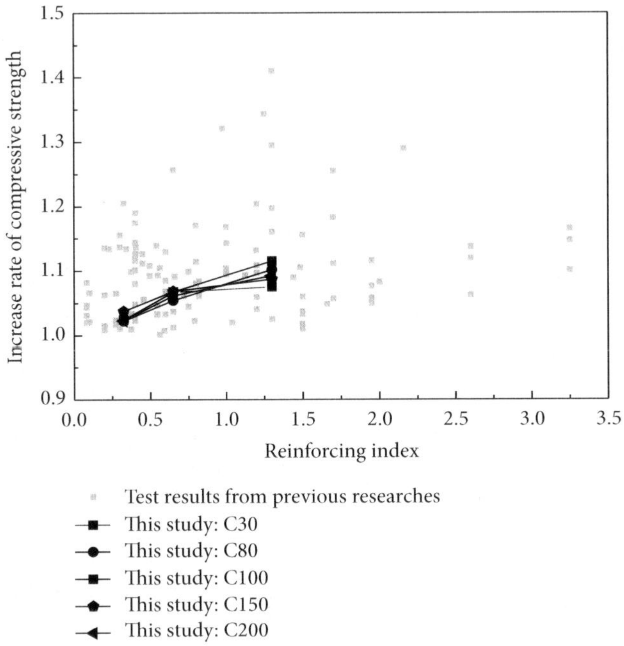

FIGURE 6: Compressive strength increase rate according to reinforcing index.

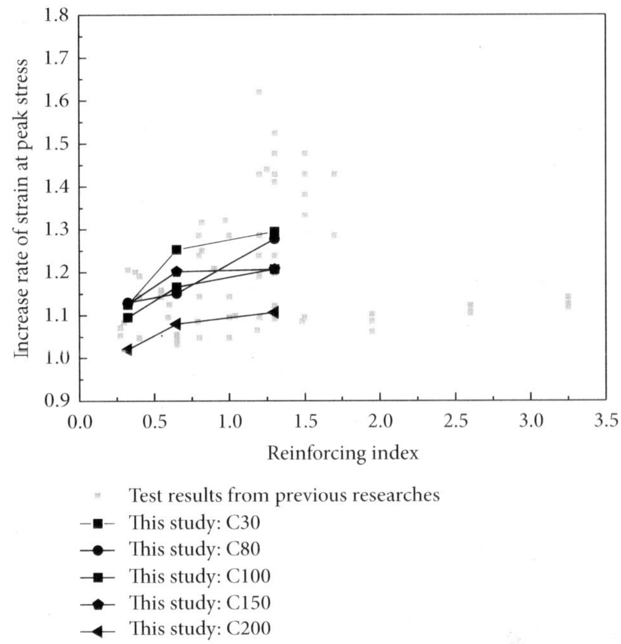

FIGURE 7: Strain at peak stress increase rate according to reinforcing index.

compressive strengths considered. A maximum compressive strength increase of 10% was verified.

3.3. Effect of Steel Fiber Contents on Strain at Peak Stress. Figure 7 illustrates the effect of the steel fiber contents on the change in the strain at the maximum stress, as indicated by the results of previous studies and by the results of this study. The change in the strain at the maximum stress exhibited a higher correlation to the contents of steel fibers than did the increase with compressive strength of matrix. The results of the experiments conducted in our study are shown in Figure 7 along with the results from previous studies. The test results indicate that, for mixes with design compressive strengths of 80 MPa, ε_{co} increases with increasing steel fiber content but, for mixes with strengths of more than 100 MPa, ε_{co} decreases with increasing fiber content. This phenomenon was particularly significant for the specimens with design compressive strengths of 200 MPa. Normal strength concrete has shown similar change according to the steel fiber inclusion.

3.4. Effect of Steel Fiber Contents on Elastic Modulus of Concrete. Higher compressive strength of concrete is achieved by homogenization of the constituent materials and of their strengths. Accordingly, the elastic modulus of concrete tends to increase with increasing concrete strength. The elastic modulus derived from test results obtained in previous studies and in our study was examined to assess the trend in the elastic modulus of fiber-reinforced reactive powder concrete with respect to the previously defined reinforcing index. Figure 8 shows the distribution of elastic modulus of concrete test results according to the reinforcing index. Because the elastic modulus is affected by the strength and strain simultaneously, the increases in strength and stress exhibit similar trends. The degree to which the elastic modulus increased with the fiber reinforcing index decreased with increasing concrete compressive strength. However, the average increase was 6.5% which is lesser than that for the strength or strain. Our test results revealed similar trends. Specimens in the C80 test group (with design compressive strengths of 80 MPa) exhibited the greatest increases among high strength concrete mixes. Specimens in test groups with design compressive strength greater than 100 MPa exhibited smaller increases and trends similar to those observed in the results from previous studies.

4. Equation for Estimating the Effect of the Steel Fiber Contents on the Mechanical Properties of Steel Fiber-Reinforced Reactive Powder Concrete under Uniaxial Compression

The effects of steel fiber on concrete strength improvement observed in the test results obtained in this study and in previous studies [30–44] were quantified for compressive strength ranges in increments of 20 MPa, as shown in Figure 9. The magnitude of the improvement within each range was determined from linear regression analysis of the data for the specimens within that range. The regression analysis was performed using the model form shown as

$$\frac{k_{cf}}{k_{co}} = 1 + \mathrm{RI} \times \alpha, \tag{2}$$

where k_{cf} is mechanical properties for steel fiber-reinforced concrete, k_{co} is mechanical properties on nonreinforced concrete, RI is reinforcing index, and α is regression coefficient.

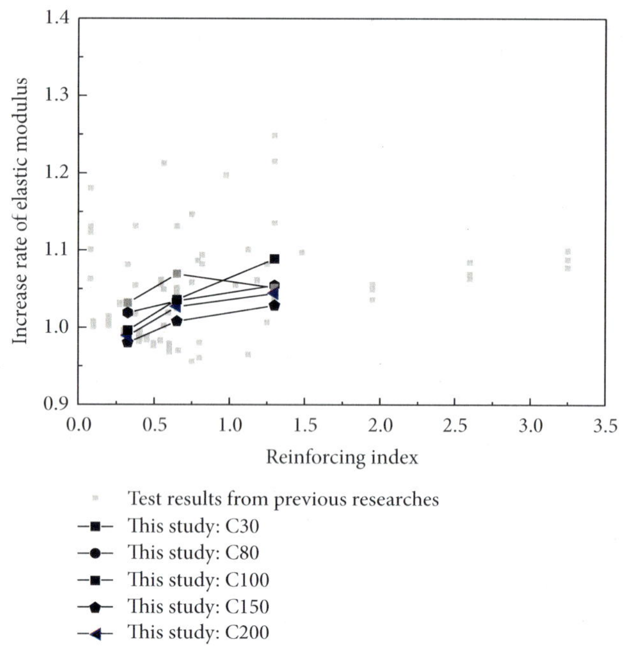

FIGURE 8: Elastic modulus increase rate according to reinforcing index.

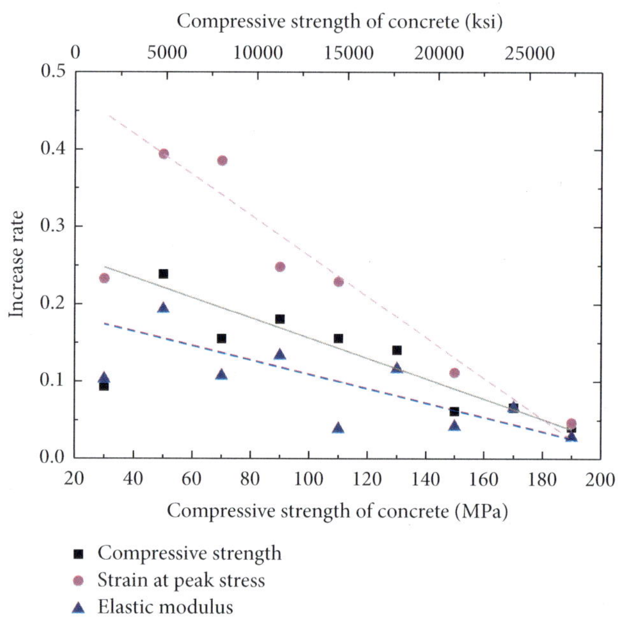

FIGURE 9: Change of statistical values and trend lines on mechanical characteristics of steel fiber-reinforced reactive powder concrete under uniaxial compression.

The increase in compressive strength with inclusion of steel fiber exhibited a clear decreasing trend with increasing concrete compressive strength. The results of the regression analysis suggest that the magnitude of the improvement in the compressive strength, which depends on the elastic modulus, the strain at the maximum stress, and the compressive strength of concrete, can be expressed by

$$\frac{k_{cf}}{k_{co}} = 1 + \mathrm{RI}\left(a_1 - b_1 f_{ck}\right). \tag{3}$$

In the above equation, a_1 and b_1 are regression coefficients. The values of a_1 and b_1 were determined to be 0.29 and 0.0013, respectively, for change in the compressive strength; 0.52 and 0.0026, respectively, for change in ε_{co}; and 0.20 and 0.00092, respectively, for change in elastic modulus.

In order to verify the applicability of proposed equations predicting mechanical properties of steel fiber-reinforced reactive powder concrete under uniaxial compression, we apply these equations to the stress-strain relation of concrete under uniaxial compression.

Most of the stress-strain relation of concrete are determined by the important mechanical properties, elastic modulus of concrete, and strain at peak stress. Stress-strain relations proposed by previous researches were investigated. There are several types of stress-strain relation of concrete. But in this study, we investigated the way to use the elastic modulus for predicting stress-strain relation of concrete, because stress-strain relations of high strength or ultrahigh strength concrete depend on the elastic modulus as we investigated in this study. The first one of stress-strain relation of concrete we investigated is suggested by Collins et al. [6] and the other is suggested by Attard and Setunge [10]. Collins et al. use the differential of plasticity as base model and Attard and Setunge [10] use the mathematical model to predict stress-strain relationship. Both of them use the elastic modulus of concrete as main variable for prediction of stress-strain relation. However, mathematical model which was suggested by Attard and Setunge [10] needs more boundary conditions such as inflection points after experiencing peak stress. Inflection points of descending curve of stress-strain relation are highly dependent on the experimental equipment. Therefore, in this study, stress-strain relation suggested by Collins et al. [6] is used for prediction of stress-strain relation by using proposed equation of main variables such as elastic modulus and strain at peak stress. The stress-strain relation suggested by Collins et al. [6] can be described using

$$\frac{f_c}{f_c'} = \frac{n\left(\varepsilon_{cf}/\varepsilon_c'\right)}{n - 1 + \left(\varepsilon_{cf}/\varepsilon_c'\right)^{nk}}, \tag{4}$$

where f_c' is peak stress obtained from cylinder test, ε_c' is strain when f_c reaches f_c', n is curve fitting factor equal to $E_c/(E_c - E_c')$, E_c is tangent stiffness when ε_{cf} is zero, E_c' is secant stiffness when ε_{cf} is ε_c', and k model the strain decay before and after experiencing peak stress. n and k are suggested by using data of high strength concrete and we decided to use these two variables without change. n and k can be calculated by using

$$n = 0.8 + \frac{f_c'}{17},$$
$$k = 0.67 + \frac{f_c'}{62}. \tag{5}$$

For the verification of the applicability of suggested equations, stress-strain relationships of concrete were constructed and compared with experimental results. As can be seen in Figure 10, predicted stress-strain relationship

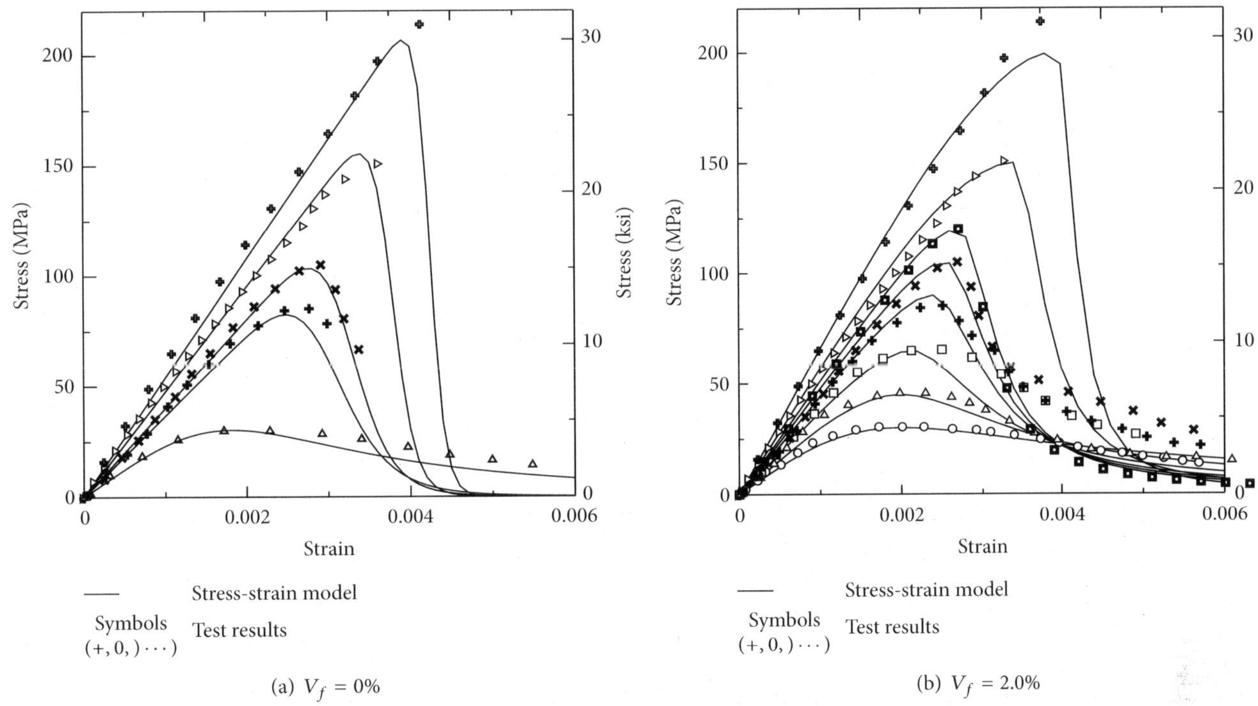

FIGURE 10: Applicability for stress-strain relationship: prediction and experimental results.

of ascending curve was well fitted into the experimental results. However, descending curve of prediction shows less accuracy in descending part of the total stress-strain relation. The inaccuracy of descending curve of stress-strain relation prediction might be caused by the difference of stiffness of experiencing equipment. Making more exact solution, energy absorption capacity under compression shall be investigated.

5. Conclusion

In this study, we conducted material testing to evaluate the properties of ultrahigh strength concrete which was made from reactive powder concrete reinforced with steel fibers under compressive loading, and we evaluated the suitability of equations developed in previous studies for use in estimating material properties of ultrahigh strength concrete. The results of this study are summarized as follows.

(1) The compression test results for non-fiber-reinforced and fiber-reinforced ultrahigh strength concrete specimens indicated that non-fiber-reinforced concrete specimens exhibited brittle fractures, whereas fiber-reinforced concrete specimens did not. The test results suggest that reinforcing fibers resisted the horizontal tensile forces induced by vertical compressive loading.

(2) The test results confirmed that the effects of steel fiber on the concrete compressive strength, strain at maximum stress, and elastic modulus exhibited linear trends regardless of the compressive strength of the mix. Compressive strength and elastic modulus were not significantly affected by the amount of reinforcing fiber in the mix. However, a relatively strong effect on the strain at the maximum stress was confirmed.

(3) The mechanical properties of the concrete, such as the elastic modulus and the strain at the peak stress, were estimated with a high degree of accuracy using the secant modulus estimation equation presented in the current design standards and by Collins et al.'s equation, respectively. Thus, the applicability of these equations to concrete in the ultrahigh strength range was confirmed.

(4) Regression analyses were performed on the test results from this study and previous studies to describe the effects of fiber reinforcement on the material properties of interest. The analysis results showed that the reinforcing effects of steel fiber can be expressed in terms of the fiber reinforcing index.

(5) Analysis of the test results confirmed that the relevant mechanical properties of concrete mix at a given strength level improve in proportion to the fiber reinforcement index. Linear regression analyses were performed for the mechanical properties using equations of the same form for all strength levels. The equation form was chosen so that different values of the regression coefficients were obtained for each strength level.

(6) Proposed equations for mechanical properties under uniaxial compression on concrete can be used as main variables of stress-strain relation. According to comparison, ascending curve of stress-strain relation can be well fitted into experimental results. However, descending part of stress-strain relation

cannot be well fitted because prediction method did not consider the energy dissipation capacity after experiencing peak stress. In order to construct full range of stress-strain relation of fiber-reinforced concrete, boundary conditions after peak stress shall be considered.

Competing Interests

The authors declare that there is no conflict of interests regarding the publication of this paper.

Acknowledgments

This research was supported by Basic Science Research Program through the National Research Foundation of Korea (NRF) funded by the Ministry of Science, ICT & Future Planning (15CTAP-C097470-01 and NRF-2014R1A1A1005444).

References

[1] ACI Committee 318, *Building Code Requirements for Structural Concrete (ACI 318-11) and Commentary*, American Concrete Institute, Farmington Hills, Mich, USA, 2011.

[2] S. Popovics, "A numerical approach to the complete stress-strain curve of concrete," *Cement and Concrete Research*, vol. 3, no. 5, pp. 583–599, 1973.

[3] M. Sargin, *Stress-Strain Relationship for Concrete and the Analysis of Structural Concrete Sections, Study 4*, Solid Mechanics Division, University of Waterloo, Waterloo, Canada, 1971.

[4] A. Tomaszewicz, Betongens Arbeidsdiagram, SINTEF, STF65, A84065, Trondheim, 1984.

[5] D. J. Carreira and K.-H. Chu, "Stress-strain relationship for plain concrete in compression," *Journal of the American Concrete Institute*, vol. 82, no. 6, pp. 797–804, 1985.

[6] M. P. Collins, D. Mitchell, and J. G. MacGregor, "Structural design considerations for high-strength concrete," *Concrete International*, vol. 15, no. 5, pp. 27–34, 1993.

[7] T. H. Wee, M. S. Chin, and M. A. Mansur, "Stress-strain relationship of high-strength concrete in compression," *Journal of Materials in Civil Engineering*, vol. 8, no. 2, pp. 70–76, 1996.

[8] P. T. Wang, S. P. Shah, and A. E. Naaman, "Stress-strain curves of normal and lightweight concrete in compression," *ACI Journal Proceedings*, vol. 75, pp. 603–611, 1978.

[9] Comité Euro-International du Béton-Fédération Internationale de la Précontrainte, "High performance concrete—recommended extensions to the model code 90, research needs," *CEB Bulletin*, no. 228, p. 60, 1995.

[10] M. M. Attard and S. Setunge, "Stress-strain relationship of confined and unconfined concrete," *ACI Materials Journal*, vol. 93, no. 5, pp. 432–442, 1996.

[11] B. A. Graybeal, "Compressive behavior of ultra-high-performance fiber-reinforced concrete," *ACI Materials Journal*, vol. 104, no. 2, pp. 146–152, 2007.

[12] A. K. H. Kwan and L. G. Li, "Combined effects of water film thickness and paste film thickness on rheology of mortar," *Materials and Structures/Materiaux et Constructions*, vol. 45, no. 9, pp. 1359–1374, 2012.

[13] F. de Larrard and T. Sedran, "Optimization of ultra-high-performance concrete by the use of a packing model," *Cement and Concrete Research*, vol. 24, no. 6, pp. 997–1009, 1994.

[14] L. G. Li and A. K. H. Kwan, "Packing density of concrete mix under dry and wet conditions," *Powder Technology*, vol. 253, pp. 514–521, 2014.

[15] KS F 2405, *Standard Test Method for Compressive Strength of Concrete*, Korean Agency for Technology and Standards, 2010.

[16] M. Ros, *Material-Technological Foundation and Problems of Reinforced Concrete*, Bericht no. 162, Eidgenössische Materialprüfungs- und Versuchsanstalt für Industrie, Bauwesen und Gewerbe, Zurich, Switzerland, 1950.

[17] A. Fafitis and S. P. Shah, "Predictions of ultimate behavior of confined columns subjected to large deformations," *Journal of the American Concrete Institute*, vol. 82, no. 4, pp. 423–433, 1985.

[18] B. De Nicolo, L. Pani, and E. Pozzo, "Strain of concrete at peak compressive stress for a wide range of compressive strengths," *Materials and Structures*, vol. 27, no. 4, pp. 206–210, 1994.

[19] European Commitee for Standardization (CEN), *Design of Concrete Structures—Part 1-1: General Rules and Rules for Buildings*, Eurocode 2, Brussels, Belgium, 2004.

[20] P. Soroushian and C. H. Lee, "Constitutive modeling of steel fiber reinforced concrete under direct tension and compression. fibre reinforced cements and concretes : recent developments," in *Proceedings of the International Conference held at the University of Wales, Collige of Cardiff, School of Engineering, United Kingdom*, London, UK, September 1989.

[21] M. C. Nataraja, N. Dhang, and A. P. Gupta, "Stress-strain curves for steel-fiber reinforced concrete under compression," *Cement and Concrete Composites*, vol. 21, no. 5-6, pp. 383–390, 1999.

[22] R. P. Dhakal, C. Wang, and J. B. Mander, "Behavior of steel fibre reinforced concrete in compression," in *Proceedings of the Nanjing: International Symposium on Innovation & Sustainability of Structures in Civil Engineering*, 2005.

[23] Korea Concrete Institute, *Concrete Design Code and Commentary*, Kimoondang Publishing Company, Seoul, Republic of Korea, 2007.

[24] Korea Concrete Institute, *Concrete Design Code and Commentary*, Kimoondang Publishing Company, Seoul, Korea, 2012.

[25] S. Martinez, A. H. Nilson, and F. Slate, "Spirally reinforced high-strength concrete columns," *ACI Journal*, vol. 81, no. 5, pp. 431–442, 1984.

[26] J. E. Cook, "10,000 PSI concrete," *Concrete International: Design and Construction*, vol. 11, no. 10, pp. 67–75, 1989.

[27] S. H. Ahmad and S. P. Shah, "Complete triaxial stress-strain curves for concrete," *Journal of the Structural Division*, vol. 108, no. 4, pp. 728–742, 1982.

[28] J. Gao, W. Sun, and K. Morino, "Mechanical properties of steel fiber-reinforced, high-strength, lightweight concrete," *Cement and Concrete Composites*, vol. 19, no. 4, pp. 307–313, 1997.

[29] S. K. Padmarajaiah, *Influence of fibers on the behavior of high strength concrete in fully/partially prestressed beams: an experimental and analytical study [Ph.D. thesis]*, Indian Institute of Science, Bangalore, India, 1999.

[30] H.-K. Choi, B.-I. Bae, and C.-S. Choi, "Mechanical characteristics of ultra high strength concrete with steel fiber under uniaxial compressive stress," *Journal of the Korea Concrete Institute*, vol. 27, no. 5, pp. 521–530, 2015.

[31] S. V. T. J. Perera, H. Mutsuyoshi, and S. Asamoto, "Properties of high-strength concrete," in *Proceedings of the 12th International Summer Symposium of Japan Society of Civil Engineers (JSCE '10)*, Funabashi, Japan, 2010.

[32] K. F. Sarsam, I. A. S. Al-Shaarbaf, and M. M. S. Ridha, "Experimental investigation of shear-critical reactive powder concrete beams without web reinforcement," *Engineering & Technology Journal*, vol. 30, no. 17, pp. 2999–3022, 2012.

[33] D. A. Pandor, *Behavior of high strength fiber reinforced concrete beams in shear [M.S. thesis]*, Massachusetts Institute of Technology, 1994.

[34] J. Thomas and A. Ramaswamy, "Mechanical properties of steel fiber-reinforced concrete," *Journal of Materials in Civil Engineering*, vol. 19, no. 5, pp. 385–392, 2007.

[35] S. Kang and G. Ryu, "The effect of steel-fiber contents on the compressive stress-strain relation of Ultra High Performance Cementitious Composites (UHPCC)," *Journal of the Korea Concrete Institute*, vol. 23, no. 1, pp. 67–75, 2011.

[36] P. Bhargava, U. K. Sharma, and S. K. Kaushik, "Compressive stress-strain behavior of small scale steel fibre reinforced high strength concrete cylinders," *Journal of Advanced Concrete Technology*, vol. 4, no. 1, pp. 109–121, 2006.

[37] Y.-C. Ou, M.-S. Tsai, K.-Y. Liu, and K.-C. Chang, "Compressive behavior of steel-fiber-reinforced concrete with a high reinforcing index," *Journal of Materials in Civil Engineering*, vol. 24, no. 2, pp. 207–215, 2012.

[38] A. Samer Ezeldin and P. N. Balaguru, "Normal- and high-strength fiber reinforced concrete under compression," *Journal of Materials in Civil Engineering*, vol. 4, no. 4, pp. 415–429, 1992.

[39] B.-W. Jo, Y.-H. Shon, and Y.-J. Kim, "The evalution of elastic modulus for steel fiber reinforced concrete," *Russian Journal of Nondestructive Testing*, vol. 37, no. 2, pp. 152–161, 2001.

[40] O. Kazuhiro, S. Shunsuke, A. Hideo et al., *An Experimental Study on the Compressive Properties of the Super-High Strength Concrete*, vol. 25, Architectural Institute of Japan, China Branch, Research Report Collection, 2002.

[41] A. R. Murthy, N. R. Iyer, and B. K. R. Prasad, "Evaluation of mechanical properties for high strength and ultrahigh strength concretes," *Advances in Concrete Construction*, vol. 1, no. 4, pp. 341–358, 2013.

[42] N. N. Meleka, A. A. Bashandy, and M. A. Arab, "Ultra high strength concrete using economical materials," *International Journal of Current Engineering and Technology*, vol. 3, no. 2, pp. 393–402, 2013.

[43] D. Moldovan and C. Măgureanu, "Stress-strain diagram for high strength concrete elements in flexure," in *Proceedings of the 3rd International Conference Advanced Composite Materials Engineering (COMAT '10)*, pp. 137–142, Transilvania University Press of Brasov, Brasov, Romania, 2010.

[44] W. I. Khalil and Y. R. Tayfur, "Flexural strength of fibrous ultra high performance reinforced concrete beams," *ARPN Journal of Engineering and Applied Sciences*, vol. 8, no. 3, pp. 200–214, 2013.

[45] R. Narayanan and I. Y. S. Darwish, "Use of steel fibers as shear reinforcement," *ACI Structural Journal*, vol. 84, no. 3, pp. 216–227, 1987.

Analysis of the Seismic Performance of Site-Bolted Beam to Column Connections in Modularized Prefabricated Steel Structures

Xuechun Liu,[1,2,3] Xiaoxiong Cui,[1] Zhiwei Yang,[1] and Xinxin Zhan[1]

[1]Beijing Engineering Research Center of High-Rise and Large-Span Prestressed Steel Structures, Beijing University of Technology, Beijing 100124, China
[2]Beijing Key Laboratory of Earthquake Engineering and Structural Retrofit, Beijing University of Technology, Beijing 100124, China
[3]Beijing Advanced Innovation Centre for Future Urban Design, Beijing 100044, China

Correspondence should be addressed to Xuechun Liu; liuxuechun@bjut.edu.cn

Academic Editor: Michael Aizenshtein

This paper proposes a site-bolted connection that is suitable for modularized prefabricated steel structures. Excellent ductility is achieved by various structural measures. Six connection specimens with different parameters were subjected to quasi-static loading tests and finite element analysis (FEA) to determine the seismic performance of the proposed connection (e.g., hysteretic behavior, skeleton curve, ductility, and failure mode). The results of the tests and FEA showed that the connection underwent sufficient plastic deformation under cyclic loading and that its ultimate rotation angle could reach 0.09 rad. A clear plastic hinge formed on the beam before the connection failed, which suggests a ductile failure mode. The connection exhibited a wide hysteresis loop, which indicated good seismic performance. The results also showed that the connection does not slip under small earthquakes and could dissipate energy through slippage in the connection region under a moderate earthquake and through slippage in the connection region as well as plastic deformation at the beam end under a severe earthquake. The number of bolts was the main parameter that affected the seismic performance of the connection. The test and FEA results demonstrated that all six specimens had excellent seismic and ductile performance and an exceptional plastic rotation capacity.

1. Introduction

Before the 1990s, welded and bolted-welded connections were used extensively due to their high load-bearing capacity, stiffness, and cost-effectiveness. However, during the Northridge earthquake in the US and the Hanshin-Awaji earthquake in Japan, brittle fracture occurred in large numbers of these types of beam-column connections, which caused many high-rise steel structures to break and even collapse. Research has shown that the cracking of weld seams under cyclic loading will lead to damage to the connection and decreases in its stiffness and strength or even brittle failure of the connection. Since these two large earthquakes, researchers around the world have extensively studied and considerably improved welded connections [1–3]. Two types of measures are used to improve these connections, namely, reducing and strengthening measures, both of which are used to prevent brittle fractures in connections by shifting the plastic hinge away from the connecting weld seam. Naimi et al. [4] proposed a new bolted-welded connection with a reduced beam section and studied the effects of the beam depth and size of the beam web section on the connection performance. Li et al. [5] and Zhao et al. [6] studied a dog bone-shaped connection, a reduced connection with a long slot, and a bolted-welded connection with enlarged weld access holes and found that good weld quality is the main factor that prevents the brittle fracture of the connections. Chen et al. [7, 8] investigated the seismic performance of a bolted-welded beam end strengthened connection and proposed a method that strengthens the beam end by forming a plastic hinge in the beam away from the column face. The proposed method could effectively improve the seismic performance of beam-column connections. However, welded connections are disadvantageous due to the relatively high welding workload, complex welding process, and difficulty in ensuring the weld quality.

Site-bolted beam-column connections can be rapidly assembled on-site. They have a good plastic rotation capacity and excellent ductility and can dissipate seismic energy through slippage and plastic deformation that occurs in the joint region, which reduces damage to the structure by an earthquake. Astaneh-Asl [9] suggested that high-strength bolted beam connections should be designed to be semirigid and allowed to slip under high seismic loads, thereby dissipating seismic energy through slippage on the friction surfaces and compression between the bolts and the bolt hole walls. There are two types of site-bolted connections: rigid connections and variable stiffness connections. The calculations and analyses for rigid connections are simple, and they are used extensively. However, rigid connections require a large number of bolts. In comparison, variable stiffness connections have a higher energy dissipation capacity and require only a small number of bolts and thus are more economical. Rigid connections can be formed using T-stubs. Ribeiro et al. [10, 11] developed and improved an analytical procedure and investigated the mechanical behavior of T-stub connections under impact loading using finite element analysis (FEA) and determined the damage scalar variable and equivalent plastic strain patterns of these connections. They compared the analytical results, three-dimensional (3D) finite element (FE) predictions, and experimental results for the connections and validated the experimental and analytical results. Hu et al. [12, 13] proposed a mechanical analysis model for T-stub connections and validated the model using both FEA and experiments.

End-plate connections are typical bolted connections with variable stiffness. Dessouki et al. [14], Prinz et al. [15], and Shi et al. [16] studied the seismic performance of end-plate beam-column connections using experiments and FEA. Tsavdaridis and Papadopoulos [17] analyzed a large number of end-plate connections in which the beam was reduced by varying degrees along its length using FEA software and determined various mechanical properties of the connections under cyclic loading. Bai et al. [18] derived a revised design method for a new end-plate connection and compared the design equations specified in various codes. They analyzed a large number of models and demonstrated that their revised design equation is highly accurate for calculating the bending moment in bolts and end-plates allowing for prying force. Li et al. [19] studied the mechanical properties of flush end-plate composite joints under the combination of axial loading and elevated temperature conditions. Rahnavard et al. [20] studied the behaviors of common types of rigid connections, including connections with an end-plate, connections with a bolted cover plate, and bolted T joints, under the effect of heat using numerical FE methods with the ABAQUS software. Fang et al. [21] proposed an extended end-plate connection that was formed using shape memory alloy bolts. They investigated various mechanical properties of this connection and provided a design method for it. Mirzaie Aliabadi et al. [22] introduced a new posttensioned T-stub connection (PTTC) for earthquake-resistant steel moment-resisting frames (MRFs) that consist of high-strength posttensioned (PT) strands and bolted T-stubs. They also investigated the theoretical behavior and proposed design methods based on parametric studies and comparisons with analytical results. Faridmehr et al. [23] proposed and studied a saddlebag connection. The results of tests and FEA showed that the proposed connection developed adequate catenary action, which is a fundamental criterion to prevent progressive collapse. Liu et al. proposed several new truss-square HSS column connections that are suitable for modularized high-rise steel structures, a fully welded connection [24] used in a module, a bolted-welded connection [25], and some fully bolted connections [26–29] used for connecting different modules, and determined the seismic performance of the proposed connections from static and quasi-static loading tests and FEA. They [30] also designed and studied a modularized prefabricated steel frame structure with inclined braces in which the beam to column and column to column connections are made by bolts on-site. However, there are still a lack of connections, especially for modularized prefabricated structures. Hu et al. [31] studied the damage behavior, bearing capacity, plastic angle, ductility, and energy dissipation capacity of end-plate bolts with a cantilever beam that is prefabricated and welded to the column in the factory. Wang et al. [32] studied the seismic performance of a new type of prefabricated connection for a concrete-filled steel tubular column and steel beam. A low-cyclic loading test was carried out for joints with different parameters. Yu et al. [33] carried out a finite element analysis of the static and hysteretic behavior of a new type of beam-column prefabricated rigid joint. The failure modes and the effect of the number of bolts, the width and thickness of the cover plate, and the length of the cantilever section on the hysteresis performance under pseudostatic loading were studied. Jiang et al. [34] proposed a self-locking joint between concrete-filled steel tubular columns and steel girders for prefabricated structures and carried out a finite element analysis. Zhang et al. [35] carried out a monotonic loading test of a new type of joint between a square HSS column and truss beam used in a prefabricated structure and analyzed the stiffness, strength, and ductility of the joint. Ma et al. [36] carried out a low-cycle repeated loading test on bolted beam-column joints with a cantilever beam prefabricated and welded to the column in the factory, in which the bolt hole is a slot for assembly convenience.

These bolted connections showed good mechanical and seismic performance and rapid construction speed. However, more new types of bolted connections, especially for prefabricated structures, need to be proposed to accelerate the construction speed and ensure structural safety, and the bolt layout, bolt distance, bolt hole size, and connecting plate thickness need more study. This study investigates a site-bolted connection that connects an H-section steel beam with a square HSS column. Six specimens were subjected to low-cyclic loading tests and FEA to study the seismic performance (hysteretic behavior, skeleton curve, ductility, rotation capacity, and stiffness degradation) of the connection and determine the key parameters that affect it.

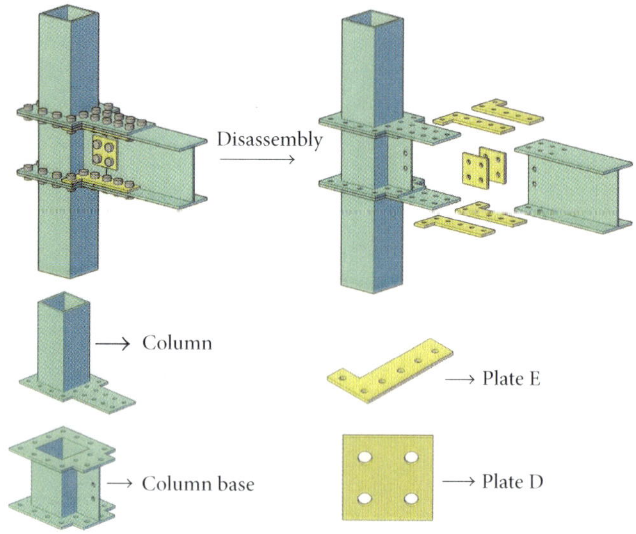

FIGURE 1: Assembly drawing of the connections.

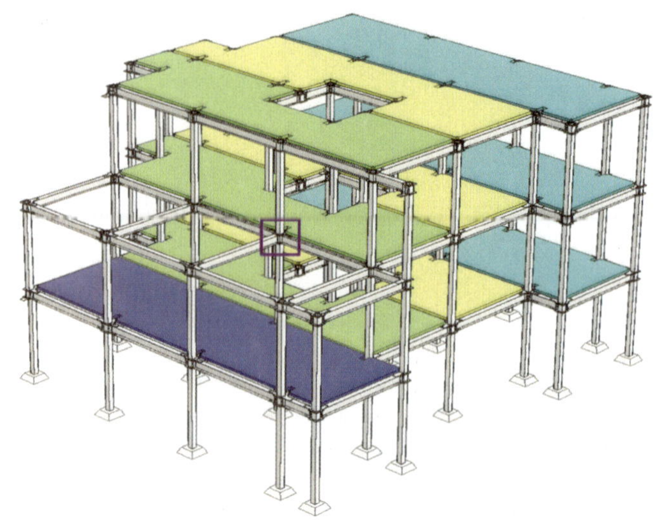

FIGURE 2: Structural system.

2. Connection Structure

Figure 1 shows the connection that is investigated in this study, which is used to connect two beam slab modules. This connection is formed by connecting two columns with flanges at the end, a steel beam, a column base, and splint plates (plate E) with bolts. Each column with a flange is composed of a steel column and a flange equipped with a cover plate. The column base is composed of flanges, a short column, and a vertical stiffener. Bolt holes are prepared in the flanges, cover plates, beam flanges, and web plate. The parts are connected with high-strength bolts on-site. Using this connection, two adjacent beam slab modules, the column above the beam slab modules, and the column below them are connected. After assembly, the columns are filled with C60 concrete. The proposed connection can be used in a modularized high-rise steel structure composed of a foundation module, column modules, and beam slab modules (Figure 2). The foundation module is made of reinforced concrete. This fabricated process can be completed at the construction site or in the factory. The prefabricated upper structures are connected to the foundation by a connection preset on the foundation. The column modules are fabricated in the factory by welding square HSS columns to flanges with cover plates. A beam slab module is composed of column bases, beams, and concrete floor slabs. The floor slabs are poured in the factory and are connected to the beams by studs. Bolt holes are prepared in the column bases, flanges, and beams.

3. Specimen Design

3.1. Specimen Design. The site-bolted beam-column connection shown in Figure 2 was selected for this study. Two 200 × 12 mm square HSS columns and a 300 × 200 × 8 × 12 mm H-section steel beam were used to form each connection specimen. Except for the beam, which was composed of Q235B steel, all of the parts of each specimen were made of Q345B steel. The nominal yield strengths of Q235B and Q345B steels are 235 MPa and 345 MPa, respectively. According to the Chinese *Code for the Seismic Design of Buildings and Technical Specification for the Steel Structure of Tall Buildings*, formula (1) is used to check the principle of strong column and weak beam [37, 38]. The plastic section modulus of beam W_{pb} is 843552 mm^3 and that of column W_{pc} is 637056 mm^3. The axial force applied on the column is 500 k N, A_c is the area of the column, and η is 1.1. The connection in this paper consists of two columns and one beam. The connection met formula (1) even before pouring the concrete, and the column will be stronger still after pouring concrete C60.

$$\sum W_{pc}\left(f_{yc}-\frac{N}{A_c}\right)\geq \eta \sum W_{pb} f_{yb}. \qquad (1)$$

Based on a previous study [39], the lengths between the respective inflection points were selected as the beam and column lengths of each specimen. Figure 3 shows the detailed dimensions of the connection. To study the effects of the number of bolts on the cover plate, the material grade of the type of plates E, and the diameter of the bolt holes on the plates E on the connection's performance, two groups of specimens (M1 and M2) were designed by varying these parameters while keeping the cross-sections of the beam and columns constant. Groups M1 and M2 each contained three specimens (hereafter referred to as the M1 specimens and the M2 specimens, resp.). Each M1 specimen contained eight bolts on the cover plates, and each M2 specimen contained six bolts on the cover plates. Table 1 lists the basic parameters of the connections.

3.2. Testing of Material Properties. According to the specifications stipulated in *Metallic Materials: Tensile Testing at Ambient Temperature* [40], the steel plate specimens were fabricated using the same batch of steel plates that were used to produce the connection specimens to determine

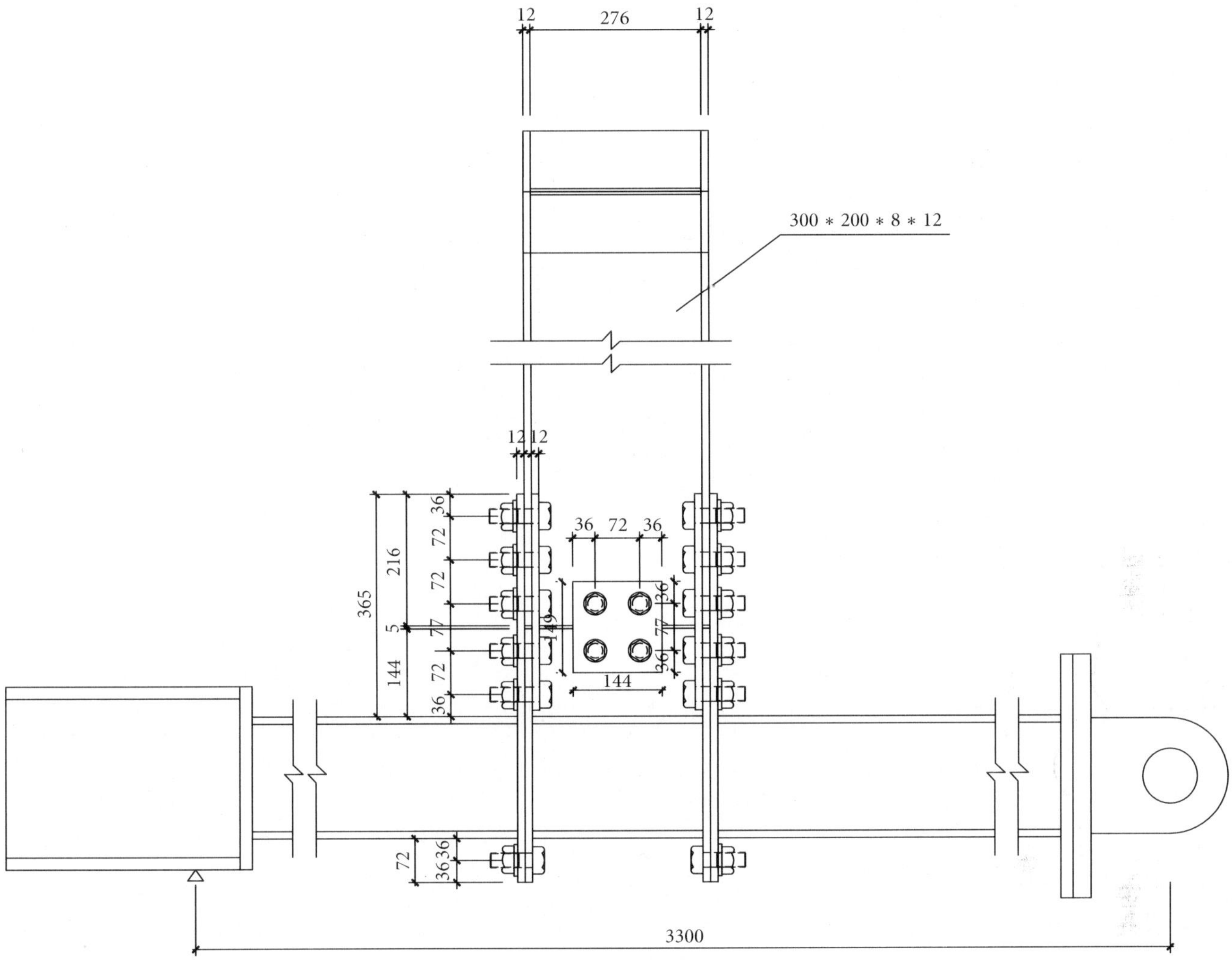

(a)

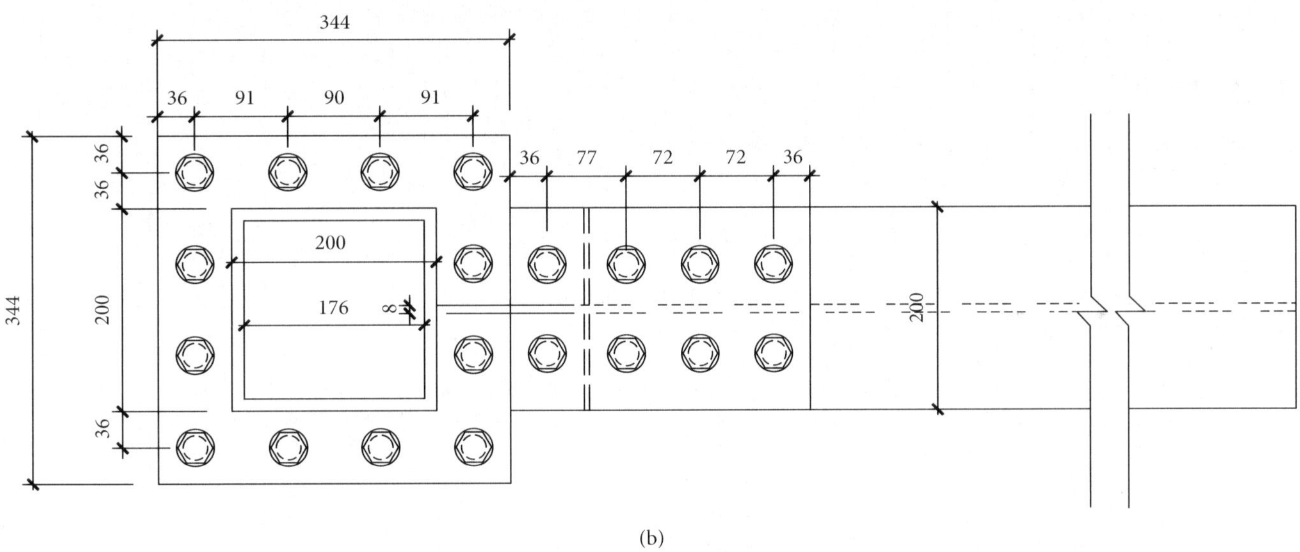

(b)

FIGURE 3: Details of the test specimens. (a) Elevation view. (b) Plan view.

Table 1: Number and parameters of the specimens.

Specimen number		Flange		Cover plate		Splint material	Splint bolt hole
		Flange thickness	Bolt	Bolt	Number of bolts		Diameter
M1	M1A	12	M20	M20	8	Q235	22
	M1B	12	M20	M20	8	Q195	22
	M1C	12	M20	M20	8	Q235	20
M2	M2A	12	M20	M20	6	Q235	22
	M2B	12	M20	M20	6	Q195	22
	M2C	12	M20	M20	6	Q235	20

Table 2: Material properties of the test specimens.

Beam		Column		S10.9 bolt	
Stress/MPa	Strain	Stress/MPa	Strain	Stress/MPa	Strain
285.4	0.00139	374.8	0.00171	995	0.0048
309.7	0.01575	400.3	0.01514	1160	0.1375
389	0.04291	448.2	0.05252	1160	0.1998
420.4	0.07082	469.5	0.08688		
435.5	0.10091	479.4	0.12272		
441.2	0.12876	483.2	0.1601		
442.1	0.17911	484	0.19972		

their material properties. Table 2 lists the nominal stress and strain of the material property tests. The stress and strain will be transformed into real stress and strain for the finite element analysis. To measure the friction coefficient between the steel plates, high-strength bolted connection specimens were fabricated based on the requirements and methods stipulated in the *Code for Acceptance of Construction Quality of Steel Structures* [41]. The friction surfaces were subjected to a sandblasting treatment, and M20 grade 10.9 high-strength bolts made of 40Cr steel were used for the connection specimens. The test results showed that the specimens had a friction coefficient of 0.4.

4. Test Scheme

4.1. Loading Device and Scheme. Each of the six specimens was subjected to a quasi-static loading test. The displacements of the inflection points of the columns were constrained, except for the displacement along the axis of the upper column. Lateral braces were used to constrain the out-of-plane movement of the beam. A constant axial pressure of 500 kN was applied at the top of the column, and a low-frequency cyclic load was applied to the end of the beam perpendicular to the axis of the beam. Figure 4 shows the loading device. A hydraulic servo system was used to apply the low-frequency cyclic load. Referencing to the American Institute of Steel Construction's *Seismic Provisions for Structural Steel Buildings* [42], the loading sequence is simplified because the connection is in the elastic state even when the rotation angle is less than 0.01 and the number of cycles in the elastic phase is reduced. Then, a variable-amplitude displacement-control loading regime was employed, and the loading was controlled through the rotation angle. Figure 5 shows the loading regime.

4.2. Arrangement of Strain Gauges and Displacement Meters. As shown in Figure 6, the measuring points on the beam-column connection specimens were determined based on the connection design and the FEA results. To measure the strains in the plates in the joint region and in the areas under relatively high stress, the strain gauges were mainly placed in this region.

Displacement meters W1 were placed at the loading point at the beam end to measure the displacement of the beam end, and the arrangement of W1 is shown in Figure 4(b). Figure 6(a) shows the arrangement of the other displacement meters. The stands of displacement transducers WS and WN were placed on the beam flange, and the pointers were placed on the glass sheet at the front of the cover plate. Displacement meters WS were used to measure the slip distances between the cover plate and the upper beam flange, and WN were used to measure the slip distances between the cover plate and the lower beam flange.

5. Test Process Analysis

5.1. Testing Process for Specimen M1. Similar phenomena were observed when applying quasi-static cyclic loads to specimens M1A, M1B, and M1C. Figure 7 shows the failure mode of specimen M1A. A sound was heard when the rotation angle of specimen M1A reached 0.02 rad cycles (32 mm in the positive direction and 37 mm in the negative direction) because slip occurred between the cover plates

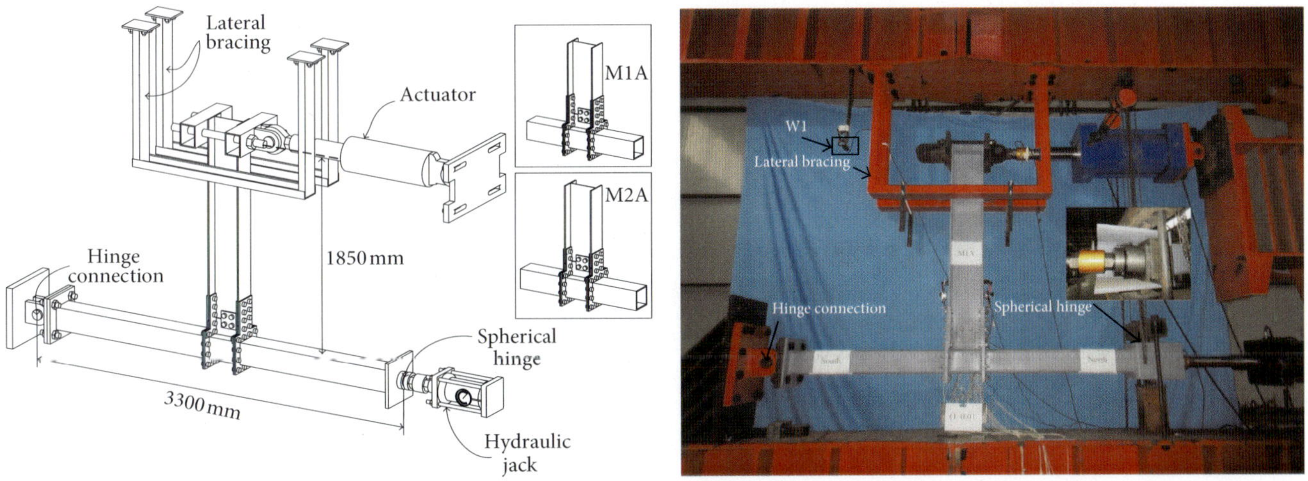

FIGURE 4: Test setup. (a) Sketch. (b) Photograph.

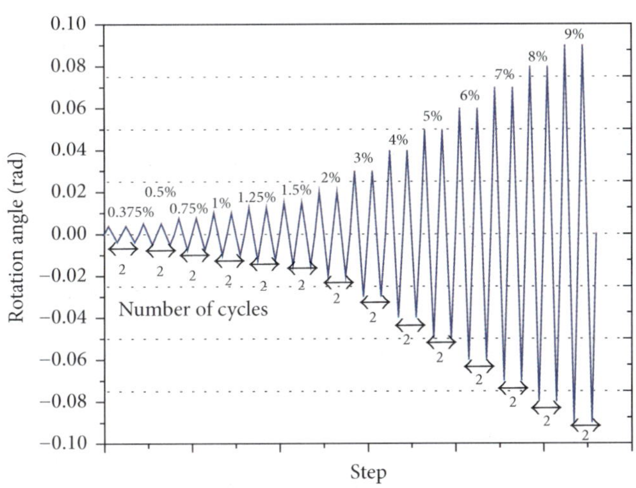

FIGURE 5: Cyclic loading history.

and beam flanges in the joint region. The entire loading process was accompanied by intermittent small sounds, which indicated that the cover plates and the beam flanges were slipping relative to each other for the duration of the loading process. The moment–rotation angle curve of specimen M1A exhibited a sawtooth shape, which shows that the load at the beam end remained nearly constant during the slipping stage. The moment is the load at the beam end multiplied by the length of the beam, which is the moment that the connection was subjected to. The rotation angle is equal to the displacement of the beam end divided by the length of the beam. When the rotation angle reached 0.08 rad cycle, clear scratches caused by the bolts could be observed on the cover plate, and the beam started to twist. The maximum slip distance between the beam flanges and the cover plates reached 18 mm, but the slip distances on the left and right sides of each cover plate were different (Figure 7(b)). In addition, the left beam flange underwent a large buckling deformation, and the beam web bulged (Figure 7(c)). The curve showed a decrease in the load-bearing capacity, and the bolt holes in the upper row underwent necking deformation. After dissembling the specimen, clear scratches were found on the contact surfaces between the beam flanges and the cover plates; moreover, the edges of the bolt holes in the upper row were found to have been pulled apart, and the bolt holes in the lower two rows were elongated (Figure 7(d)).

A sound was heard when the rotation angle of specimen M1B reached 0.02 rad cycle (46.5 mm in the positive direction and 38 mm in the negative direction). An examination of the joint region showed that slip had occurred between the cover plates and the beam flanges. A sound was also heard when the rotation angle of specimen M1C reached 0.015 rad cycle (32 mm in the positive direction and 27 mm in the negative direction). Slip was also found to have occurred between the cover plates and the beam flanges.

5.2. Testing Process for Specimen M2. Similar phenomena were observed when applying quasi-static cyclic loads to specimens M2A, M2B, and M2C. Figure 8 shows the failure mode of specimen M2A. A sound was heard when the rotation angle of specimen M2A reached 0.015 rad cycle (23 mm in the positive direction and 21 in the negative direction), and slippage occurred between the cover plates and the beam flanges. Moreover, the entire loading process was accompanied by intermittent small sounds, which indicates that the cover plates and the beam flanges were slipping relative to each other for the duration of the loading process. The moment–rotation angle curve of specimen M2A exhibited a sawtooth shape, and the load at the beam end remained approximately constant during the slipping stage. When the rotation angle at the beam end reached 0.09 rad cycle, clear scratches caused by the bolts were observed on the cover plate, but the beam was not twisted. In addition, the maximum slip distance between the beam flanges and the cover plates in the joint region reached 19 mm, and the

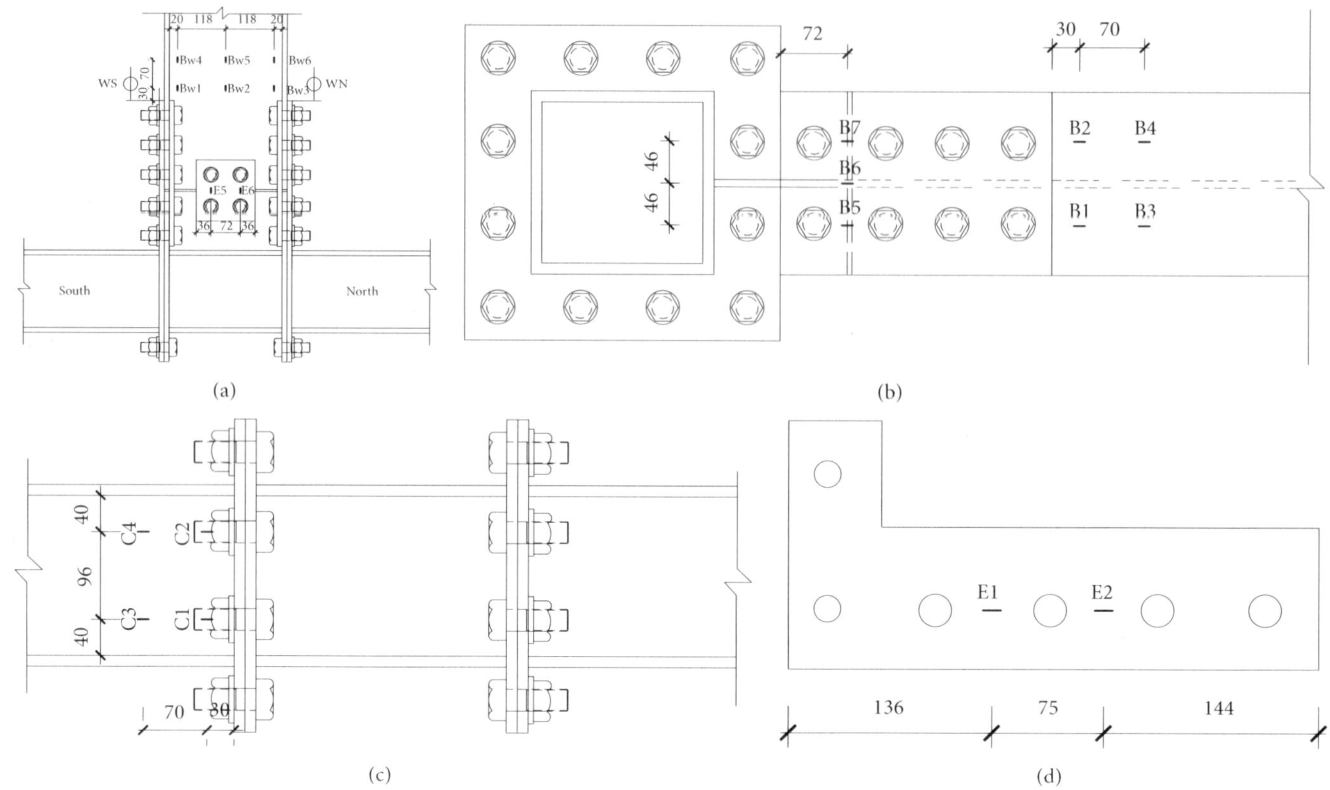

FIGURE 6: Arrangements of strain gauges and displacement meters. (a) Front view. (b) Left-side view. (c) Top view. (d) Plate E.

slip distances on the left and right sides of the cover plate were the same (Figure 8(b)). Moreover, the left beam flange underwent a slight buckling deformation, but the web did not bulge. There was a large gap between the beam web and the column base (Figure 8(c)). The curve showed no decrease in the load-bearing capacity. The bolt holes in the upper row were found to have undergone necking deformation. After dissembling the specimen, clear scratches were found on the contact surfaces between the beam flanges and the cover plates; moreover, the edges of the bolt holes were found to have been pulled apart and greatly elongated (Figure 8(d)); the elongated bolt holes of M2 were larger than those of M1.

A sound was heard when the rotation angle at the beam end of specimen M2B reached 0.015 rad cycle (17.8 mm in the positive direction and 18 mm in the negative direction). By examining the joint region, slip was found to have occurred between the cover plates and the beam flanges. A sound was also heard when the rotation angle at the beam end of specimen M2C reached 0.015 rad cycle (25 mm in the positive direction and 25 mm in the negative direction). Slip was found to have occurred between the cover plates and the beam flanges. The sound occurred because of the slip between the steel parts and could be eliminated by adding a thin copper sheet between them, so that a stable slipping energy dissipation could be obtained [43].

6. Test Results and Analysis

6.1. Moment–Rotation Angle Hysteretic Curves. Figure 9 shows the moment–rotation angle hysteretic curves of the six specimens. The moment is the load at the beam end multiplied by the length of the beam, which is the moment that the connection is subjected to. The rotation angle is equal to the displacement of the beam end divided by the length of beam. Specimens M1A, M1B, M1C, M2A, and M2B were loaded until a rotation angle of 0.09 rad was reached, and specimen M2C was loaded until a rotation angle of 0.07 rad was reached. The curves obtained from the FEA closely match those of the tests, and their maximum bearing capacity in each cycle is almost the same. Based on Figure 9, all of the specimens exhibited a relatively good energy dissipation capacity, ductility, and load-bearing capacity. Due to the relative slip that occurred in the joint region, the hysteretic curves of all of the specimens exhibited a Z-shape, which indicates that the pinching of the hysteretic curves was a result of slip and the bearing capacity could still increase after the slip. The fewer bolts there were, the smaller the slip-load was, and the more significantly the hysteretic curve was pinched. Each hysteretic loop had an approximately horizontal slipping segment. The slip-load continuously decreased as the number of loading cycles increased. This was mainly due to two reasons: the pretension of the bolts decreased because the plates were plastically compressed by the bolt gaskets during the loading process, and the slip coefficient of the friction surfaces decreased as the number of loading cycles increased due to wear and tear. When the slip stopped, the bolt shanks came in contact with the bolt holes, and the bolt holes started to bear the load and undergo deformation; in addition, the beam started to dissipate energy through plastic

FIGURE 7: Failure modes of specimen M1A.

deformation. Although this character makes the hysteretic curve look like it is not very wide, it could make the structure avoid collapse because the bearing capacity does not decrease after the slipping dissipation. A comparison shows that the M1 specimens had wider hysteretic curves than the M2 specimens. This is because the M1 specimens had more bolts; thus, their joint regions were stronger, and more energy was dissipated through buckling at the beam end after the bolts slipped. This theory is corroborated by a photograph that shows the deformation of the beam after loading. The size of the bolt holes and the material of the plates E had no significant impact on the connection performance. All of the M1 specimens ultimately failed due to the occurrence of significant buckling at the beam end; that is, their failures were ductile. For the M2 specimens, the beam was not considerably buckled when the load was removed. After dissembling each specimen, the bolt holes were found to have been greatly elongated, and clear cracks had formed near the bolt holes. The moment–rotation angle hysteretic test curves of all of the M1 specimens have decreasing segments after the maximum moment, whereas those of the M2 specimens did not demonstrate such. This is consistent with the phenomena observed during the tests; a plastic hinge formed at the beam end of each M1 specimen, whereas such did not occur in the M2 specimens.

6.2. Performance Indicators. Several performance indicators were obtained by processing the test data, including the moment at which significant slip occurred (M_s), the corresponding rotation angle of the connection (θ_s), the peak moment at which the load at the beam end reached its maximum (M_u), the ultimate rotation angle at which the displacement reached its maximum (θ_u), the displacement ductility factor ($\mu = \theta_u/\theta_s$), the plastic rotation angle ($\theta_p = \theta_u - \theta_s$), and the total energy dissipated by the specimen E_{total}.

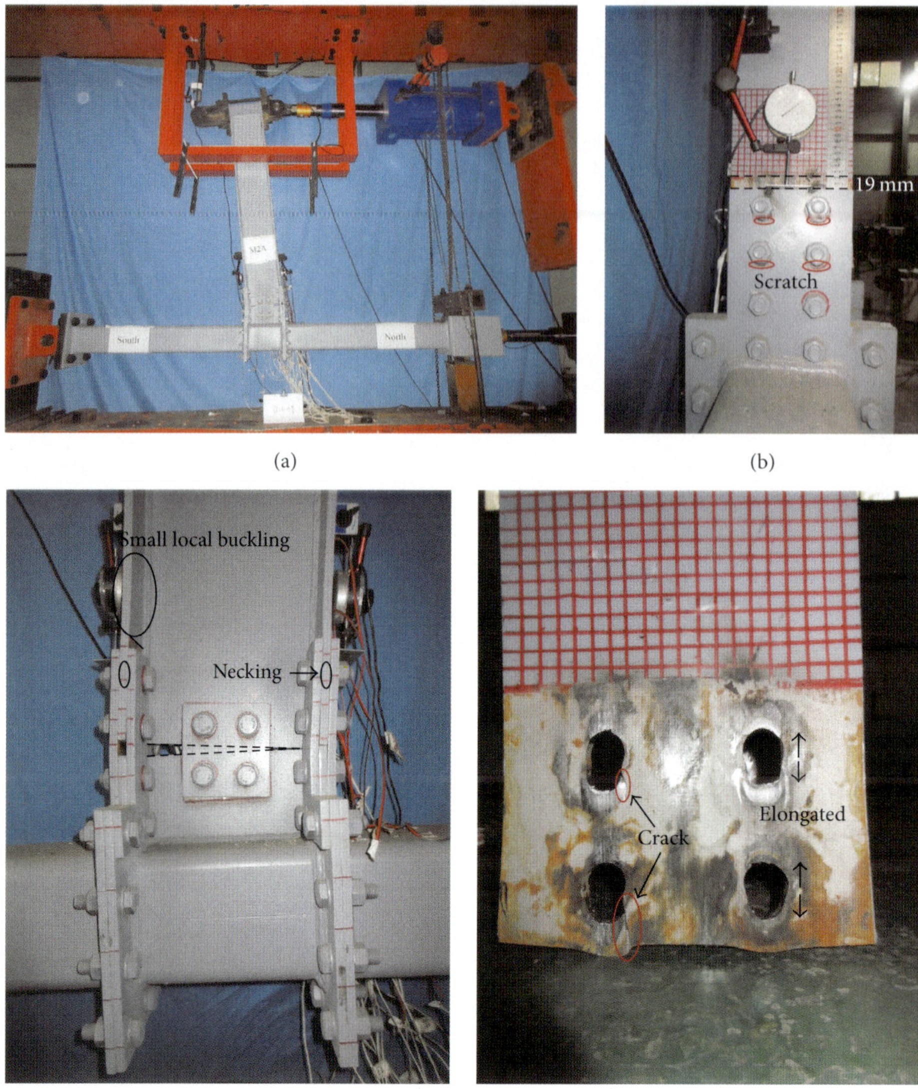

Figure 8: Failure modes of specimen M2A.

Table 3 lists the performance indicators of the six specimens obtained from quasi-static loading tests. The results show that significant slip occurred in the M1 specimens when the rotation angle reached approximately 0.02 rad and in the M2 specimens when the rotation angle reached approximately 0.015 rad. The slip moment M_s when significant slip occurred in specimens M1 and M2 was approximately 300 kN·m and 200 kN·m, respectively. This suggests that the more bolts there are on the cover plate, the greater the slip moment is. The rotation angle is small when the slip occurred, which indicates that the connection can dissipate energy through slip during a moderate earthquake and the slip moment can be controlled by varying the number of bolts on the cover plate.

The ultimate moment M_u of the M1 specimens reached approximately 400 kN·m, whereas that of the M2 specimens reached approximately 350 kN·m. This indicates that for the specimens that contained more bolts, the frictional contact surface has a larger friction-resistance, and the bolt shank and bolt hole surface bear larger pressures after the conclusion of the slip when the rotation angle was very large, which resulted in a higher load-bearing capacity. Moreover, the connection dissipated more seismic energy through plastic deformation at the beam end. The μ values of the specimens were symmetrical in the positive and negative directions. The M1 specimens had μ values of approximately 5, whereas the M2 specimens had μ values of approximately 6. The μ values of the M1 specimens were smaller than those of the M2 specimens. The rotation angle of the M2 specimens when the slip occurred was smaller than that of the M1 specimens because they contained fewer bolts, leading to a relatively large μ value but a relatively low energy dissipation capacity. As shown in Table 3, μ is larger than 4.0 [44] and θ_p is larger than 0.04 [42], and thus the connection specimens had relatively good ductility.

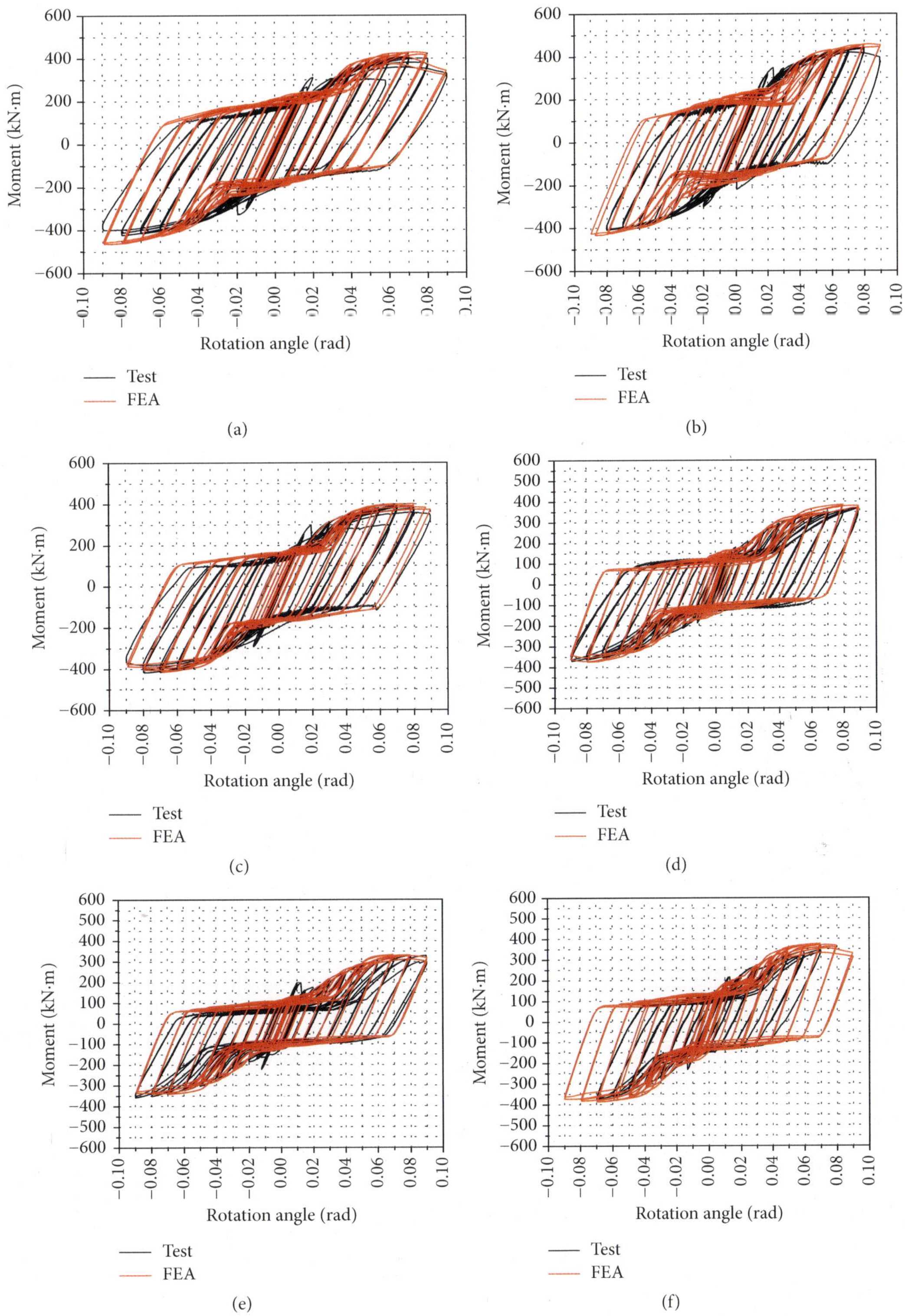

FIGURE 9: Moment–rotation angle hysteretic curves. (a) Specimen M1A. (b) Specimen M1B. (c) Specimen M1C. (d) Specimen M2A. (e) Specimen M2B. (f) Specimen M2C.

Compared to the slip moment, the ultimate strength was greatly improved. This indicates that the bolted connections between the cover plates and the flanges are transformed from friction-type to bearing-type as the connection dissipates energy, and the load-bearing capacity increases after the slip is finished. This process is accompanied by continuous increases in the load-bearing capacity and ductility of the connection. The data shown in Table 3 demonstrate that the

TABLE 3: Primary performance indicators during the quasi-static tests.

Specimen	Loading direction	M_s (kN·m)	θ_s (% rad)	M_u (kN·m) (% rad)	θ_u (% rad)	μ	θ_p (% rad)	E_{total}
M1A	Positive	310.7	1.98	408.4 (7)	9	4.5	7.02	391.3 kJ
	Negative	298.4	1.51	412.1 (7)	8.99	6	7.48	
M1B	Positive	318.4	2	431.1 (8)	8.95	4.5	6.95	374.8 kJ
	Negative	300.1	2	402.9 (8)	8.01	4	6.01	
M1C	Positive	252.3	1.5	388.1 (7)	8.99	6	7.49	353.3 kJ
	Negative	290.1	1.5	404.8 (7)	9	6	7.50	
M2A	Positive	179.5	1.53	362.5 (8)	8.9	5.8	7.37	388 kJ
	Negative	185.8	1.52	357.9 (8)	8.95	5.9	7.43	
M2B	Positive	160.4	1.49	325.7 (8)	8.99	6	7.5	292 kJ
	Negative	171.2	1.51	348 (8)	8.99	6	7.48	
M2C	Positive	179.1	1.51	342 (7)	6.96	4.6	5.45	222.1 kJ
	Negative	179.1	1.5	366.1 (7)	6.99	4.7	5.49	

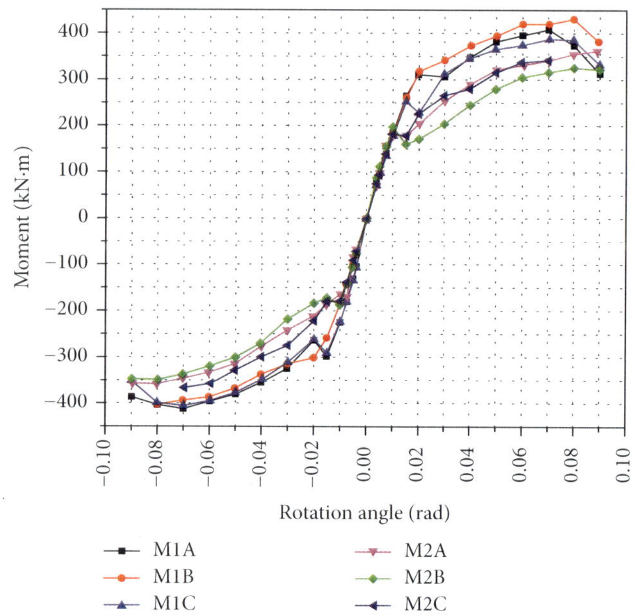

FIGURE 10: Skeleton curves of the specimens.

connection that dissipates energy through slipping has excellent seismic performance and a high load-bearing capacity after the slip is finished.

6.3. Skeleton Curves. The points of the maximum values of the loads in each cycle of the hysteresis curve are extracted and connected in turn to obtain the skeleton curve. Figure 10 shows a comparison of the moment–rotation angle skeleton curves for the six specimens. The three M1 specimens had similar skeleton curves. For these three specimens, slip occurred when the rotation angle reached 0.02 rad, which corresponds to a sudden decrease in the skeleton curves of the connection. Afterwards, the bolt shanks came in contact with the bolt hole walls, and the bolt hole walls started to bear the load and undergo deformation, which resulted in an increase in the load-bearing capacity of the connection. Subsequently, the beam end started to buckle in several local areas, which corresponds to the descending segment of the moment–rotation angle skeleton curve. The M1 specimens reached their peak load-bearing capacity of approximately 400 kN·m when the rotation angle of the beam reached 0.08 rad. The three M2 specimens also had similar skeleton curves. In these three specimens, slip occurred when the rotation angle reached 0.015 rad, which corresponds to a sudden decrease in the skeleton curves of the connection. Afterwards, the bolt shanks came in contact with the bolt hole walls, and the bolt hole walls started to bear the load and deform, which resulted in an increase in the load-bearing capacity of the connection. However, the beam did not undergo significant local buckling deformation at its end before the load was removed. The M2 specimens reached their peak load-bearing capacity of approximately 350 kN·m when the rotation angle of the beam reached 0.09 rad. Table 3 lists the ultimate load-bearing capacities and other performance indicators in detail. Figure 10 shows that the positive and negative directions of the skeleton curves of the specimens were approximately symmetrical. The skeleton test curves of all of the M1 specimens have decreasing segments after the maximum moment, whereas those of the M2 specimens did not have such. This is consistent with the phenomena observed during the tests, in which a plastic hinge formed at the beam end of each M1 specimen, whereas no plastic hinge formed at the beam end of each M2 specimen.

6.4. Stiffness Degradation. On the basis of the skeleton curves, the stiffness of each point is calculated, the ratio of the stiffness of each point to the initial stiffness is calculated, and the point with the ratio of each point as the vertical coordinate and the rotation angle as the horizontal coordinate is obtained. Drawing a line between these points, the equivalent stiffness degradation coefficient curve is obtained, which shows the law of the stiffness degradation of the connection. Figure 11 shows a comparison of the equivalent stiffness degradation coefficient curves for the six specimens. The curve of each specimen had similar trends in the positive and negative directions. The curves contained three segments in each direction, including a horizontal linear segment, a

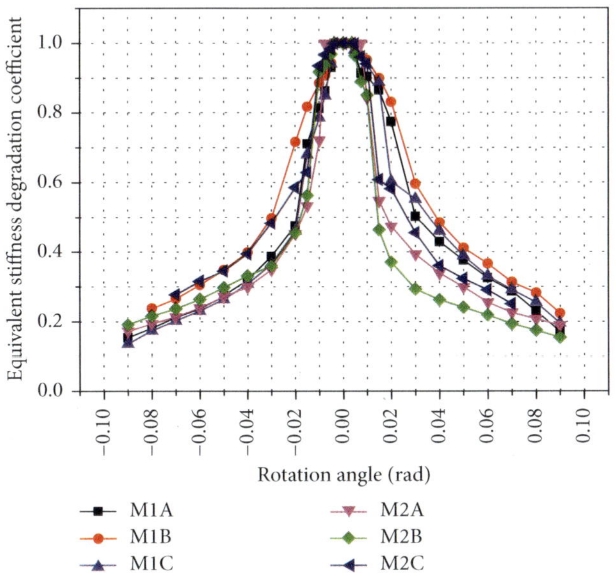

FIGURE 11: Equivalent stiffness degradation coefficient curves of the specimens.

linearly descending segment, and a curved descending segment. Each specimen was in an elastic state during the initial loading stage, and the segment of its curve that corresponds to the initial loading stage was a horizontal straight line. After significant slip occurred, the stiffness of each specimen started degrading at a higher rate, and the corresponding segment of its curve exhibited a sudden linear descent. When the bolts contacted with the bolt hole walls, the slip stopped, at which time the connection was strengthened, and the stiffness degradation slowed down significantly. During the entire loading process, the equivalent stiffness degradation coefficients of the M1 specimens were significantly higher than those of the M2 specimens, which indicates that the stiffness of the specimens that contained fewer bolts degraded at a higher rate.

6.5. Energy Dissipation Capacity. Figure 12 shows the amounts of energy that were dissipated by the six specimens with respect to each rotation angle of connection. The amounts of energy dissipated by the M1 and M2 specimens began to increase significantly in the seventh loading cycle (±2% rad) and the sixth loading cycle (±1.5% rad), respectively. This indicates that specimens M2A, M2B, and M2C were no longer in an elastic state and started to slip and strengthen starting in the sixth loading cycle. In specimens M1A, M1B, and M1C, this occurred in the seventh loading cycle. The number of bolts had a significant impact on the slip-load of the connection. The specimens that contained fewer bolts (i.e., the M2 specimens) started to slip sooner than those that contained more bolts (i.e., the M1 specimens). After slip occurred, the beam of each specimen underwent plastic deformation, and energy was dissipated through plastic deformation in addition to slipping, which resulted in a rapid increase in the amount of energy dissipated. The total amounts of energy dissipated by specimens M1A, M1B, M1C, M2A, M2B, and M2C were 391.3 kJ (0.08 rad), 374.8 kJ (0.08 rad), 353.3 kJ (0.08 rad), 388 kJ (0.09 rad), 292 kJ (0.09 rad), and 222.1 kJ (0.07 rad), respectively. Figure 12(a) shows that the energy dissipation capacity of the M1 specimens was significantly higher than that of the M2 specimens when the connection reached the same rotation angle, which indicates that the more bolts there were, the higher the capacity of the connection to dissipate energy was, and the effects of the bolt hole size and the grade of the splint material on the total energy dissipation were small on the basis of the analysis of Figures 9 and 12.

In this study, the equivalent viscous damping coefficient h_e in each hysteretic loop is used to describe the energy dissipation [45]. Figure 12(b) shows that the equivalent viscous damping coefficient of the M1 specimens increased with the increasing rotation angle. This indicates that each specimen of M1 dissipated energy only through slip during the initial stage and thus had a relatively low energy dissipation capacity. Later, the energy dissipation capacity of the specimen increased due to the compressive deformation of the bolts' hole walls and the plastic deformation at the beam end. Because there were fewer bolts at the beam end of the M2 specimens than in the M1 specimens, the M2 specimens slip at smaller rotation angles, and their equivalent viscous damping coefficients were larger than those of M1 at the early stage. But, with the process of slipping, the friction of M2 dropped significantly because the bolt tension drops drastically. Thus, the energy dissipation dropped, as did the equivalent viscous damping coefficient. With the development of plastic deformation on the bolt hole and beam end, the energy dissipation increases again, and then the equivalent viscous damping coefficient increases, too. Figure 12 shows that the connection dissipates energy by different modes at different stages. There are three stages. During the first stage, the connection dissipates energy mainly through slip. During the second stage, the connection dissipates energy mainly through slip and the compressive deformation of the bolt hole walls. During the third stage, the connection dissipates energy mainly through slip and plastic deformation at the beam end. The energy dissipation capacity of the M1 specimens is higher than that of the M2 specimens because the bolt constraints (friction and bearing pressure) of the M1 specimens are stronger and the beam end undergoes more plastic deformation. Except for M2B, when the rotation angle is greater than 0.06, the equivalent viscous damping coefficients of the connections are greater than 0.2. This shows that the connections have a high energy dissipation capacity, except M2B.

6.6. Strain Analysis. Figure 13 shows the strain–rotation angle curves for the six specimens with respect to four locations: the beam flanges in the joint region, the beam web in the joint region, the cover plates in the joint region, and plate E. The two black horizontal lines represent the yield strain of Q235 steel ($\varepsilon_y = \pm 1.383 \times 10^{-3}$), and the two red horizontal lines represent the yield strain of Q345 steel ($\varepsilon_y = \pm 1.785 \times 10^{-3}$). The specimens remained elastic in these four locations before reaching the yield strain. In the M1 specimens, the

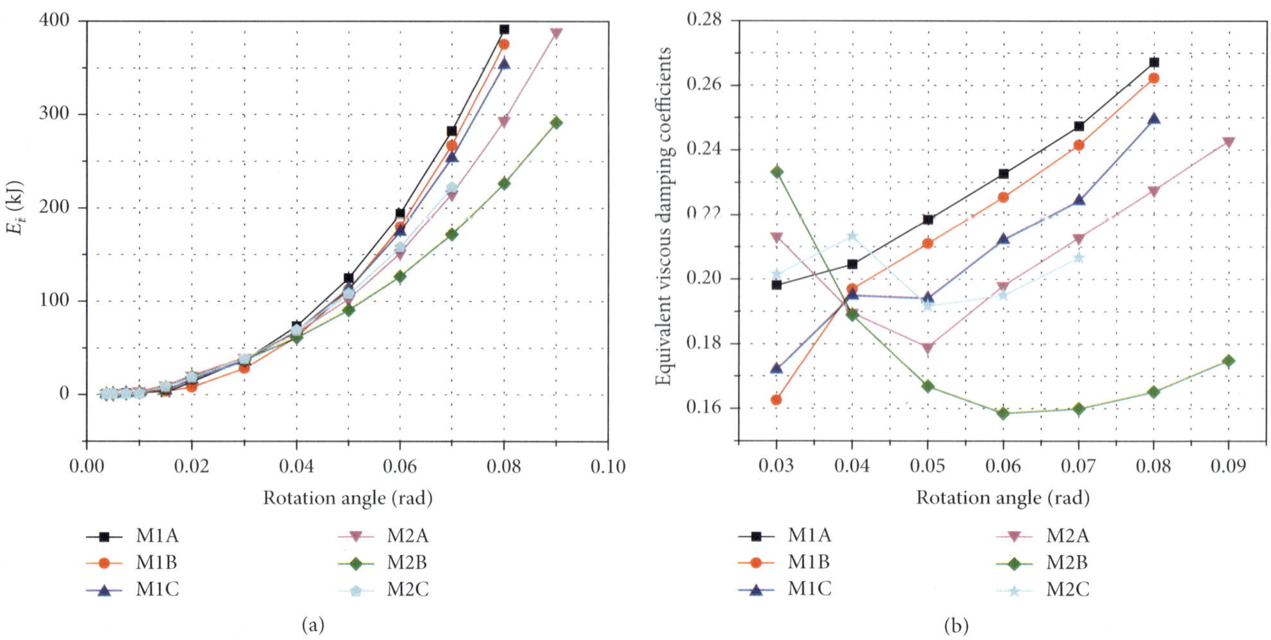

FIGURE 12: Energy dissipation capacity for each rotation angle. (a) Energy dissipation capacity curves of the specimens. (b) Equivalent viscous damping coefficient curves of the specimens.

beam flanges in the joint region yielded first, followed by the cover plates in the joint region, the splints, and the beam web in the joint region. In the M2 specimens, the cover plates in the joint region yielded first, followed by the beam flanges in the joint region, the splints, and the beam web in the joint region.

7. Finite Element Analysis

7.1. Basic Information of the FE Model. Numerical simulations were performed using the finite element analysis software ABAQUS. The six connections were subjected to quasi-static FEA. The FE models were established using the same dimensions as those of the test specimens. The steel beam, steel columns, and high-strength bolts were all simulated using 3D solid elements (C3D8R). In addition, the mesh was refined in the areas where the stress is relatively large based on the results of the tests and the theoretical analysis. The friction coefficient used in the modeling was obtained from the antislip tests. To determine the constitutive relations of the materials, nominal stress–strain curves were obtained from the tests (shown in Table 2) and converted to true stress–strain curves for the FEA. The kinematic hardening of the materials is used for the seismic finite element analysis. The interactions between the bolts, splint plate, steel beam, and column are all built, including the positive-pressure contact relationships and frictional contact relationships. The lateral surface of the bolt shank has a potential positive-pressure contact relationship with the wall of the bolt hole. The friction coefficient between the bolt and steel plates is 0.05 and that between the steel plates is 0.35. Based on the actual boundary conditions that were used during the tests, the end of each square HSS column was simulated as a hinged end. To be consistent with the tests, lateral braces were simulated to constrain the lateral displacement of the H-section steel beam. A low-frequency cyclic load was applied at the beam end, and the same loading regime that was used in the tests was used in the simulations. Figure 14 shows the FE model of the connection.

7.2. Failure Modes. A comparison of the hysteretic curves in Figure 9 obtained from the tests and the FEA shows good agreement between them. A comparison of the failure modes in Figure 15 obtained from the tests and the FEA similarly shows that the deformation and strain of the FEA agreed well with the test results, which verifies the FE model. Figures 15(a) and 15(b) show the deformation and strain in the joint regions of specimens M1A and M2A, respectively, and Figures 15(c) and 15(d) show the deformation and strain in the cover plate regions of specimens M1A and M2A, respectively. The bolt holes on the beam flanges and cover plates experienced a very high strain. Figures 15(e) and 15(f) show the deformation and strain in the beam end of specimens M1A and M2A, respectively, indicating that the bolt holes on the beam end are elongated. The color contours in the FEA results show the strain of the components. The deeper the color in the picture is, the greater the strain is.

The FEA results were thus consistent with the test results. The FEA results for specimen M1A show that after loading, a plastic hinge formed at the beam end in the joint region, the web bulged, and only the bolt holes in the upper row were significantly elongated. The areas under relatively high stress were mainly located near the beam end and on the cover plate. The FEA results for specimen M2A show that a small buckling deformation only occurred in some areas at the beam end, the beam web did not bulge, and all of the bolt holes at the beam

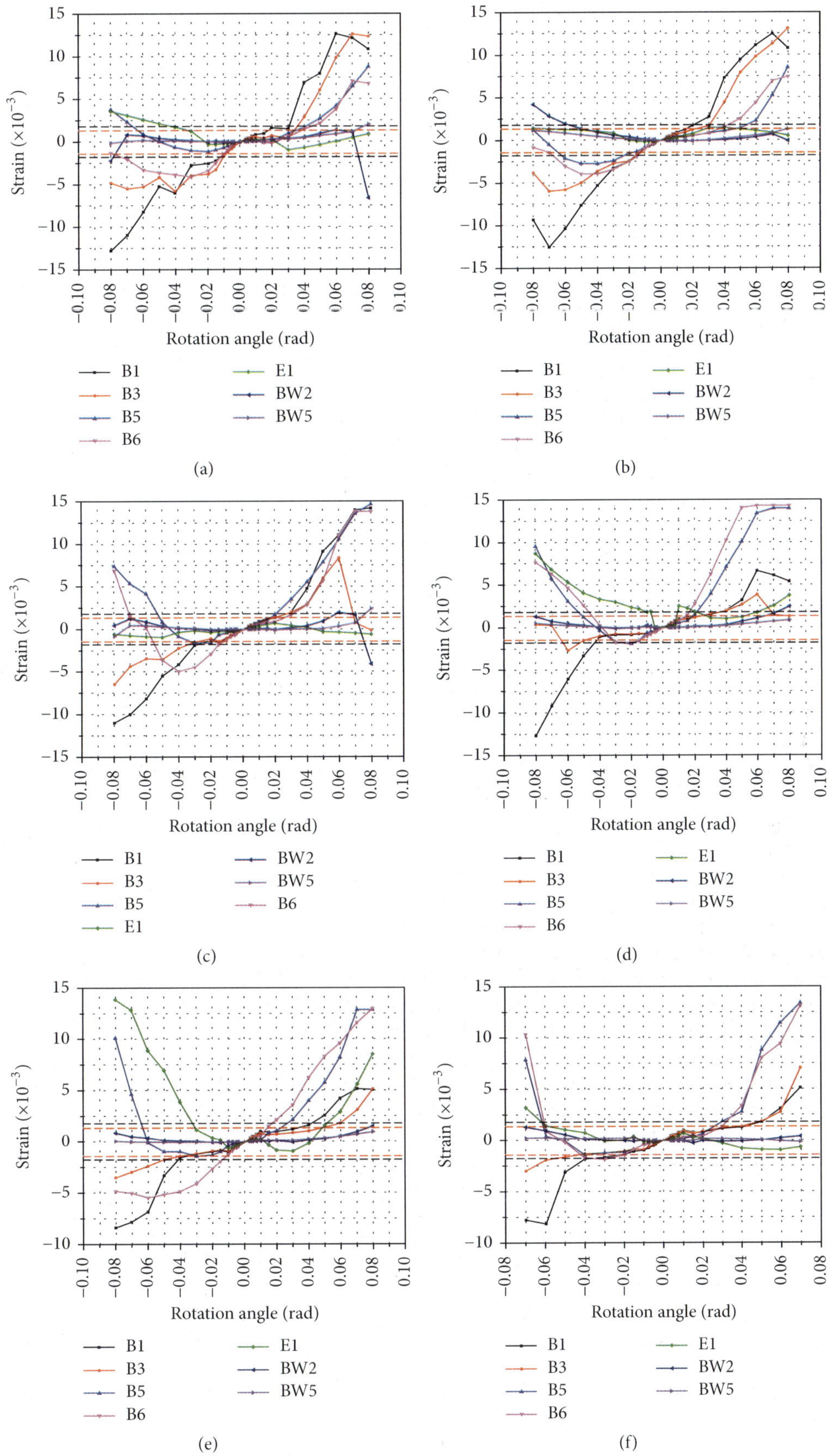

FIGURE 13: Strain under cyclic loading. (a) Specimen M1A. (b) Specimen M1B. (c) Specimen M1C. (d) Specimen M2A. (e) Specimen M2B. (f) Specimen M2C.

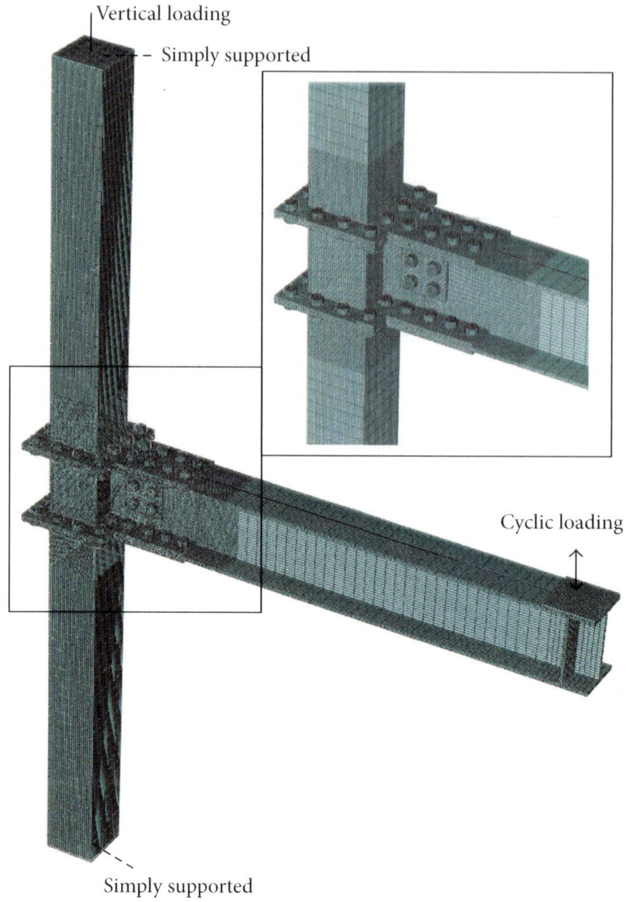

FIGURE 14: Finite element model.

end were greatly elongated. The areas under relatively high stress were located near the bolt holes on the cover plate and the bolt holes at the beam end. A comparison of the stress distributions from the FEA and the strain curves obtained from the tests shows that the areas under relatively high stress in the simulations were the same as those areas where large strains and relatively large deformations occurred in the tests.

8. Conclusion

A site-bolted beam-column connection that is suitable for modularized prefabricated steel structures was proposed. Six connection specimens were subjected to quasi-static cyclic loading tests and FEA. The FE model was validated by comparing the results of the FEA and tests. The following conclusions are drawn from this study.

The M1 specimens dissipated energy by the following mechanism. During the initial loading stage, energy was mainly dissipated by slip. After the bolt hole wall started to bear the load, energy was mainly dissipated by slip as well as the plastic deformation of the bolt holes. When the rotation angle was excessively large, the seismic energy was dissipated mainly by slip, the plastic deformation of the bolt holes, and the plastic deformation at the beam end. The M2 specimens dissipated energy by a similar mechanism; however, when the rotation angle was excessively large, seismic energy was still dissipated by slip and the deformation of the bolt holes, and only slight plastic deformation was observed in the joint region. The more bolts there were, the larger the buckling deformation of the flange at the beam end was, and the greater the energy dissipation capacity of the connection was.

The M1 specimens failed under quasi-static cyclic loading due to the relatively large local buckling deformation of the flange at the beam end. The M2 specimens failed under quasi-static cyclic loading due to cross-sectional rupturing at the bolt holes in the innermost row on the flange at the beam end and the slight buckling deformation of the cantilever beam flange near the joint region. The connection specimens underwent sufficient plastic deformation before failure, and the ultimate rotation angle reached 0.09 rad; thus, their failure was ductile. The observations revealed that the fewer bolts there were in the joint region, the earlier the slip occurred. Therefore, the ductility factor of the M2 specimens was greater than that of the M1 specimens. Increasing the number of bolts can effectively increase the slip moment of the connection and cause an outward shift of the plastic hinge.

The strains on the sections at the beam flange near the joint region, at the bolt holes in the innermost row on the cover plate, and at the bolt holes in the outermost row on the beam end were greater than those at other locations.

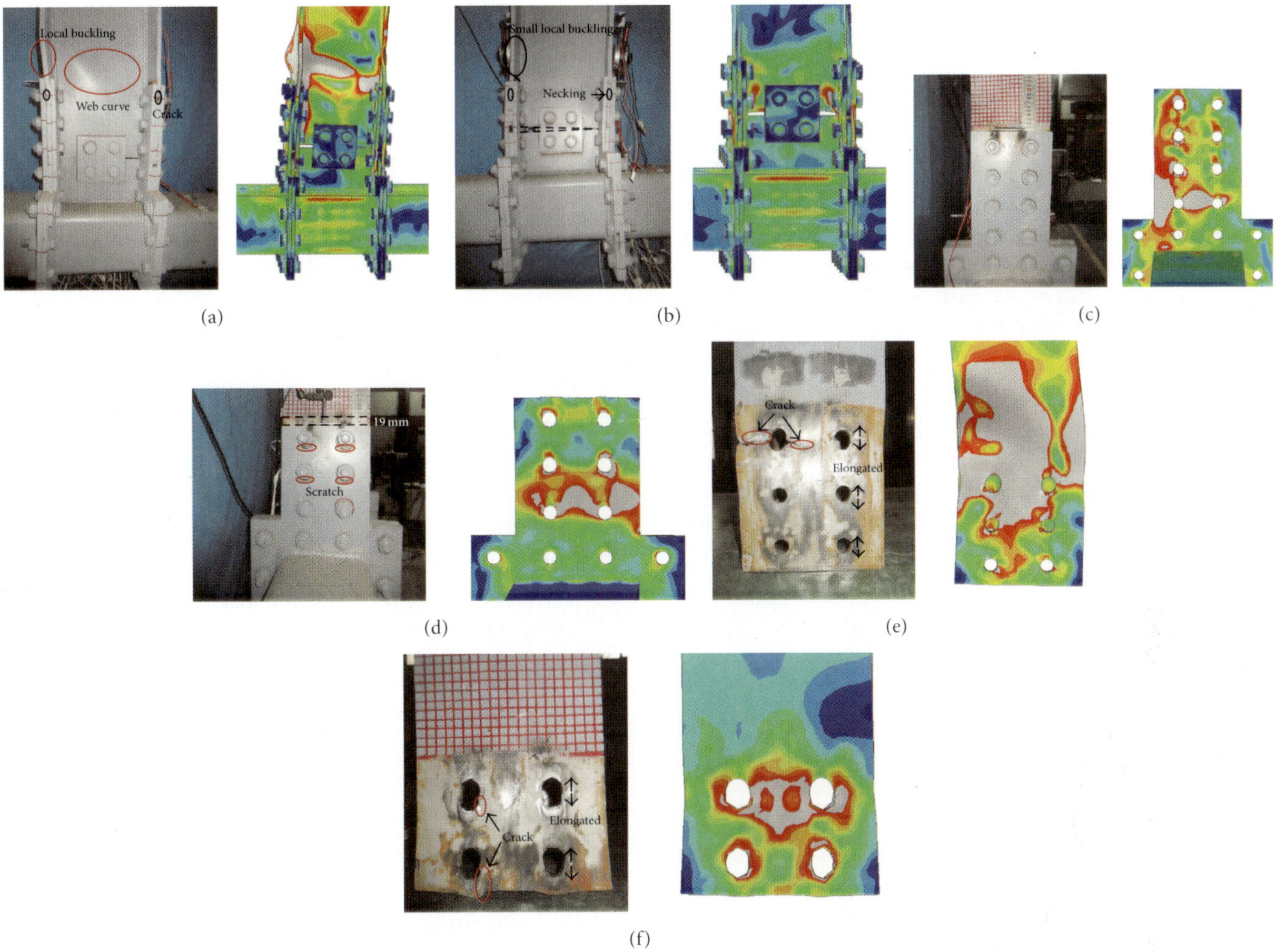

FIGURE 15: Comparison of the failure modes of the specimens in the experiments and the FEM analyses. (a) Joint region of specimen M1A. (b) Joint region of specimen M2A. (c) Cover plate of specimen M1A. (d) Cover plate of specimen M2A. (e) Beam end of specimen M1A. (f) Beam end of specimen M2A.

These three sections are weak sections and thus require safety checks during the connection design process.

The comparative analysis showed that the number of bolts in the joint region was a major parameter that affected the connection's performance. The more bolts there were in the joint region, the wider the hysteretic curve of the connection was, the greater the deformation of the beam flange near the joint region was, and the greater the energy dissipation capacity of the connection was. The material of the splint plates and the size of the bolt holes at the beam end had no significant impact on the connection's performance. The test and FEA results showed that the proposed connection exhibited reasonable seismic performance (e.g., hysteretic behavior, energy dissipation capacity, ductility, and plastic rotation capacity) while meeting the requirements of easy on-site installation and fast assembly.

Conflicts of Interest

The authors declare that there are no conflicts of interest regarding the publication of this paper.

Acknowledgments

The authors gratefully acknowledge the support for this work, which was funded by the National Key Research and Development Program of China (016YFC0701503), the National Natural Science Foundation of China (51678010), the Beijing Municipal Natural Science Foundation (8172009), and the Beijing Municipal Science and Technology Plan Program (Z171100002217017).

References

[1] D. K. Miller, "Lessons learned from the Northridge earthquake," *Engineering Structures*, vol. 20, no. 4-6, pp. 249–260, 1998.

[2] M. Nakashima, K. Inoue, and M. Tada, "Classification of damage to steel buildings observed in the 1995 Hyogoken-Nanbu earthquake," *Engineering Structures*, vol. 20, no. 4-6, pp. 271–281, 1998.

[3] E. P. Popov, T.-S. Yang, and S.-P. Chang, "Design of steel MRF connections before and after 1994 Northridge earthquake," *Engineering Structures*, vol. 20, no. 12, pp. 1030–1038, 1998.

[4] S. Naimi, M. Celikag, and A. A. Hedayat, "Ductility enhancement of post-northridge connections by multilongitudinal voids in the beam web," *The Scientific World Journal*, vol. 2013, Article ID 515936, 2013.

[5] Z. F. Li, Y. J. Shi, H. Chen, and Y. Q. Wang, "Nonlinear element analysis of improved beam-column joint of steel structure," *Building Structure*, vol. 32, no. 9, pp. 015–018, 2002 (Chinese).

[6] D. W. Zhao, Y. J. Shi, and H. Chen, "Experimental study of beam-column joint under low cycling load," *Building Structure*, vol. 30, no. 9, pp. 003–007, 2000 (Chinese).

[7] C.-C. Chen, C.-C. Lin, and C.-H. Lin, "Ductile moment connections used in steel column-tree moment-resisting frames," *Journal of Constructional Steel Research*, vol. 62, no. 8, pp. 793–801, 2006.

[8] C.-C. Chen and C.-C. Lin, "Seismic performance of steel beam-to-column moment connections with tapered beam flanges," *Engineering Structures*, vol. 48, pp. 588–601, 2013.

[9] A. Astaneh-Asl, *Seismic Design Steel Column-Tree Moment Resisting Frames*, Structural Steel Educational Council, Berkeley, Calif, USA, 1997.

[10] J. Ribeiro, A. Santiago, C. Rigueiro, and L. S. Da Silva, "Analytical model for the response of T-stub joint component under impact loading," *Journal of Constructional Steel Research*, vol. 106, pp. 23–34, 2015.

[11] J. Ribeiro, A. Santiago, C. Rigueiro, P. Barata, and M. Veljkovic, "Numerical assessment of T-stub component subjected to impact loading," *Engineering Structures*, vol. 106, pp. 450–460, 2016.

[12] J. W. Hu, R. T. Leon, and T. H. Park, "Mechanical modeling of bolted T-stub connections under cyclic loads part I: stiffness modeling," *Journal of Constructional Steel Research*, vol. 67, no. 11, pp. 1710–1718, 2011.

[13] J. W. Hu, R. T. Leon, and T. Park, "Mechanical models for the analysis of bolted T-stub connections under cyclic loads," *Journal of Constructional Steel Research*, vol. 78, pp. 45–57, 2012.

[14] A. K. Dessouki, A. H. Youssef, and M. M. Ibrahim, "Behavior of I-beam bolted extended end-plate moment connections," *Ain Shams Engineering Journal*, vol. 4, no. 4, pp. 685–699, 2013.

[15] G. S. Prinz, A. Nussbaumer, L. Borges, and S. Khadka, "Experimental testing and simulation of bolted beam-column connections having thick extended endplates and multiple bolts per row," *Engineering Structures*, vol. 59, pp. 434–447, 2014.

[16] Y. Shi, G. Shi, and Y. Wang, "Experimental and theoretical analysis of the moment-rotation behaviour of stiffened extended end-plate connections," *Journal of Constructional Steel Research*, vol. 63, no. 9, pp. 1279–1293, 2007.

[17] K. D. Tsavdaridis and T. Papadopoulos, "A FE parametric study of RWS beam-to-column bolted connections with cellular beams," *Journal of Constructional Steel Research*, vol. 116, pp. 92–113, 2016.

[18] R. Bai, S.-L. Chan, and J.-P. Hao, "Improved design of extended end-plate connection allowing for prying effects," *Journal of Constructional Steel Research*, vol. 113, pp. 13–27, 2015.

[19] G.-Q. Li, L.-Z. Chen, J.-T. Li, and G.-B. Lou, "Modeling of end-plate bolted composite connection in fire considering axial force effects," *Journal of Constructional Steel Research*, vol. 76, pp. 133–143, 2012.

[20] R. Rahnavard, N. Siahpolo, M. Naghavi, and A. Hassanipour, "Analytical study of common rigid steel connections under the effect of heat," *Advances in Civil Engineering*, vol. 2014, Article ID 692323, 10 pages, 2014.

[21] C. Fang, M. C. H. Yam, A. C. C. Lam, and L. Xie, "Cyclic performance of extended end-plate connections equipped with shape memory alloy bolts," *Journal of Constructional Steel Research*, vol. 94, pp. 122–136, 2014.

[22] M. Mirzaie Aliabadi, M. R. Bahaari, and S. Torabian, "Design and analytical evaluation of a new self-centering connection with bolted T-stub devices," *Advances in Materials Science and Engineering*, vol. 2013, Article ID 163021, 2013.

[23] I. Faridmehr, Y. Ahmad, M. M. Tahir, and M. H. Osman, "Cyclic and explosive evaluation of new proposed steel joint," *Advances in Civil Engineering*, vol. 2016, Article ID 4975097, 2016.

[24] X. C. Liu, A. X. Xu, A. L. Zhang, Z. Ni, H. X. Wang, and L. Wu, "Static and seismic experiment for welded joints in modularized prefabricated steel structure," *Journal of Constructional Steel Research*, vol. 112, pp. 183–195, 2015.

[25] X. C. Liu, S. H. Pu, A. L. Zhang et al., "Static and seismic experiment for bolted-welded joint in modularized prefabricated steel structure," *Journal of Constructional Steel Research*, vol. 115, pp. 417–433, 2015.

[26] X. C. Liu, S. H. Pu, A. X. Xu, Z. Ni, A. L. Zhang, and Z. W. Yang, "Experimental study on static and seismic performance of bolted joint in modularized multi-layer and high-rise prefabricated steel structures," *Journal of Building Structures*, vol. 36, no. 12, pp. 43–51, 2015 (Chinese).

[27] X. C. Liu, Z. W. Yang, H. X. Wang, A. L. Zhang, S. H. Pu, and L. Wu, "Seismic performance study on beam-to-column joint of bolted assembled multi-high-level steel structure," *Journal of Building Structures*, vol. 38, no. 3, pp. 43–51, 2017 (Chinese).

[28] X. Liu, Z. Yang, H. Wang et al., "Seismic performance of H-section beam to HSS column connection in prefabricated structures," *Journal of Constructional Steel Research*, vol. 138, pp. 1–16, 2017.

[29] X. C. Liu, S. H. Pu, and A. L. Zhang, "Performance analysis and design on bolted connections in modularized prefabricated steel structures," *Journal of Constructional Steel Research*, vol. 19, no. 9, pp. 183–195, 2017.

[30] X. Liu, A. Zhang, J. Ma, Y. Tan, and Y. Bai, "Design and model test of a modularized prefabricated steel frame structure with inclined braces," *Advances in Materials Science and Engineering*, vol. 2015, Article ID 291481, 12 pages, 2015.

[31] F. X. Hu, G. Shi, Y. J. Shi et al., "Experimental study on seismic performance of specifically pre-fabricated beam-column connections," *Journal of Building Structures*, vol. 35, no. 7, pp. 034–043, 2014 (Chinese).

[32] Q. J. Wang, Y. N. Zhang, C. Y. Xu et al., "Seismic test for fabricated connection of concrete-filled steel tubular column and H-shaped steel beam," *Journal of Shenyang University of Technology*, vol. 36, no. 3, pp. 355–360, 2014 (Chinese).

[33] Y. S. Yu, L. Wei, and G. X. Fu, "Nonlinear static analysis of a new beam-column assembly rigid connection," *World Earthquake Engineering*, vol. 30, no. 1, pp. 050–056, 2014 (Chinese).

[34] L. X. Jiang, Y. L. Guo, Q. T. Lu et al., "Behavior of a new type of self-locked joint of concrete-filled steel tubular column and steel beam," *Construction Technology*, vol. 38, no. 3, pp. 067–069, 2009 (Chinese).

[35] A. Zhang, L. Zhao, X. Liu, and Y. Sun, "Monotonic experimental study of prefabricated square tubular column to truss beam connections," *China Civil Engineering Journal*, vol. 47, pp. 169–174, 2014 (Chinese).

[36] R. L. Ma, Y. Yang, Q. S. Chen et al., "Seismic performance testing study on high strength bolt connections with slotted holes,"

Journal of Building Structures, vol. 30, no. 1, pp. 101–106, 2009 (Chinese).

[37] GB 50011, *Code for Seismic Design of Buildings*, China Planning Press, Beijing, China, 2010.

[38] JGJ 99, *Technical Specification for Steel Structure of Tall Building*, China Architecture & Building Press, Beijing, China, 2015.

[39] M. Gerami, H. Saberi, V. Saberi, and A. S. Daryan, "Cyclic behavior of bolted connections with different arrangement of bolts," *Journal of Constructional Steel Research*, vol. 67, no. 4, pp. 690–705, 2011.

[40] GB/T 288. 1. Metallic Materials: Tensile Testing at Ambient Temperature, China Standards Press, Beijing, China, [In Chinese] 2010.

[41] GB50205, *Code for Acceptance of Construction Quality of Steel Structures*, China Planning Press, Beijing, China, 2001.

[42] ANSI/AISC341, *Seismic Provisions for Structural Steel Buildings*, American Institute of Steel Construction, Chicago, Ill, USA, 2010.

[43] A. L. Zhang, Q. X. Quan, X. X. Zhan, and X. F. Ma, "Force mechanism analysis of fabricated friction dissipation re-centering brace with zero initial cable force," *Journal of Southeast University (Natural Science Edition)*, vol. 47, no. 1, pp. 142–149, 2017.

[44] A. M. Girão Coelho, L. S. D. Silva, and F. S. K. Bijlaard, "Ductility analysis of bolted extended end plate beam-to-column connections in the framework of the component method," *Steel and Composite Structures*, vol. 6, no. 1, pp. 33–53, 2006.

[45] Li. Lm, Z. H. Chen, and N. Li, "Experimental study on seismic capability of diaphragm-through style beam-column joint," *Earthquake Engineering and Engineering Vibration*, vol. 27, no. 1, pp. 46–53, 2007 (Chinese).

17

Effect of Concrete Age and Creep on the Behavior of Concrete-Filled Steel Tube Columns

HaiYang Wang,[1,2] XiaoXiong Zha,[2] and Wei Feng[2]

[1] *School of Environment and Energy, Peking University Shenzhen Graduate School, Shenzhen 518055, China*
[2] *School of Civil and Environment Engineering, Harbin Institute of Technology Shenzhen Graduate School, Shenzhen 518055, China*

Correspondence should be addressed to XiaoXiong Zha; zhahero@126.com

Academic Editor: Doo-Yeol Yoo

The influence of concrete age and creep on the ultimate axial loading capacity of concrete-filled steel tube (CFST) columns is experimentally and numerically investigated. After validation of numerical models, a parametric study is conducted and the results are used to formulate empirical formulas for predicting the ultimate axial load-bearing capacity of the columns. Formulas are also proposed for predicting both the composite creep and aging coefficients of the CFST columns, which consider the confinement action of steel tubes on concrete. Then, the proposed formulas are validated independently by comparing their predictions with existing test results performed by other researchers. The comparisons show that the empirical formulas have the potential to be used in the practical design of CFST columns.

1. Introduction

Concrete-filled steel tube (CFST) columns have been extensively used in civil engineering structures over the past few decades. Extensive experimental and theoretical investigations on CFST components have been conducted and are well documented in the literature. The experimental work includes tests reported by O'Shea and Bridge [1] and Giakoumelis and Lam [2], who conducted research on the bearing capacity of short CFST columns. Zhong [3] and Zha [4] also reported their works on short and long CFST columns. Han et al. [5] and Wang [6] also studied property of the CFST under long-time load, while these results cannot be implemented conveniently considering the new code (GB50936-2014). Although significant work has been conducted on the properties of matured CFST columns, less emphasis has been placed on the influence of concrete age. Tan and Qi [7] have worked systematically by testing the effect of creep on columns under axially applied eccentric compressive load. Terry et al. [8] and Luo et al. [9] tested the creep of CFST columns and plain concrete in order to investigate the influence of steel tube reinforcement. Ichinose et al. [10] also reported a series of similar tests to obtain creep coefficients. Naguib and Mirmiran [11] and Liu et al. [12] developed an algorithm for the time-dependent behavior of CFST using the rate of flow method and the double power law functions for the creep of concrete. Shrestha and Chen [13] worked on the aging and creep coefficients of confined concrete, but the effect of steel tube was excluded in their formulas.

This study investigates the effect of concrete age and creep by experimental test and FEM modeling. Then, empirical formulas are proposed by treating a CFST column as a structural component made of a single material through homogenization.

2. Simulation and Methods

After pouring concrete into the steel tube, the elastic modulus of concrete increases with age. At the early stage, the steel tube is subjected to higher values of load than that specified by the design code. Thus, the ultimate bearing capacity of the CFST column is lower than the designed value. In this section, laboratory tests and FEM model are used to obtain the time-dependent load-bearing capacity. An empirical formula is established to provide a simple form of calculation for estimating the age-related ultimate load-bearing capacity.

TABLE 1: Geometrical and mechanical properties of the specimens.

D (mm)	T (mm)	L (mm)	f_y (MPa)	f_{ck} (MPa)	E_s (MPa)
219	4.5	766	345	40.7	2.06e5

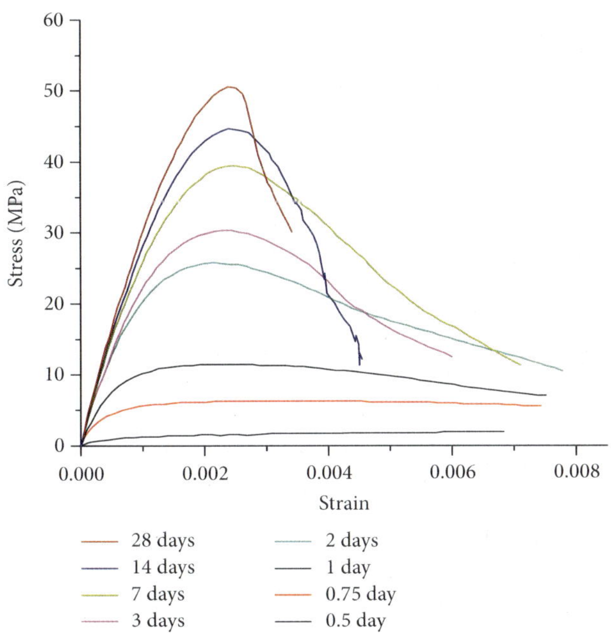

FIGURE 1: Stress strain relations of concrete at different ages.

2.1. FEM Modeling of CFST Columns. The numerical model is implemented in ABAQUS eight-node quadrilateral in-plane continuum shell elements which are used for the steel tube whereas the concrete is modeled employing eight-node linear brick element. The bilinear material model is adopted for the steel, with tangent modulus being 5% of the elastic modulus of 2.06e5 MPa. The material model used for the concrete is the damaged plasticity model, and hardening of concrete at different ages follows the curves proposed by Yi et al. [14], as shown in Figure 1. Other mechanical and geometrical parameters of the column are given in Table 1. FE meshes of both concrete and steel are shown in Figure 2. The contact between the steel tube and concrete is assumed to be perfect; here the nodes on the interface between solid and shell elements are tied.

Different meshing schemes are tested in order to assess the convergence of the FEM model, and the details are listed in Table 2, where *N*_Ring, *N*_Radius, and *N*_Axial are, respectively, the numbers of elements in the circumferential, radial, and axial directions. *E*_Section and *E*_Total are the total number of elements on a cross section and of the whole column, respectively. *N*_3 d and *N*_28 d are the ultimate axial loads of the column at ages of 3 days and 28 days.

Figure 3 shows terms of load against relative displacement of the different meshing FE model for concrete at the age 3 of and 28 days. With the increase in the number of elements, from M1 to M6, Table 2 shows the convergence of the ultimate load. When the elements number is more than that of M4, the difference between the different meshing schemes is nonsignificant. Therefore, meshing scheme M4 will be used in the following numerical simulations.

Laboratory tests have been conducted (see Figure 4) on the column specified in Table 1 in order to verify the FEM model. Load-bearing capacity tests are conducted at ages equal to 3, 7, 14, 21, 28, and 49 days.

The experimental test results of the time-dependent elastic modulus at different ages are shown in Table 3, and the load-bearing capacity of the CFST column is shown in Figure 5 as a function of the considered ages. Comparison between experimental and numerical results shows that the FE model can clearly predict the age-dependent load-bearing capacity.

2.2. Empirical Formula of the Age-Related Load-Bearing Capacity. Parametric studies on a variety of CFST columns presented in Table 4 are conducted to derive an empirical formula for the load-bearing capacity of the CFST columns, whose numerical results are presented in Table 5.

Supposing that the age-strength relationship of concrete in the steel tube has the same form as when it is unconfined, according to ACI-209 [15], the following equation can be established:

$$\frac{N_{\mathrm{ut}} - N_k}{N_{u28} - N_k} = \frac{t}{a + bt}, \tag{1}$$

where the two unknown constants in the equation can be estimated by fitting FE results presented in Table 5. Then, $a = 1.243$ and $b = 0.977$ are obtained with the maximum error of 3.4%. Therefore, the ultimate load-bearing capacity of the CFST column is obtained as follows:

$$N_{\mathrm{ut}} = N_k + \frac{(N_{u28} - N_k)\,t}{1.243 + 0.977t}. \tag{2}$$

The previously presented equation is validated through its application on the specimen defined in Table 1, and a comparison between experimental and numerical results is shown in Figure 6. F(exp) and F(FEM) are obtained applying (2) to the experimental and numerical results, respectively, shown in Figure 5. The predicted ultimate loading capacities obtained from the formula are consistent with those obtained from the experimental tests and FE modeling.

3. Influence of Creep on CFST Columns

In this section, a composite creep coefficient of CFST columns is proposed, in order to consider steel confinement, and the factor of the effective aging coefficient is obtained by regression on the FEM results.

3.1. Creep Coefficient of the CFST Column. Yi et al. [14] proposed the age-adjusted effective modulus method, also known as BT theory, which represents an effective method to analyze the long-term effect of concrete creep. An aging coefficient was introduced in order to consider the effect of

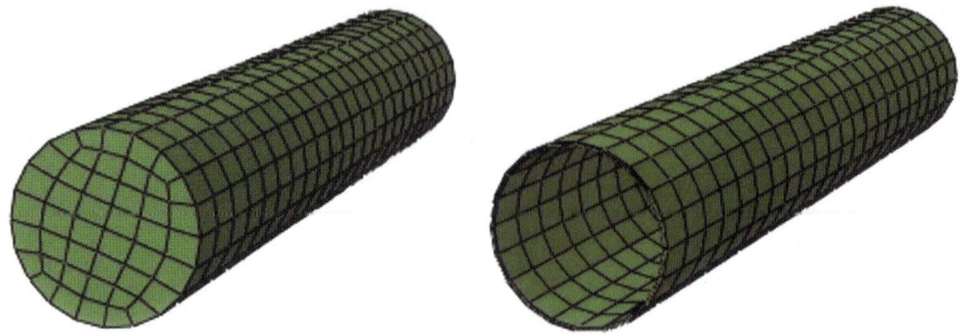

(a) (b)

Figure 2: Finite element mesh of (a) concrete and (b) steel tube.

Table 2: Meshes parameter and results.

Model	N_Ring	N_Radius	N_Axial	E_Section	E_Total	N_3 d (kN)	N_28 d (kN)
M1	12	2,3	18	56	1120	2240	2868
M2	12	3,4	18	68	1224	2256	2904
M3	16	3,4	18	96	1728	2285	2942
M4	20	3,4	18	104	1872	**2358**	**2965**
M5	24	3,4	18	132	2376	2372	2972
M6	28	3,4	18	160	2880	2377	2979

concrete aging on the ultimate creep, and the integral equation for creep was transformed into an algebraic equation by employing the integral mean value theorem.

When stress varies with time, the corresponding total strain at time t can be expressed as follows:

$$\varepsilon_c^c(t) = \frac{\sigma_{c1}^0(\tau_0)}{E_c(\tau_0)}\left[1+\varphi(t,\tau_0)\right] + \frac{1}{E_c(\tau_0)}\int_{\tau_0}^{t}\frac{d\sigma_{c1}(\tau)}{d\tau}\left[1+\varphi(t,\tau)\right]d\tau, \quad (3)$$

where $\varepsilon_c^c(t)$ is the strain of concrete when creep is with no steel confinement at age t days, $\sigma_{c1}^0(\tau_0)$ is the stress of the concrete at τ_0 days, $E_c(\tau_0)$ is the elastic modulus of concrete at τ_0 days, and $\varphi(t,\tau_0)$ is the creep coefficient at t when loaded at τ_0 days.

Applying the integral mean value theorem on formula (3) leads to

$$\varepsilon_c^c(t) = \frac{\sigma_{c1}^0(\tau_0)}{E_c(\tau_0)}\left[1+\varphi(t,\tau_0)\right] + \frac{\sigma_{c1}(t)-\sigma_{c1}^0(\tau_0)}{E_\varphi}. \quad (4)$$

Consider

$$E_\varphi = \frac{E_c(\tau_0)}{1+\rho\varphi(t,\tau_0)}, \quad (5)$$

where $\sigma_{c1}(t)$ is the stress of concrete at age of t days. E_φ is the age-adjusted effective modulus. ρ is the aging coefficient of concrete that is calculated as follows:

$$\rho = \frac{E_c(\tau_0)}{1-e^{-\varphi(t,\tau_0)}} - \frac{1}{\varphi(t,\tau_0)}. \quad (6)$$

The following calculation is based on the assumption of plane section and superposition principle of creep regardless of the toroidal shrinkage and strain. The axial strains of concrete and steel tube at age of τ_0 are denoted as ε_{c1}^0 and ε_{s1}^0, respectively, whereas σ_{c1}^0 and σ_{s1}^0 represent the related stresses. The creep of concrete with no steel confinement is $\varepsilon_c^c = \varepsilon_c^0\varphi(t,\tau_0)$ at the age of t. As a result of the effect of the steel tube confinement, the concrete creep is reduced to ε_{c1}^c, as shown in Figure 7.

It is clear that the strain increment of the steel tube is the actual concrete creep. Hence,

$$\begin{aligned} \varepsilon_{sc1}^c &= \varepsilon_{s1}^c, \\ \varepsilon_{s1}^c &= \varepsilon_c^c - \varepsilon_{c1}^c, \\ \varepsilon_{s1}^0 &= \varepsilon_{c1}^0, \end{aligned} \quad (7)$$

where ε_{sc1}^c is the axial creep strain of the CFST column, ε_{s1}^c is the axial strain of steel tube due to concrete creep, ε_c^c is the creep strain of the concrete with no steel confinement, ε_{c1}^c is the reduced strain of the concrete considering the steel tube confinement, ε_{s1}^0 is the strain of the steel tube when loaded at τ_0 days, and ε_{c1}^0 is the strain of the concrete when loaded at τ_0 days.

The total strain of CFST, or concrete in the steel tube, is derived as follows:

$$\varepsilon(t) = \varepsilon_{c1}^0 + \varepsilon_c^c - \varepsilon_{c1}^c. \quad (8)$$

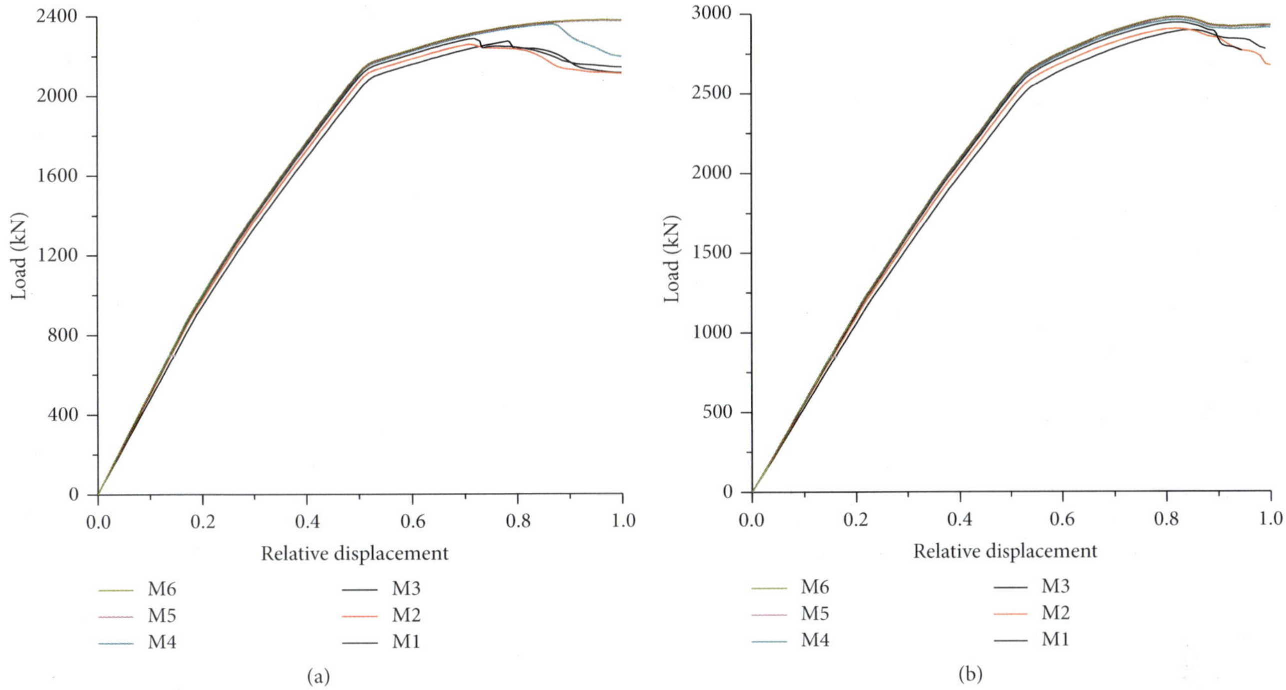

FIGURE 3: The load curve for different meshes at the (a) age of 3 days and (b) age of 28 days.

From (3), it becomes

$$\begin{aligned}\varepsilon(t) &= \frac{\sigma_{c1}^{0}}{E_c(\tau_0)}\left[1+\varphi(t,\tau_0)\right] \\ &\quad + \frac{1}{E_c(\tau_0)}\int_{\tau_0}^{t}\frac{d\sigma_{c1}(\tau)}{d\tau}\left[1+\varphi(t,\tau)\right]d\tau, \\ \varepsilon_{c1}^{0} &= \frac{\sigma_{c1}^{0}}{E_c(\tau_0)}, \\ \varepsilon_c^{c} &= \frac{\sigma_{c1}^{0}}{E_c(\tau_0)}\varphi(t,\tau_0).\end{aligned} \tag{9}$$

From (8) and (9), it is obtained that

$$\varepsilon_{c1}^{c} = -\frac{1}{E_c(\tau_0)}\int_{\tau_0}^{t}\frac{d\sigma_{c1}(\tau)}{d\tau}\left[1+\varphi(t,\tau)\right]d\tau \tag{10}$$

and by considering (4) and (5)

$$\begin{aligned}\varepsilon_{c1}^{c} &= -\frac{\left[1+\rho\varphi(t,\tau_0)\right]\left[\sigma_{c1}(t)-\sigma_{c1}^{0}\right]}{E_c(\tau_0)} \\ &= -\frac{\left[1+\rho\varphi(t,\tau_0)\right]\sigma_{c1}^{c}}{E_c(\tau_0)},\end{aligned} \tag{11}$$

where σ_{c1}^{c} is the stress change of the concrete due to creep. In spite of extensive microcracks that will be induced in the core as reported by previous researchers and splitting cracks that can occur at 30% to 40% stress [16–18], the cracks could close if they are subject to an outer sustained compression load at an early age. The confining stress would delay the formation of splitting cracks and restrict the widening of splitting cracks [19, 20], and the confinement effect appears based on the splitting cracks formation [21]. There will be no or little splitting cracks in the core concrete under the maximum sustained load after seven stages at an early age that is 40% bearing capacity of the specimens in the experiment and this had been contained naturally in the total creep. Therefore, the effect of splitting cracks on the creep was not considered separately here.

FIGURE 4: Load-bearing capacity test of the CFST column.

TABLE 3: Elastic modulus of the concrete at different age.

Age (days)	3	7	14	21	28
E_c (MPa)	30680	32319	34053	35101	35905

TABLE 4: Axially compressed short columns.

Number	$D \times T \times L$ (mm)	f_y (MPa)	f_{ck} (MPa)	α	ξ
1	$1000 \times 12 \times 3000$	345	40.7	0.050	0.422
2	$1000 \times 18 \times 3000$	345	40.7	0.076	0.645
3	$1000 \times 24 \times 3000$	345	40.7	0.103	0.876
4	$1000 \times 30 \times 3000$	345	40.7	0.132	1.117
5	$1000 \times 36 \times 3000$	345	40.7	0.161	1.366

α is the area ratio of the steel with respect to that of concrete ($\alpha = A_s/A_c$), whereas ξ is the hoop effect coefficient ($\xi = A_s f_y / A_c f_{ck}$).

The equilibrium force condition in the axial direction satisfies the following equations:

$$\sigma_{c1}^c A_c + \sigma_{s1}^c A_s = 0,$$
$$\sigma_{c1}^c = -\frac{A_s \sigma_{s1}^c}{A_c} = -\alpha \sigma_{s1}^c, \tag{12}$$

where σ_{s1}^c is the stress change of steel tube due to creep.

Not considering the creep of the steel tube and taking steel as elastic modulus value a constant give

$$\varepsilon_{s1}^c = \frac{\sigma_{s1}^c}{E_s} = -\frac{\sigma_{c1}^c}{\alpha E_s} \tag{13}$$

and (11) becomes

$$\frac{\sigma_{c1}^0}{E_c(\tau_0)} \varphi(t, \tau_0) + \frac{[1 + \rho\varphi(t, \tau_0)]\, \sigma_{c1}^c}{E_c(\tau_0)} = -\frac{\sigma_{c1}^c}{\alpha E_s}; \tag{14}$$

letting $n = E_s/E_c(\tau_0)$ in (14), the change in stresses on concrete and steel tube because of the concrete creep is expressed as follows:

$$\sigma_{c1}^c = -\frac{\alpha n \sigma_{c1}^0 \varphi(t, \tau_0)}{1 + \alpha n [1 + \rho(t, \tau_0)\varphi(t, \tau_0)]}, \tag{15a}$$

$$\sigma_{s1}^c = \frac{n \sigma_{c1}^0 \varphi(t, \tau_0)}{1 + \alpha n [1 + \rho(t, \tau_0)\varphi(t, \tau_0)]}. \tag{15b}$$

Consider the following equations:

$$\sigma_{s1}^c = \varepsilon_{s1}^c E_s = \varepsilon_{sc1}^c E_s,$$
$$\sigma_{c1}^0 = E_c(\tau_0)\, \varepsilon_{sc1}^0 = \frac{E_s \varepsilon_{sc1}^0}{n}. \tag{16}$$

Equations (16) are introduced into (15a) and (15b) to obtain the total creep strain ε_{sc1}^c of the CFST column, as follows:

$$\varepsilon_{sc1}^c = \frac{\varepsilon_{sc1}^0 \varphi(t, \tau_0)}{1 + \alpha n [1 + \rho(t, \tau_0)\varphi(t, \tau_0)]}. \tag{17}$$

Here the creep prediction function $\varphi(t, \tau_0)$ employed the equation from ACI-209 [15]. Then, the composite creep coefficient of the CFST, $\varphi_{sc}(t, \tau_0) = \varepsilon_{sc1}^c / \varepsilon_{sc}^0$, is obtained as follows:

$$\varphi_{sc}(t, \tau_0) = \frac{\varphi(t, \tau_0)}{1 + \alpha n [1 + \rho_{sc}(t, \tau_0)\varphi(t, \tau_0)]}, \tag{18a}$$

$$\varphi(t, \tau_0) = \frac{(t - \tau_0)^{0.6}}{10 + (t - \tau_0)^{0.6}} \varphi_{cu}, \tag{18b}$$

$$\varphi_{cu} = 2.35 \gamma_{la} \gamma_h \gamma_\psi \gamma_s \gamma_\lambda \gamma_a, \tag{18c}$$

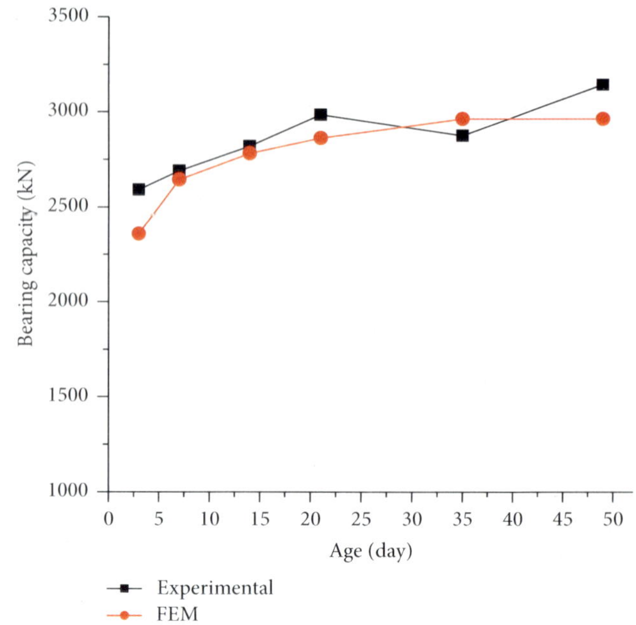

FIGURE 5: Load-bearing capacity of the CFST column at different ages.

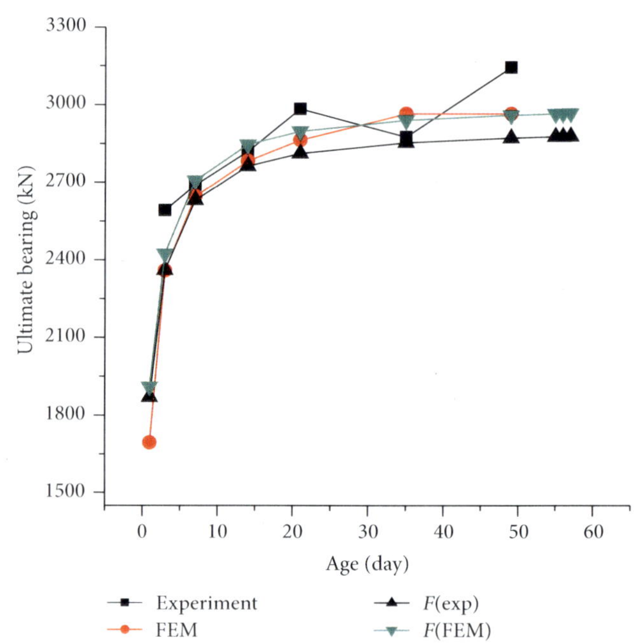

FIGURE 6: Comparison with the experimental and numerical results.

TABLE 5: Ultimate load-bearing capacity of the axially compressed columns.

	N_k (kN)	N_{ut} ($\times 10^3$ kN)					$(N_{\text{ut}} - N_k)/(N_{u28} - N_k)$			
Concrete age (day)	0	1	3	7	14	28	1	3	7	14
Number 1	13.070	30.876	43.497	49.731	52.161	54.934	0.425	0.727	0.876	0.934
Number 2	19.235	38.841	51.250	57.432	59.979	63.839	0.440	0.718	0.856	0.913
Number 3	25.238	45.828	58.000	64.098	66.483	70.591	0.454	0.722	0.857	0.909
Number 4	31.084	52.171	64.129	70.131	72.343	76.533	0.464	0.727	0.859	0.908
Number 5	36.770	58.070	69.795	75.706	77.777	81.960	0.471	0.731	0.862	0.907

$N_k = A_s f_y$; N_{u28} is the load-bearing capacity when the concrete age is equal to 28 days.

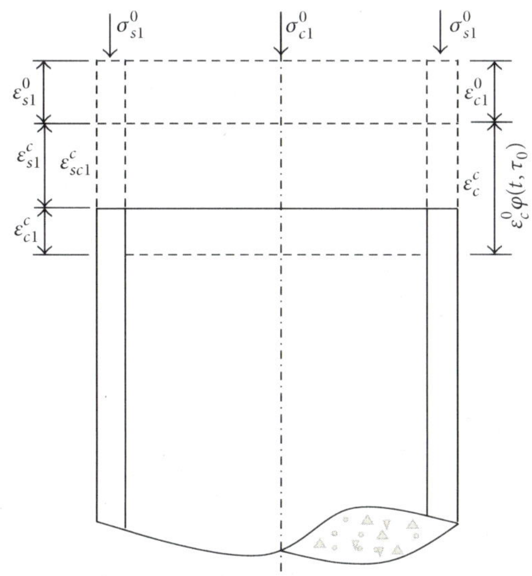

FIGURE 7: Creep strain of CFST member under axial compression.

FIGURE 8: Experimental testing devices.

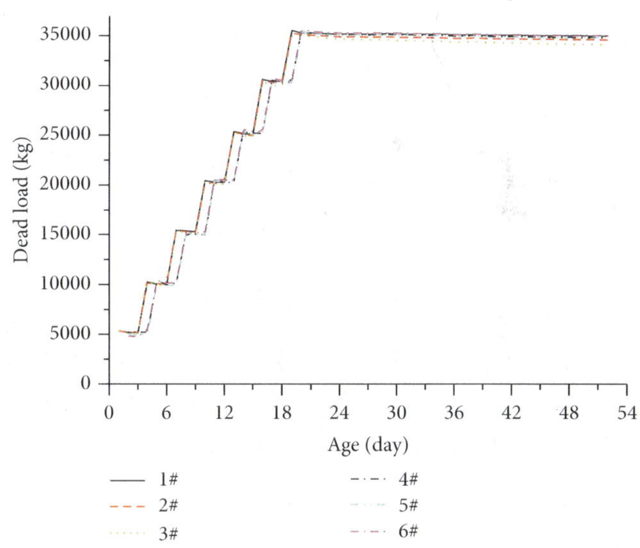

FIGURE 9: The dead load applied at related time.

where φ_{cu} is the ultimate creep of concrete; γ_{la} is the influence coefficient of loading age; γ_h is influence coefficient of average thickness or the ratio of volume and area of the component; γ_ψ is influence coefficient of the fine aggregate content; γ_s is influence coefficient of the concrete slump; γ_λ is influence coefficient of the ambient humidity; γ_a is influence coefficient of the air content.

In this study, $\rho_{sc}(t, \tau_0)$ represents the aging coefficient that introduce the effect of not only concrete age but also the steel tube confinement when the CFST column is considered as a single homogeneous material, which is distinguished by $\rho(t, \tau_0)$, the aging coefficient of concrete.

Experimental test and FEM creep analysis of the CFST column are conducted in order to determine the aging coefficient, $\rho_{sc}(t, \tau_0)$.

3.2. Experimental Test and FEM Simulation of the Creep Model during Construction. A special loading device was designed and manufactured in order to simulate the loading process during construction, as shown in Figure 8. Six columns, divided into two groups, were tested. The three columns in the first group are filled with concrete with recycled aggregates, whereas the other three columns in the second group are filled with concrete with conventional aggregates. The volume percentage of the expansive agent is 0%, 5%, and 10% for the three columns in each group, respectively. The design details of the columns are shown in Table 6.

The ultimate bearing capacity of the columns is designed as 1,100 kN, whose 40% represents the maximum dead load of 350 kN reached at the final stage of loading. The entire loading process is divided into seven stages and the load at each stage is recorded by the force sensor, as shown in Table 7 and Figure 9. An initial load of approximately 50 kN is applied to all columns of the two groups when concrete is three days

TABLE 6: Geometrical and mechanical properties of tests columns.

Number	D (mm)	T (mm)	H (mm)	f_y (MPa)	f_{ck} (MPa)	Aggregate kind	Percent of expansive
1#	133	4.5	400	345	30.4	Recycled	0
2#	133	4.5	400	345	30.4	Recycled	5
3#	133	4.5	400	345	30.4	Recycled	10
4#	133	4.5	400	345	30.4	Conventional	0
5#	133	4.5	400	345	30.4	Conventional	5
6#	133	4.5	400	345	30.4	Conventional	10

TABLE 7: Registered load at each stage (kN).

Number	1st stage	2nd stage	3rd stage	4th stage	5th stage	6th stage	7th stage
1#	53.05	102.45	154.42	204.31	253.60	306.10	355.20
2#	53.31	103.40	153.87	203.30	252.75	305.05	354.28
3#	53.54	102.70	152.94	202.90	252.58	303.37	349.58
4#	54.48	103.40	151.44	208.40	253.92	305.79	355.70
5#	52.90	103.16	150.71	204.46	253.00	304.43	354.28
6#	49.40	103.50	152.39	205.95	255.11	306.54	355.99

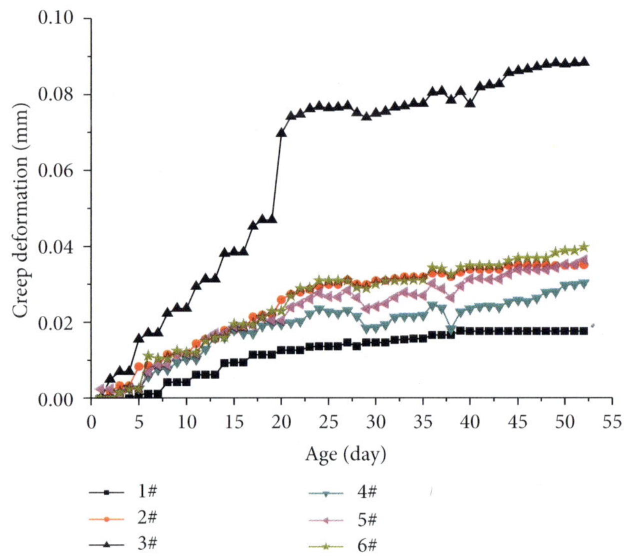

FIGURE 10: Longitudinal creep deformation curve obtained by FE model.

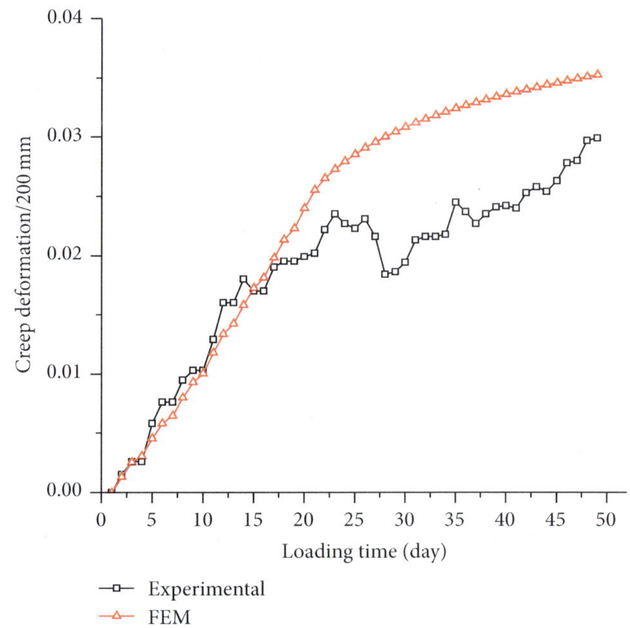

FIGURE 11: Experimental and numerical longitudinal creep deformation curves of specimen 4#.

old. Dead load increases by approximately 50 kN at each stage with a three-day interval, until it reaches 350 kN.

The creep deformation of the columns is measured by excluding the instantaneous displacements from the total displacement, as shown in Figure 10. Instantaneous displacements from the dead load are recorded immediately after such a load is applied, at each stage of the loading process. From Figure 10, the longitudinal deformation of the columns caused by the creep is notable and an increase in the expansion agent results in a larger creep deformation.

Column 4 is modeled by ABAQUS using a user-supplied subroutine of the creep effect. The creep model of ACI-209 [15], which considers the moist-cured condition, is adopted in the subroutine, where the progression method proposed by Gao et al. [22] is used to consider the change in concrete elastic modulus with age.

Creep deformation of column 4 obtained from the FE analysis is compared with the experimental results shown in Figure 11. In general, the FE solutions are consistent with the experimental test results.

After validating the FE analysis, the model is used to evaluate the aging coefficient $\rho_{sc}(t, \tau_0)$ of the columns specified in Table 8; here the composite creep coefficient $\varphi_{sc}(t, \tau_0)$ is obtained by ratio between creep and instantaneous displacements from the numerical result, shown in Figure 12.

TABLE 8: Parameters of column needed in calculation.

D (mm)	T (mm)	H (mm)	f_y (MPa)	f_{ck} (MPa)	τ_0 (day)	TL (day)	φ_{cu}
1000	12~36	4000	345	40.7	7	50	2.12

TL is the time of the load duration; φ_{cu} is the ultimate creep coefficient of concrete according to Terry et al. [8].

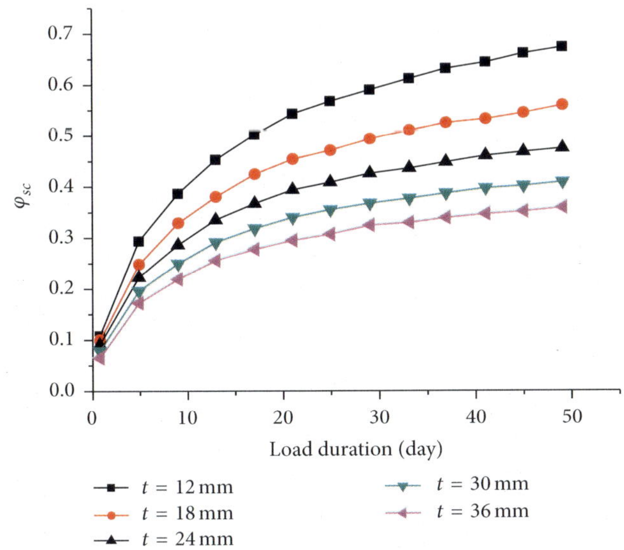

FIGURE 12: Composite creep coefficient.

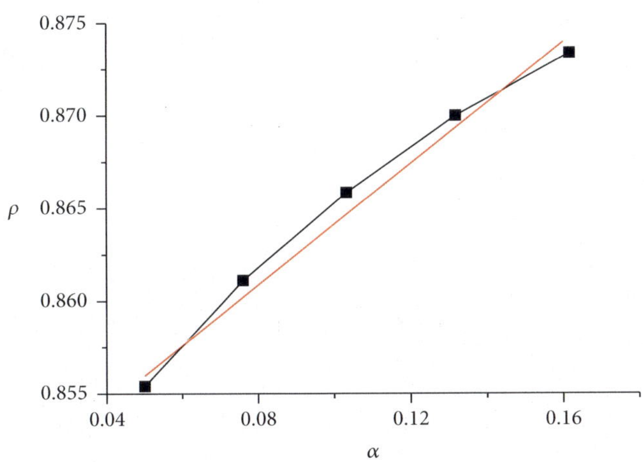

FIGURE 13: Aging coefficient of CFST column.

$\rho_{sc}(t,\tau_0)$ can be back calculated employing (18a), (18b), and (18c) with known $\varphi_{sc}(t,\tau_0)$ in Figure 12. Figure 13 shows the resulting aging coefficients, which are almost linear for the steel ratio α ranging from 0.05 to 0.20, representing the typical reinforcement ratio of practical designs. Thus, the aging coefficient within this range can be approximately computed as follows:

$$\rho_{sc}(t,\tau_0) = 0.848 + 0.16\alpha. \quad (19)$$

Then, the composite creep coefficient $\varphi_{sc}(t,\tau_0)$ of a CFST column can be calculated employing (18a), (18b), (18c), and (19).

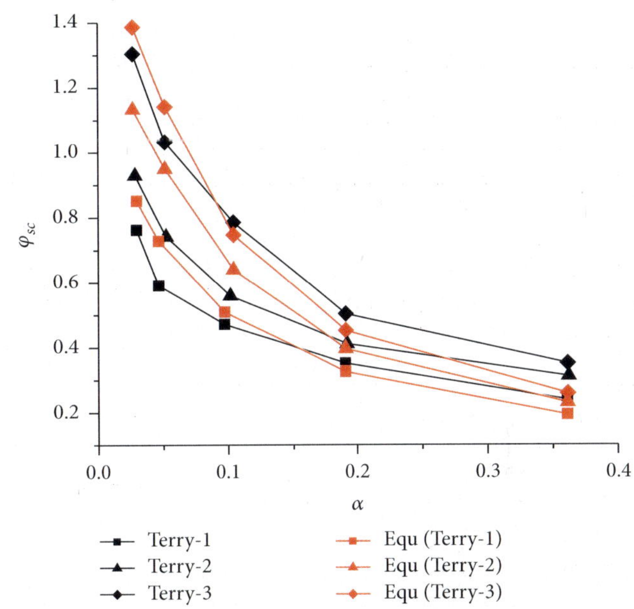

FIGURE 14: Composite creep coefficient compared with results of Terry et al. [8].

Predictions by (18a), (18b), and (18c) are compared with the test results from the studies of Terry et al. [8], in order to verify such equations, as shown in Figure 14, where Equ(Terry-1) is the equation result obtained by (18a), (18b), and (18c) related to the experimental test of Terry-1. A quite agreement can be seen, with some differences due to the different value adoptions of the ultimate creep of concrete and the loading duration.

4. Conclusion

In this study, the effect of concrete age and creep on the ultimate loading capacity of CFST columns had been experimentally and numerically investigated. The validated numerical models were employed in order to propose a set of empirical equations that can be used to predict the axial loading capacity of CFST columns with different concrete ages and composite creep coefficients.

According to the analysis of formula derivation and verification by experimental test and FE method, the ultimate bearing capacity considering the age of the CFST column can be calculated using (2). When treating the CFST column as a single material, both the unity creep and the aging coefficient, determined by employing (18a), (18b), (18c), and (19), respectively, can be used for long-term property analysis.

Equations in this study will be simplified to a certain degree on the design and time-dependent analysis of the CFST column. More studies considering the step upload will be conducted in the future.

Symbols

$E_c(t)$: The elastic modulus of concrete at time t
E_c^{28}: The elastic modulus of concrete at 28th day
E_s: The elastic modulus of the steel tube
E_{sc}: The elastic modulus of the CFST column
A_c: The area of the cross section of the concrete
A_s: The area of the cross section of the tube
α: Ratio between cross areas of steel and concrete
n: The ratio between the elastic modulus of the steel and the concrete at t time
N_{ut}: The ultimate bearing capacity of CFST column at t time (days)
N_k: The bearing capacity of hollow steel tube
N_r: The stability of the CFST column
N_{u28}: The ultimate bearing capacity of CFST column at 28th day
$E_c(\tau_0)$: The elastic modulus of concrete at loading age τ_0
$\varphi(t, \tau_0)$: The concrete creep coefficient
$\xi = \alpha f_y / f_{ck}$: The hoop effect coefficient
D: The outer diameter of the steel tube
T: The thickness of the steel tube
L: The length of the steel tube
t: The age of the concrete
e: The eccentricity of the load
λ: The slenderness of the column
$\varepsilon_c^c(t)$: The total strain of concrete when creep is with no steel confinement at age t days
$\sigma_{c1}^0(\tau_0)$: The initial stress of the concrete when loaded at τ_0 days
$E_c(\tau_0)$: The elastic modulus of concrete at τ_0 days
$\varphi(t, \tau_0)$: The creep coefficient at t when loaded at τ_0
$\sigma_{c1}(t)$: The stress of concrete at age t days
ρ: The aging coefficient of concrete
E_φ: The age-adjusted effective modulus
ε_{sc1}^c: The axial creep strain of the CFST column
ε_{s1}^c: The axial strain of steel tube due to concrete creep
ε_c^c: The creep strain of the concrete with no steel confinement
ε_{c1}^c: The reduced strain of the concrete due to the effect of the steel tube confinement
ε_{sc1}^0: The initial strain of CFST when loading
ε_{s1}^0: The initial strain of the steel tube when loaded
ε_{c1}^0: The initial strain of the concrete when loaded
σ_{c1}^c: The stress change of concrete due to creep
σ_{s1}^c: The stress change of steel tube due to creep
$\varphi_{sc}(t, \tau_0)$: The unity creep coefficient of CFST
$\rho_{sc}(t, \tau_0)$: The aging coefficient of CFST.

Competing Interests

The authors declare that they have no competing interests.

References

[1] M. D. O'Shea and R. Q. Bridge, "Design of circular thin-walled concrete filled steel tubes," *Journal of Structural Engineering*, vol. 126, no. 11, pp. 1295–1303, 2000.

[2] G. Giakoumelis and D. Lam, "Axial capacity of circular concrete-filled tube columns," *Journal of Constructional Steel Research*, vol. 60, no. 7, pp. 1049–1068, 2004.

[3] S. T. Zhong, *The Concrete-Filled Steel Tubular Structures*, Tsinghua University Press, 3rd edition, 2005 (Chinese).

[4] X. X. Zha, *Hollow and Solid Concrete-Filled Steel Tube Columns Structure*, Science Press, 2011 (Chinese).

[5] L.-H. Han, W. Li, and R. Bjorhovde, "Developments and advanced applications of concrete-filled steel tubular (CFST) structures: members," *Journal of Constructional Steel Research*, vol. 100, pp. 211–228, 2014.

[6] Y. F. Wang, *Creep of the Concrete-Filled Steel Tubular Structures*, Science Press, 2006.

[7] S. J. Tan and J. L. Qi, "Experimental investigation of the effect on strenth of concrete filled steel tubular compressive members under long-time load," *Journal of Harbin Engineering University*, no. 2, pp. 10–24, 1987 (Chinese).

[8] P. J. Terry, M. A. Bradford, and R. I. Gilbert, "Creep and shrinkage in concrete filled steel tubes," in *Proceedings of the 6th International Symposium in Tubular Structures Melbourne*, pp. 293–298, Melbourne, Australia, 1994.

[9] K. Luo, Y.-L. Pi, W. Gao, M. A. Bradford, and D. Hui, "Investigation into long-term behaviour and stability of concrete-filled steel tubular arches," *Journal of Constructional Steel Research*, vol. 104, pp. 127–136, 2015.

[10] L. H. Ichinose, E. Watanabe, and H. Nakai, "An experimental study on creep of concrete filled steel pipes," *Journal of Constructional Steel Research*, vol. 57, no. 4, pp. 453–466, 2001.

[11] W. Naguib and A. Mirmiran, "Creep modeling for concrete-filled steel tubes," *Journal of Constructional Steel Research*, vol. 59, no. 11, pp. 1327–1344, 2003.

[12] H. Liu, Y. Wang, M. He, Y. Shi, and H. Waisman, "Strength and ductility performance of concrete-filled steel tubular columns after long-term service loading," *Engineering Structures*, vol. 100, pp. 308–325, 2015.

[13] K. M. Shrestha and B. C. Chen, "Aging coefficient, creep coefficient and extrapolating aging coefficient from short term test for sealed concrete," *Journal Wuhan University of Technology, Materials Science Edition*, vol. 26, no. 1, pp. 154–159, 2011.

[14] S.-T. Yi, J.-K. Kim, and T.-K. Oh, "Effect of strength and age on the stress-strain curves of concrete specimens," *Cement and Concrete Research*, vol. 33, no. 8, pp. 1235–1244, 2003.

[15] ACI, "Prediction of creep, shrinkage, and temperature effects in concrete structures," ACI 209, ACI Special Publication, 1992.

[16] P. Shokouhi, A. Zoëga, and H. Wiggenhauser, "Nondestructive investigation of stress-induced damage in concrete," *Advances in Civil Engineering*, vol. 2010, Article ID 740189, 9 pages, 2010.

[17] J. Newman, "Strength and failure of concrete under short-term, cyclic and sustained loading," in *Advanced Concrete Technology—Concrete Properties*, J. Newman and B. S. Choo, Eds., Elsevier, 2003.

[18] G. T. Liu, H. Gao, and F. Q. Chen, "Microstudy on creep of concrete at early age under biaxial compression," *Cement and Concrete Research*, vol. 32, no. 12, pp. 1865–1870, 2002.

[19] M. M. Attard and S. Setunge, "Stress-strain relationship of confined and unconfined concrete," *ACI Materials Journal*, vol. 93, no. 5, pp. 432–442, 1996.

[20] I. Imran and S. J. Pantazopoulou, "Experimental study of plain concrete under triaxial stress," *ACI Materials Journal*, vol. 93, no. 6, pp. 589–601, 1996.

[21] A. Kwan, C. Dong, and J. Ho, "Axial and lateral stress–strain model for concrete-filled steel tubes," *Journal of Constructional Steel Research*, vol. 122, pp. 421–433, 2016.

[22] Z. G. Gao, D. G. Huang, and G. F. Zhao, "A method for creep stess analysis of concrete structures," *China Civil Engineering Journal*, vol. 34, no. 4, pp. 10–14, 2001 (Chinese).

Estimation of Oxidation Kinetics and Oxide Scale Void Position of Ferritic-Martensitic Steels in Supercritical Water

Li Sun and Weiping Yan

School of Energy Power and Mechanical Engineering, North China Electric Power University, Yong Hua Bei No. 619, Lian Chi District, Baoding 071003, China

Correspondence should be addressed to Li Sun; sunli1001315@163.com

Academic Editor: Pavel Lejcek

Exfoliation of oxide scales from high-temperature heating surfaces of power boilers threatened the safety of supercritical power generating units. According to available space model, the oxidation kinetics of two ferritic-martensitic steels are developed to predict in supercritical water at 400°C, 500°C, and 600°C. The iron diffusion coefficients in magnetite and Fe-Cr spinel are extrapolated from studies of Backhaus and Töpfer. According to Fe-Cr-O ternary phase diagram, oxygen partial pressure at the steel/Fe-Cr spinel oxide interface is determined. The oxygen partial pressure at the magnetite/supercritical water interface meets the equivalent oxygen partial pressure when system equilibrium has been attained. The relative error between calculated values and experimental values is analyzed and the reasons of error are suggested. The research results show that the results of simulation at 600°C are approximately close to experimental results. The iron diffusion coefficient is discontinuous in the duplex scale of two ferritic-martensitic steels. The simulation results of thicknesses of the oxide scale on tubes (T91) of final superheater of a 600 MW supercritical boiler are compared with field measurement data and calculation results by Adrian's method. The calculated void positions of oxide scales are in good agreement with a cross-sectional SEM image of the oxide layers.

1. Introduction

According to the characteristic of supercritical water [1], supercritical and ultrasupercritical boiler power plants have been constructed for a long time in order to improve efficiency of units in some countries. Under the supercritical water (the critical point of water, 374°C, 22.1 MPa) environment, the corrosion resistance of candidate materials is one of the key requirements for these materials safe application in units. Two ferritic-martensitic steels of candidate structural materials for superheaters and reheaters, such as HCM12A and NF616, have been attracted increasing attention by researchers, because they have good high-temperature strength, high thermal conductivity, creep resistance, and low thermal expansion coefficients [2]. Exfoliation of oxide scales from high-temperature heating surfaces of power boilers threatened the safety of supercritical power generating units, and prediction of oxidation kinetics and void formation of oxide scales are desired.

Generally, good corrosion resistance of the steels with higher Cr contents is presented compared to the steels with lower Cr contents [3, 4]. In order to understand the oxidation mechanism of candidate materials under supercritical water environment, the oxidation tests of ferritic-martensitic steels are performed and the effects of SCW temperature and dissolved oxygen on the oxidation rate are analyzed. The diffusion processes in oxide in terms of diffusion species and mechanisms are determined and the structure of the various oxide phases and their formation are understood [5]. In situ electrical and electrochemical measurements during oxidation of ferritic steel P91 and austenitic steel AISI 316L (NG) in ultrasupercritical water (500–700°C, 30 MPa) have been carried out. The surface and in-depth composition and structure of the oxides formed on ferritic and austenitic steels in ultrasupercritical water are broadly analogous to those obtained by wet air oxidation in the same temperature range [6]. Angell et al. [7] studied the effect of pressure on the steam oxidation of 9Cr-1Mo steels at 500–600°C and the oxidation of the steels followed a parabolic kinetics with the oxide scale thickness increasing with the increasing pressure. The corrosion of the ferritic-martensitic steel P92 exposed to supercritical water at 550°C under 25 MPa with the dissolved oxygen contents of 100, 300, and 2000 ppb was investigated. The results indicated that the weight gain increased with the dissolved oxygen content [8]. The oxidation behavior of ferritic/martensitic steel HCM12A exposed in SCW at 500°C, 25 MPa, and two levels of oxygen partial pressure were investigated. Oxygen partial pressure effects on oxide structure and porosity in oxide scale are rationalized in terms of thermodynamics and diffusion mechanism. A desirable continuous protective oxide layer structure with limited pores may be obtained by controlling the oxygen partial pressure [9]. Various growth mechanisms of the oxide scale of 9–12% Cr F-M steels exposed to SCW were proposed by Ampornrat and Was [5], Bischoff and Motta [10], and Zhong et al. [11]. The oxide films on F-M steels followed parabolic oxidation kinetics under supercritical water. Microstructure of oxide formed on four F-M steels consisting of two oxide layers and a transition layer. The outer oxide consisted of dense columnar grains of magnetite (Fe_3O_4). The inner oxide consisted of small equiaxed grains of Fe-Cr spinel oxide $Fe_{3-x}Cr_x\ O_4$, where x (0.7–1) depends on alloy type. The transition layer consisted of grain boundary oxides of chromia and chromite and fine oxide grains of a spinel structure precipitated inside laths [5]. Martinelli et al. [12] focused on the growth kinetics simulation of a duplex oxide scale of a Fe-9Cr-1Mo martensitic steel (T91) in static liquid Pb-Bi. The magnetite layer growth is controlled by iron diffusion inside the Fe-Cr spinel and the magnetite lattice. An analytical simulation is matched with the experimental kinetics. Rouillard and Martinelli [13] have suggested that 9Cr-1Mo steel forms in CO_2 at 550°C a duplex oxide layer containing an outer magnetite scale and an inner Fe-Cr rich spinel scale. The inner spinel oxide layer is formed according to a void-induced oxidation mechanism. The kinetics of the total oxide growth are simulated from the proposed oxidation model.

To sum up, the research of oxidation kinetics of two ferritic-martensitic steels is focused on experiments; simulation about oxidation kinetics is relatively few. In the present study, the growth kinetics of oxide films of ferritic-martensitic steel under supercritical water are simulated. Considering the presence of Fe-Cr rich spinel oxide layer, the oxidation rates of magnetite layer and Fe-Cr rich spinel oxide layer are calculated simultaneously. With thermodynamic analysis of reaction of materials and calculation of oxygen activity at the interface, the growth kinetics of the inner layer and the outer layer are simulated. The results of simulation of oxidation of two ferritic-martensitic steels under supercritical water are compared to the experimental data in Tan et al. [14]. An example of one power plant is given. The calculated thickness and quantitative location of voids formation of oxide scale are compared with the experimental results.

2. Assumptions and Formulas of Simulation

Oxidation of different Fe-Cr steels under steam, CO_2, and LBE leads to the formation of a duplex oxide scale, and the characteristic of oxide scales is similar. Microstructure of oxide formed on the two steels consisted of two oxide layers, which are a Fe-Cr spinel oxide and a magnetite layer. The oxide thickness ratio is close to 1, and the oxidation rate constant of the two layers follows approximately the parabolic law. The "available space model" is proposed to explain the oxidation mechanism of the duplex oxide scale [15]. In the "available space" model, the thickness ratio between the oxide layers formed is constant with exposure time, because the growth mechanism of both oxide layers is linked. Furthermore, the kinetics of oxidation and especially the growth of the inner layer are not limited by the diffusion of oxygen, which reaches the oxide-metal interface through short-circuits such as microchannels. In general, the original oxide-metal interface corresponds to the outer-inner layer interface during the oxidation of alloys under supercritical water. The outer layer is formed by outward migration of iron while the inner layer is formed by inwards migration of the oxidant. Consequently, the limiting step in this mechanism is the diffusion of iron, which enables the formation of the available space for the oxide to grow in. So the growth of oxide scales is dependant on the diffusion of iron in the double layers.

In order to simulate the oxide growth on ferritic-martensitic steels surface during exposure to supercritical water, the following assumptions are made:

① The oxide scale formed on ferritic-martensitic steels in supercritical water contains a dual structure. The outer layer, in contact with supercritical water, is a magnetite layer which grows at the magnetite/supercritical water interface involving an iron diffusion from the steel to this external interface. The inner layer, in contact with the steel, consisted of Fe-Cr spinel layer involving a transport of oxygen from supercritical water to the steel/oxide interface. The iron flux across the Fe-Cr spinel is equal to the iron flux across the magnetite layer. A steady state is reached at the interface between the two oxide layers.

② Thermodynamic equilibrium exists at both the steel/oxide and oxide/supercritical water interfaces.

③ The composition concentrations are constant at the interfaces.

④ The growth of oxide scales follows parabolic laws.

⑤ The oxygen activity is constant at interfaces.

⑥ The Fe-Cr spinel oxide slightly deviates from the stoichiometry.

⑦ The oxide scale is compact and adherent and voids and cracks in the oxide may be neglected.

⑧ The dissociation of oxide scales is neglected

Generally, the thickness of consumption of ferritic-martensitic steel is equal to the thickness of the Fe-Cr spinel layer. The relation between the thickness of Fe-Cr spinel layer and that of the magnetite layer can be described as follows [12]:

$$h_{\text{spinel}} = \frac{C_{\text{Fe(magnetite)}}}{C_{\text{Fe(steel)}} - C_{\text{Fe(spinel)}}} h_{\text{magnetite}}, \quad (1)$$

where h_{spinel} and $h_{\text{magnetite}}$ are the thicknesses of the Fe-Cr spinel layer and the magnetite layer (μm). $C_{\text{Fe(magnetite)}}$, $C_{\text{Fe(spinel)}}$, and $C_{\text{Fe(steel)}}$ are the concentrations of Fe in magnetite, spinel oxide, and steel substrate (mol/cm^3), respectively.

For a given time, the formula of the thickness of oxide scale is as follows:

$$h^2 = k_p t, \quad (2)$$

where h is the thickness of the entire oxide or the thickness of the spinel layer or the thickness of the magnetite layer and k_p is the oxidation rate constant (cm^2/s), respectively. t is time (s).

According to Wagner's theory of oxidation and considering that magnetite is formed by outward migration of Fe ions, the oxidation rate constant of oxide obeys the law as proposed by Atkinson et al. [16]:

$$k_p = \frac{4}{3} \int_{a'_{O_2}}^{a''_{O_2}} f^{-1} D^* d \ln a_{O_2}, \quad (3)$$

where a'_{O_2} and a''_{O_2} are the oxygen activities at oxide/environment interface and steel/oxide interface and f is the correlation factor for the diffusion mechanism and is approximately equal to 0.5. D^* is the tracer diffusion coefficient of Fe in oxide. $a'_{O_2} = P'_{O_2}/P^0_{O_2}$, P'_{O_2} is the oxygen partial pressure (atm) at the interface, $P^0_{O_2} = 1$ atm.

The diffusion coefficient of Fe in oxide is very dependent on different types of defects that are present in the oxide lattice. The concentration of defects is linked to the oxygen activity in oxide scales and the diffusion coefficient of Fe ions is dependent on the oxygen activity in two oxide layers. The diffusion coefficient of Fe ions in magnetite can be expressed by oxygen activity and temperature, and the equation can be written [17]:

$$D_{\text{Fe,mag}} = \frac{A a_{O_2}^{2/3}}{\left(1 + 2K_V a_{O_2}^{2/3}\right)} + B a_{O_2}^{-2/3}, \quad (4)$$

where $A = D_V K_V/12$, $B = 4D_I K_I/3$, D_V and D_I are self-diffusion coefficients of cation vacancies and iron interstitials, and K_V and K_I are the equilibrium constants for the formation of cation vacancies and iron interstitials in magnetite. The first term and the second term as a function of temperature and oxygen activity represent the iron diffusion via vacancies and interstitials. The diffusion of Fe in the outer oxide layer at high-oxygen activities is governed by cation vacancies and cation interstitials at low-oxygen activities.

The diffusion coefficient of Fe in Fe-Cr spinel oxide layer can be expressed by the following relation [13]:

$$D_{\text{Fe,spinel}} = R_V \frac{A a_{O_2}^{2/3}}{\left(1 + 2K_V a_{O_2}^{2/3}\right)} + R_I B a_{O_2}^{-2/3}$$

$$R_V = \frac{D^{\text{topfer}}_{\text{spinel},V}}{D^{\text{topfer}}_{\text{mag},V}} \quad (5)$$

$$R_I = \frac{D^{\text{topfer}}_{\text{spinel},I}}{D^{\text{topfer}}_{\text{mag},I}},$$

where $D^{\text{topfer}}_{\text{mag},V}$ and $D^{\text{topfer}}_{\text{mag},I}$ are the partial cation diffusion coefficients of the cation diffusion via vacancies (V) and interstitials (I) in Fe_3O_4. $D^{\text{topfer}}_{\text{spinel},V}$ and $D^{\text{topfer}}_{\text{spinel},I}$ are the partial cation diffusion coefficients of the cation diffusion via vacancies (V) and interstitials (I) in Fe-Cr spinel oxide layer. R_V and R_I are the ratios of D_V and D_I of Fe-Cr spinel oxide to that of Fe_3O_4 by Töpfer et al.'s data [18].

According to assumption ①, the iron flux across the Fe-Cr spinel is equal to the iron flux across the magnetite layer. The oxygen activity at the interface of ferritic-martensitic steel/Fe-Cr spinel layer is determined from the oxygen partial pressure of the phase boundary at different temperatures. The oxygen activity at the interface of magnetite/supercritical water is calculated by the thermodynamics equilibrium between Fe_3O_4 and Fe_2O_3. The oxygen activity at the interface of magnetite/Fe-Cr spinel is unknown. According to (1)–(5), the oxidation rates and thickness of magnetite layer and Fe-Cr spinel layer and the oxygen activity at the interface of magnetite/Fe-Cr spinel can be evaluated.

3. Determination of the Parameters

Two types of ferritic-martensitic steels HCM12A and NF616 were evaluated in this study. The oxidation rates of these steels are calculated in supercritical water at 400°C, 500°C, and 600°C under 25 MPa. A summary of steel composition in weight percent is presented in Table 1 [14].

The densities of F-M steels and magnetite are 7.77 mg/cm^3 and 5.17 mg/cm^3, respectively. The mole volume of Fe-Cr spinel oxide approximately equals that of magnetite because the lattice parameter of Fe-Cr spinel oxide is close to that of magnetite. According to chemical compositions of the ferritic-martensitic steels HCM12A and NF616 and the densities of scales, the concentrations of Fe in magnetite, spinel oxide, and steel substrate, $C_{\text{Fe(magnetite)}}$, $C_{\text{Fe(steel)}}$, and $C_{\text{Fe(spinel)}}$, are approximately evaluated. The values of $C_{\text{Fe(magnetite)}}/(C_{\text{Fe(steel)}} - C_{\text{Fe(spinel)}})$ for NF616 and HCM12A are 0.985 and 0.932, respectively.

3.1. Determination of Iron Diffusion Coefficient in Magnetite and Fe-Cr Spinel Oxide. The iron diffusion coefficient in magnetite at lower temperatures can be extrapolated from the data of the iron diffusion coefficient for temperatures between 900 and 1400°C, which is given by Backhaus-Ricoult and

TABLE 1: Chemical composition of HCM12A and NF616 (in wt%).

Steel	C	Mn	P	S	Si	Ni	Cr	Mo	Cu	N	Nb	Al	V	W
HCM12A	0.11	0.64	0.016	0.002	0.27	0.39	10.83	0.3	1.02	0.063	0.054	0.001	0.19	1.89
NF616	0.109	0.45			0.102	0.174	8.82	0.468			0.064		0.194	1.87

Dieckmann [17]. For the iron diffusion coefficient in spinel oxide, the data at 1200°C given by Töfpfer et al. [18] is used and extrapolated towards lower temperature. Assuming the same temperature dependence for iron diffusion in magnetite and in Fe-Cr spinel, the ratio between iron diffusion coefficient in magnetite and in spinel oxide is a function of oxygen activity.

The mean values of the ratio between iron diffusion coefficient via vacancies or interstitials in spinel oxide and in magnetite for different experimental points at 1200°C are calculated, namely, R_V and R_I. For NF616, composition of inner oxide scale is $Fe_{2.4}Cr_{0.6}O_4$ [19], R_V equals 0.644, and R_I equals 0.31. For HCM12A, composition of inner oxide scale is Fe_2CrO_4 [19], R_V equals 0.5, and R_I equals 0.039.

3.2. Estimation of the Oxygen Activities of Magnetite/Supercritical Water Interface and Steel Substrate/Fe-Cr Spinel Interface. Formation of oxide phase depends on the oxygen potential and local chemistry, which can be described by the diffusion path in Fe-Cr-O phase diagrams [19]. The oxygen partial pressure at this interface could be determined from the phase boundary in the predominance diagram. The oxygen partial pressures at steel substrate/Fe-Cr spinel interface are approximately 3×10^{-35} atm, 1×10^{-29} atm, and 2×10^{-25} atm for 400°C, 500°C, and 600°C, respectively.

The real oxygen concentration at the magnetite/atmosphere interface is not the oxygen concentration in air or water. Zurek et al. [20], who studied oxidation of 10% Cr ferritic steel in Ar-H_2O mixture, showed that the oxide scale thickness does not depend on the H_2O content in the gas mixture (Ar-4% H_2O or Ar-50% H_2O) if the oxygen activity is sufficient to allow the thin hematite layer formation. The estimated oxygen partial pressure using Henry's constant of SCW is lower than that of steam. The values also are still quite high compared to a dissociation P_{O_2} of surface oxide phase (magnetite) observed. It can be assumed that the dissolved oxygen was totally removed from water under deaerated condition, and oxygen was miscible under SCW. Therefore, the oxygen partial pressure at specimen surface was very low and resulted from dissociation of water molecules only [19]. The oxidation rate of oxide scale is related to oxygen activity at magnetite/supercritical water. Furukawa et al. [21] suggested that the oxide growth kinetics do not depend on the nature of the oxidizing medium but just on the oxygen activity in the oxidizing medium.

Fe transports through the spinel oxide layer and reacts with H_2O; the Fe_3O_4 layer becomes thick gradually as exposure time increases. At magnetite/supercritical water interface, Fe_2O_3 is formed through reaction (6) at the grain boundary of magnetite. With outward transport of Fe and inward transport of oxidation species along the oxide grain boundaries and short-circuit paths, both the Fe_3O_4 layer and Fe-Cr spinel layer became thicker. The real oxygen partial pressure at magnetite/supercritical water interface is not dissolved oxygen partial pressure, but the equivalent oxygen pressure which corresponds to the thermodynamic equilibrium of the following.

$$2Fe_3O_4 + H_2O \longrightarrow 3Fe_2O_3 + H_2 \quad (6)$$

Using Gaskell's data [22], the standard free energy for the formation of reaction (6) is given by

$$\Delta G_T = -3440 + 81.7T. \quad (7)$$

At 600°C, $\Delta G_T = 67884$ J, and the equilibrium P_{H_2O}/P_{H_2} ratio is 11532. According to dissociation equation, Gibbs free energy, and water dissociation reaction rate, the equivalent oxygen pressure at magnetite/supercritical water interface can be evaluated. The equivalent oxygen partial pressure magnetite/supercritical water interface is as follows at 400°C, 500°C, and 600°C in Table 2.

TABLE 2: Thermodynamic data of oxidation reaction (6) at different temperatures.

Temperature (°C)	ΔG_T (J)	P_{H_2O}/P_{H_2} ratio	Equivalent oxygen partial pressure P_{O_2} (atm)
400	51544	10016	3.46×10^{-25}
500	59714	10846	3.5×10^{-20}
600	67884	11532	2.57×10^{-16}

3.3. Relation between Weight Gain and Thickness of Inner and Outer Layers of Oxide Scale. The weight gain of oxide scales is related to oxide thickness as a linear function because of oxygen absorption during exposure. The density of oxide scales is nearly constant, and the thickness of inner and outer layers of oxide scales is becoming thicker with exposure time increasing. The relation between weight gain and thickness of inner and outer layers of oxide scale was estimated by the following equation [8]:

$$\Delta w = \rho_{\text{outer}} h_{\text{outer}} \left(\frac{4M_o}{M_{Fe_3O_4}} \right) + \rho_{\text{inner}} h_{\text{inner}} \left(\frac{4M_o}{M_{Fe_3O_4\text{-}FeCr_2O_4}} \right), \quad (8)$$

where Δw is the amount of absorbed oxygen per unit area in mg/cm^2, ρ_{outer} and ρ_{inner} are the outer and inner layer density (mg/cm^3), respectively, h_{outer} and h_{inner} are the outer and inner layer thickness (cm), respectively, $4M_o$ are the mole mass of oxygen in oxide layer, and $M_{Fe_3O_4}$ and $M_{Fe_3O_4\text{-}FeCr_2O_4}$ are the mole mass of outer and inner layer, respectively.

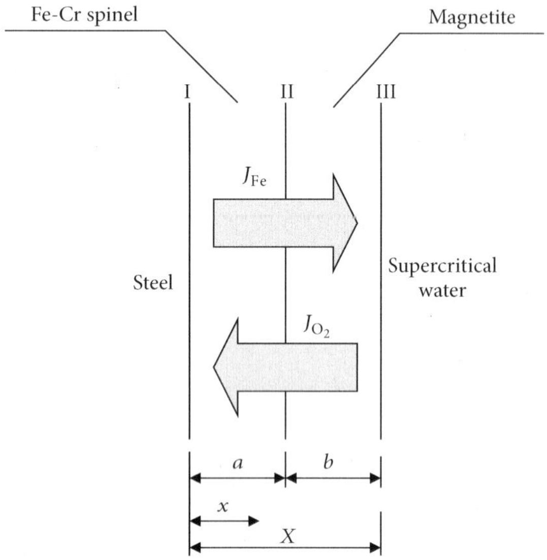

FIGURE 1: Schematic illustration of diffusion of oxygen and iron in the duplex scale in the magnetite and Fe-Cr spinel formation process.

3.4. The Description of the Oxygen Activity, Iron Diffusion Coefficient, the Flux of Oxide Ion, and Its Divergence in Oxide as a Function of the Location (x/X) in the Duplex Scale. Figure 1 shows a schematic illustration of diffusion of oxygen and iron in the duplex scale formed on ferritic-martensitic steel. The iron migrates outward during the growth of the duplex scale, while the oxidant migrates inwards.

The flux of iron in the Fe-Cr spinel layer formed on NF616 is described as [23]

$$J_{Fe} = \frac{2C_{Fe(spinel)}D_{Fe,spinel}}{2.4}\frac{d\ln a_{O_2}}{dx}. \quad (9)$$

The integration is performed over the entire Fe-Cr spinel layer.

$$J_{Fe}a = \frac{2C_{Fe(spinel)}}{2.4}\int_{a_{O_2}(I)}^{a_{O_2}(II)} D_{Fe,spinel}d\ln a_{O_2}. \quad (10)$$

Equation (9) is integrated from $x = 0$ to $x = x$:

$$J_{Fe}x = \frac{2C_{Fe(spinel)}}{2.4}\int_{a_{O_2}(I)}^{a_{O_2}} D_{Fe,spinel}d\ln a_{O_2}. \quad (11)$$

The values of a, b, and X are the thickness of Fe-Cr spinel layer, magnetite layer, and entire duplex scale layer, respectively; that is, $a + b = X$. The relationship between the oxygen activity and the location (x/X) in the Fe-Cr spinel layer can be derived as follows:

$$\frac{x}{X} = \frac{\int_{a_{O_2}(I)}^{a_{O_2}} D_{Fe,spinel}d\ln a_{O_2}}{\int_{a_{O_2}(I)}^{a_{O_2}(II)} D_{Fe,spinel}d\ln a_{O_2}}\frac{a}{a+b} \quad 0 \le x \le a. \quad (12)$$

According to assumption ①, the iron flux across the Fe-Cr spinel is equal to the iron flux across the magnetite layer. With the above method, the relationship between the oxygen activity and the location (x/X) in the magnetite layer can be expressed as

$$\frac{x}{X} = \frac{\int_{a_{O_2}(II)}^{a_{O_2}} D_{Fe,mag}d\ln a_{O_2}}{\int_{a_{O_2}(II)}^{a_{O_2}(III)} D_{Fe,mag}d\ln a_{O_2}}\frac{b}{a+b} + \frac{a}{a+b} \quad (13)$$

$$a \le x \le b.$$

3.5. Process of the Calculation. The oxygen partial pressure at the magnetite/SCW interface could be determined from Fe-Cr-O phase diagrams [19]. The oxygen partial pressure at the steel/Fe-Cr spinel interface can be calculated by the thermodynamical equilibrium of reaction (6) from Section 3.2.

The oxygen activities (i.e., $a_{O_2}(I)$ and $a_{O_2}(III)$) at the magnetite/SCW and the steel/Fe-Cr spinel interfaces are known. The only unknown variable is thus the oxygen activity ($a_{O_2}(II)$) at the Fe-Cr spinel/magnetite interface. With (1)–(5), the oxygen activity ($a_{O_2}(II)$) is quantified. Then the magnetite layer oxidation rate, k_{mag}, and the Fe-Cr spinel layer oxidation rate, k_{spinel}, can be evaluated. And the thickness (h_{mag}) of the magnetite layer and the thickness (h_{spinel}) of Fe-Cr spinel layer can be calculated also. The weight gain of oxide scale can be estimated by thickness of inner and outer layers of oxide scale with (7).

4. Results and Discussion

4.1. Iron Diffusion Coefficient in Magnetite and Fe-Cr Spinel Oxide. According to (4) and (5), the iron diffusion coefficient in magnetite and Fe-Cr spinel oxide is evaluated at different temperatures. The relation between oxygen activity and iron diffusion coefficient in the oxides for HCM12A and NF616 is expressed as a function of logarithm of diffusion coefficients and logarithm of oxygen activity on Figure 2. The different kinds of diffusion regimes of iron diffusion coefficient in magnetite, vacancy, and interstitial are also specified on Figure 2.

The oxide growth kinetics depend on the oxygen activity in the interfaces. The self-iron diffusion has a strong dependence on the oxygen activity in the oxide. The value of the oxygen activity fixes the iron diffusion mechanism, which can follow either a vacancy mechanism for high-oxygen activities or an interstitial mechanism for low-oxygen activities.

In fact, the oxygen activity is low and the iron diffuses through the interstitial mechanism at the steel/oxide interface. The iron diffusion coefficient jumps from the Fe-Cr spinel to the magnetite one at the Fe-Cr spinel/magnetite interface. The diffusion mechanism at this site is still not predictable because the oxygen activity is needed to be calculated. The oxygen activity is high and the iron diffuses through the vacancy mechanism at the magnetite/supercritical interface.

4.2. Comparison between Simulation and Experiments of Oxide Weight Gain and Error Analysis. Using (1)–(5), (8), the simulation is performed for HAM12A and NF616 exposed to

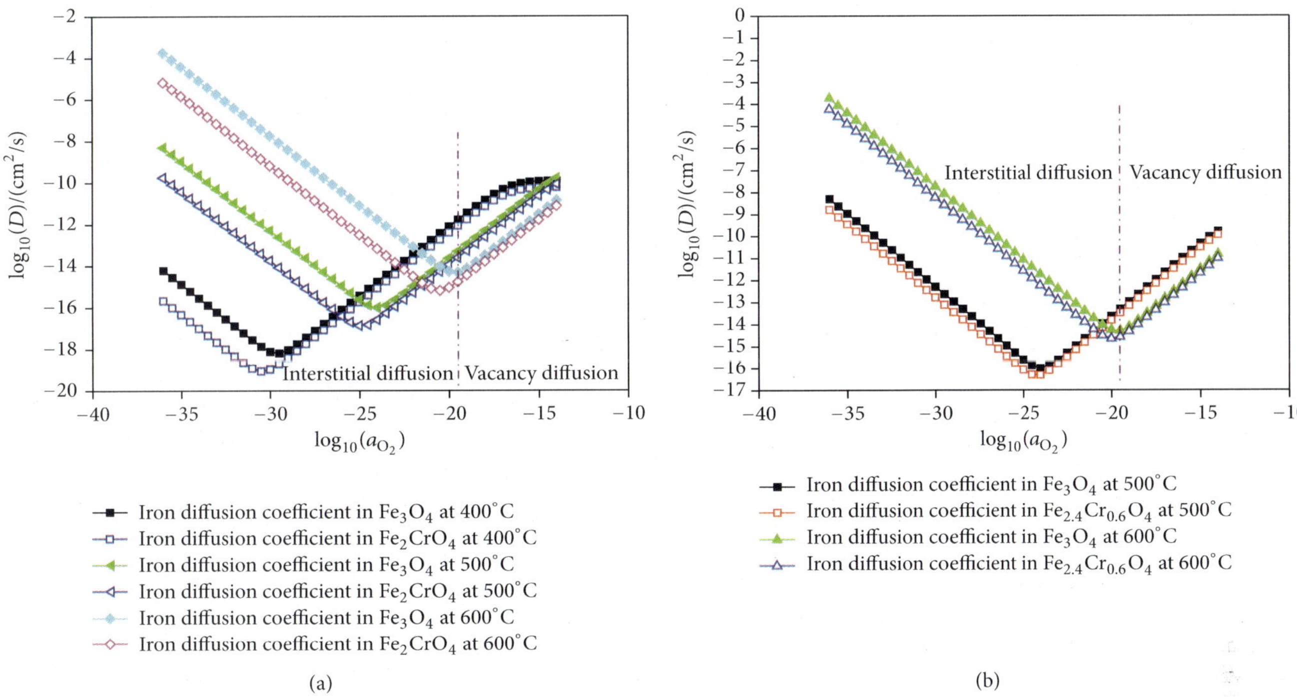

FIGURE 2: Iron diffusion coefficient of the duplex scale formed on HCM12A (a) and NF616 (b) as a function of oxygen activity in supercritical water at different temperatures.

supercritical water at 400°C, 500°C, and 600°C under 25 MPa. The comparison between simulation and experiments of oxide weight gain are presented in Figure 3.

In [14], tests are performed on NF616 exposed to a SCW at 400, 500, and 600°C and 25 MPa with an oxygen concentration of less than 25 ppb in weight for time periods from 1 to 3000 h.

For HCM12A, the simulation of oxidation weight gain is lower than experimental value at 500°C and 600°C. The maximum difference between calculation and experiment of weight gain at 600°C is 104 mg/dm^2 and is equivalent to thickness of about 7.5 μm according to formula (8). The maximum difference between calculation and experiment of weight gain at 500°C is 181 mg/dm^2 and is equivalent to thickness of about 13 μm. The maximum difference between calculation and experiment of weight gain at 400°C is 2 mg/dm^2 and is equivalent to thickness of about 1.57 μm.

For NF616, the simulation of oxidation weight gain is higher than the experimental values at 600°C. The maximum difference between calculation and experiment of weight gain at 600°C for 3000 hours is 306 mg/dm^2 and is equivalent to thickness of about 22 μm. The reason is that the hematite phase was not detected in outer oxide layer [8], and the selective thermodynamic equilibrium oxygen partial pressure of reaction (6) is larger than actual pressure during calculation. The simulation of oxidation weight gain is lower than the experimental values at 500°C. The maximum difference between calculation and experiment of weight gain for 3000-hour exposure is 76 mg/dm^2 and is equivalent to thickness of about 5.4 μm.

For 3000 hours of exposure, relative error of simulated weight gain by oxidation of HCM12A and NF616 in supercritical water at different temperatures compared to the experiments is shown in Figure 4.

The relative error of oxide weight gain between simulation and experiments on NF616 at 500°C and 600°C and HCM12A at 600°C is less than 20%, and the calculated values are close to experimental values. The relative error of oxide weight gain between simulation and experiments on HCM12A at 400°C and 500°C is above 40%.

4.3. Comparison and Analysis of Oxidation Rates. In [24], the temperature dependence of the parabolic rate constant is given by

$$k_{\text{EPRI}} = 2 \cdot A \cdot \exp\left(-\frac{Q}{RT}\right), \tag{14}$$

where Q is the activation energy for the rate-controlling process, T is the metal temperature (absolute), A is the Arrhenius constant, and R is the Universal Gas Constant. The oxidation parameters used for alloy NF616 are $Q = 5.13 \times 10^{10}\ \mu\text{m}^2/\text{hr}$ and $A = 169$ kJ/mole. The oxidation parameters used for alloy HCM12A are $Q = 5.6 \times 10^4\ \mu\text{m}^2/\text{hr}$ and $A = 75$ kJ/mole.

According to the above equations, the oxidation rate of inner layer and outer layer and the global oxidation rate which is compared to the data in [24] are evaluated. These values are presented in Table 3.

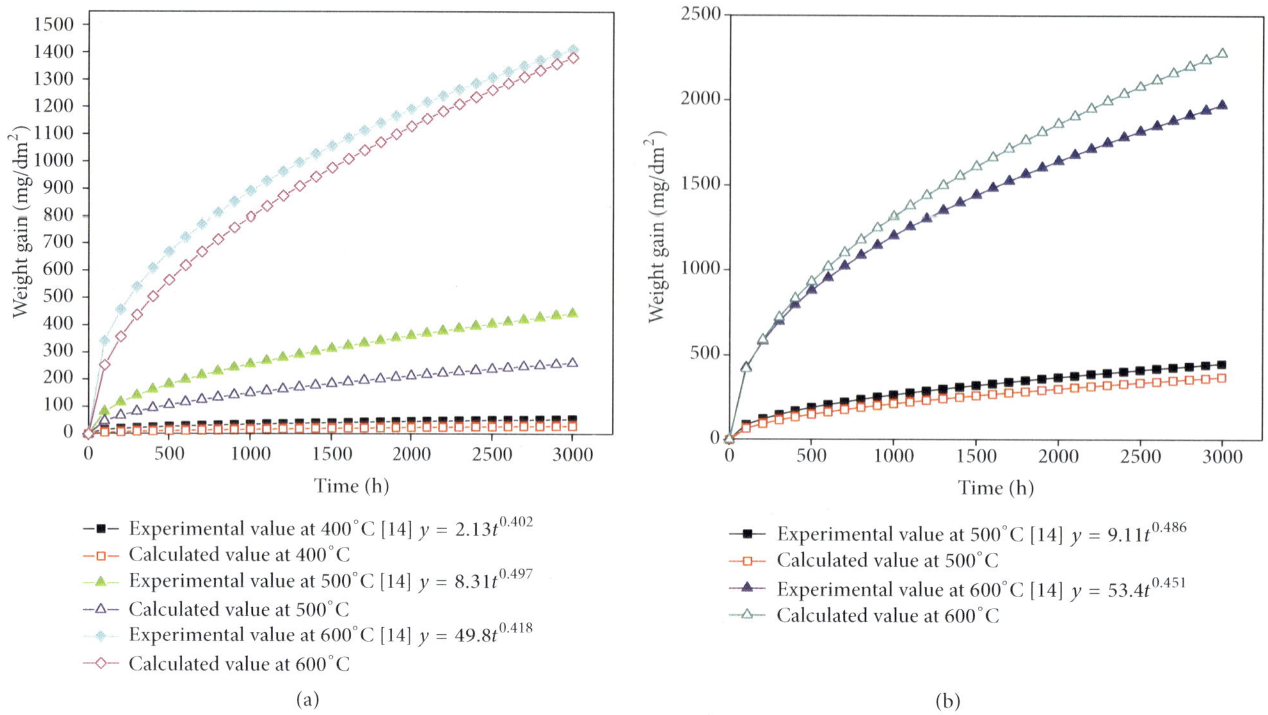

FIGURE 3: Experimental weight gain by oxidation of HCM12A (a) and NF616 (b) in supercritical water at different temperatures compared to the simulation.

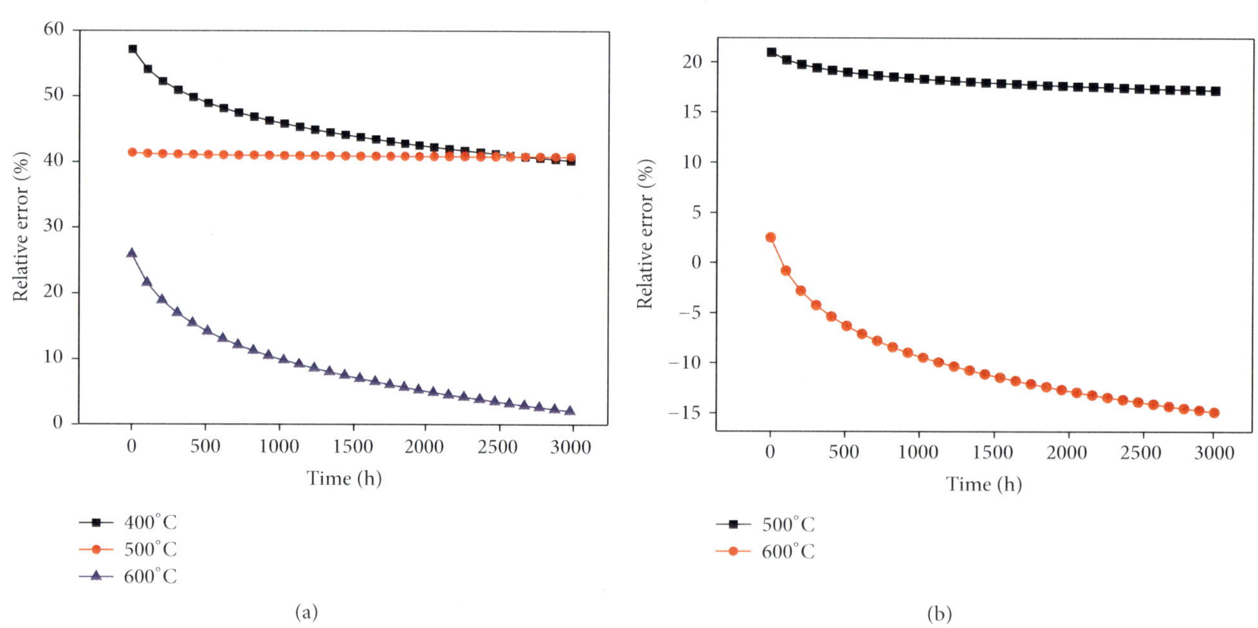

FIGURE 4: Relative error of simulated weight gain by oxidation of HCM12A (a) and NF616 (b) in supercritical water at different temperatures compared to the experiments.

The simulation of oxidation rates for NF616 and HCM12A at 550°C and 600°C are the same order of magnitude as one of the data in [24].

The simulation of oxidation rate for NF616 at 600°C is nearly two orders of magnitude higher than that at 500°C, while the simulation of oxidation rate for HCM12A at 600°C is nearly one order of magnitude higher than that at 500°C. The oxidation rate of Fe-Cr spinel layer is lower than that of magnetite layer because of the presence of Cr in Fe-Cr spinel layer. HCM12A with higher Cr contents show superior oxidation resistance as compared to NF616 with lower Cr contents at the same temperature.

TABLE 3: Oxidation rate of inner and outer layer oxides, k_{mag} and k_{spinel}, the global oxidation rate, k_{oxide}, and oxygen partial pressure of interfaces.

Steel	Temperature (°C)	oxidation rate (cm^2/s)				Oxygen partial pressure at interfaces (atm)		
		k_{mag}	k_{spinel}	k_{oxide}	k_{EPRI} [24]	$p_{alloy/spinel}$	$p_{spinel/mag}$	$p_{mag/scw}$
NF616	500	1.66×10^{-13}	1.61×10^{-13}	6.54×10^{-13}	—	1×10^{-29}	3.3×10^{-21}	3.5×10^{-20}
	550	8.84×10^{-13}	8.57×10^{-13}	3.45×10^{-11}	5.35×10^{-12}	3.5×10^{-27}	3.97×10^{-19}	3.9×10^{-18}
	600	6.33×10^{-12}	6.15×10^{-12}	2.5×10^{-11}	2.22×10^{-11}	2×10^{-25}	1.41×10^{-23}	2.57×10^{-16}
HCM12A	400	1.3×10^{-15}	1.13×10^{-15}	4.85×10^{-15}	—	3×10^{-35}	1.66×10^{-25}	3.46×10^{-25}
	500	8.85×10^{-14}	7.7×10^{-14}	3.3×10^{-13}	—	1×10^{-29}	1.53×10^{-20}	3.5×10^{-20}
	550	4.8×10^{-13}	4.17×10^{-13}	1.8×10^{-12}	5.4×10^{-12}	3.5×10^{-27}	1.7×10^{-18}	3.9×10^{-18}
	600	2.46×10^{-12}	2.14×10^{-12}	9.2×10^{-12}	10.1×10^{-12}	2×10^{-25}	9.59×10^{-17}	2.57×10^{-16}

4.4. The Distribution of Oxygen Activities and Iron Diffusion Coefficient in Duplex Oxide Scale. The oxygen partial pressures at interfaces of oxidation of HCM12A and NF616 at different temperatures are listed in Table 4. For the oxidation of NF616 at 600°C, the oxygen activity in the Fe-Cr spinel layer evolves from 2×10^{-25} to 1.41×10^{-23}, and the diffusion process follows an interstitial mechanism; the oxygen activity in the magnetite layer evolves from 1.41×10^{-23} to 2.57×10^{-16}, and the diffusion process varies from a interstitial mechanism to a vacancy mechanism. For the oxidation of HCM12A at 600°C, the oxygen activity in the Fe-Cr spinel layer evolves from 2×10^{-25} to 9.59×10^{-17}, and the diffusion process varies from a interstitial mechanism to a vacancy mechanism; the oxygen activity in the magnetite layer evolves from 9.59×10^{-17} to 2.57×10^{-16}, and the diffusion process follows a vacancy mechanism.

The iron diffusion coefficient in the duplex scale of two ferritic-martensitic steels exposed to supercritical water for 3000 h is expressed as a function of normalized position (x/X). Figure 5 shows calculations of the oxygen activity distribution in the Fe-Cr spinel layer and the magnetite scale formed on NF616 and HCM12A at 600°C and 500°C. Figure 6 shows the iron diffusion coefficient distribution along thickness direction of the duplex scale formed on NF616 and HCM12A at 600°C and 500°C. In the duplex scale of two ferritic-martensitic steels, the oxygen activity is continuous, while the iron diffusion coefficient is discontinuous at the inner layer/outer layer interface.

For the oxidation of NF616 at 600°C, the oxygen activity drastically increases at the normalized position (x/X = 0.6), while iron diffusion coefficient becomes minimum. For the oxidation of HCM12A at 600°C, the oxygen activity drastically increases at the normalized position (x/X = 0.2), while iron diffusion coefficient becomes minimum. There are two diffusion mechanisms of iron in the oxide, including interstitial mechanism and vacancy mechanism. When the vacancy concentration in oxide is high enough, the vacancies may collapse into voids. From Figure 2, with increasing of the oxygen activity, the diffusion mechanism changes from interstitial to vacancy. The vacancy diffusion mode starts to act a primary role at the location of the minimum diffusion coefficient of iron. So void is expected to form at the location of the minimum iron diffusion coefficient.

As shown in Figure 6, the iron diffusion coefficient in the Fe-Cr spinel layer on NF616 is higher than that in the Fe-Cr spinel layer on HCM12A, while the iron diffusion coefficient in the magnetite layer on NF616 is close to that in the magnetite layer on HCM12A. The oxygen partial pressure at the inner layer/outer layer interface formed on NF616 and HCM12A at 500°C is 3.3×10^{-21} and 1.53×10^{-20}, respectively (Table 4). The normalized position of the minimum iron diffusion coefficient in the duplex scale is about 0.4 and 0.1 (Figure 6), so the location of void formation in duplex scale formed on NF616 is close to the inner layer/outer layer interface compared to that formed on HCM12A. Similar result, which is that the minimum iron diffusion coefficient of low-oxygen exposed sample occurs at the location closer to the surface compared to high-oxygen exposed sample, has been reported in HCM12A oxidized in supercritical water [25].

4.5. Model Validation of an Example of One Power Plant. In the previous section, the model of oxidation kinetics of the ferritic-martensitic steel is validated with experimental data of Tan et al. [14]. In this section, the tubes (T91 steel) of the final superheater in one power plant are selected as research objects, and the thickness of steam side oxide scales of the final superheater is estimated by the above method. The simulation results are compared with field measurement data of the oxide scale thickness and calculation results by the EPRI's oxidation kinetics parameters.

Since its service, exfoliation of the oxide scale has become an acute problem in a 600 MW supercritical unit of one power plant, whose boiler type is SG2080/25.4-M969. The main thermodynamic parameters of the boiler, for example, calculation, are listed in Table 4. The thicknesses of oxide scales of the outlet part of the final superheater are measured by ultrasonic technology after 13000 h, 21000 h, 25000 h, and 36000 h periods of service. The range of values of thickness of oxide scales is listed in Table 5.

In order to verify the simulation of oxidation kinetics of ferritic-martensitic steels, an oxide scale growth model of superheater tubes is established. The empirical formula correlating scale thickness with the simulation oxidation rates is combined to the heat transfer theory. The model involves forced convections on the inner surface due to the

TABLE 4: Main thermodynamic parameters of a boiler, for example, calculation.

Item	Unit	BRL	BMCR
Superheated steam mass flow	t/h	1979	2080
Pressure of superheated steam outlet	MPa	25.28	25.4
Temperature of superheated steam outlet	°C	571	571
Reheated steam flow	t/h	1638	1764
Pressure of reheated steam inlet	MPa	4.38	4.6
Pressure of reheated steam outlet	MPa	3.28	3.66
Temperature of reheated steam inlet	°C	313	318
Temperature of reheated steam outlet	°C	569	569

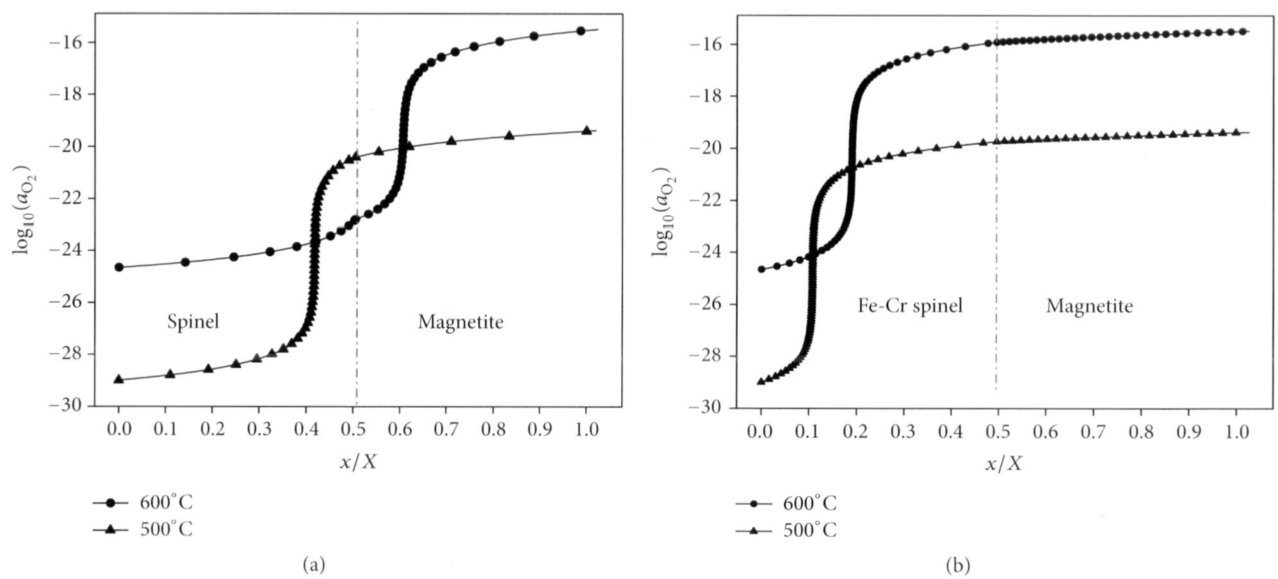

FIGURE 5: The calculation oxygen activity in the duplex scale formed on NF616 (a) and HCM12A (b) as a function of normalized position.

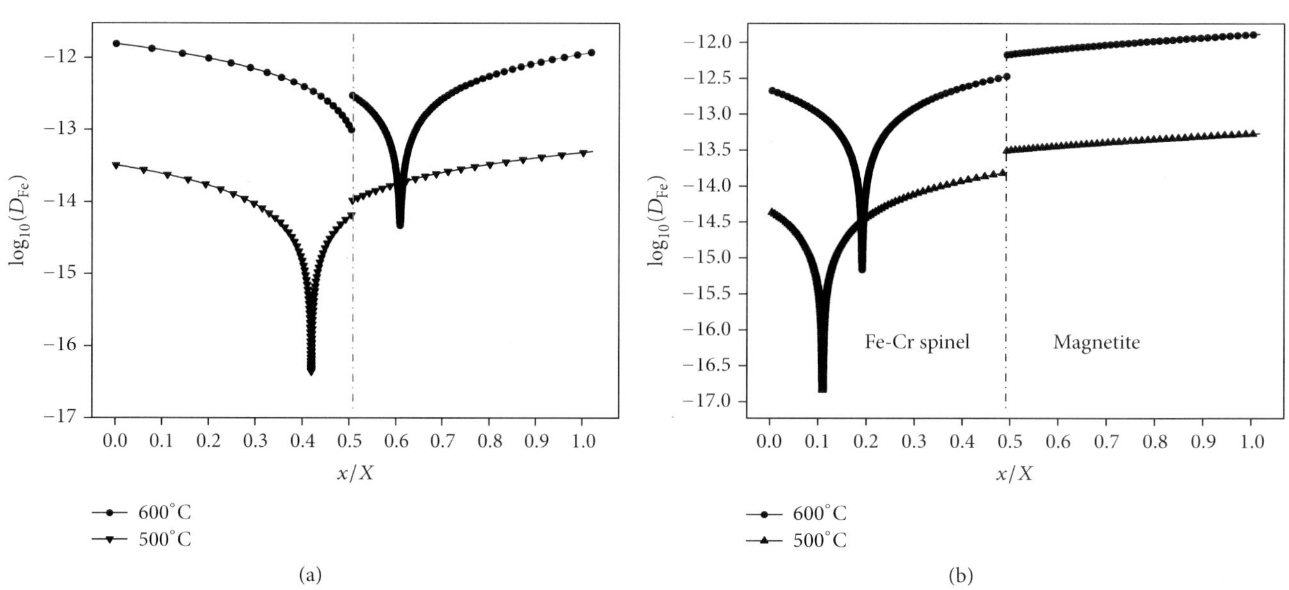

FIGURE 6: The diffusion coefficient of iron in the duplex scale formed on NF616 (a) and HCM12A (b) as a function of the normalized position.

TABLE 5: Comparison between field measurement data and simulation results of oxide scale thickness.

Exposure time/h	Field measurement data of oxide scale thickness/μm	Simulation results of oxide scale thickness/μm	Relative error compared to the maximum value/%
10000	116–148	177.5	19.9
21000	188–209	225.6	7.9
25000	187–225	246.1	9.4
36000	144–258	295.4	14.5

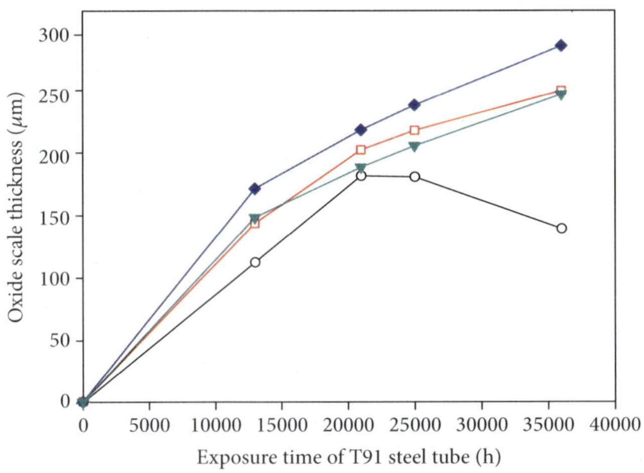

FIGURE 7: Comparison of the thickness of oxide scale of final superheater tubes during the oxidation of T91 for different exposure time among field measurement data, simulation results, and calculation results according to Sabau and Wright's method [26].

turbulent flow of steam and on the outer surface due to cross flow of the hot flue gas over ash tubes. The oxidation rates of ferritic-martensitic steels at different temperatures are calculated and carried out by a quadratic multinomial fitting. An iterative procedure is used to determine scale thickness as both temperature and time increase [26]. The results are presented in Figure 7.

Based on the parabolic law of high-temperature oxidation, Arrhenius equation, and related basic theories of heat transfer, an oxide scale growth model of boiler superheater tubes is proposed through iterative technique [26]. Related parameters are from the EPRI report [24]. According to this algorithm, the calculation results of oxide scale thickness of boiler superheater tubes for different exposure time are presented in Figure 7.

Figure 7 shows the predicted oxide scale thickness, the field measurement data, and calculated results by Sabau and Wright's method. From this figure, it can be concluded that the prediction results are credible. The calculation results of this method and Sabau and Wright's method are comparatively close to field measurement data. Because exfoliation of oxide scales probably emerged during the operation of the unit, the maximum value of the field measurement data is lower than calculation results of oxide thickness of boiler superheater tubes. The parameters in Adrian's iterative method are extrapolated from field measurement data, so the calculated results are closer to the actual experimental data.

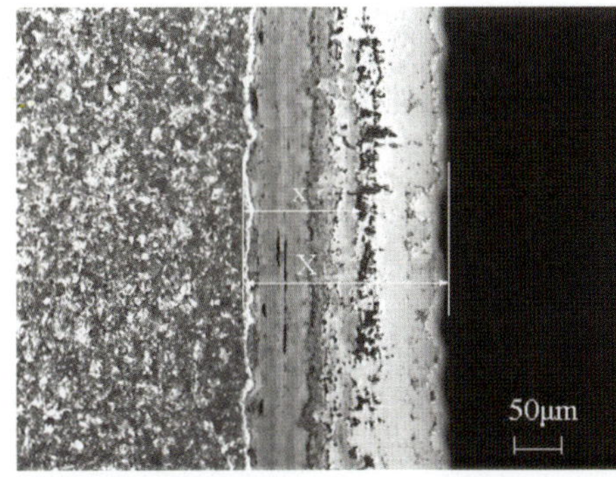

FIGURE 8: A cross-sectional SEM image of the oxide layer formed on T91 used in a superheater tube for 25000 hours.

To verify the location of voids, the scanning electron microscope image of the oxide layer formed on T91 used in a superheater tube for 25000 hours is captured and shown in Figure 8. The experimental thickness of oxide scale is 225.34 μm, and the distance is 125.2 μm from the steel/Fe-Cr spinel interface to the location of void formation. The calculated quantitative location of voids formation (x/X = 0.52) is in good agreement with the experimental result (x/X = 0.56) with minor difference.

5. Conclusion

Based on available space model, the oxidation kinetics of two ferritic-martensitic steels are developed to predict in supercritical water at 400°C, 500°C, and 600°C. The iron diffusion coefficients in magnetite and Fe-Cr spinel, which are extrapolated from studies of Backhaus-Ricoult and Dieckmann [17] and Töpfer et al. [18], are used for calculation of oxidation kinetics of steels under supercritical water environment. The main conclusions are as follows.

(i) The simulation of oxidation of NF616 at 500°C and 600°C and HCM12A at 600°C is approximately close to the experimental weight gain during the oxidation of NF616 and HCM12A in supercritical water at 400–600°C [14], which proves the validity of the simulation assumptions.

(ii) The simulation value of the oxidation rate depends on oxygen partial pressure at the steel/spinel interface and magnetite/supercritical water interface. The equilibrium oxygen partial pressure of reaction (6) is used as oxygen partial pressure at magnetite/supercritical water interface, and the results of calculation are in agreement with the experimental weight gain.

(iii) The relative error of oxide weight gain between simulation and experiments of HCM12A at 400°C and 500°C is above 40%, and the reason is that magnetite/supercritical water oxygen partial pressure is not the equilibrium oxygen partial pressure of reaction (6) or probably the error of diffusion coefficient when it extrapolated from high temperature to low temperature according to Töpfer et al.'s data [18].

(iv) In the duplex scale of two ferritic-martensitic steels, the oxygen activity is continuous, while the iron diffusion coefficient is discontinuous at the inner layer/outer layer interface. The oxygen activity drastically increases at one normalized position (x/X), while iron diffusion coefficient becomes minimum. Void is expected to form at the location of the minimum iron diffusion coefficient.

(v) At the same oxidation temperature, the iron diffusion coefficient in the Fe-Cr spinel layer on NF616 is higher than that in the Fe-Cr spinel layer on HCM12A, while the iron diffusion coefficient in the magnetite layer on NF616 is close to that in the magnetite layer on HCM12A.

Conflicts of Interest

The authors declare that there are no conflicts of interest regarding the publication of this paper.

Acknowledgments

This paper was supported by the Fundamental Research Funds for the Central Universities (2014MS107).

References

[1] P. Kritzer, "Corrosion in high-temperature and supercritical water and aqueous solutions: a review," *Journal of Supercritical Fluids*, vol. 29, no. 1-2, pp. 1–29, 2004.

[2] R. L. Klueh and A. T. Nelson, "Ferritic/martensitic steels for next-generation reactors," *Journal of Nuclear Materials*, vol. 371, no. 1-3, pp. 37–52, 2007.

[3] S. R. J. Saunders, M. Monteiro, and F. Rizzo, "The oxidation behaviour of metals and alloys at high temperatures in atmospheres containing water vapour: a review," *Progress in Materials Science*, vol. 53, no. 5, pp. 775–837, 2008.

[4] W. J. Quadakkers, J. Zurek, and M. Hänsel, "Effect of water vapor on high-temperature oxidation of FeCr alloys," *JOM Journal of the Minerals Metals and Materials Society*, vol. 61, no. 7, pp. 44–50, 2009.

[5] P. Ampornrat and G. S. Was, "Oxidation of ferritic-martensitic alloys T91, HCM12A and HT-9 in supercritical water," *Journal of Nuclear Materials*, vol. 371, no. 1-3, pp. 1–17, 2007.

[6] I. Betova, M. Bojinov, P. Kinnunen et al., "Composition, structure, and properties of corrosion layers on ferritic and austenitic steels in ultrasupercritical water," *Journal of the Electrochemical Society*, vol. 153, no. 11, Article ID 020611JES, pp. B464–B473, 2006.

[7] M. G. Angell, S. K. Lister, and A. Rudge, "The effect of steam pressure on the oxidation behavior of annealed 9Cr1Mo boiler tubing materials," in *Proceedings of the in 15th International Conference on the Properties of Water and Steam (ICPWS XV)*, pp. 1–9, Berlin, Germany, 2008.

[8] N.-Q. Zhang, H. Xu, B.-R. Li, Y. Bai, and D.-Y. Liu, "Influence of the dissolved oxygen content on corrosion of the ferritic-martensitic steel P92 in supercritical water," *Corrosion Science*, vol. 56, pp. 123–128, 2012.

[9] L. Tan, Y. Yang, and T. R. Allen, "Oxidation behavior of iron-based alloy HCM12A exposed in supercritical water," *Corrosion Science*, vol. 48, no. 10, pp. 3123–3138, 2006.

[10] J. Bischoff and A. T. Motta, "Oxidation behavior of ferritic-martensitic and ODS steels in supercritical water," *Journal of Nuclear Materials*, vol. 424, no. 1-3, pp. 261–276, 2012.

[11] X. Zhong, X. Wu, and E.-H. Han, "Effects of exposure temperature and time on corrosion behavior of a ferritic-martensitic steel P92 in aerated supercritical water," *Corrosion Science*, vol. 90, pp. 511–521, 2015.

[12] L. Martinelli, F. Balbaud-Célérier, G. Picard, and G. Santarini, "Oxidation mechanism of a Fe-9Cr-1Mo steel by liquid Pb-Bi eutectic alloy (Part III)," *Corrosion Science*, vol. 50, no. 9, pp. 2549–2559, 2008.

[13] F. Rouillard and L. Martinelli, "Corrosion of 9Cr steel in CO_2 at intermediate temperature III: Modelling and simulation of void-induced duplex oxide growth," *Oxidation of Metals*, vol. 77, no. 1-2, pp. 71–83, 2012.

[14] L. Tan, X. Ren, and T. R. Allen, "Corrosion behavior of 9-12% Cr ferritic-martensitic steels in supercritical water," *Corrosion Science*, vol. 52, no. 4, pp. 1520–1528, 2010.

[15] L. Martinelli, F. Balbaud-Célérier, A. Terlain, S. Bosonnet, G. Picard, and G. Santarini, "Oxidation mechanism of an Fe-9Cr-1Mo steel by liquid Pb-Bi eutectic alloy at 470°C (Part II)," *Corrosion Science*, vol. 50, no. 9, pp. 2537–2548, 2008.

[16] A. Atkinson, M. L. O'Dwyer, and R. I. Taylor, "55Fe diffusion in magnetite crystals at 500°C and its relevance to oxidation of iron," *Journal of Materials Science*, vol. 18, no. 8, pp. 2371–2379, 1983.

[17] M. Backhaus-Ricoult and R. Dieckmann, "Defects and cation diffusion in magnetite (VII): diffusion controlled formation of magnetite during reactions," *Berichte der Bunsengesellschaft für physikalische Chemie*, vol. 90, no. 8, pp. 690–698, 1986.

[18] J. Töpfer, S. Aggarwal, and R. Dieckmann, "Point defects and cation tracer diffusion in $(Cr_xFe_{1-x})_{3-\delta}O_4$ spinels," *Solid State Ionics*, vol. 81, no. 3-4, pp. 251–266, 1995.

[19] P. Ampornrat, *Determination of oxidation mechanisms of ferritic - martensitic steels in supercritical water*, The University of Michigan, 2011.

[20] J. Zurek, M. Michalik, F. Schmitz, T.-U. Kern, L. Singheiser, and W. J. Quadakkers, "The effect of water-vapor content and gas flow rate on the oxidation mechanism of a 10%Cr-ferritic steel in Ar-H_2O mixtures," *Oxidation of Metals*, vol. 63, no. 5-6, pp. 401–422, 2005.

[21] T. Furukawa, G. Müller, G. Schumacher, A. Weisenburger, A. Heinzel, and K. Aoto, "Effect of oxygen concentration and temperature on compatibility of ODS steel with liquid, stagnant Pb45Bi55," *Journal of Nuclear Materials*, vol. 335, no. 2, pp. 189–193, 2004.

[22] D. R. Gaskell, *Introduction to Metallurgical Thermodynamics*, Taylor, Francis, London, 2nd edition, 1981.

[23] T. Maruyama, N. Fukagai, M. Ueda, and K. Kawamura, "Chemical potential distribution and void formation in magnetite scale formed in oxidation of iron at 823K," *Materials Science Forum*, vol. 461-464, no. II, pp. 807–814, 2004.

[24] I. G. Wright and P. F. Tortorelli, "Program on technology innovation: oxide growth and exfoliation on alloys exposed to steam," Tech. Rep., Electric Power Research Institute, PaloAlto, Calif, USA, 2007.

[25] L. Tan, Y. Yang, and T. R. Allen, "Porosity prediction in supercritical water exposed ferritic/martensitic steel HCM12A," *Corrosion Science*, vol. 48, no. 12, pp. 4234–4242, 2006.

[26] A. S. Sabau and I. G. Wright, "On the estimation of thermal strains developed during oxide growth," *Journal of Applied Physics*, vol. 106, no. 2, Article ID 023503, 8 pages, 2009.

19

Effect of Deformation Temperature on Microstructure Evolution and Mechanical Properties of Low-Carbon High-Mn Steel

Adam Grajcar, Aleksandra Kozłowska, Santina Topolska, and Mateusz Morawiec

Institute of Engineering Materials and Biomaterials, Silesian University of Technology, 18a Konarskiego Street, 44-100 Gliwice, Poland

Correspondence should be addressed to Adam Grajcar; adam.grajcar@polsl.pl

Academic Editor: Akihiko Kimura

This work addresses the influence of deformation temperature in a range from −40°C to 200°C on the microstructure evolution and mechanical properties of a low-carbon high-manganese austenitic steel. The temperature range was chosen to cope at the time during sheet processing or car crash events. Experimental results show that yield stress and ultimate tensile strength gradually deteriorate with an increase in the tensile testing temperature. The dominant mechanism responsible for the strain hardening of steel changes as a function of deformation temperature, which is related to stacking fault energy (SFE) changes. When the deformation temperature rises, twinning decreases while a role of dislocation slip increases.

1. Introduction

Among new steel grades developed for application to car bodies [1–5], high-manganese austenitic steels are attractive materials for the automotive industry because of their unique combination of high strength, ductility, and formability. High-manganese steels are characterized by a high work-hardening rate resulting from TRIP (transformation-induced plasticity), TWIP (twinning-induced plasticity), or SIP (shear-induced plasticity) effects [6, 7]. A lot of reports concerning the mechanical behavior of high-Mn TRIP or TWIP steels have already been available in the literature. Most of them characterize the influence of heat treatment parameters on microstructure, mechanical properties [8–11], and corrosion behavior of the steels [12, 13].

The dominant mechanism of strengthening occurring in these steels strongly depends on the chemical composition, particularly on carbon, manganese, aluminum, and silicon contents. The deformation mechanism depends also on stacking fault energy, strain rate, and temperature. TRIP effect is dominant when SFE is lower than 25 mJ/m^2—some fraction of austenite transforms into ε or α' martensites. When SFE value is in a range of 25–60 mJ/m^2, deformation twins occur (TWIP effect). If SFE value is higher than 60 mJ/m^2, the steel is strengthened mostly by work hardening. Aluminum strongly increases SFE, whereas silicon causes an opposite effect [14–16].

Deformation temperature changes affect significantly the SFE, which in turn influences the character of the hardening. Shterner et al. reported [17] that the increase of deformation temperature of the Fe-0.6C-18Mn-1Al TWIP steel caused a decrease of yield strength, ultimate tensile strength, and total and uniform elongation. The highest mechanical properties of the TWIP steel were detected at room temperature. They decrease gradually with the increase in deformation temperature. They also reported [17] that the work-hardening behavior was attributed to complex dynamic strain-induced microstructural changes including dynamic recovery, dislocation dissociation, stacking fault formation, mechanical twinning, and dynamic strain aging.

Asghari et al. [18] classified the strengthening mechanisms depending on the characteristic temperature regimes in unidirectional compression tests: 25–300°C, 300–700°C, and 700–1000°C. The Fe-0.07C-18Mn-2Si-2Al steel showed transformation-induced plasticity effect (TRIP effect) as the major deformation mechanism from 25°C to 200°C, whereas deformation twinning (TWIP effect) started from 200°C to 300°C. Above 700°C the dynamic restorations, recovery and

Table 1: Chemical composition of the investigated steel (wt.%).

C	Mn	Si	Al	S	P	Nb	Ti	N	O
0.065	26.00	3.08	2.87	0.013	0.002	0.034	0.010	0.0028	0.0006

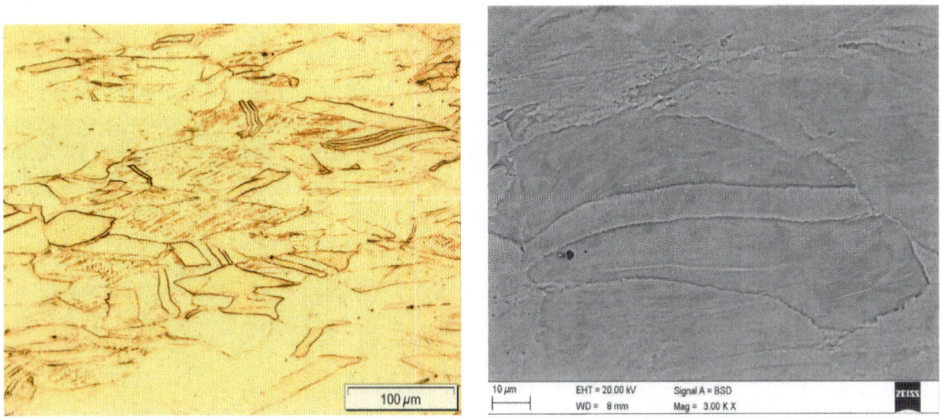

(a) (b)

Figure 1: Optical micrograph (a) and SEM image (b) of the X6MnSiAlNbTi26-3-3 steel before tensile test.

recrystallization play a key role in microstructural evolution of the experimental steel in accordance with the diminishing SFE. Similar results have been shown by Salas-Reyes et al. [19] for the Fe-0.1C-21Mn-2.5Si-1.6Al steel during compression tests in a temperature range of 25°C to 1000°C. The local composition of steel may be changed during hot deformation, for instance due to the strain-induced precipitation or decomposition, thereby resulting in SFE changes [19].

Eskandari et al. [20] studied the hot ductility behavior of the Fe-0.45C-22Mn-1.5Al-1.5Si steel in uniaxial hot tensile tests in the temperature range of 700–1100°C under a constant strain rate. They indicated that the peak stress decreases as the temperature increases, from 250 MPa at 700°C up to 30 MPa at 1100°C. The highest ductility was detected at the intermediate temperature range (800–900°C). It was caused by dynamic recrystallization.

So far, the mechanical behavior of low-carbon high-Mn steels in a temperature range of 20–200°C and below room temperature has been investigated rarely [18, 19]. Most publications concern the mechanical properties of high-carbon high-manganese steels at elevated temperatures: 700–1000°C, which simulate the hot-rolling conditions. Therefore, the aim of this paper is to determine the effect of deformation temperature on the microstructure evolution and mechanical properties of low-carbon high-manganese steel in a temperature range −40°C to 200°C that is chosen to cope at the time during sheet stamping or car crashes.

2. Experimental Procedure

2.1. Material. The chemical composition of the X6MnSiAlNbTi26-3-3 steel used is given in Table 1. Carbon and manganese are major austenite stabilizers, whereas silicon and aluminum were added for providing solid solution strengthening. Small amounts of Nb and Ti were added for precipitation strengthening and grain refinement. The chemical composition of steel significantly affects SFE, which is related to the strengthening mechanisms [14–16].

A steel ingot (25 kg) was prepared by vacuum melting. Then, it was hot-forged using a high-speed hydraulic press with forging pressure of 300 tons in a temperature range from 1200°C to 900°C. The flat bars after forging were hot-rolled to a thickness of 4.5 mm. The thermomechanical processing included hot rolling of the flat samples in 3 passes (at temperatures: 1050°C, 950°C, and 850°C) to a final thickness of ~2 mm obtained at the finishing rolling temperature of 850°C. The hot rolling was conducted using a reversible rolling mill at a circumferential speed of rollers of 0.65 $m{\cdot}s^{-1}$. A solution heat treatment at 900°C for 20 min was followed by a rapid water cooling to room temperature.

2.2. Tensile Tests. 2 mm thick tensile specimens were machined from the hot-rolled sheet along the rolling direction to investigate the effect of the deformation temperature on the microstructure evolution and mechanical properties. The tensile tests were performed at −40°C, 20°C, 80°C, 140°C, and 200°C at a strain rate of $5 \times 10^{-3}\,s^{-1}$ using an INSTRON 4505 universal testing machine.

2.3. Microstructure Characterization. The microstructural analysis was performed using Zeiss Axio Observer Z1m optical microscope. The microstructural details were revealed with a scanning electron microscope Zeiss SUPRA 25 operating at 20 kV. The specimens at the initial state and after tensile tests were mechanically ground with SiC paper up to 1500 grid, polished with Al_2O_3 and then etched using 5% nital to observe microstructures.

3. Results and Discussion

3.1. Microstructural Evolution. Microstructure of the X6MnSiAlNbTi26-3-3 steel in the initial state is shown in

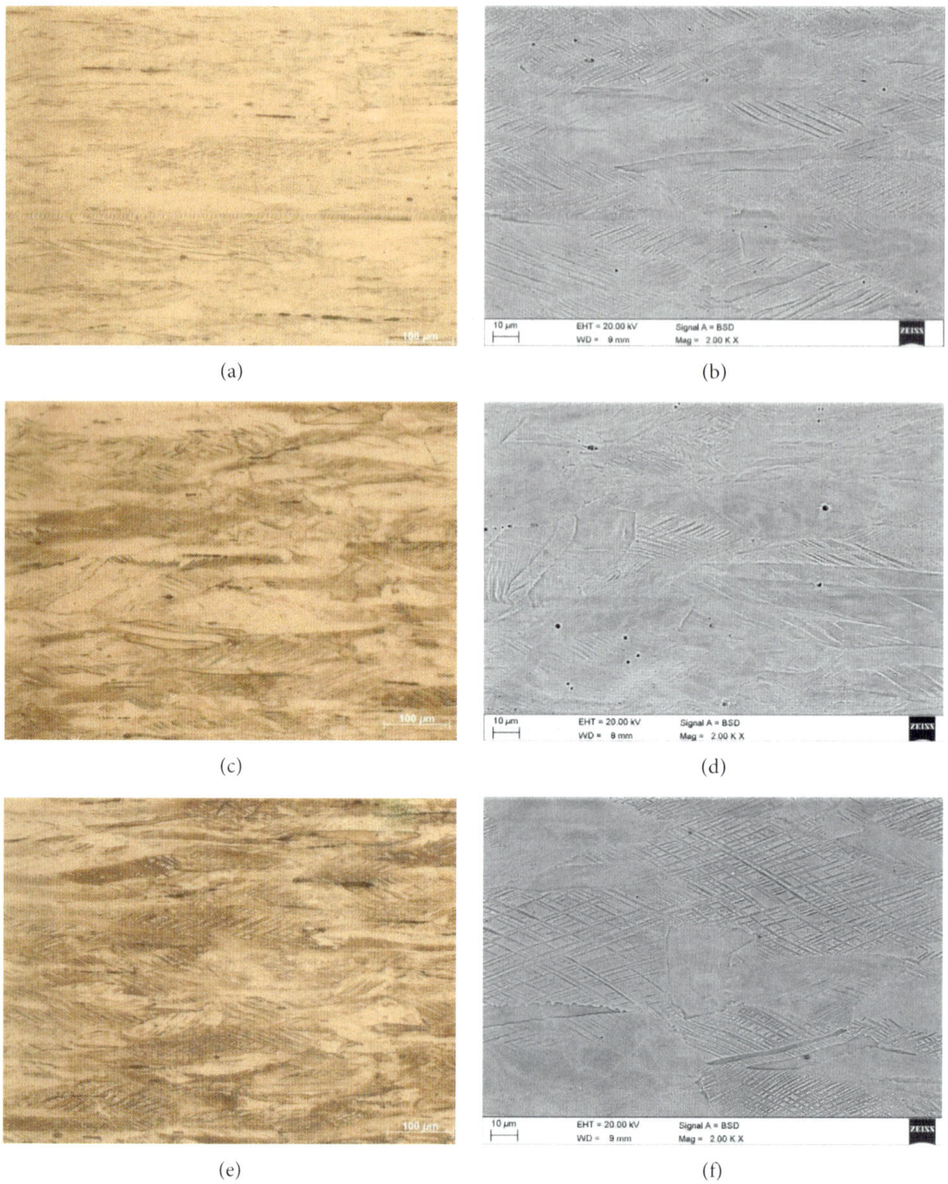

Figure 2: Microstructures of the X6MnSiAlNbTi26-3-3 steel deformed at a temperature of −40° ((a) LM; (b) SEM), 80°C ((c) LM; (d) SEM), and 200°C ((e) LM; (f) SEM) observed by a light microscope (LM), and a scanning electron microscope (SEM).

Figure 1. Austenite grains are relatively coarse and elongated along rolling direction. Annealing twins were also observed.

The results obtained from the tensile tests indicate a relationship among the test temperature, microstructure, and mechanical properties. The microstructure of the steel deformed in a static tensile test at −40°C is characterized by the presence of austenitic grains elongated according to the direction of applied tensile loading. With the reduction of deformation temperature the SFE decreases [18], therefore mechanical twinning occurred as a main strengthening mechanism (Figure 2(a)). The low SFE results in a reduction of stress levels necessary to initiate the twinning process. SEM image revealed also the presence of slip bands (Figure 2(b)). Shterner et al. reported [17] that, in the Fe-0.6C-18Mn-1Al steel, mechanical twins nucleated at the grain boundaries and then propagated across the grain interior, as the degree of deformation increased. Amount of deformation twins in the microstructure is related to the increasing dislocation density during deformation process. The microstructure of the steel deformed at room temperature is similar to the one obtained at −40°C. This concerns the amount of crossing slip lines and bands as wells as numerous deformation twins. The occurrence of the transformation-induced plasticity effect (TRIP) or twinning mechanism (TWIP) is common at relatively low test temperature. Shterner et al. [17] reported that, at a strain level of 0.4, the volume fraction of mechanical twins reduced from 11.2% at room temperature to 5% and 2.9% at 100°C and 200°C, respectively.

The microstructure of the sample deformed at 80°C indicates a dominant role of crossed slip lines and bands (Figures 2(c) and 2(d)). The amount of deformation twins is

TABLE 2: Mechanical properties of the tensile tested steel at each temperature.

Temperature (T)		Tensile strength (UTS) (MPa)	Yield stress ($YS_{0.2}$) (MPa)	$YS_{0.2}$/UTS	Total elongation (TE) (%)
1	−40°C	834	577	0.69	43.4
2	20°C	743	554	0.75	43.8
3	80°C	676	527	0.78	41.8
4	140°C	617	511	0.83	43.4
5	200°C	604	509	0.84	22.4

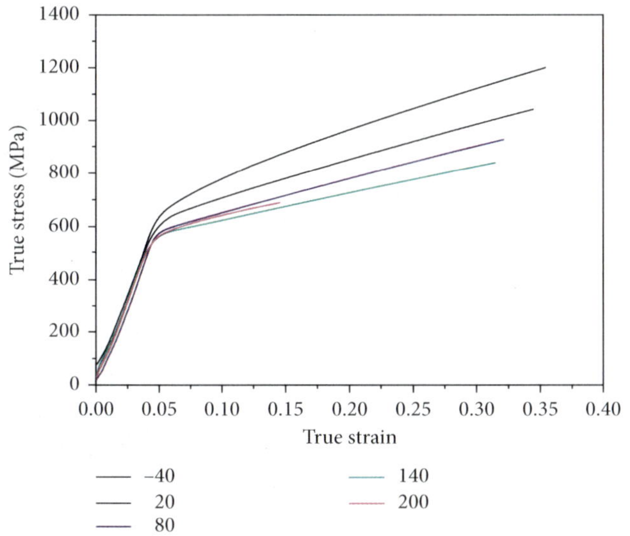

FIGURE 3: True stress-true strain curves in a uniform elongation range of the X6MnSiAlNbTi26-3-3 steel at various deformation temperatures.

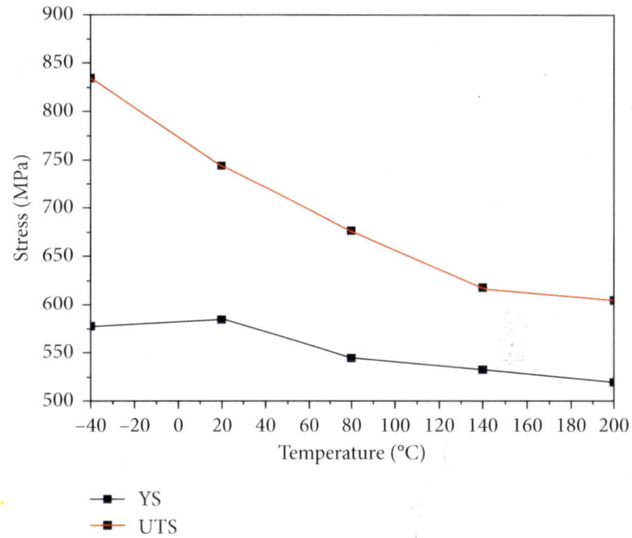

FIGURE 4: Mechanical properties of the X6MnSiAlNbTi26-3-3 steel: yield stress (YS) and ultimate tensile strength (UTS) at various tensile test temperatures.

smaller than in the samples deformed at −40°C and 20°C. Increasing the test temperature to 80°C results in an increase of SFE, which limits the twinning occurrence. The deformation of the sample at 140°C favors dislocation slip as a main deformation mechanism, while the formation of twins is strongly limited. The prevailing mechanism of deformation at 200°C appears to be dislocation slip (Figures 2(e) and 2(f)). Chen et al. [21] confirmed that deformation bands can be observed in pancaked austenite deformed in a non-recrystallization region, which bridge the grains by roughly parallel lines. No ε or α' martensites were observed in the temperature range −40°C to 200°C. Such behavior is typical for steels characterized by SFE > 20 mJ/m^2, resulting from the chemical composition.

3.2. Mechanical Properties. The mechanical properties of the steel tensile tested at each temperature is shown in Table 2, indicating a strong temperature dependence.

Figure 3 shows the true stress-true strain curves in the temperature range of −40°C to 200°C. The X6MnSiAlNbTi26-3-3 steel showed the best mechanical properties at −40°C, having a tensile strength (UTS) of 834 MPa, yield stress (YS) of 577 MPa, and total elongation of 43.4% (Figures 3–5). At −40°C, the dominant mechanism responsible for the high strain hardening of the steel is twinning (Figures 2(a) and 2(b)). The YS/UTS ratio in this case is 0.69, which indicates relatively high strengthening potential (Figure 5).

The steel deformed at 20°C is characterized by slightly lower mechanical properties (YS = 584 MPa; UTS = 743 MPa) than the one deformed at −40°C (Figure 4). The YS/UTS ratio rises to 0.75 (Figure 5). It shows the gradual lowering of strengthening intensity with the increase of deformation temperature. Total elongation remains at the same level ~43.8% (Figure 5). The combination of high strength and ductility both at −40°C and at room temperature is mostly related to twinning mechanism. At 80°C, the mechanical properties decreased: YS = 527 MPa and UTS = 676 MPa (Figure 4), which is related to a decrease in the density of deformation twins (Figures 2(c) and 2(d)) and a gradual increase of the SFE. The YS/UTS ratio increased to 0.78 (Figure 5). When the temperature increases, the role of slip lines and bands in the plastic deformation increases. However, the reduction of total elongation is relatively small (~2%) when compared to the specimen deformed at 20°C (Figure 5). Such a small reduction is due to the combination of mechanical twinning and slip and overall high plasticity of the austenite phase.

The specimen deformed at 140°C shows further lower strengthening effect than the samples deformed in a lower temperature range (Figure 3). YS and UTS are 511 MPa and 617 MPa, respectively. The YS/UTS ratio in this temperature

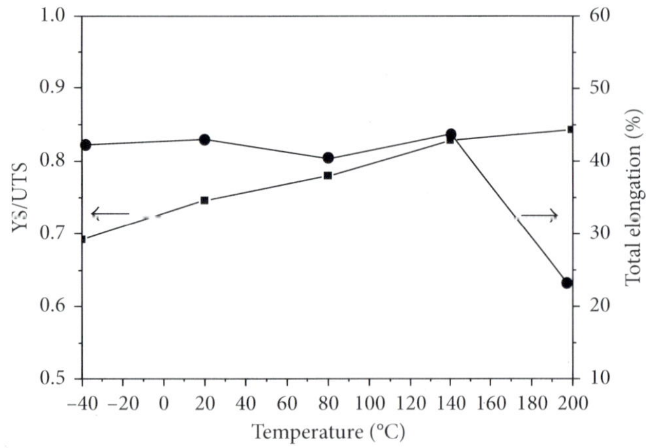

FIGURE 5: YS/UTS ratio (square) and total elongation (circle) of the X6MnSiAlNbTi26-3-3 steel at various tensile test temperatures.

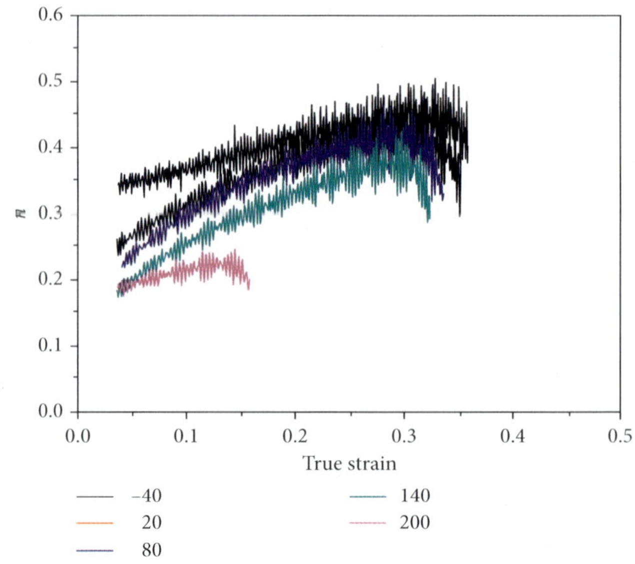

FIGURE 6: Work-hardening exponent n changes as a function of true strain at various tensile test temperatures.

corresponds to 0.83 (Figure 5). It indicates a significant decrease in the intensity of the strengthening. It is also reflected in a slope of the σ-ε curve (Figure 3). The temperature of 140°C favors dislocation slip as a main deformation mechanism. However, the change of dominant deformation mechanism from twinning to dislocation slip is not still too detrimental for total elongation (43.4%), but these conditions strongly affect the strength properties.

The mechanical properties determined for the sample deformed at 200°C are the lowest compared to the results obtained at the lower temperatures (Table 2). YS and UTS are 509 MPa and 604 MPa, respectively (Figure 4), and YS/UTS = 0.84 (Figure 5). The slight slope of the σ-ε curve shown in Figure 3 indicates the minimum intensity of the strengthening. The specimen deformed at 200°C is characterized by a rapid decrease of plasticity expressed as the lowest total elongation = 22.4% (Figure 5). The plasticity is almost twice lower when compared to the other samples. The significant decrease of plasticity is due to disappearance of mechanical twinning and increased importance of diffusion processes activated at elevated temperatures.

Asghari et al. [18] reported that the mechanical twinning occurring inside the austenite grains by the variation of deformation temperature strongly affects the mechanical behavior of steel. The mechanical twins subdivide the austenite grains and reduce the dislocations mean free path. Therefore, the twin boundaries act as strong barriers to dislocation motion. Hence, higher mechanical properties reflect the occurrence of mechanical twinning. Mechanical twins were observed in the microstructures of samples deformed in the temperature range: −40°C–80°C (Figures 2(a)–2(d)). Eskandari et al. [20] observed mechanical twins in a temperature range 150°C–600°C during compression of the Fe-0.1C-21Mn-2.5Si-1.6Al-0.02Nb-0.02Ti-0.01V steel. Shterner et al. [17] showed that mechanical properties such as YS and UTS of the Fe-0.6C-18Mn-1A steel decreased with increasing temperature of tensile test: from YS = 500 MPa and UTS = 1000 MPa at room temperature to YS = 430 MPa and UTS = 840 MPa at 200°C. Total elongation was high between 20°C and 100°C (60–65%) and gradually decreased above 100°C (e.g., 47% at 200°C).

The steel shows the TWIP effect most intensively at −40°C. The mechanical twins were detected up to ~80°C. Asghari et al. [18] and Eskandari et al. [20] observed the TWIP effect in a temperature range 200–300°C. It seems that the steel shows an optimal TWIP effect at lower temperatures. In order to obtain a better TWIP effect at room temperature and at increased temperatures, some modification of the chemical composition of the steel would be required. It is expected that TWIP effect could be enhanced by reducing the aluminum and manganese contents, when the SFE at room temperature would be high.

The mechanical twinning is believed to significantly influence the work hardening of high-Mn steels. Presence of mechanical twins in the microstructure blocks the movement of dislocation glide [17, 21]. The gradual increase in the amount of mechanical twins and the increase of deformation level prevent the localization of deformation. The work-hardening exponent n gradually rises as the degree of deformation increases (Figure 6). The highest value of n exponent = 0.46 was obtained at −40°C (Figure 7). It is related to the dominant character of twinning mechanism, which decreases with increasing the deformation temperature. With increasing test temperature, the strengthening potential expressed as n values gradually decreases. Shterner et al. [17] suggested that the interactions of glide dislocations with stacking faults have greater contributions to the work-hardening behavior (plateau region) compared with the mechanical twinning at room temperature. The present steel has a relatively small grain size (high grain boundary area), which significantly enhances the interactions between the dislocations and the grain boundaries. However, at 200°C grain boundaries start to lose their reinforcing character due to the initiation of diffusion processes.

The deformation temperature strongly affects the work-hardening rate (Figure 8). The shape of the curves is typical; that is, the $d\sigma/d\varepsilon$ values decrease rapidly at an initial range of

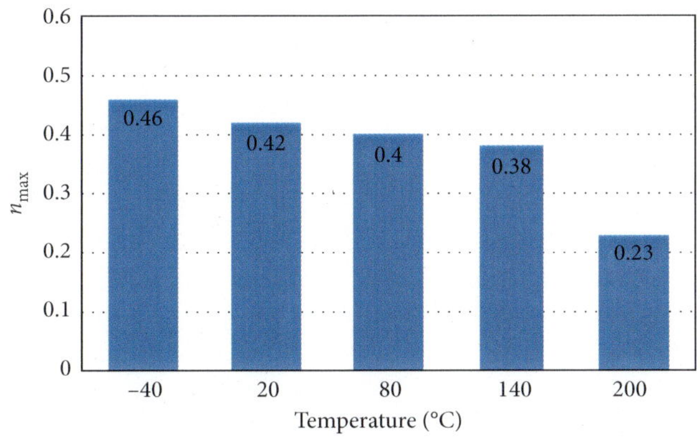

FIGURE 7: Maximum values of work-hardening exponent n_{max} at various tensile test temperatures.

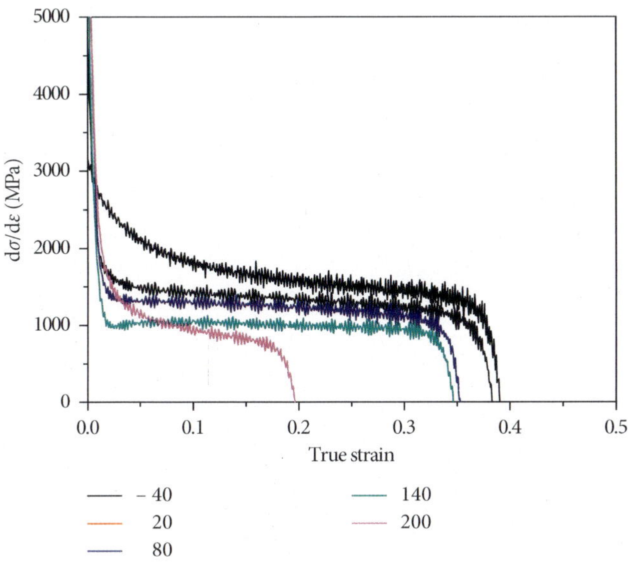

FIGURE 8: Work-hardening rate as a function of true strain at different tensile test temperatures.

deformation. Then, the work-hardening rate stabilizes at a constant level for the samples deformed between −40°C and 140°C due to the combined occurrence of deformation twinning and deformation slip. A various level of the deformation twinning at various temperatures is reflected in a different constant value of $d\sigma/d\varepsilon$. It is circa 1900–1800 MPa for the temperature of −40°C and about 1000 MPa for a sample deformed at 140°C. The different behavior is registered for a sample deformed at the highest deformation temperature, where a continuous decrease of the work-hardening rate can be observed due to a lack of deformation twinning.

4. Conclusions

The present study investigated the effect of deformation temperature on the microstructure evolution and mechanical properties of low-carbon high-Mn steel. The steel showed the best mechanical properties at −40°C due to the TWIP effect. Yield stress and ultimate tensile strength gradually deteriorated with an increase in the deformation temperature from −40°C to 200°C. The reduction of the mechanical properties was related to the decrease in the contribution of twinning, whereas that of dislocation slip increased. The significant reduction of the plasticity occurred at 200°C due to a lack of mechanical twinning, and diffusion processes activated at elevated temperatures.

The friction occurring during steel sheet forming or a crash event can lead very often to an increase in the temperature of a formed metal piece to ~100°C. To maximize the potential application of the high-Mn steel, its chemical composition should be modified to reduce its SFE in a desired temperature range from 20°C to 100°C. As a result, the optimum window of mechanical properties would shift to higher temperatures. One of the options is to reduce the aluminum content.

Conflicts of Interest

The authors declare that there are no conflicts of interest regarding the publication of this paper.

References

[1] E. I. Galindo-Nava and P. E. J. Rivera-Díaz-del-Castillo, "Understanding martensite and twin formation in austenitic steels: a model describing TRIP and TWIP effects," *Acta Materialia*, vol. 128, pp. 120–134, 2017.

[2] A. Grajcar and R. Kuziak, "Softening kinetics in Nb-microalloyed TRIP steels with increased Mn content," *Advanced Materials Research*, vol. 314–316, pp. 119–122, 2011.

[3] H. Jirkova, L. Kucerova, D. Aisman, and B. Masek, "Optimization of the Q-P process parameters for low alloyed steels with 0.2% C," *Archives of Metallurgy and Materials*, vol. 59, no. 3, pp. 1205–1210, 2014.

[4] R. A. Mesquita, R. Schneider, K. Steineder, L. Samek, and E. Arenholz, "On the austenite stability of a new quality of twinning induced plasticity steel, exploring new ranges of Mn and C," *Metallurgical and Materials Transactions A*, vol. 44, no. 9, pp. 4015–4019, 2013.

[5] K. Steineder, R. Schneider, D. Krizan, C. Beal, and C. Sommitsch, "Investigation on the microstructural evolution in a medium-Mn steel (X10Mn5) after intercritical annealing," *HTM Journal of Heat Treatment and Materials*, vol. 70, no. 1, pp. 19–25, 2015.

[6] R. Kuziak, R. Kawalla, and S. Waengler, "Advanced high strength steels for automotive industry," *Archives of Civil and Mechanical Engineering*, vol. 8, no. 2, pp. 103–117, 2008.

[7] B. C. De Cooman, K. Chin, and J. Kim, *High-Mn TWIP Steels for Automotive Applications*, Intech, Open Access Publisher, Rijeka, Croatia, 2011.

[8] L. A. Dobrzański, M. Czaja, W. Borek, K. Labisz, and T. Tański, "Influence of hot-working conditions on a structure of X11MnSiAl17-1-3 steel for automotive industry," *International Journal of Materials and Product Technology*, vol. 51, no. 3, pp. 264–280, 2015.

[9] M. Jabłońska and A. Śmiglewicz, "A study of mechanical properties of high manganese steels after different rolling conditions," *Metalurgija*, vol. 54, no. 4, pp. 619–622, 2015.

[10] J. Kowalska, W. Ratuszek, M. Witkowska, A. Zielińska-Lipiec, and M. Kowalski, "Microstructure and texture evolution

during cold-rolling in the Fe-23Mn-3Si-3Al alloy," *Archives of Metallurgy and Materials*, vol. 60, no. 3, pp. 1789–1794, 2015.

[11] X. Yuana, L. Chena, Y. Zhao, H. Di, and F. Zhu, "Influence of annealing temperature on mechanical properties and microstructures of a high manganese austenitic steel," *Journal of Materials Processing Technology*, vol. 217, pp. 278–285, 2015.

[12] A. Grajcar, A. Płachcińska, S. Topolska, and M. Kciuk, "Effect of thermomechanical treatment on the corrosion behavior of Si- and Al- containing high-Mn austenitic steel with Nb and Ti micro-additions," *Materiali in Tehnologije*, vol. 49, no. 6, pp. 889–894, 2015.

[13] A. Grajcar, M. Kciuk, S. Topolska, and A. Płachcińska, "Microstructure and corrosion behavior of hot-deformed and cold-strained high-Mn steels," *Journal of Materials Engineering and Performance*, vol. 25, no. 6, pp. 2245–2254, 2016.

[14] E. Mazancova and K. Mazanec, "The stacking fault energy evaluation of the TWIP and TRIPLEX alloys," *Kovove Materialy*, vol. 47, no. 6, pp. 353–358, 2009.

[15] D. Kuc, E. Hadasik, G. Niewielski et al., "Structural and mechanical properties of laboratory rolled steels high-alloyed with manganese and aluminium," *Archives of Civil and Mechanical Engineering*, vol. 12, no. 3, pp. 312–317, 2012.

[16] J. S. Jeong, W. Woo, K. H. Oh, S. K. Kwon, and Y. M. Koo, "In situ neutron diffraction study of the microstructure and tensile deformation behavior in Al-added high manganese austenitic steels," *Acta Materialia*, vol. 60, no. 5, pp. 2290–2299, 2012.

[17] V. Shterner, I. B. Timokhina, and H. Beladi, "On the work-hardening behaviour of a high manganese TWIP steel at different deformation temperatures," *Materials Science and Engineering A*, vol. 669, pp. 437–446, 2016.

[18] A. Asghari, A. Zarei-Hanzaki, and M. Eskandari, "Temperature dependence of plastic deformation mechanisms in a modified transformation-twinning induced plasticity steel," *Materials Science and Engineering A*, vol. 579, pp. 150–156, 2013.

[19] A. E. Salas-Reyes, I. Mejía, A. Bedolla-Jacuinde, A. Boulaajaj, J. Calvo, and J. M. Cabrera, "Hot ductility behavior of high-Mn austenitic Fe–22Mn–1.5Al–1.5Si–0.45C TWIP steels microalloyed with Ti and V," *Materials Science and Engineering A*, vol. 611, pp. 77–89, 2014.

[20] M. Eskandari, A. Zarei-Hanzaki, J. A. Szpunar, M. A. Mohtadi-Bonab, A. R. Kamali, and M. Nazarian-Samani, "Microstructure evolution and mechanical behavior of a new microalloyed high Mn austenitic steel during compressive deformation," *Materials Science and Engineering A*, vol. 615, pp. 424–435, 2014.

[21] L. Chen, H. Kim, S. K Kim, and B. C. De Cooman, "Localized deformation due to Portevin–Le Chatelier effect in 18Mn–0.6C TWIP austenitic steel," *ISIJ International*, vol. 47, no. 12, pp. 1804–1812, 2007.

20

Experimental and Numerical Study of Mild Steel Behaviour under Cyclic Loading with Variable Strain Ranges

Paulina Krolo, Davor Grandić, and Željko Smolčić

Department of Structural Engineering and Technical Mechanics, Faculty of Civil Engineering, University of Rijeka, Radmile Matejčić 3, 51000 Rijeka, Croatia

Correspondence should be addressed to Paulina Krolo; paulina.krolo@uniri.hr

Academic Editor: Paolo Ferro

To simulate the effect of variable strains on steel grades S275 and S355, an experimental displacement control test of plate specimens was performed. Specimens were tested under monotonic and cyclic loading according to the standard loading protocol of SAC 2000. During experimental testing, strain values were measured with an extensometer at the tapered part of the specimen. Strains obtained by the experimental tests are disproportional to the applied displacements at the ends of the specimens. This phenomenon occurs due to the imperfections of the specimen, hardening of the material, and the buckling behaviour that appears in real structures due to the high deformation experienced during earthquakes. Due to the relative simplicity and wide applicability of the Chaboche hardening model of steel, the calibration of hardening parameters based on experimental test results was conducted. For the first time, calibration of steel hardening parameters was performed following the Chaboche procedure to define the cyclic behaviour with variable strain ranges. The accuracy of the hardening model with variable strain ranges, which were simulated using ABAQUS software, was verified using the experimental results.

1. Introduction

Seismic resistant steel structures designed as a dissipative structure must allow for plastic deformation to develop in its specific members [1]. The common practice is to increase the hysteretic energy as much as possible through inelastic behaviour using the ductile properties of the structure. "Plastic" members consisting of mild carbon steel (S235 to S355) in terms of ductility, strength, and stiffness will dissipate seismic energy. Under extreme seismic action, structural steel members, especially the dissipative elements, have to resist enormous cyclic displacement, which is classified as low-cycle fatigue and characterized by repeated inelastic strain leading to material failure. The seismic resistance of the structure is estimated on the basis of structural displacements and preservation of its integrity at the largest displacements that are expected for the earthquake [2, 3]. The structure displacements in earthquake engineering are expressed as the maximum displacement of structures in the form of interstorey drift and rotation of structural member ends or their connections [4]. The response of these elements mainly depends on the geometric dimensions and hysteretic behaviour of the material [5].

In the structures that are subject to earthquake action, the occurrence of large displacements results in inelastic deformation of the material from which the structure is built. Under cyclic loading, structural steel exhibits complicated mechanical behaviour, which includes the Bauschinger effect as well as hardening behaviour. The effect of loading history on the cyclic behaviour of different types of structural steel is given in [5–7], demonstrating that the responses of structural steel under cyclic and monotonic loading are quite different. With increasing cyclic loops, structural steel exhibits cyclic hardening behaviour, and the hysteresis stress-strain curve is much higher than the monotonic stress-strain curve after steel yielding. The local strain of the material in the plastic deformation areas of structural members and connections of steel frames are not proportional to the displacements. This suggests that varying the symmetric cycle of displacement in the structure leads to unsymmetrical and variable local strain of materials [8]. This occurs due to the local hardening of materials, imperfections in the structural elements, and local buckling of the structure [9]. In experimental tests focusing on the connections, the relationship between the stresses and variable strains of materials under cyclic changes of the displacements cannot be determined [8, 9], even though it is possible to measure the local strains because the local stresses in the material cannot be measured.

To simulate the effect of variable strains that are not proportional to displacements, an experimental programme is selected in which the test specimens are subjected to uniaxial cyclic loading using displacement control by moving crosshead. A total of 24 specimens of the S275 and S355 steel are tested under the monotonic (9 specimens) and cyclic (15 specimens) loading, according to the standard loading protocol of SAC 2000 [10]. During experimental testing, strain values were measured with an extensometer at the tapered part of the specimen. Additionally, the axial force in the specimens and displacements of the moving crosshead were measured and recorded. Based on the measured axial forces, the stresses in the materials at a tapered part of the specimen were determined.

Due to the high cost of these experimental tests, the numerical simulation technique has been widely used, and it is a very powerful tool in the field of structural design. To accurately simulate the behaviour of a given structure, the behaviour of material should be carefully defined in a structural model. Experimental studies provide basic knowledge about the seismic performance of structural steel; however, those experimental results also need to be defined in constitutive form for further implementation in numerical simulation. Most of the calculations are performed using a standard tensile test, which is based on uniaxial loading and the stress-strain relationship, commonly defined as bilinear or multilinear. Such test results are useful only for simple elastic problems or elastoplastic problems with low plastic deformation. However, these models cannot provide an accurate simulation of steel material under cyclic loading. Many researchers have proposed constitutive models to simulate material under cyclic loading. Ramberg and Osgood [11] proposed a three-parameter stress-strain constitutive model of the skeleton curve, which is widely used in metal materials. The application of this model is shown [7, 12] on structural steel. Chaboche [13] proposed a cyclic constitutive model that includes the isotropic and kinematic hardening and is used to simulate the inelastic behaviour of materials that are subjected to cyclic loading. This model is applicable in most finite element software in which the nonlinear combined hardening model is supported. Calibration of hardening parameters of materials is usually conducted on specimens exposed to symmetrical cycles, with a constant strain range [14–21]. The resulting parameters are then used in the numerical simulation for solving different engineering problems. However, the structure under the influence of the earthquake action is exposed to variable strain ranges. The complex hardening law of steel is not practical for application due to changes in the strain ranges of the materials that are used in the structure when exposed to an earthquake of random nature. Due to the relative simplicity and wide applicability of the Chaboche hardening model of steel in existing software packages, the calibration of hardening

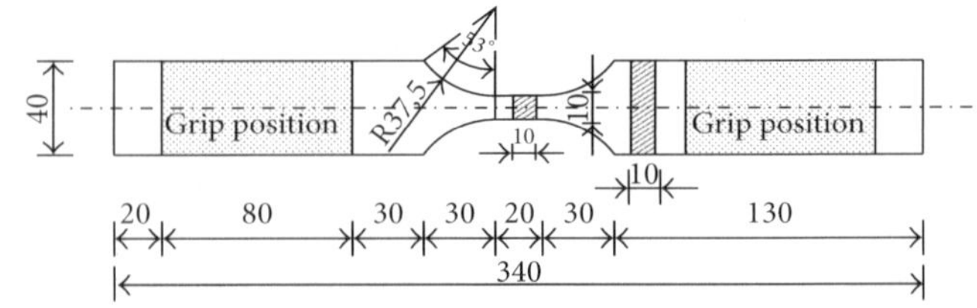

FIGURE 1: Shape and size of the specimen (mm).

parameters based on experimental test results is conducted. To verify the accuracy of the hardening model with variable strain ranges, experimental results will be simulated using ABAQUS software [22].

2. Experimental Study

2.1. Description of Specimens and the Device. As the plated elements are widely used in engineering structures, the plate specimens are adopted in the experimental study instead of test pieces with round cross sections. The materials used in this research were European mild hot-rolled structural steel S275 and S355. A total of 24 plate specimens were tested; the specimen shapes and sizes are shown in Figure 1. The loading device used was the Zwick/Roell Z600, which is designed to perform the tensile and compression tests (see Figure 2). The strains were measured with a gauge length of 20 mm. The minimum yield strength was specified as 275 MPa and 355 MPa for steel grades S275 and S355, respectively, for thicknesses below 16 mm. The steel S275 should exhibit the ultimate tensile strength within the range of 430 MPa and 580 MPa for thicknesses below 16 mm, whereas the steel S355 should exhibit a value within the range of 470 MPa and 630 MPa for thicknesses below 16 mm, according to the EN 10025 standard [23]. The chemical composition of the liquid alloy provided by the processing factory is shown in Table 1. M specimens are used for the monotonic tests, whereas C specimens are used for cyclic tests.

During the experimental tests, in addition to the strain value, the axial force in the specimens and displacements of the moving crosshead were also recorded. Stress is defined as the ratio between the axial force and the initial cross section area at a tapered part of the specimen. Thus, the stress-strain curves for monotonic and cyclic loading (hysteresis curve) were obtained. Test management and registration of the data were conducted using testXpert II software [24].

2.2. Monotonic Test. To verify the mechanical properties of steel grades S275 and S355, the uniaxial tensile tests were performed according to the standard for metallic materials EN ISO 6892-1:2009 [25] for 9 specimens at room temperature. The strains were measured by the extensometer on the gauge length of 20 mm. Test values of Young's modulus E, yield stresses f_y, ultimate stresses f_u, yield strains ε_y, ultimate strains ε_u, and fracture strains ε_{u1} are summarized in Table 2, while the monotonic loading stress-strain curves are shown in Figure 3. The results of the steel S275 and steel S355 used in this research demonstrate that the yield stress f_y and

FIGURE 2: Details of the device.

TABLE 1: Chemical composition of liquid alloys of S275 and S355 (%).

	Steel grade	C	Mn	Si	P	S	N	Cu	Ni	Cr	Mo	V	Al	Ti	Nb
EN 10025 standard	S275	<0.18	<1.5	<0.55	<0.03	<0.03	<0.012	<0.55	—	—	—	—	—	—	—
	S355	<0.2	<1.6	>0.14 <0.25	<0.03	<0.03	<0.012	<0.55	—	—	—	—	—	—	—
Measured value	S275	0.173	1.02	0.021	0.016	0.005	0.0051	0.008	0.012	0.014	0.001	0.001	0.05	0.001	0.001
	S355	0.16	1.36	0.192	0.015	0.007	0.005	0.04	0.02	0.07	0.01	0.01	0.041	<0.001	0.03

TABLE 2: Main mechanical properties of specimens for S275 and S355 obtained by monotonic loading.

Type	E [GPa]	f_y [MPa]	f_u [MPa]	ε_y [%]	ε_u [%]	ε_{u1} [%]
M_S275_01	213	323	471	0.187	19.04	50.2
M_S275_02	203	309	470	0.188	18.92	50
M_S275_03	209	318	471	0.175	18.93	50.1
Mean	208	317	471	0.183	18.96	50.1
M_S355_01	208	385	542	0.221	15.69	47.5
M_S355_02	195	380	553	0.222	15.05	47.3
M_S355_03	200	373	546	0.244	15.17	47.2
M_S355_04	199	399	556	0.228	14.99	46.8
M_S355_05	205	386	569	0.289	15.40	47.4
M_S355_06	188	381	548	0.243	15.32	47.4
Mean	199	384	552	0.288	15.27	47.3

ultimate stress f_u are in the range specified in the standard. The elongation value of steel shown in Table 2 indicates the high ductility of S275 and S355.

2.3. Cyclic Test. The purpose of the cyclic experimental test is to evaluate the seismic performance of material as an important part of the structure. A series of cyclic loading tests were programmed according to the standard loading protocol by SAC 2000 [10] (see Figure 4). The SAC programme was originally developed as a basic loading protocol for the beam to column connections in steel moment resisting frames. The basic loading history is the multiple step test, in which the loading (deformation) history consists of stepwise increasing cycles. The displacement parameter, which is used to control the loading history, is the interstorey drift angle. Since the material is an important component of the structure, the SAC protocol, which simulates the impact of earthquakes, was chosen for an experimental test of steel as a displacement control mode. A total of 15 specimens with grades of S275 and S355 were tested for the same loading protocol and five different maximum strain values (see Table 3). Maximum displacements δ_G are shown in Table 3, corresponding to the strain of steel at which significant strains of members were developed, and seismic energy was dissipated. The

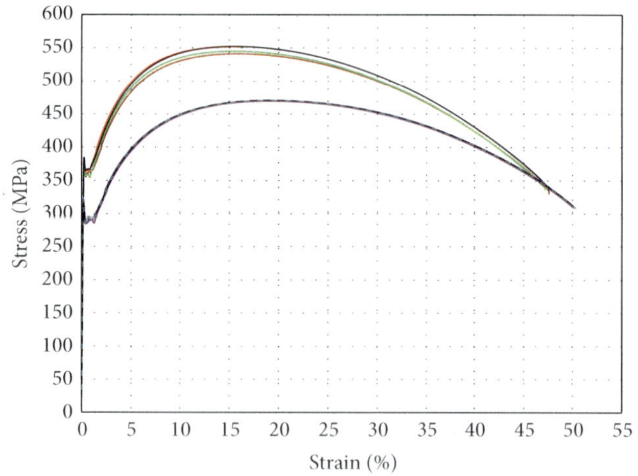

FIGURE 3: Stress-strain curves for monotonic loading.

maximum displacement of 5.23 mm (Lo. 1), according to the monotonic test (see Figure 4), corresponds to a strain of 15%, the displacement of 3.08 mm (Lo. 2) corresponds to a strain of 8%, the displacement of 2.02 mm (Lo. 3) corresponds to a strain of 5%, the displacement of 2.09 mm (Lo. 4) corresponds to a strain of 5%, and the displacement of 1.73 mm (Lo. 5) corresponds to a strain of 4%. To reduce the problems of buckling behaviour due to compression strain, the test specimens were shortened, compared to the standard uniaxial tension test specimen. Zhou et al. [5] verified the catted shape of the specimen, and monotonic stress-strain curves obtained from the catted specimen were compared with that using a standard specimen.

The strain was measured using extensometers with a gauge length of 20 mm, and the obtained values were used to control the experimental results. The loading frequency was 0.025 Hz (40 s) per each cycle, which includes one tension and one compression half cycle. The first three specimens C_S275_01, C_S275_02, and C_S275_03 were tested under the loading protocol Lo. 1 to maximum displacement values of 5.23 mm (total of 32 cycles), 4.36 mm (total of 30 cycles), and 3.42 mm (total of 28 cycles), respectively. In all three cases, the buckling occurred after 27 cycles, whereas C_S275_01 was fractured in the 32 tensile half cycles. As a result of the testing, strain ε at time t is obtained and is disproportional to the applied displacements δ_G, which are subjected at the ends of the specimen (see Figure 5). This phenomenon occurs due to the imperfections of the specimen and hardening of the material, and buckling occurs in the actual structures due to the large deformation under earthquake action. Specimen C_S275_04 was tested under the loading protocol Lo. 2 to the maximum displacement value of 3.08 mm (total of 32 cycles), whereas specimens C_S275_05 and C_S275_06 were tested to the maximum strain value of 2.52 mm (total of 30 cycles). After the last cycle, the specimens were tensile and stretched to fracture (Figure 5).

TABLE 3: Loading protocol details.

Load step	Number of cycles	δ_G [mm] Lo. 1	Lo. 2	Lo. 3	Lo. 4	Lo. 5
1	6	0.60	0.45	0.40	0.47	0.46
2	6	0.67	0.50	0.42	0.51	0.48
3	6	0.94	0.59	0.49	0.56	0.53
4	4	1.25	0.70	0.54	0.63	0.57
5	2	1.82	1.00	0.67	0.78	0.68
6	2	2.38	1.33	0.85	0.97	0.82
7	2	3.42	1.94	1.25	1.35	1.12
8	2	4.36	2.52	1.64	1.73	1.43
9	2	5.23	3.08	2.02	2.09	1.73

Capacity reduction due to buckling behaviour occurs only in the specimen C_S275_04. Specimens C_S275_07, C_S275_08, and C_S278_09 were tested under the loading protocol Lo. 3 to the maximum displacement value of 2.02 mm (total of 32 cycles). No capacity reduction occurred due to the buckling behaviour. Specimens C_S355_01, C_S355_02, and C_S355_03 were tested under the loading protocol Lo. 4 to the maximum displacement value of 2.09 mm (total of 32 cycles; Figure 5), whereas specimens C_S355_04, C_S355_05, and C_S355_06 were tested under loading protocol Lo. 5 to the maximum displacement value of 1.73 mm (total of 32 cycles; Figure 5). After the last cycle, specimen C_S355_01 to 06 was tensile and stretched to fracture.

The stress σ and strain ε relationships present the hysteresis curve shown in Figure 6, obtained for both steel grades using five different loading protocols with variable strain ranges.

A comparison between the monotonic stress-strain curve and hysteresis stress-strain curve obtained by cyclic loading is shown in Figure 7. A monotonic stress-strain curve is often used as a constitutive model of the material in seismic calculations. However, hysteresis and monotonic constitutive stress-strain curves are quite different under cyclic loading. Damage accumulation due to cyclic loading leads to ductility reduction, as we can see in the strain results for ultimate strength and fracture strength. The ultimate strength value was almost equal for the monotonic and cyclic loading case. The strain at ultimate strength ε_u and strain at fracture strength ε_{u1} for the monotonic and cyclic curve indicate a significant difference. With increasing cyclic loops, the structural steel exhibits cyclic hardening behaviour, and the hysteresis stress-strain curve is much higher than the monotonic stress-strain curve after steel yielding. The fracture strain of steel S275 is maximally reduced by 43.47%, and S355 is reduced by 27.84% after cyclic loading.

3. Numerical Study of Cyclic Behaviour

To simulate the accurate behaviour of materials, the model of material should be carefully defined in the finite element model. Under cyclic loading, with increasing cyclic loops, the steel exhibits hardening behaviour. By repeating the cycle, the

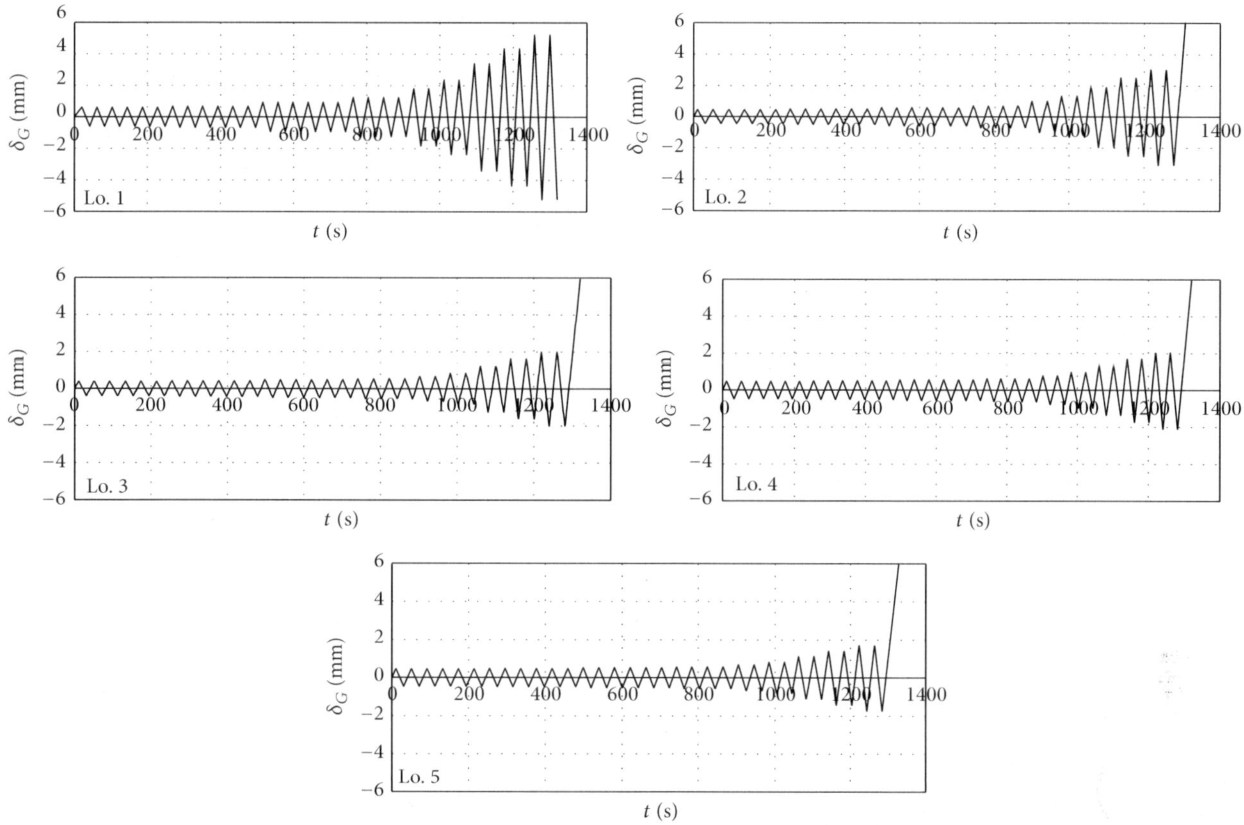

FIGURE 4: Cyclic loading histories according to SAC 2000 [10].

steel includes the Bauschinger effect, which is characterized by reduced yield stress when the direction of strain is changed and decreases with continued cyclic loading. The accurate constitutive model of steel under cyclic loading is quite important to the numerical simulation. Due to the relative simplicity and wide applicability of the Chaboche hardening model [13] of steel, in most software packages, calibration of hardening parameters based on experimental test results is performed. The accuracy of the hardening model with variable strain ranges was verified using the experimental results that were simulated using ABAQUS software [22]. The nonlinear isotropic/kinematic hardening model was used in the simulation for the von Mises flow rule. The Chaboche hardening model consists of two components: an isotropic hardening component in which the yield surface remains the same shape but expands with increasing stress and a kinematic hardening component in which the yield surface remains in the same shape and size but is translated in the stress space. The cyclic hardening behaviour of steel is described by the isotropic hardening law, whereas the Bauschinger effect is described by the kinematic hardening law.

3.1. Constitutive Model. The yield surface is defined by the function, using the von Mises yield criterion:

$$f = J_2(\sigma - \alpha) - \sigma^0 = 0, \tag{1}$$

where σ^0 is the yield stress and $J_2(\sigma - \alpha)$ is the equivalent von Mises stress with respect to the backstress α, defined as

$$J_2(\sigma - \alpha) = \sqrt{\frac{3}{2}\left(S - \alpha^{\text{dev}}\right) : \left(S - \alpha^{\text{dev}}\right)}, \tag{2}$$

where σ is the stress tensor, S is the deviatoric stress tensor, and α^{dev} is the deviatoric part of the backstress tensor.

The isotropic hardening behaviour defines the evolution of the yield surface size σ^0 as a function of the equivalent plastic strain $\bar{\varepsilon}^{\text{pl}}$ and can be expressed as the simple exponential law:

$$\sigma^0 = \sigma|_0 + Q_\infty\left(1 - e^{-b\bar{\varepsilon}^{\text{pl}}}\right), \tag{3}$$

where $\sigma|_0$ is the initial yield stress at zero plastic strain and Q_∞ and b are material parameters. Q_∞ is the maximum change in the size of the yield surface, and b defines the rate at which the size of the yield surface changes as plastic strain develops.

The kinematic hardening behaviour is defined as an additive combination of a purely kinematic term (linear Ziegler hardening law) and relaxation term, which introduces the nonlinearity and can be expressed as follows:

$$d\alpha_i = \frac{C}{\sigma^0}(\sigma - \alpha)\, d\bar{\varepsilon}^{\text{pl}} - \gamma\alpha_i d\bar{\varepsilon}^{\text{pl}}, \tag{4}$$

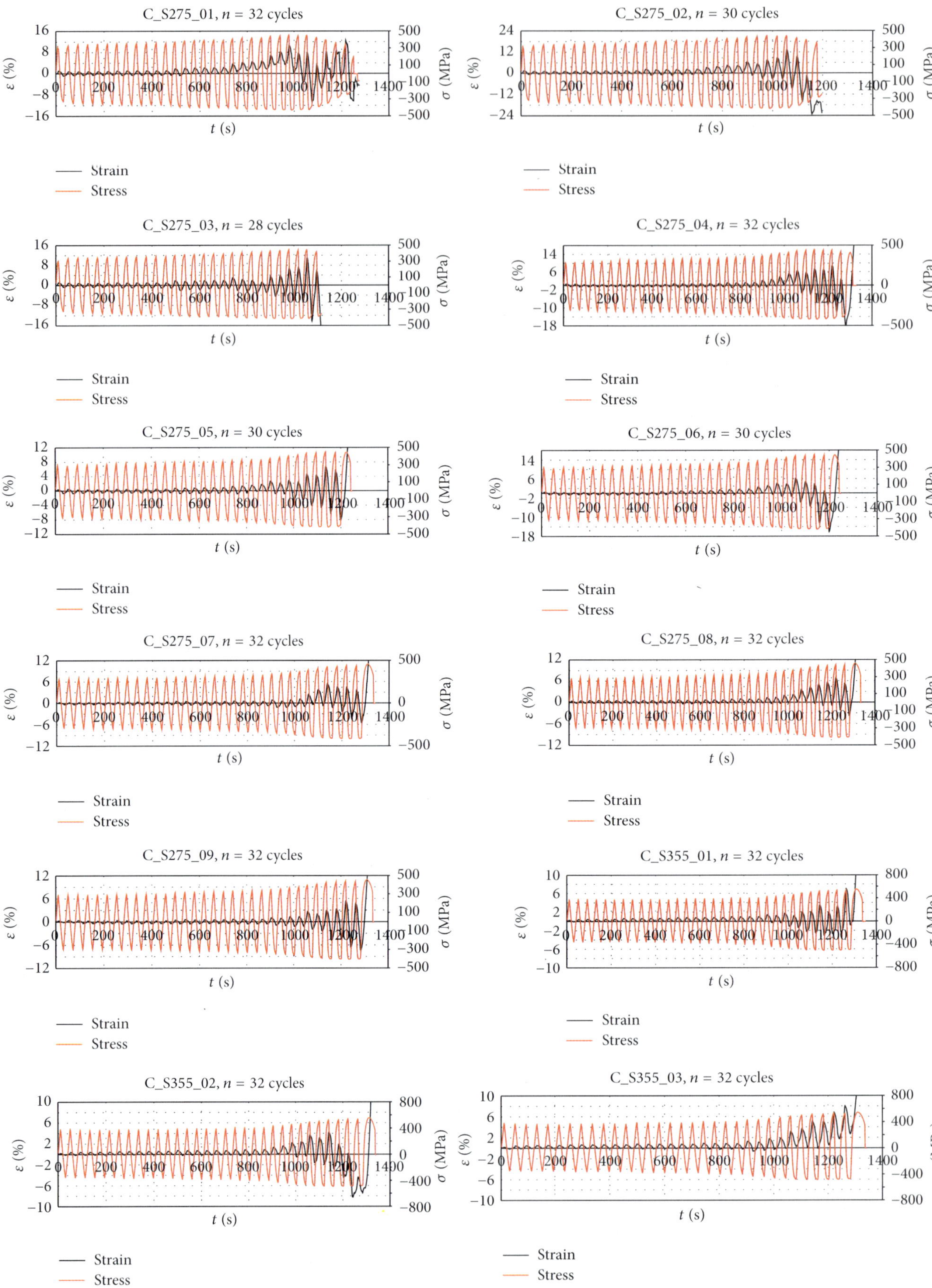

FIGURE 5: Continued.

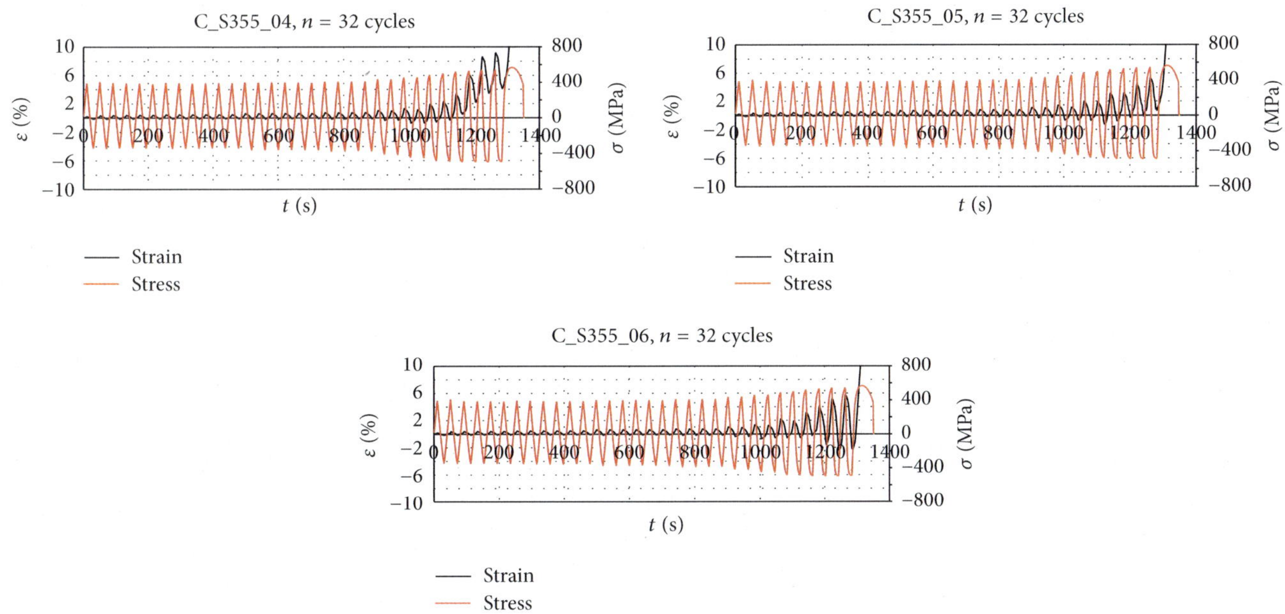

FIGURE 5: Strain and stress records for 15 specimens of steel grades S275 and S355.

where C and γ are the material parameters that are calibrated from cyclic test data. C is the initial kinematic hardening module, whereas γ determines the rate at which the kinematic hardening module decreases with increasing plastic deformation.

The overall backstress is computed from the following expression:

$$\alpha = \sum_{i=1}^{N} \alpha_i, \quad (5)$$

where N is the number of backstresses; Chaboche proposed using three backstresses ($N = 3$).

3.2. Calibration of the Material Parameters. For the first time, the Chaboche model [13] was used for calibration of isotropic and kinematic hardening parameters (Q_∞, b, C, and γ) of steel subjected to cyclic behaviour with variable strain ranges. The experimental test results of specimens C_S275_09 and C_S355_06 were used for calibration of isotropic and kinematic hardening parameters. Those two specimens show the most stabilized cycles and minimum effect of buckling behaviour of steel grades S275 and S355, respectively. The numerical model for simulation of the cyclic behaviour of steel is modelled as a unit beam element. That unit beam finite element model represents the tapered part of the tested specimen in which the strains were measured. Parameters E (Young's modulus), ν (Poisson's ratio), and $\sigma|_0$ (initial yield stress at zero plastic strain), which describe the elastic behaviour of steel, were defined as the mean values from the experimental tests. The hardening parameters for each steel grade are identified by comparison of the numerically simulated hysteresis stress-strain curve and experimental hysteresis stress-strain curve using the "trial and error method." Strain values ε, which were obtained from the experimental test of specimens C_S275_09 and C_S355_06 (Figure 5), were used as loading protocol in the numerical simulation. The last half cycle of the experimental stabilized hysteresis stress-strain curve is used for calibration of the kinematic hardening parameters ($C_1, C_2, C_3, \gamma_1, \gamma_2, \gamma_3$). Figure 8 shows the comparison between the obtained numerical simulations and experimental test of cyclic behaviour for both calibrated specimens. The numerical model does not contain the damage model. The described model shows the behaviour of steel up to the strain value, which corresponds to the ultimate strength obtained by the experimental tests. Calibrated isotropic and kinematic hardening parameters for the steel grades S275 and S355 using previously described protocol and test results of specimens C_S275_09 and C_S355_06 are shown in Table 4.

Numerical simulations performed using the calibrated isotropic and kinematic hardening parameters from Table 4 were also conducted for the other 13 strain protocols (Figure 5) that were not used in the parameter calibration procedure. For example, the numerical simulations of steel specimen C_S275_01 were subjected to the C_S275_01 protocol with 32 cycles of strain amplitudes, as shown in Figure 5, and specimen C_S275_02 was subjected to the C_S275_02 protocol with 30 cycles of strain amplitudes, also shown in Figure 5. The results and discussion of the material behaviour obtained by conducting the numerical simulations are presented in the next chapter.

4. Results and Discussion

The comparison between the experimental test and numerical simulation for steel S275 and steel S355 is presented in Figures 9 and 10. Table 5 summarizes the mechanical properties of the experimental test and numerical simulation results for both grades of steel. Table 5 includes the values

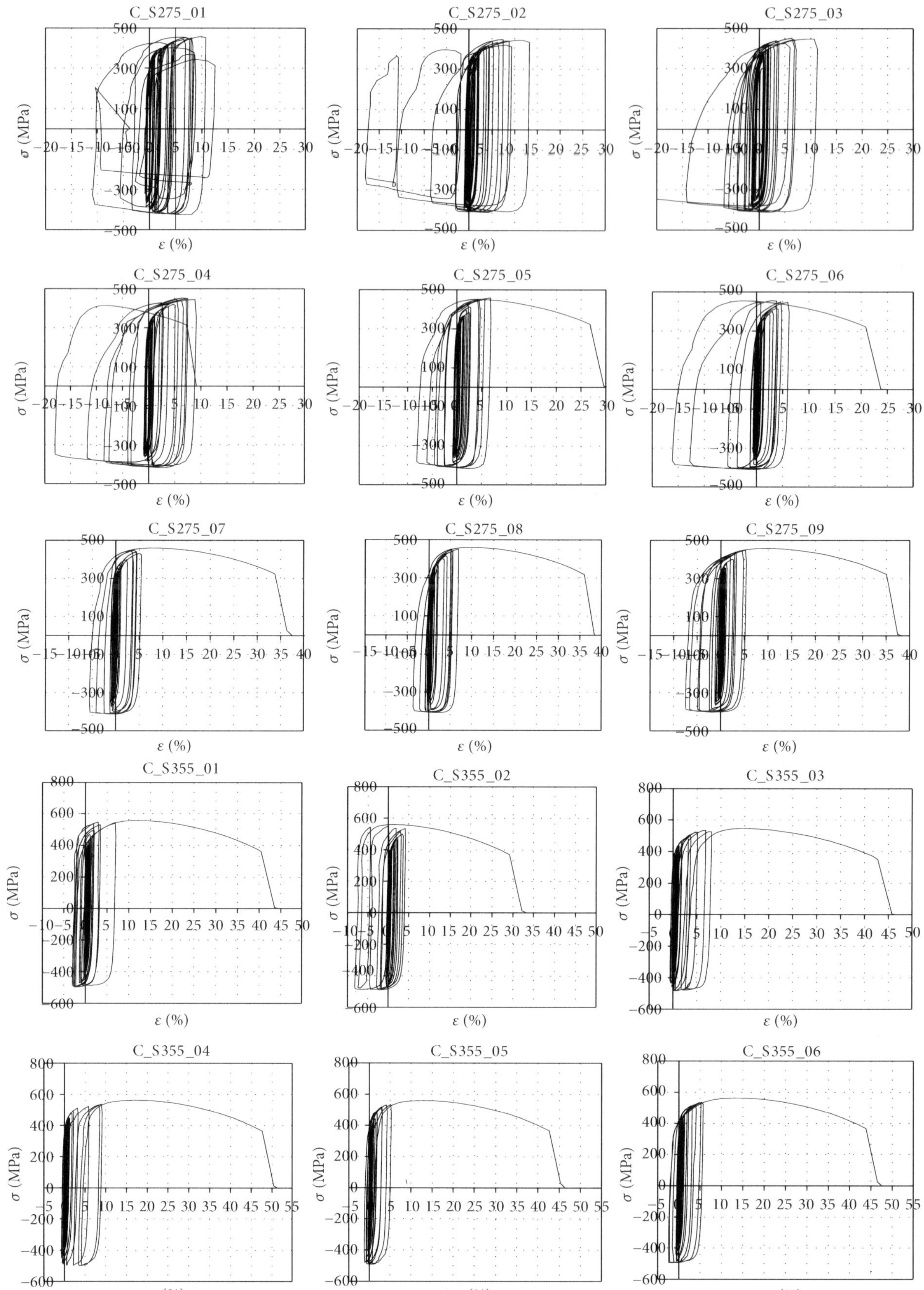

FIGURE 6: The hysteresis stress-strain curves for steel grades S275 and S355.

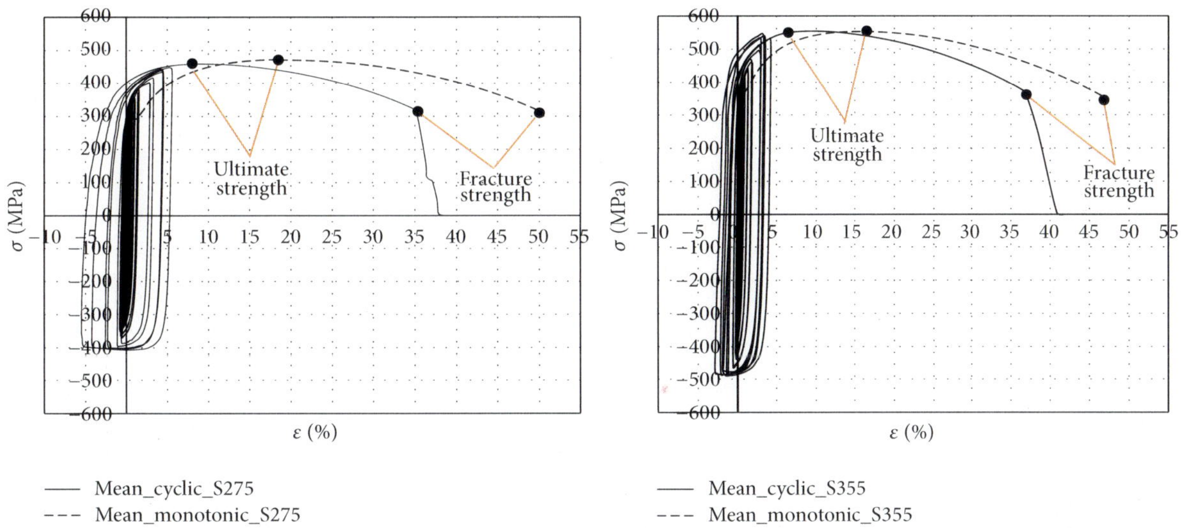

FIGURE 7: Comparison between monotonic loading stress-strain curves and hysteresis stress-strain curves.

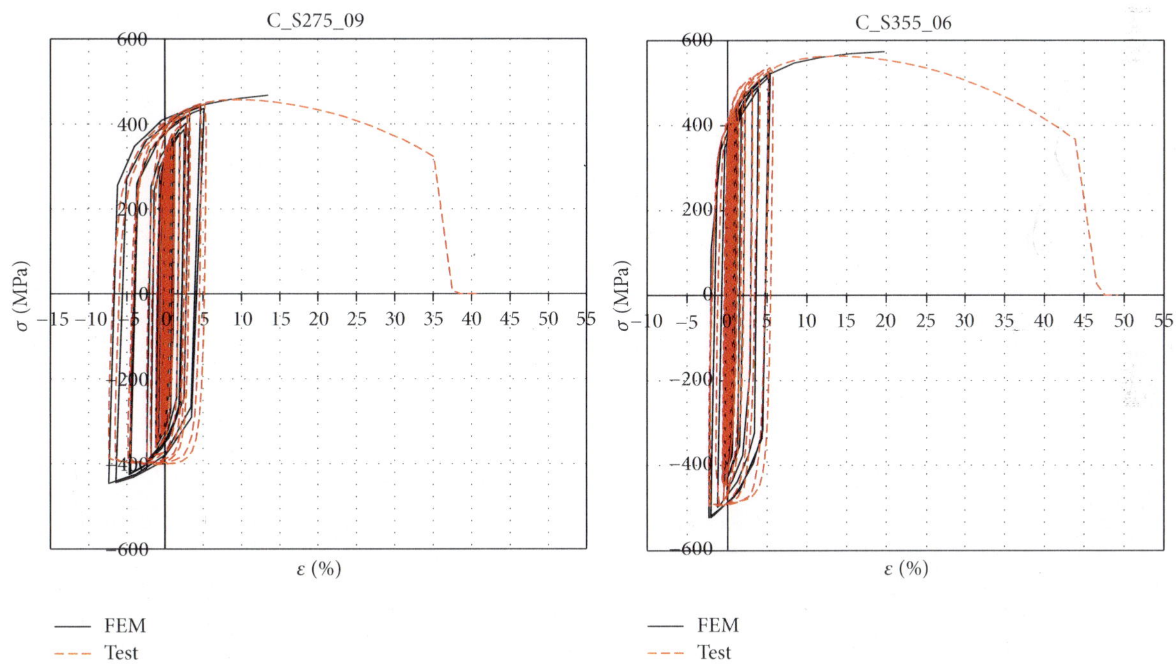

FIGURE 8: Comparison of the numerical simulation and experimental test results for the calibrated specimens.

of Young's modulus E, yield stress f_y, ultimate stress f_u, yield strain ε_u, fracture strain ε_{u1}, and Δf_u, which represents the absolute deviation of the ultimate stress value obtained numerically in relation to the ultimate stress value obtained by the experimental test. Deviation is calculated according to the following equation:

$$\Delta f_u = \left| \left(\frac{f_u\,(\text{FEM})}{f_u\,(\text{Test})} - 1 \right) \cdot 100 \right|. \tag{6}$$

The numerical model does not contain the damage model; hence, the fracture strains were obtained only from experimental test data. Hysteresis stress-strain curves obtained by numerical simulation (FEM) fit the hysteresis stress-strain curves obtained by the experimental test (Test) in cycles in which no buckling behaviour was observed (see Figures 9 and 10). Numerical simulations show the visible difference in describing the last few cycles in compression when the buckling occurred. In these cycles, the satisfactory agreement between numerical and experimental results is achieved in tension. The largest buckling behaviour of the tested specimens occurred in the first six tested specimens (C_S275_01 to C_S275_06), which were subjected to loading protocols with the highest level of strain (13.3% to −23.2%). Significantly less buckling behaviour occurred in the tested

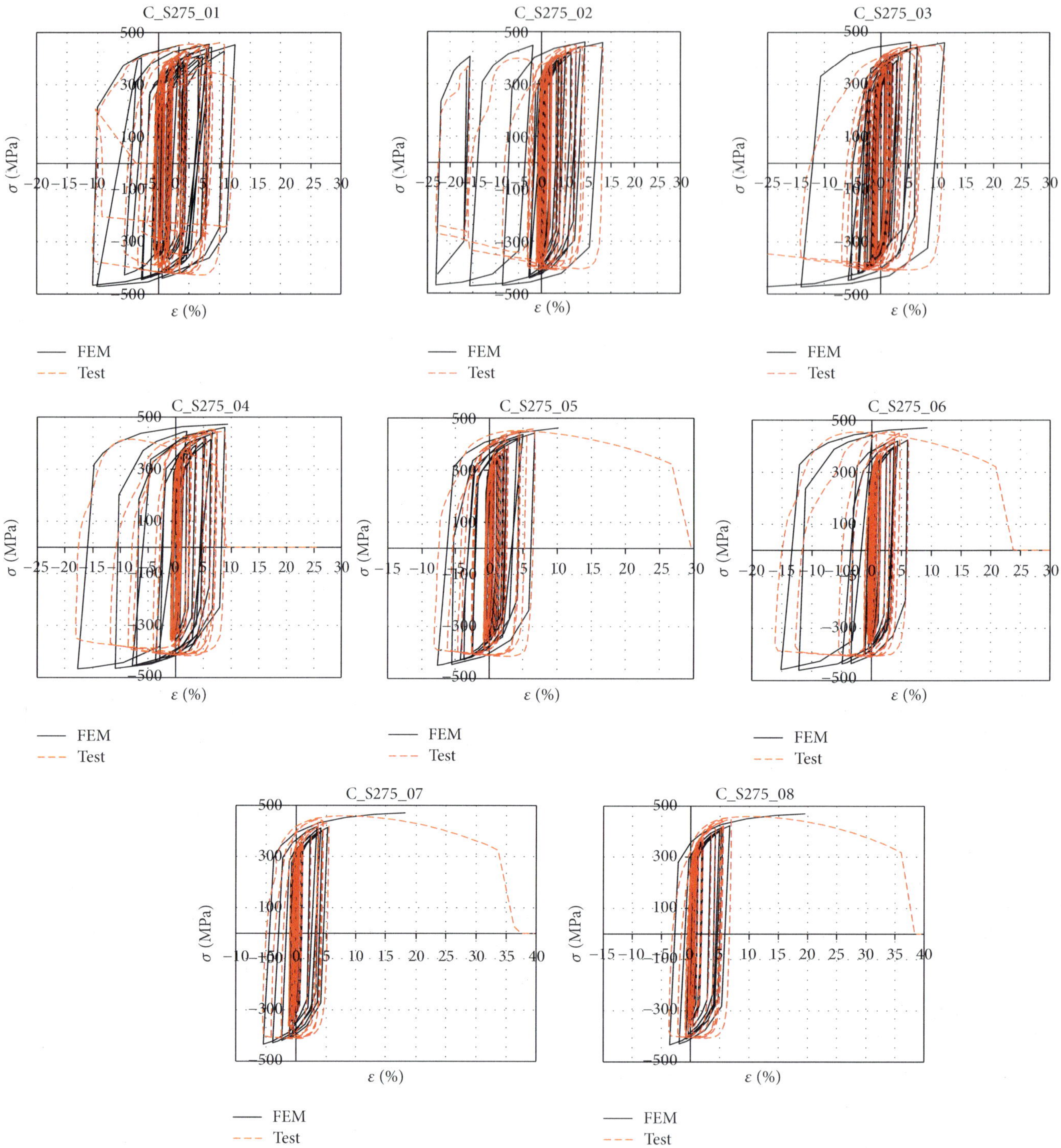

FIGURE 9: Comparison of the numerical simulation and experimental test for steel S275.

specimens C_S275_07 to C_S275_09, which were subjected to a loading protocol with strain levels in the range of 6.9% to −6.5% and C_S355_01 to C_S355_06 with strain levels of 9.2% to −7.5%. The buckling problems in the conducted experimental tests occur for all tested specimens only in the last few cycles at each level of strain greater than 2% for steel grade S275 and 2.8% for steel grade S355, respectively.

Ultimate stresses obtained by numerical simulation were determined for the strain value that corresponds to the ultimate stresses obtained by the experimental test. As can be observed from Table 5, the differences between the ultimate stresses obtained by the numerical simulation and experimental test for specimens of steel grade S275 show values between 0.44% and 3.49%, with a mean value of 0.86%.

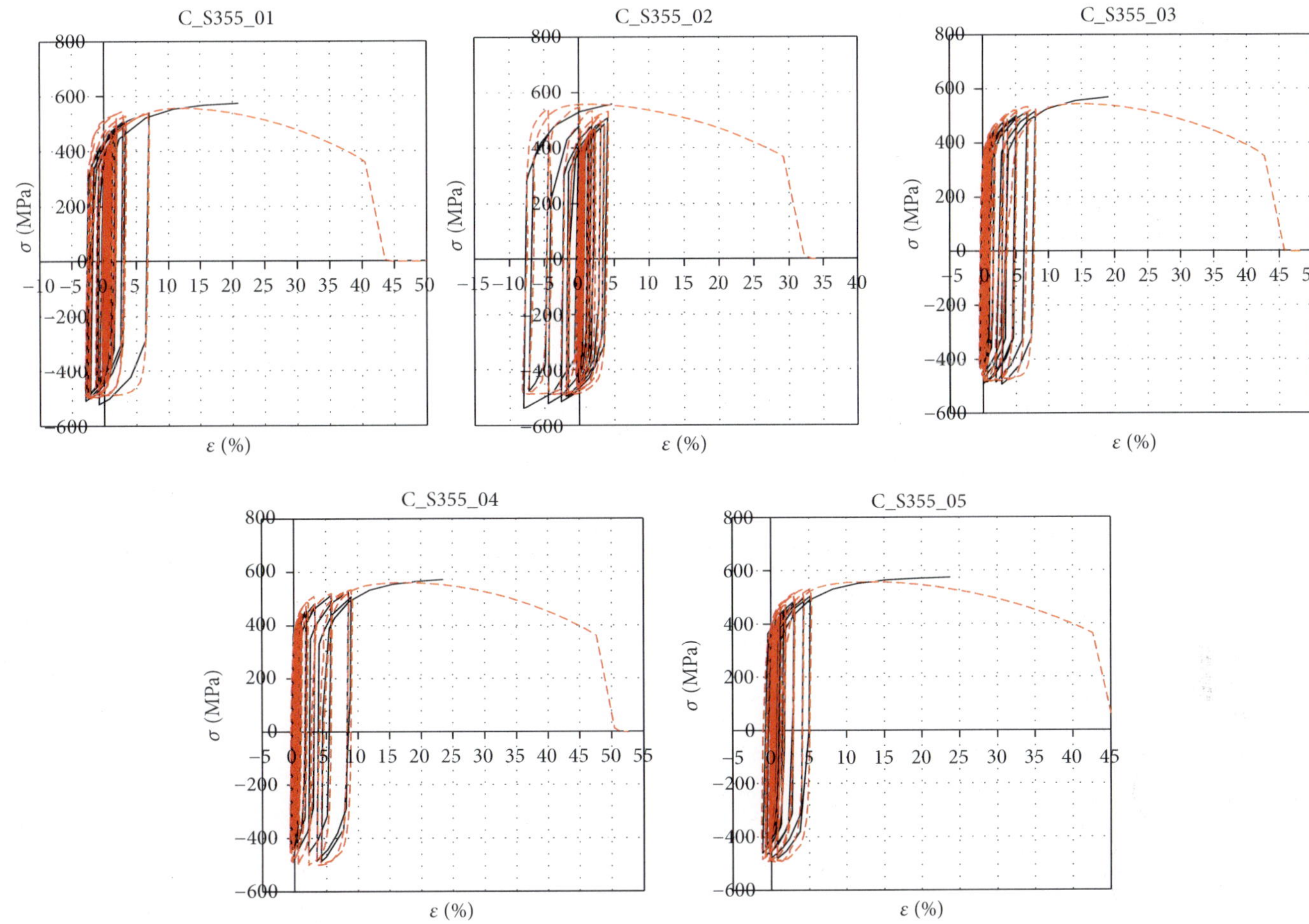

FIGURE 10: Comparison of the numerical simulation and experimental test results for steel S355.

The differences of the ultimate stresses between numerical simulation and experimental test of steel grade S355 take values between 0.18% and 1.46%, with a mean value of 0.36%.

Based on previous analysis, it can be concluded that the determined hardening parameters obtained by the Chaboche model effectively describe the behaviour of steel grades S275 and S355 under the influence of cyclic loading with variable strain ranges.

5. Conclusions

A series of experimental tests were conducted on structural mild carbon steel of S275 and S355 under monotonic and cyclic loading. To simulate real seismic action on the material, a loading protocol was used in experimental testing with variable strain ranges that were not proportional to the applied displacements. During the experimental testing, strain on the specimen was measured with an extensometer. The strain values obtained from the experimental test were then used as loading protocol for a numerical simulation. An accurate constitutive model of steel under cyclic loading is quite important for numerical simulation. For this purpose, the Chaboche hardening model was used for the calibration of the hardening parameters based on the experimental test data. The fundamental contribution of this research is a new approach to the calibration procedure in which the experimental values of strain are used for numerical simulation of steel behaviour. Calibration of hardening parameters of steel according to the Chaboche procedure was applied for the first time to define the cyclic behaviour of steel with variable strain ranges. The hardening model includes isotropic and kinematic hardening. The accuracy of the hardening model with variable strain ranges was verified using the experimental results that were simulated in the ABAQUS software. The analysis includes the stress-strain response, hardening behaviour, and damage evaluation. Based on this work, the following conclusions can be drawn:

(1) The behaviour of S275 steel and S355 steel under cyclic loading and monotonic loading is quite different. Hysteresis stress-strain curves lie much higher than monotonic stress-strain curves between the steel yielding and ultimate strength of cyclically loaded specimens.

(2) Increasing the cyclic loops and strain amplitudes directly affects the material ductility. Ultimate strain and fracture strain are reduced if the cyclic and monotonic stress-strain curves are compared.

TABLE 4: Calibrated parameters of mild steel grade S275 and S355.

Material	Elastic behaviour		Plastic behaviour								
			Kinematic hardening							Isotropic hardening	
	E (MPa)	ν	$\sigma\|_0$ (MPa)	C_1 (MPa)	γ_1	C_2 (MPa)	γ_2	C_3 (MPa)	γ_3	Q_∞ (MPa)	b
S275	207000	0.3	285	13921	765	4240	52	1573	14	25.6	4.4
S355	185000	0.3	386	5327	75	1725	16	1120	10	20.8	3.2

TABLE 5: Summary of experimental and numerical test results of steel specimens of grade S275 and S355.

Type	E [GPa]	f_y [MPa]	f_u [MPa]		Δf_u [%]	ε_u [%]	ε_{u1} [%]
	Test	Test	Test	FEM		Test	Test
C_S275_01	206	334	459	454	1.09	9.88	—
C_S275_02	187	332	447	449	0.45	6.40	—
C_S275_03	195	279	453	447	1.32	6.38	—
C_S275_04	199	287	452	449	0.66	6.38	—
C_S275_05	195	322	459	443	3.49	6.73	—
C_S275_06	180	287	456	458	0.44	2.86	—
C_S275_07	218	289	459	453	1.31	8.10	33.67
C_S275_08	199	282	460	464	0.86	10.40	36.05
C_S275_09	207	284	458	460	0.44	9.41	35.03
C_S355_01	191	380	558	555	0.54	11.64	40.43
C_S355_02	190	384	559	558	0.18	3.61	29.20
C_S355_03	193	386	546	554	1.46	14.91	42.80
C_S355_04	192	379	563	560	0.53	17.61	47.62
C_S355_05	179	382	560	559	0.18	12.92	42.64
C_S355_06	185	386	563	564	0.18	13.97	43.82

(3) The loading history has a significant effect on the cyclic behaviour of steel; the cyclic stress-strain curves are not unique for the same material due to variable strain.

(4) Numerical stress-strain curves using Chaboche hardening parameters agree well with the stress-strain curves obtained by the experimental test.

(5) The obtained hardening parameters can be used in engineering practice and finite element analysis of the cyclic behaviour of steel structures in seismic situations.

(6) The numerical model of steel in this study does not consider the damage model; hence, the model does not account for the decrease in strength after reaching the ultimate strength. This will be analysed in future work.

Competing Interests

The authors declare that there are no competing interests regarding the publication of this paper.

Acknowledgments

This work has been supported in part by Ministry of Science, Education and Sports of the Republic of Croatia under the project Research Infrastructure for Campus-based Laboratories at the University of Rijeka (no. RC.2.2.06-0001). The project has been cofunded by the European Fund for Regional Development (ERDF). This research was performed within the scientific project Grant no. 13.05.1.1.01, supported by the University of Rijeka.

References

[1] D. Grecea, N. Muntean, and D. Dubina, "Control of bolted beam-to-column connections in moment joints by T-stub properties," in *Proceedings of the 7th International Conference on Behaviour of Steel Structures in Seismic Areas*, pp. 249–254, CRC Press, Santiago, Chile, January 2012.

[2] EN, "Eurocode 8: design of structures for earthquake resistance—Part 1: general rules, seismic actions and rules for buildings," EN 1998-1, European Committee for Standardization (CEN), Brussels, Belgium, 2004.

[3] FEMA, "Improvement of nonlinear static seismic analysis procedure," FEMA 440, Prepared by the Applied Technology Council (ATC-55 Project), Washington, DC, USA, 2005.

[4] V. Gioncu and F. M. Mazzolani, *Ductility of Seismic Resistant Steel Structures*, Spon Press, London, UK, 2002.

[5] F. Zhou, Y. Chen, and Q. Wu, "Dependence of the cyclic response of structural steel on loading history under large inelastic strains," *Journal of Constructional Steel Research*, vol. 104, pp. 64–73, 2015.

[6] Y. Shi, M. Wang, and Y. Wang, "Experimental and constitutive model study of structural steel under cyclic loading," *Journal of Constructional Steel Research*, vol. 67, no. 8, pp. 1185–1197, 2011.

[7] G. Shi, M. Wang, Y. Bai, F. Wang, Y. Shi, and Y. Wang, "Experimental and modeling study of high-strength structural steel under cyclic loading," *Engineering Structures*, vol. 37, pp. 1–13, 2012.

[8] C.-M. Uang, Q.-S. Yu Kent, S. Noel, and J. Gross, "Cyclic testing of steel moment connections rehabilitated with RBS or welded haunch," *Journal of Structural Engineering*, vol. 126, no. 1, pp. 57–68, 2000.

[9] G. Shi, Y. Shi, and Y. Wang, "Behaviour of end-plate moment connections under earthquake loading," *Engineering Structures*, vol. 29, no. 5, pp. 703–716, 2007.

[10] SAC-SteelProject, *Protocol for Fabrication, Inspection, Testing and Documentation of Beam-Column Connection Tests and Other Experimental Specimens*, FEMA, Oakland, Calif, USA, 1997.

[11] W. Ramberg and W. R. Osgood, "Description of stress-strain curves by three-parameters," Tech. Rep. 902, National Advisory for Aeronautics, 1943.

[12] P. Dusicka, A. M. Itani, and I. G. Buckle, "Cyclic response of plate steels under large inelastic strains," *Journal of Constructional Steel Research*, vol. 63, no. 2, pp. 156–164, 2007.

[13] J. L. Chaboche, "Time-independent constitutive theories for cyclic plasticity," *International Journal of Plasticity*, vol. 2, no. 2, pp. 149–188, 1986.

[14] K. M. Zhao and J. K. Lee, "Finite element analysis of the three-point bending of sheet metals," *Journal of Materials Processing Technology*, vol. 122, no. 1, pp. 6–11, 2002.

[15] S. M. H. Kabir, T.-I. Yeo, and S.-H. Kim, "Characterization of material parameters," in *Proceedings of the World Congress on Engineering (WCE '09)*, vol. 2, London, UK, July 2009.

[16] M. Rezaiee-Pajand and S. Sinaie, "On the calibration of the Chaboche hardening model and a modified hardening rule for uniaxial ratcheting prediction," *International Journal of Solids and Structures*, vol. 46, no. 16, pp. 3009–3017, 2009.

[17] S. J. Zakavi, M. Zehsaz, and M. R. Eslami, "The ratchetting behavior of pressurized plain pipework subjected to cyclic bending moment with the combined hardening model," *Nuclear Engineering and Design*, vol. 240, no. 4, pp. 726–737, 2010.

[18] H. Wang, Y. Yan, M. Wan, and X. Wu, "Experimental investigation and constitutive modeling for the hardening behavior of 5754O aluminum alloy sheet under two-stage loading," *International Journal of Solids and Structures*, vol. 49, no. 26, pp. 3693–3710, 2012.

[19] A. Kalnins, J. Rudolph, and A. Willuweit, "Using the nonlinear kinematic hardening material model of Chaboche for elastic-plastic ratcheting analysis," *Journal of Pressure Vessel Technology, Transactions of the ASME*, vol. 137, no. 3, Article ID 031006, 2015.

[20] V. N. Van Do, C.-H. Lee, and K.-H. Chang, "A constitutive model for uniaxial/multiaxial ratcheting behavior of a duplex stainless steel," *Materials and Design*, vol. 65, pp. 1161–1171, 2015.

[21] M. Zehsaz, F. V. Tahami, and H. Akhani, "Experimental determination of material parameters using stabilized cycle tests to predict thermal ratchetting," *UPB Scientific Bulletin, Series D: Mechanical Engineering*, vol. 78, no. 2, pp. 17–30, 2016.

[22] ABAQUS, *Analysis User'S Manual I_V, Version 6.12*, Dassault Systemes, Fremont, Calif, USA, 2012.

[23] C. European Standard, *EN 10025: Hot Rolled Products of Structural Steel*, European Committee for Standardization, Brussels, Belgium, 2004.

[24] TestXpertII, *Instruction Manual, Version 2.2*, Zwick GmbH&Co. KG, Ulm, Germany, 2015.

[25] C. European Standard, *EN ISO 6892-1: Metallic Materials-Tensile Testing-Part1: Method of Test at Room Temperature*, European Committee for Standardization, Brussels, Belgium, 2010.

21

Effects of Deposition Potentials on the Morphology and Structure of Iron-Based Films on Carbon Steel Substrate in an Alkaline Solution

Soon-Hyeok Jeon, Geun-Dong Song, and Do Haeng Hur

Nuclear Materials Safety Research Division, Korea Atomic Energy Research Institute, 989-111 Daedeok-daero, Yuseong-gu, Daejeon 305-353, Republic of Korea

Correspondence should be addressed to Do Haeng Hur; dhhur@kaeri.re.kr

Academic Editor: Daniel Guay

The purpose of this work is to investigate the effect of electrochemical deposition potential on the morphology and structure of iron-based films on the carbon steel in an alkaline Fe(III)-triethanolamine solution. The deposition potentials were controlled in the range from −1.05 to −1.23 V_{SCE} for 1800 s at 80°C. Total amount of electric charge for electrodeposition process was increased with increasing deposition potential in negative direction. Pure magnetite films with a columnar and defect-free structure were deposited in the potential range from −1.05 V_{SCE} to −1.11 V_{SCE}. However, petal-like magnetite film containing ferrihydrite and iron was formed at −1.17 V_{SCE}. At more negative potential of −1.23 V_{SCE}, two distinct layers were observed: a porous outer layer containing ferrihydrite and goethite and a compact inner layer consisting of columnar metallic iron. In the potential range from −1.05 to −1.11 V_{SCE}, the pure magnetite films gradually increased the thickness and decreased the surface roughness with an increase of the overpotential. The magnetite film deposited at −1.11 V_{SCE} showed the most thick layer and smooth surface state.

1. Introduction

Magnetite is one of the principal corrosion products formed on the carbon steel surfaces of piping and equipment of the secondary coolant system in pressurized water reactor (PWR) of nuclear power plant. Magnetite is released from carbon steel surfaces into the high temperature feed water and is transported to the steam generator [1, 2]. Flow accelerated corrosion (FAC) leads to wall thinning of carbon steel piping and components exposed to flowing water or wet steam in the secondary system of PWR. The process of FAC is related to the dissolution of magnetite on the surface of carbon steel under turbulent flow [3, 4]. In addition, magnetite deposits on the outside of the steam generator tubes and tube support structure. There has been an intense interest in magnetite deposits on the nuclear materials because its removal improves heat transfer and reduces the corrosion of steam generator tube and other parts of the steam generator system [5, 6]. In order to evaluate the corrosion behavior of the magnetite covering the nuclear materials, a rigid method to deposit magnetite on the corresponding material substrates should be established. Therefore, it is of high interest to find a simple and efficient method to produce the magnetite film.

Recently, electrodeposition method has extensively been investigated to produce magnetite films [7–12]. There are two basic approaches to deposit the magnetite by anodic deposition in Fe(II) systems [7–10] and cathodic deposition in Fe(III) systems [11, 12]. Most of the previous work was conducted in the Fe(II) systems, where Fe(II) ions were complexed with organic compounds and were electrochemically oxidized to form magnetite film in an inert atmosphere at an elevated temperature. A problem with this method is that the complexed Fe(II) can be oxidized by molecular oxygen in the air; thus special attention must be paid to exclude molecular oxygen.

Sapieszko and Matijević [13] developed the hydrothermal method to produce the magnetite powder in Fe(III)-triethanolamine (TEA) solution agents. Switzer's group have modified the Fe(III)-TEA solution and developed the efficient method to produce the magnetite film on the stainless steel substrate [14], single crystal Au [15], and Ni [16] at 80°C. This method is not necessary to eliminate molecular oxygen from the deposition environment due to the good stability of Fe(III)-TEA complex [14–16]. Recently, Switzer's group showed that the polycrystalline magnetite films electrodeposited on stainless steels exhibit multistate resistance switching and resistance switching can be tuned by controlling the composition of the films by varying the deposition potential [17, 18]. Goujon et al. [19, 20] showed the characterization of magnetite films electrodeposited on the Ni-based alloys and established the optimal temperature and agitation condition for electrodeposition to grow magnetite films in the Fe(III)-TEA solution. Duan et al. [21] also presented that the concentration of Fe(III) ions had an impact on the electrodeposition rate, while the deposition temperature and time strongly affected the morphology of the magnetite films.

However, there is a few researches for the electrodeposition of iron-based films on the carbon steel substrate in the Fe(III)-TEA solution. In particular, the effect of deposition potentials on the morphology, structure, thickness, and surface roughness of the electrodeposited films on the carbon steel substrate has not been verified.

The aim of this study is to elucidate the effects of deposition potential on the morphology, structure, thickness, and surface roughness of the electrodeposited films on the carbon steel in Fe(III)-TEA solution, thus establishing the optimal deposition potential to form dense and thick magnetite films. It is considered that these results are useful for simulating the deposited corrosion products on the surface of carbon steel materials in the secondary system of nuclear power plant.

2. Experimental

2.1. Material Preparation. Carbon steel specimens for magnetite deposition were machined into 10 × 5 × 1 mm from the SA106Gr.B pipe material. The carbon steel substrates were mechanically ground up to #1100 grit silicon carbide paper and then ultrasonically cleaned in acetone and deionized water. The chemical composition of SA106Gr.B is given in Table 1.

Table 1: Chemical composition of the carbon steel (wt.%).

C	Cr	Ni	Mo	Si	Cu	Mn	S	P	Fe
0.19	0.4	0.2	0.1	0.23	0.1	1.05	0.005	0.012	Bal.

2.2. Electrodeposition of Films. The iron-based films were electrodeposited from an alkaline solution of $Fe_2(SO_4)_3$ complexed with TEA. The deposition bath for magnetite was prepared by dissolving 26 g of Fe(III) sulfate hydrate in 150 mL of 1 M TEA. It was diluted with 500 mL of deionized water. The solution was then added to a second solution of 120 g of NaOH pellets in 850 mL of deionized water. The resulting deposition solution was gray-green colored with pH 12.5 at 25°C.

The electrochemical deposition was carried out in a three-electrode cell. A saturated calomel electrode (SCE) and a platinum wire were used as the reference and counter electrode, respectively. During the deposition process, the solution was maintained at 80°C and stirred at 200 rpm. The deposition potentials were as follows: $-1.05\ V_{SCE}$, $-1.08\ V_{SCE}$, $-1.11\ V_{SCE}$, $-1.17\ V_{SCE}$, and $-1.23\ V_{SCE}$. The electrodeposition time was 1800 s at all the deposition potentials. To ensure a deposition as homogeneous as possible, the carbon steel substrate was fixed in front of the counter electrode. After electrodeposition, the electrodeposited films were carefully rinsed with deionized water and dried in air.

2.3. Characterization of Electrodeposited Films. The electrodeposited films were analyzed by scanning electron microscope-energy dispersive X-ray spectroscopy (SEM-EDS) and X-ray diffraction (XRD). In order to analyze the morphology, thickness, and chemical composition of the electrodeposited films on the carbon steel substrate, the electrodeposited films were milled by a focused ion beam (FIB) toward a vertical direction of the electrodeposited films. Adherence of electrodeposited films was confirmed by FIB-SEM. The values of the surface roughness were measured using a noncontacting surface profiler. In this paper, the root mean square roughness (R_q) is used. The thickness and roughness of electrodeposited films were measured at least three times to confirm the reproducibility. The values were presented by the average with error bars, and good reproducibility was confirmed.

The chemical species of the electrodeposited films were analyzed using X-ray photoelectron spectroscopy (XPS). An XPS analysis was carried out using a Thermo Fisher Scientific (Theta Probe AR-XPS) X-ray photoelectron spectrometer with an Al Kα X-ray source (1486.6 eV) operated at 15 kV and 150 W under a base pressure of 2.7×10^{-7} Pa. The binding energies of all peaks were corrected by the reference C 1s peak at 284.5 eV.

3. Results and Discussion

3.1. Electrodeposition of the Films. Figure 1 shows linear sweep voltammogram of the carbon steel substrate in the Fe(III)-TEA bath at 80°C. The potential was swept from the open circuit potential to $-1.30\ V_{SCE}$ at a scan rate of 10 mV/s. The electrochemical reduction of Fe(III)-TEA started at potentials more negative than $-0.95\ V_{SCE}$. The first reduction wave of this linear sweep was measured approximately between -0.95 and $-1.20\ V_{SCE}$. The second reduction wave was also observed in the potential range from -1.20 to $-1.30\ V_{SCE}$.

The electrodeposition of magnetite film in Fe(III)-TEA solution can be simplified by two steps. The first step is that the Fe(III)-TEA solution is electrochemically reduced to an Fe(II)-TEA complex. The reaction, which may occur, is expressed as reaction (1). The second step is that Fe(III)-TEA complex would then chemically react with Fe(II)-TEA to produce magnetite at the surface of the substrate in the following reaction (2). The overall electrochemical reaction leading to the formation of magnetite deposit on the substrate

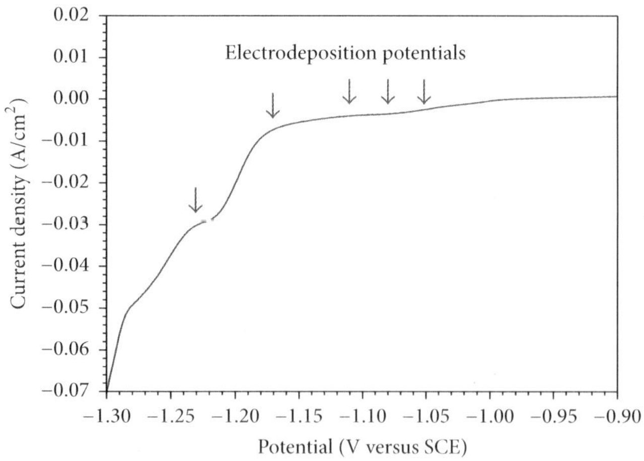

FIGURE 1: Linear sweep voltammogram of the carbon steel substrate in Fe(III)-TEA solution at 80°C.

is reaction (3). The proposed mechanism is expressed in the following reactions (1) to (3) [14–18]:

$$Fe(OH)_4TEA_2^- + e^- \longrightarrow Fe(TEA)_2^{2+} + 4OH^- \quad (1)$$

$$2Fe(OH)_4TEA_2^- + Fe(TEA)_2^{2+} \longleftrightarrow Fe_3O_4 + 6TEA + 4H_2O \quad (2)$$

$$3Fe(OH)_4TEA_2^- + e^- \longrightarrow Fe_3O_4 + 6TEA + 4H_2O + 4OH^- \quad (3)$$

Therefore, the first reduction wave observed between −0.95 and −1.20 V_{SCE} can be attributed to a one electron reaction corresponding to the above reaction (1) [14, 17]. On the contrary, the second reduction wave between −1.20 and −1.30 V_{SCE} seems to be due to a two-electron process corresponding to the reduction of Fe(II) to Fe [15–18].

Figure 2 shows the surface morphologies of the films formed at different deposition potentials. It is obvious that the morphology depends on the electrochemical potential. The films deposited at −1.05 and −1.08 V_{SCE} have highly faceted morphologies (Figures 2(a) and 2(b)). In this range of potentials, the degree of faceting and particle size of the deposits decrease as the overpotential increases, whereas the film formed at −1.11 V_{SCE} is flat morphologies (Figure 2(c)). As shown in Figure 2(d), the film deposited at −1.17 V_{SCE} has a petal-like microstructure. The film deposited at −1.23 V_{SCE} is similar to that deposited at −1.17 V_{SCE} (Figure 2(e)). These morphologies of the five films are homogeneous on the whole surface of the deposits.

At low overpotentials (−1.05, −1.08, and −1.11), the surface concentration of Fe(III) is nearly the same as the bulk concentration because the ion diffusion is sufficient. Hence, the structure is determined by the preferential growth of favorably oriented crystal faces. As a result, the deposited film formed transforms to a dense and crack-free morphology with tightly packed, well-faceted crystallites. Whereas at high overpotentials (−1.17, −1.23 V_{SCE}) the current becomes mass-transport controlled, the surface concentration of Fe(III) approaches zero because the ion diffusion is insufficient. As a result, the resultant structure is dominated by open boundaries and the deposit exhibits a petal-like structure.

Figure 3 shows the FIB-SEM images on a cross section of the electrodeposited films. The films deposited in a range from −1.05 to −1.11 V_{SCE} grew with a columnar morphology. The columns are very compact and adherent to the carbon steel substrate. No defects such as hole, gap, and crack can also be observed at the interface between the deposited films and carbon steel substrate, which confirms that the deposited films are tightly bonded to carbon steel substrate. However, the film deposited at −1.17 V_{SCE} shows a very defective structure containing numerous holes and gaps, indicating that the deposited films are bonded weakly to the carbon steel substrate. The film deposited at −1.23 V_{SCE} consists of two distinct layers: a porous external layer and a compact columnar inner layer. The numerous defects are observed in the external layer, similar to the film at −1.17 V_{SCE}. However, no defects such as hole and gap can be observed in the columns and at the interface between the inner layer and carbon steel substrate. The porous cross-sectional structure of the films formed at −1.17 V_{SCE} and −1.23 V_{SCE} is well consistent with the top view images shown in Figures 2(d) and 2(e).

Additional analysis was performed to evaluate the double layer observed from the film deposited at −1.23 V_{SCE}. Figure 4 shows the EDS element profile. Only iron without oxygen was detected in the inner columnar layer and the content of iron gradually increased as departing from the interface. It is therefore believed that the inner layer is composed of pure metallic iron.

3.2. Characterization of the Deposited Films. Figure 5 shows the XRD patterns of the deposited films at the various deposition potentials. In the range from −1.05 to −1.11 V_{SCE}, the XRD pattern shows that deposited films are all crystalline and only have peaks corresponding to magnetite. The very weak iron peak from the film at −1.08 V_{SCE} is probably originated from the carbon steel substrate. The intensity of magnetite diffraction peaks is almost equivalent. At the electrodeposition potential of −1.17 V_{SCE}, new XRD peaks corresponding to ferrihydrite and iron appeared, indicating the formation of a mixture of three structures: magnetite, ferrihydrite, and iron. When the electrodeposition potential has been moved to −1.23 V_{SCE}, additional XRD peak was detected and indexed to goethite. It should be noted that the peak intensity of iron became higher than that of magnetite. Consequently, this film consisted of a mixture of complex phases: magnetite, ferrihydrite, goethite, and iron. The above XRD results indicate that the deposition potential strongly affects not only the morphology but also the structure of the deposited films.

The detection of metallic iron is in good agreement with the observation of the inner column layer consisting of metallic iron shown in Figure 4. Ferrihydrite was also detected by Mossbauer spectroscopy and XRD on the films deposited in the same electrochemical solution [15, 19]. However, goethite and metallic iron were not reported. Since the driving force for reduction became larger at more negative

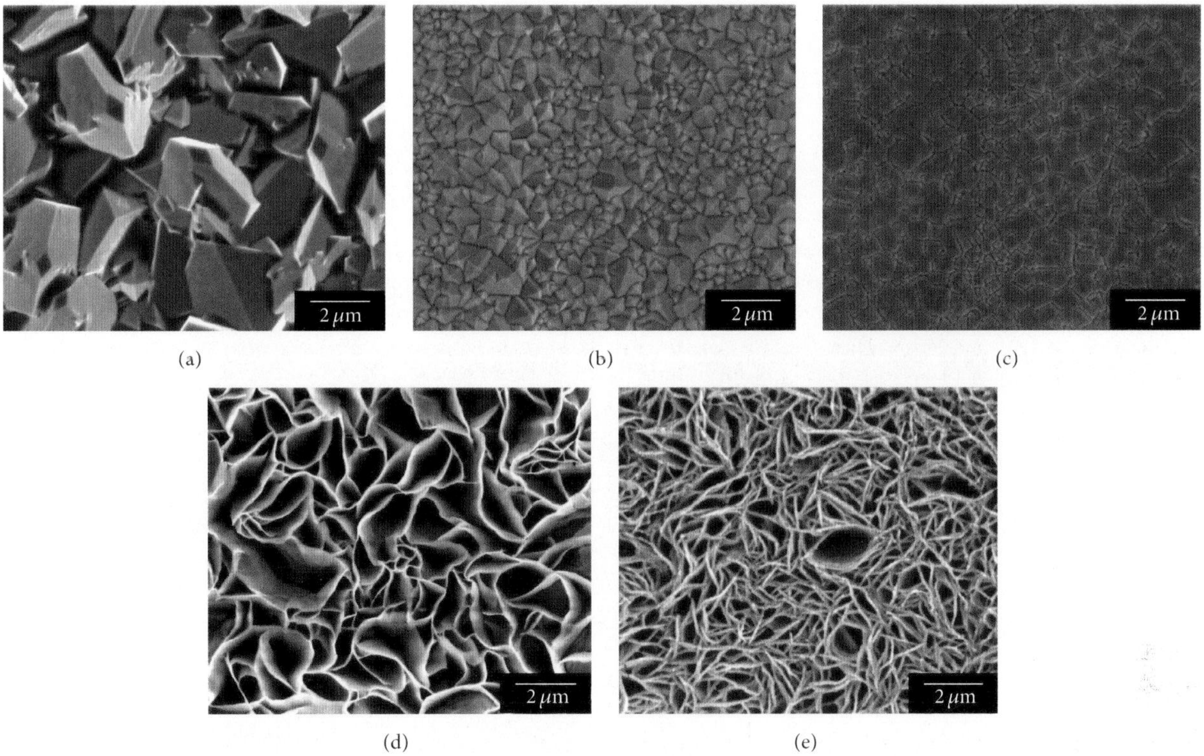

FIGURE 2: SEM surface images of electrodeposited films on the carbon steel substrate at the various deposition potentials in Fe(III)-TEA solution for 1800 s at 80°C: (a) –1.05, (b) –1.08, (c) –1.11, (d) –1.17, and (e) –1.23 V_{SCE}.

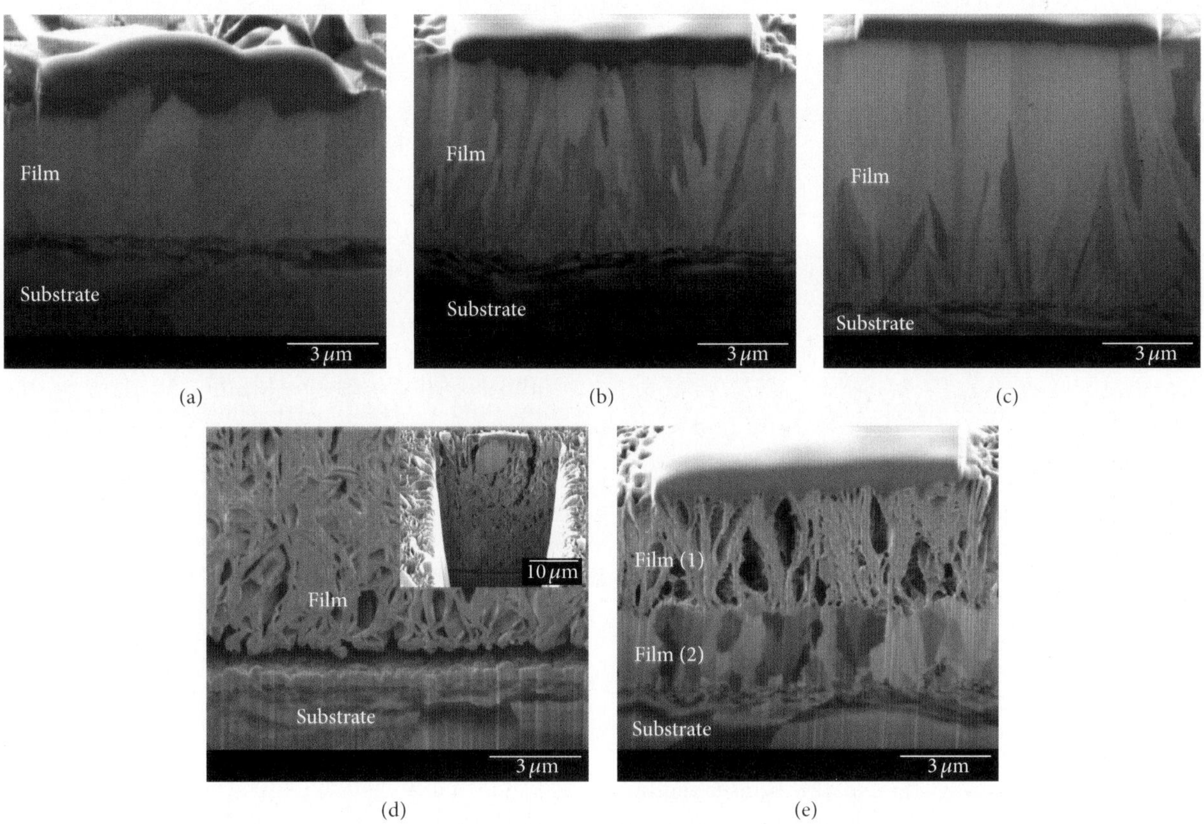

FIGURE 3: FIB-SEM images of the cross section of the electrodeposited films on the carbon steel substrate at the various deposition potentials in Fe(III)-TEA solution for 1800 s at 80°C: (a) –1.05, (b) –1.08, (c) –1.11, (d) –1.17, and (e) –1.23 V_{SCE}.

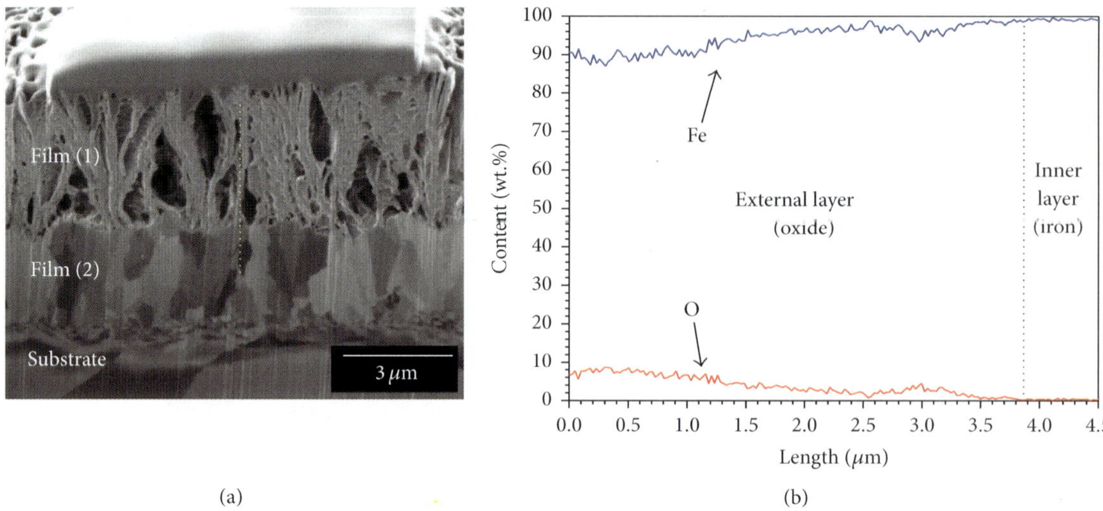

FIGURE 4: Cross-sectional EDS analysis of the film electrodeposited at $-1.23\ V_{SCE}$ on the carbon steel substrate in Fe(III)-TEA solution for 1800 s at 80°C: (a) SEM image and (b) line analysis.

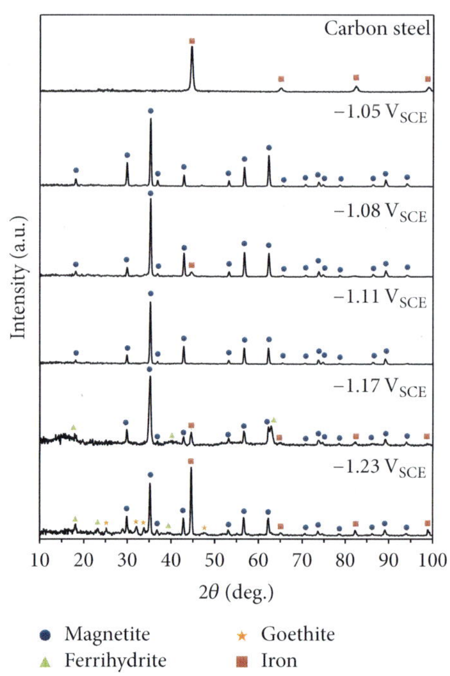

FIGURE 5: XRD analysis of the electrodeposited films on the carbon steel substrate at the various deposition potentials in Fe(III)-TEA solution for 1800 s at 80°C.

potential, Fe(III) in the solution can be reduced to metallic iron. In addition to the reduction force, the mass transfer rate of Fe(II) in the solution to the substrate will be fast at more negative potential. This will increase the deposition rate of magnetite through reactions (1) and (2). Consequently, the presence of ferrihydrite and goethite is unexpected at a lower electrochemical potential. Although the formation of ferrihydrite can be attributed to the aerial oxidation of a film in basic conditions [15], further study is needed.

Figure 6 shows Fe 2p and O 1s XPS spectra of the films deposited at the various potentials. As shown in Figure 6(a), Fe 2p core levels were split into $2p_{3/2}$ and $2p_{1/2}$ doublet. The binding energy of Fe $2p_{3/2}$ and Fe $2p_{1/2}$ of the film deposited at $-1.05\ V_{SCE}$ was 710.8 and 724.0 eV, respectively, which is in good agreement with the reported values for Fe_3O_4 [22, 23]. However, the peak positions for Fe $2p_{1/2}$ and Fe $2p_{3/2}$ slightly shifted to the higher binding energy side with an increase of the overpotential. This indicates an increase in the proportion of Fe(III) species in the films. The binding energy of ferrihydrite and goethite is higher than that of pure magnetite [24, 25]. Therefore, the shift of the peak positions to the higher binding energy is attributed to the formation of ferrihydrite and goethite. The binding energies of O 1s are found at 530.0 eV corresponding to O^{2-} bonds and at 531.1 eV corresponding to OH^- bonds (Figure 6(b)). In the O 1s spectrum, the film deposited at $-1.23\ V_{SCE}$ has its maximum at 531.1 eV due to OH^- bonds.

This is well consistent with the XRD analysis shown in Figure 4. It should be noted that the above Fe 2p spectra represent the chemical states of the outermost surface film, because the spectra were obtained without sputtering. Therefore, the above XPS results do not entirely reflect the XRD analysis.

Figure 7 shows the thickness variation of the deposited films with the deposition potentials. The thickness of the deposited films gradually increased with an increase of the overpotential in the range from −1.05 to $-1.17\ V_{SCE}$, whereas the thickness sharply decreased at $-1.23\ V_{SCE}$. The largest thickness of the deposited film at $-1.17\ V_{SCE}$ was about 33.2 μm. The double layered films at $-1.23\ V_{SCE}$ consisted of about 4.6 μm of the porous external film and about 4.6 μm of the compact inner film for a total of about 8.1 μm.

Figure 8 shows the variation of deposition current density at each potential during the deposition. It is clear that the more negative deposition potential is, the higher current density is, resulting in a thick film. In particular, the abrupt

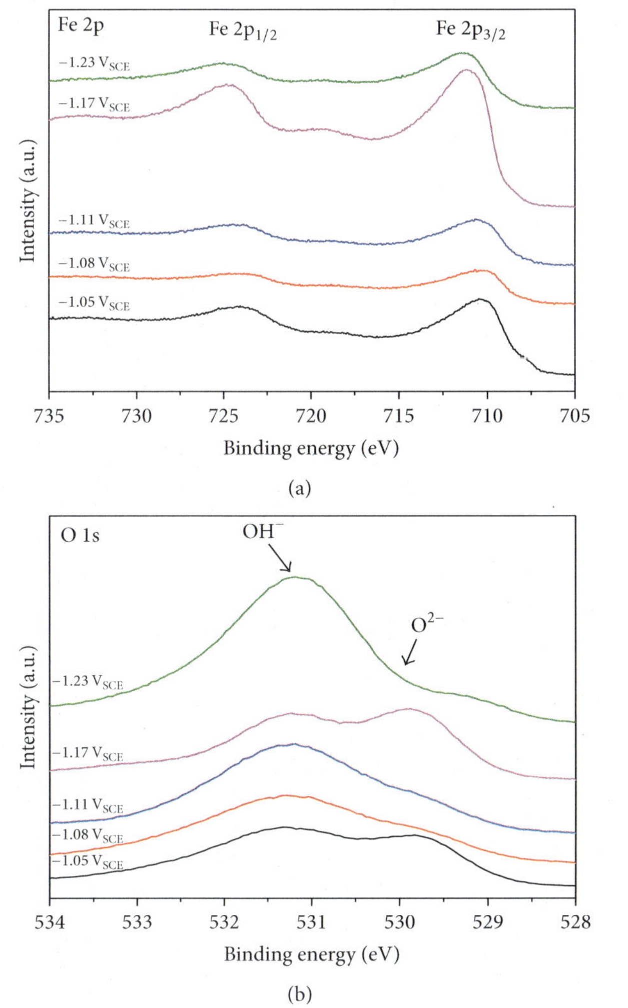

FIGURE 6: XPS spectra of the electrodeposited films on the carbon steel substrate at the various potentials in Fe(III)-TEA solution for 1800 s at 80°C: (a) Fe 2p and (b) O 1s.

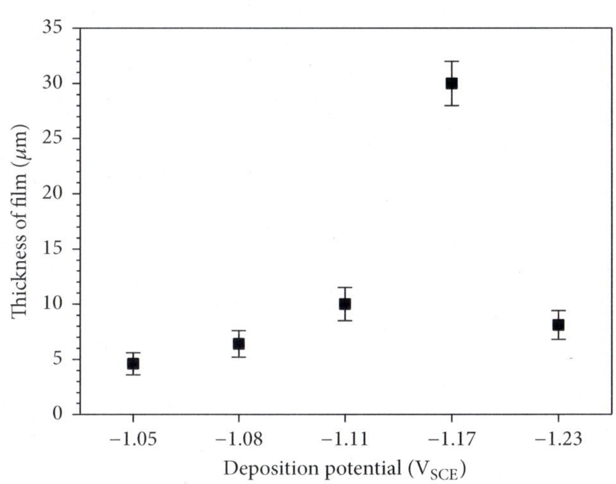

FIGURE 7: Thickness of the electrodeposited films on the carbon steel substrate at the various deposition potentials in Fe(III)-TEA solution for 1800 s at 80°C.

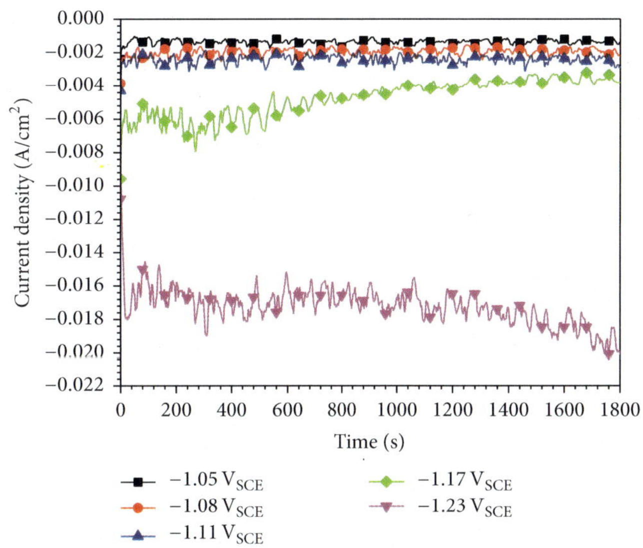

FIGURE 8: Current density variation at different potentials during electrochemical deposition in Fe(III)-TEA solution for 1800 s at 80°C.

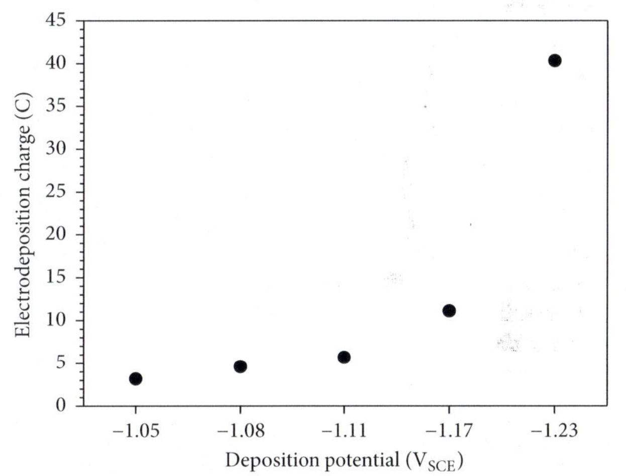

FIGURE 9: Electrodeposition charge at different potentials during electrochemical deposition in Fe(III)-TEA solution for 1800 s at 80°C.

increase of the film at $-1.17\,V_{SCE}$ can be attributed to 2-dimensional growth perpendiculars to the substrate and the resulting porous structure as shown in Figures 2(d) and 3(d). Comparing the thickness of the films, the film deposited at $-1.23\,V_{SCE}$ is thinner than that deposited at $-1.11\,V_{SCE}$, although the current density at $-1.23\,V_{SCE}$ is about three times higher than that at $-1.11\,V_{SCE}$. This is because the three electrons are consumed to reduce Fe(III) ion to a metallic ion.

Figure 9 shows the electrodeposition charge was measured at the various potentials during electrochemical deposition. Total amount of electric charge was increased with increasing deposition potential in negative direction.

Figure 10 shows the surface roughness of films on the carbon steel substrate deposited at the various potentials using a noncontacting surface profiler. The roughness variation of

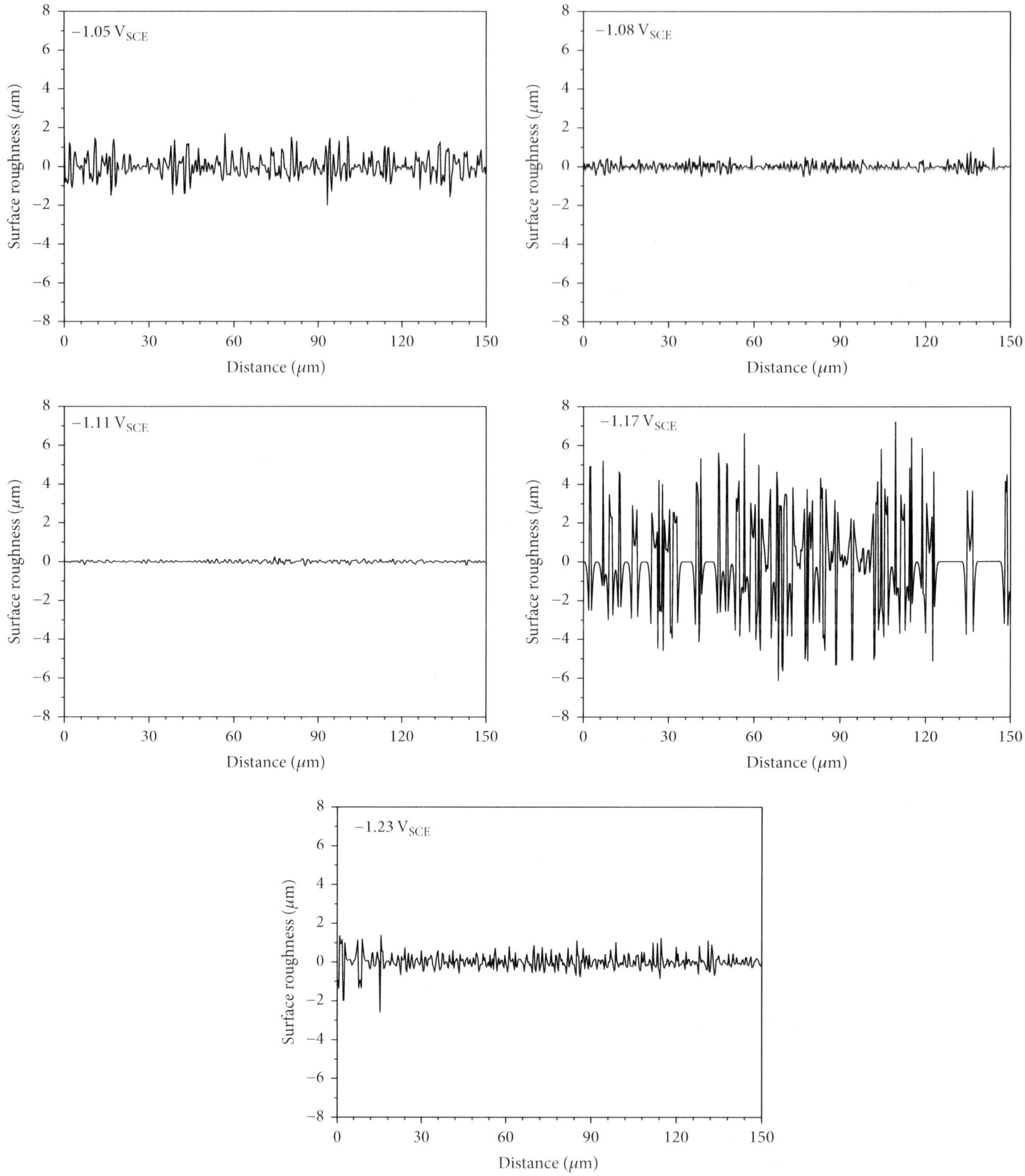

FIGURE 10: Surface roughness profile of electrodeposited films on the carbon steel substrate at the various deposition potentials in the Fe(III)-TEA solution for 1800 s at 80°C.

films is well associated with the surface morphology of the films shown in Figure 2. The roughness of the deposited films gradually decreased with an increase of the overpotential in the range from −1.05 to −1.11 V_{SCE} due to the decrease of particle size. The steep increase in the roughness at −1.17 V_{SCE} is owing to the porous petal-like morphology. At the deposition potential of −1.23 V_{SCE}, the roughness decreased in accordance with the more dense surface morphology of the film. As shown in Figure 11, the value of R_q of the deposited films on the carbon steel substrate is obtained from the

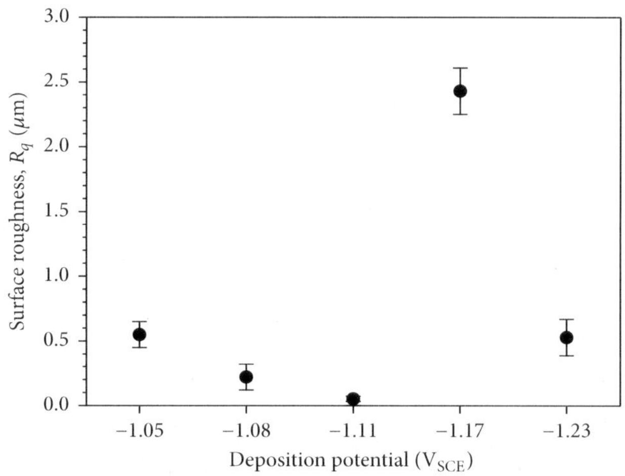

FIGURE 11: Surface roughness of the electrodeposited films on the carbon steel substrate at the various deposition potentials in Fe(III)-TEA solution for 1800 s at 80°C.

surface roughness curve. The surface roughness value of the deposited films at −1.11 V_{SCE} was the smallest with 0.06 μm.

Based on the results obtained in this work, it is confirmed that the electrochemical potentials affect the morphology, structure, thickness, and surface roughness of the deposited films.

In actual secondary system of nuclear power plants, porous magnetite is the main corrosion product formed on the surface of carbon steel piping, steam generator tubing, and other support structures due to the alkaline and reducing condition. Though the corrosion products are mainly made of magnetite, other corrosion products such as ferrihydrite and goethite are also present [26]. And thickness of real corrosion products on carbon steel piping, steam generator tubing, and other support structures exceeds several tens of micrometers. The film deposited at −1.17 V_{SCE} has porous morphology and several tens of micrometers thickness. Furthermore, this film contains the magnetite and ferrihydrite like the actual corrosion products in secondary system of nuclear power plants. Based on these results, the deposition potential at −1.17 V_{SCE} is appropriate to simulate the real corrosion products formed in the secondary system of nuclear power plants. However, the deposition potential range from −1.05 V_{SCE} to −1.11 V_{SCE} is proper deposition condition to use specimens for the electrochemical tests due to the formation of pure adhesive homogeneous magnetite. In this range, the film deposited at −1.11 V_{SCE} showed the most thick layer and smooth surface state. Therefore, the optimal potential for the electrodeposition of pure magnetite on the carbon steel substrate was determined to be −1.11 V_{SCE}.

To simulate the electrochemical corrosion behavior of the magnetite covering the materials, a rigid method to deposit magnetite on the corresponding material substrates should be established. In this respect, it is considered that these results are useful for simulating the deposited magnetite on the surface of carbon steel materials in the secondary system.

4. Conclusions

Magnetite films were electrodeposited on carbon steel substrate at different potentials in Fe(III)-TEA solution at 80°C. The electrochemical potentials affected the morphology, structure, thickness, and surface roughness of the deposited films. Pure magnetite films with a compact and defect-free structure were electrodeposited in the potential range from −1.05 V_{SCE} to −1.11 V_{SCE}. Porous magnetite film containing ferrihydrite and iron was obtained at −1.17 V_{SCE}. At a more negative potential at −1.23 V_{SCE}, double layered film was formed: a porous outer layer containing ferrihydrite and goethite and an inner columnar layer consisting of metallic iron. Based on these results, the optimal potential for the electrodeposition of pure magnetite was determined to be −1.11 V_{SCE}. These results are useful to prepare magnetite specimens for the electrochemical corrosion tests of magnetite itself and materials in contact with magnetite.

Conflict of Interests

The authors declare that there is no conflict of interests regarding the publication of this paper.

Acknowledgment

This work was supported by the National Research Foundation of Korea (NRF) grant funded by the Korean government (MSIP).

References

[1] C. Ramesh, N. Murugesan, A. A. M. Prince, S. Velmurugan, S. V. Narasimhan, and V. Ganesan, "Application of polymer electrolyte based hydrogen sensor to study corrosion of carbon steel in acid medium," *Corrosion Science*, vol. 43, no. 10, pp. 1865–1875, 2001.

[2] A. A. M. Prince, S. Velmurugan, S. V. Narasimhan et al., "Dissolution behaviour of magnetite film formed over carbon steel in dilute organic acid media," *Journal of Nuclear Materials*, vol. 289, no. 3, pp. 281–290, 2001.

[3] C. Hales, K. J. Stevens, P. L. Daniel, M. Zamanzadeh, and A. D. Owens, "Boiler feedwater pipe failure by flow-assisted chelant corrosion," *Engineering Failure Analysis*, vol. 9, no. 2, pp. 235–243, 2002.

[4] X. Jiang, Y. G. Zheng, and W. Ke, "Effect of flow velocity and entrained sand on inhibition performances of two inhibitors for CO_2 corrosion of N80 steel in 3% NaCl solution," *Corrosion Science*, vol. 47, no. 11, pp. 2636–2658, 2005.

[5] S. Odar, "Water chemistry measures to improve steam generator performance," in *Proceedings of the 14th International Conference on the properties of Water and Steam*, pp. 531–538, Kyoto, Japan, August 2004.

[6] D. H. Hur, M. S. Choi, U. C. Kim, and J. H. Han, "Magnetite dissolution and corrosion behavior in high temperature EDTA solvents," *Nuclear Engineering and Design*, vol. 220, no. 1, pp. 11–16, 2003.

[7] M. P. Nikiforov, A. A. Vertegel, M. G. Shumsky, and J. A. Switzer, "Epitaxial electrodeposition of Fe_3O_4 on single-crystal Au(111)," *Advanced Materials*, vol. 12, no. 18, pp. 1351–1353, 2000.

[8] D. Carlier, C. Terrier, C. Arm, and J.-P. Ansermet, "Preparation and magnetic properties of Fe_3O_4 nanostructures grown by electrodeposition," *Electrochemical and Solid-State Letters*, vol. 8, no. 3, pp. C43–C46, 2005.

[9] S. Peulon, H. Antony, L. Legrand, and A. Chausse, "Thin layers of iron corrosion products electrochemically deposited on inert substrates: synthesis and behaviour," *Electrochimica Acta*, vol. 49, no. 17-18, pp. 2891–2899, 2004.

[10] S.-Y. Wang, K.-C. Ho, S.-L. Kuo, and N.-L. Wu, "Investigation on capacitance mechanisms of Fe_3O_4 electrochemical capacitors," *Journal of the Electrochemical Society*, vol. 153, no. 1, pp. A75–A80, 2006.

[11] C.-L. Teng and M. P. Ryan, "A morphological study of nanocrystalline magnetite electrodeposited onto polycrystalline copper substrates," *Electrochemical and Solid-State Letters*, vol. 10, no. 10, pp. D108–D112, 2007.

[12] S. Mitra, P. Poizot, A. Finke, and J.-M. Tarascon, "Growth and electrochemical characterization versus lithium of Fe_3O_4 electrodes made by electrodeposition," *Advanced Functional Materials*, vol. 16, no. 17, pp. 2281–2287, 2006.

[13] R. S. Sapieszko and E. Matijević, "Preparation of well-defined colloidal particles by thermal decomposition of metal chelates. I. Iron oxides," *Journal of Colloid And Interface Science*, vol. 74, no. 2, pp. 405–422, 1980.

[14] H. M. Kothari, E. A. Kulp, S. J. Limmer, P. Poizot, E. W. Bohannan, and J. A. Switzer, "Electrochemical deposition and characterization of Fe_3O_4 films produced by the reduction of Fe(III)-triethanolamine," *Journal of Materials Research*, vol. 21, no. 1, pp. 293–301, 2006.

[15] E. A. Kulp, H. M. Kothari, S. J. Limmer et al., "Electrodeposition of epitaxial magnetite films and ferrihydrite nanoribbons on single-crystal gold," *Chemistry of Materials*, vol. 21, no. 21, pp. 5022–5031, 2009.

[16] R. V. Gudavarthy, S. Gorantla, G. Mu et al., "Epitaxial electrodeposition of Fe_3O_4 on single-crystal Ni(111)," *Chemistry of Materials*, vol. 23, no. 8, pp. 2017–2019, 2011.

[17] J. A. Switzer, R. V. Gudavarthy, E. A. Kulp, G. Mu, Z. He, and A. J. Wessel, "Resistance switching in electrodeposited magnetite superlattices," *Journal of the American Chemical Society*, vol. 132, no. 4, pp. 1258–1260, 2010.

[18] R. V. Gudavarthy, A. S. Miller, E. W. Bohannan, E. A. Kulp, Z. He, and J. A. Switzer, "Resistance switching in electrodeposited polycrystalline Fe_3O_4 films," *Electrochimica Acta*, vol. 56, no. 28, pp. 10550–10556, 2011.

[19] C. Goujon, T. Pauporté, C. Mansour, S. Delaunary, and J. L. Bretelle, "Fouling of steam generator tubes in nuclear power plants: laboratory tests to reproduce oxides deposition," in *Proceedings of the International Conference on Heat Exchanger Fouling and Cleaning*, pp. 101–107, Budapest, Hungary, June 2013.

[20] C. Goujon, T. Pauporté, C. Mansour, S. Delaunay, and J. Bretelle, "Electrochemical deposition of thick iron oxide films on nickel based superalloy substrates," *Electrochimica Acta*, vol. 176, pp. 230–239, 2015.

[21] H. Duan, X. Chen, B. Li, and J. Liang, "Growth morphology study of cathodically electrodeposited Fe_3O_4 thin films at elevated temperatures," *Materials Research Bulletin*, vol. 45, no. 11, pp. 1696–1702, 2010.

[22] Y. Jia, X.-Y. Yu, T. Luo, M.-Y. Zhang, J.-H. Liu, and X.-J. Huang, "Two-step self-assembly of iron oxide into three-dimensional hollow magnetic porous microspheres and their toxic ion adsorption mechanism," *Dalton Transactions*, vol. 42, no. 5, pp. 1921–1928, 2013.

[23] L. Martinez, D. Leinen, F. Martín et al., "Electrochemical growth of diverse iron oxide (Fe_3O_4, α-FeOOH, and γ-FeOOH) thin films by electrodeposition potential tuning," *Journal of the Electrochemical Society*, vol. 154, no. 3, pp. D126–D133, 2007.

[24] H. Abdel-Samad and P. R. Watson, "An XPS study of the adsorption of lead on goethite (α-FeOOH)," *Applied Surface Science*, vol. 136, no. 1-2, pp. 46–54, 1998.

[25] P. Ghods, O. B. Isgor, F. Bensebaa, and D. Kingston, "Angle-resolved XPS study of carbon steel passivity and chloride-induced depassivation in simulated concrete pore solution," *Corrosion Science*, vol. 58, pp. 159–167, 2012.

[26] H. Namduri, *Formation and quantification of corrosion deposits in the power industry [Ph.D. dissertation]*, University of North Texas, Denton, Tex, USA, 2007.

Permissions

All chapters in this book were first published in AMSE, by Hindawi Publishing Corporation; hereby published with permission under the Creative Commons Attribution License or equivalent. Every chapter published in this book has been scrutinized by our experts. Their significance has been extensively debated. The topics covered herein carry significant findings which will fuel the growth of the discipline. They may even be implemented as practical applications or may be referred to as a beginning point for another development.

The contributors of this book come from diverse backgrounds, making this book a truly international effort. This book will bring forth new frontiers with its revolutionizing research information and detailed analysis of the nascent developments around the world.

We would like to thank all the contributing authors for lending their expertise to make the book truly unique. They have played a crucial role in the development of this book. Without their invaluable contributions this book wouldn't have been possible. They have made vital efforts to compile up to date information on the varied aspects of this subject to make this book a valuable addition to the collection of many professionals and students.

This book was conceptualized with the vision of imparting up-to-date information and advanced data in this field. To ensure the same, a matchless editorial board was set up. Every individual on the board went through rigorous rounds of assessment to prove their worth. After which they invested a large part of their time researching and compiling the most relevant data for our readers.

The editorial board has been involved in producing this book since its inception. They have spent rigorous hours researching and exploring the diverse topics which have resulted in the successful publishing of this book. They have passed on their knowledge of decades through this book. To expedite this challenging task, the publisher supported the team at every step. A small team of assistant editors was also appointed to further simplify the editing procedure and attain best results for the readers.

Apart from the editorial board, the designing team has also invested a significant amount of their time in understanding the subject and creating the most relevant covers. They scrutinized every image to scout for the most suitable representation of the subject and create an appropriate cover for the book.

The publishing team has been an ardent support to the editorial, designing and production team. Their endless efforts to recruit the best for this project, has resulted in the accomplishment of this book. They are a veteran in the field of academics and their pool of knowledge is as vast as their experience in printing. Their expertise and guidance has proved useful at every step. Their uncompromising quality standards have made this book an exceptional effort. Their encouragement from time to time has been an inspiration for everyone.

The publisher and the editorial board hope that this book will prove to be a valuable piece of knowledge for researchers, students, practitioners and scholars across the globe.

List of Contributors

Imed Miraoui
Department of Mechanical Engineering, College of Engineering, Aljouf University, Sakaka 2014, Saudi Arabia
College of Sciences, Research Unit MEER, University of Gafsa, 2100 Gafsa, Tunisia

Mohamed Boujelbene
Department of Industrial Engineering, College of Engineering, Hail University, Hail, Saudi Arabia

Mouna Zaied
College of Sciences, Research Unit MEER, University of Gafsa, 2100 Gafsa, Tunisia

Jonathan Parker and John Siefert
Electric Power Research Institute, Charlotte, NC, USA

Shunbo Zhao, Changyong Li, Mingshuang Zhao and Xiaoyan Zhang
School of Civil Engineering and Communication, North China University of Water Resources and Electric Power, No. 36 Beihuan Road, Zhengzhou 450045, China

Vineet Shibe
Research Scholar, IKG Punjab Technical University, Kapurthala, Punjab 144603, India

Vikas Chawla
Ferozepur College of Engineering and Technology, Ferozepur, Punjab 142052, India

Agnieszka E. Kochmánska
West Pomeranian University of Technology in Szczecin, Piastow 17, 70-310 Szczecin, Poland

Liang Liu, Ting-Ting Cao, Qi-Wei Zhang and Chong-Wei Cui
Harbin Institute of Technology, School of Environment, Harbin 150090, China

Ladislav Falat, Viera Homolová, Lucia Hiripová and Peter Ševc
Institute of Materials Research, Slovak Academy of Sciences, Watsonova 47, 04001 Kŏsice, Slovakia

Milan Svoboda
Institute of Physics of Materials, Academy of Sciences of the Czech Republic, Žižkova 22, 61662 Brno, Czech Republic

J. C. Velázquez
Departamento de Ingeniería Química Industrial, ESIQIE, Instituto Politécnico Nacional, Zacatenco, México City 07738, Mexico
Gerencia de Ingeniería y Costos, Pemex Transformación Industrial, Piso 6, Edificio B1, Centro Administrativo Pemex, Marina Nacional 329, Col. Petróleos Mexicanos, México City 11311, Mexico

F. Caleyo
Departamento de Ingeniería Metalúrgica, ESIQIE, Instituto Politécnico Nacional, Zacatenco, México City 07738, Mexico

R. Cabrera-Sierra, G. Teran, S. Capula-Colindres and C. C. Ortiz-Herrera
Departamento de Ingeniería Química Industrial, ESIQIE, Instituto Politécnico Nacional, Zacatenco, México City 07738, Mexico

E. Hernandez-Sanchez
Instituto Politécnico Nacional, UPIBI, Avenida Acueducto s/n Barrio La Laguna Ticomán, México City 07340, Mexico

H. Herrera-Hernández
Universidad Autónoma del Estado de México, IIN-Lab. de Electroquímica Aplicada y Corrosión de Materiales Industriales, Blvd. Universitario s/n, Atizapán de Zaragoza, Estado de México 54500, Mexico

Zhi-chao Li
Collaborative Innovation Center of Steel Technology, University of Science and Technology, Beijing 100083, China

Ning Dang and Zhen-li Mi
Institute of Engineering Technology, University of Science and Technology, Beijing 100083, China

Jari Aromaa, Antero Pehkonen, Sönke Schmachtel, Istvan Galfi and Olof Forsén
Department of Chemical Engineering and Metallurgy, School of Chemical Technology, Aalto University, Aalto 00076, Finland

Xu Yan, Li-min Liu, Jin-peng Zhang, Yan-hui Li and He Wang
College of Mining and Safety Engineering, Shandong University of Science and Technology, Qingdao 266590, China

Hai Fang, Yue Wang and Yujun Qi
College of Civil Engineering, Nanjing Tech University, Nanjing, China

Huiyuan Shi
School of Civil Engineering, Southeast University, Nanjing, China

Weiqing Liu
Advanced Engineering Composites Research Center, Nanjing Tech University, Nanjing, China

H. S. Ho, D. L. Li, E. L. Zhang and P. H. Niu
School of Mechanical Engineering and Henan Key Engineering Laboratory of Anti-Fatigue Manufacturing Technology, Zhengzhou University, Science Road 100, Zhengzhou 450001, China

Abdel Monem Masmoudi
Department of Civil Engineering, LA2MP Laboratory, ENIS, Université de Sfax, Sfax, Tunisia

Jamel Bouaziz
Department of Industrial Chemistry, National School of Engineers of Sfax, Sfax, Tunisia

Baek-Il Bae
Research Institute of Industrial Science, Hanyang University, 17 Haengdang-Dong, Seongdong-Gu, Seoul 04763, Republic of Korea

Hyun-Ki Choi, Bong-Seop Lee and Chang-Hoon Bang
Department of Fire and Disaster Prevention Engineering, Kyungnam University, Gyeongsangnam-do 51767, Republic of Korea

Xuechun Liu
Beijing Engineering Research Center of High-Rise and Large-Span Prestressed Steel Structures, Beijing University of Technology, Beijing 100124, China
Beijing Key Laboratory of Earthquake Engineering and Structural Retrofit, Beijing University of Technology, Beijing 100124, China
Beijing Advanced Innovation Centre for Future Urban Design, Beijing 100044, China

Xiaoxiong Cui, Zhiwei Yang and Xinxin Zhan
Beijing Engineering Research Center of High-Rise and Large-Span Prestressed Steel Structures, Beijing University of Technology, Beijing 100124, China

Hai Yang Wang
School of Environment and Energy, Peking University Shenzhen Graduate School, Shenzhen 518055, China
School of Civil and Environment Engineering, Harbin Institute of Technology Shenzhen Graduate School, Shenzhen 518055, China

Xiao Xiong Zha and Wei Feng
School of Environment and Energy, Peking University Shenzhen Graduate School, Shenzhen 518055, China

Li Sun and Weiping Yan
School of Energy Power and Mechanical Engineering, North China Electric Power University, Yong Hua Bei No. 619, Lian Chi District, Baoding 071003, China

Adam Grajcar, Aleksandra Kozłowska, Santina Topolska and Mateusz Morawiec
Institute of Engineering Materials and Biomaterials, Silesian University of Technology, 18a Konarskiego Street, 44-100 Gliwice, Poland

Paulina Krolo, Davor Grandić and Željko Smolčić
Department of Structural Engineering and Technical Mechanics, Faculty of Civil Engineering, University of Rijeka, Radmile Matejčić 3, 51000 Rijeka, Croatia

Soon-Hyeok Jeon, Geun-Dong Song and Do Haeng Hur
Nuclear Materials Safety Research Division, Korea Atomic Energy Research Institute, 989-111 Daedeok-daero, Yuseong-gu, Daejeon 305-353, Republic of Korea

Index